# Reliability-Based Analysis and Design of Structures and Infrastructure

**Taylor & Francis Series in Resilience and Sustainability in Civil, Mechanical, Aerospace, and Manufacturing Engineering Systems**

Series Editor: Mohammad Noori

*Cal Poly San Luis Obispo*

*Published Titles*

**Resilience of Critical Infrastructure Systems**
Emerging Development and Future Challenges
*Zhishen Wu, Xilin Lu, Mohammad Noori*

**Experimental Vibration Analysis for Civil Structures**
Testing, Sensing, Monitoring, and Control
*Jian Zhang, Zhishen Wu, Mohammad Noori, and Yong Li*

**Reliability and Safety of Cable-Supported Bridges**
*Naiwei Lu, Yang Liu, and Mohammad Noori*

**Reliability-Based Analysis and Design of Structures and Infrastructure**
*Ehsan Noroozinejad Farsangi, Mohammad Noori, Paolo Gardoni, Izuru Takewaki, Humberto Varum, and Aleksandra Bogdanovic*

**Sustainable Development for the Americas: Science, Health, and Engineering Policy and Diplomacy**
*W. William Colglazier, Hassan A. Vafai, Kevin E. Lansey, Molli D. Bryson*

**Seismic Analysis and Design using the Endurance Time Method**
*Homayoon E. Estekanchi, Hassan A. Vafai*

For more information about this series, please visit: https://www.routledge.com/Resilience-and-Sustainability-in-Civil-Mechanical-Aerospace-and-Manufacturing/book-series/ENG

# Reliability-Based Analysis and Design of Structures and Infrastructure

Edited by

Ehsan Noroozinejad Farsangi
*Graduate University of Advanced Technology, Iran*

Mohammad Noori
*California Polytechnic State University, USA*

Paolo Gardoni
*University of Illinois at Urbana-Champaign, USA*

Izuru Takewaki
*Kyoto University, Japan*

Humberto Varum
*University of Porto, Portugal*

Aleksandra Bogdanovic
*Institute of Earthquake Engineering and Engineering Seismology,
North Macedonia*

CRC Press
Taylor & Francis Group
Boca Raton London New York

CRC Press is an imprint of the
Taylor & Francis Group, an **informa** business

First edition published 2022
by CRC Press
6000 Broken Sound Parkway NW, Suite 300, Boca Raton, FL 33487-2742
and by CRC Press

2 Park Square, Milton Park, Abingdon, Oxon, OX14 4RN

© 2022 selection and editorial matter, Ehsan Noroozinejad Farsangi, Mohammad Noori, Paolo Gardoni, Izuru Takewaki, Humberto Varum, Aleksandra Bogdanovic; individual chapters, the contributors

CRC Press is an imprint of Taylor & Francis Group, LLC

---

**Library of Congress Cataloging-in-Publication Data**

---

Names: Farsangi, Ehsan Noroozinejad, editor.
Title: Reliability-based analysis and design of structures and infrastructure / edited by Ehsan Noroozinejad Farsangi, Graduate University of Advanced Technology, Iran, Mohammad Noori, California Polytechnic State University, USA, Paolo Gardoni, University of Illinois at Urbana-Champaign, USA, Izuru Takewaki, Kyoto University, Japan, Humberto Varum, University of Porto, Portugal, Aleksandra Bogdanovic, Inst. of Earthquake Engineering and Engineering Seismology, North Macedonia.
Description: First edition. | Boca Raton, FL : CRC Press, 2021. | Series: Resilience and sustainability in civil, mechanical, aerospace and manufacturing engineering systems | Includes bibliographical references and index.
Identifiers: LCCN 2021006002 (print) | LCCN 2021006003 (ebook) | ISBN 9780367758080 (hardback) | ISBN 9781032047720 (pbk) | ISBN 9781003194613 (ebook)
Subjects: LCSH: Structural design. | Structural analysis (Engineering) | Structural failures–Prevention. | Buildings–Reliability. | Public works–Reliability.
Classification: LCC TA658 .R45 2021 (print) | LCC TA658 (ebook) | DDC 624.1/7–dc23
LC record available at https://lccn.loc.gov/2021006002
LC ebook record available at https://lccn.loc.gov/2021006003

---

ISBN: 978-0-367-75808-0 (hbk)
ISBN: 978-1-032-04772-0 (pbk)
ISBN: 978-1-003-19461-3 (ebk)

DOI: 10.1201/9781003194613

Typeset in Times LT Std
by KnowledgeWorks Global Ltd.

# Contents

# Preface

Increasing demand for improving the resilience of modern structures and infrastructure requires more critical and complex designs that take into account the recent developments in uncertainty quantification and reliability measures. Hence, the need for accurate and efficient approaches to assess uncertainties in loads, geometry, material properties, manufacturing processes, and operational environments has increased significantly. For problems in which randomness is relatively small, deterministic models can be used rather than probabilistic ones. However, when the level of uncertainty is high and cannot be ignored, probabilistic approaches are necessary for system analysis and design.

Reliability-based techniques help develop more accurate initial guidance for robust designs. They can also be used to identify the sources of significant uncertainties in structures and infrastructure. The necessity for incorporating reliability-based techniques in the design of structures and infrastructure has resulted in the need to identify where further research, testing, and quality control could increase safety and efficiency.

The current edited volume, which includes contributions from recognized scholars in the relevant fields, provides a practical and comprehensive overview of reliability and risk analysis, and design techniques. Since reliability analysis and reliability-based design are multidisciplinary subjects, the scope is not limited to any particular engineering discipline; rather, the material applies to most engineering subjects. This book also intends to provide engineers and researchers with an intuitive appreciation for probability theory, statistical methods, and reliability analysis methods, such as Monte Carlo sampling, Latin hypercube sampling, first- and second-order reliability methods, stochastic finite element method, and stochastic optimization. The reliability-based design is the key focus of the book, which is currently a relatively unknown topic for practicing engineers.

We hope that this book will serve as a comprehensive guide and reference for practicing engineers, researchers, educators, and recent graduates entering the engineering profession by assuring them that they have discovered an exciting world of challenges and opportunities.

**Ehsan Noroozinejad Farsangi**
*Iran*
**Mohammad Noori**
*USA*
**Paolo Gardoni**
*USA*
**Izuru Takewaki**
*Japan*
**Humberto Varum**
*Portugal*
**Aleksandra Bogdanovic**
*North Macedonia*

# Editors

**Dr. Ehsan Noroozinejad Farsangi** is an Assistant Professor in the Department of Earthquake and Geotechnical Engineering at the Graduate University of Advanced Technology, Iran. Dr. Noroozinejad is the Founder and Chief Editor of the *International Journal of Earthquake and Impact Engineering*, the Associate Editor of the *ASCE Practice Periodical on Structural Design and Construction*, the Associate Editor of the *IET Journal of Engineering*, the Associate Editor of *Frontiers in Built Environment: Earthquake Engineering Section*, the Associate Editor of the *Open Civil Engineering Journal*, the Associate Editor of the *Journal of Applied Science and Engineering*, the Editor of the *Journal of Reliability Engineering and Resilience*, and the Engineering Editor of *ASCE Natural Hazards Review*. He is also the Core member of the FIB Commission on Resilient RC Structures, the ASCE Objective Resilience Committee, the ASCE Risk and Resilience Measurements Committee, the ASCE Civil Infrastructure and Lifeline Systems Committee, the ASCE Structural Health Monitoring & Control Committee, and also the IEEE P693 Committee on Seismic Design for Substations. He has authored more than 70 journal papers in indexed journals and has published four books with reputed publishers in his field of expertise. His main research interests include structural and geotechnical earthquake engineering, structural dynamics, computational dynamics, smart structures, solid and fracture mechanics, resilience-based design, reliability analysis, artificial intelligence, soil-structure interaction (SSI), and vibration control in structures and infrastructure under extreme loading.

**Dr. Mohammad Noori** is a Professor of Mechanical Engineering at California Polytechnic State University, San Luis Obispo, a Fellow of the American Society of Mechanical Engineering, and a recipient of the Japan Society for Promotion of Science Fellowship. Dr. Noori's work in nonlinear random vibrations, seismic isolation, and application of artificial intelligence methods for structural health monitoring is widely cited. He has authored more than 250 refereed papers, including more than 100 journal articles, 6 scientific books, and has edited 25 technical and special journal volumes. Dr. Noori has supervised more than 90 graduate students and post-doc scholars, and has presented more than 100 keynotes, plenary, and invited talks. He is the founding executive editor of an international journal and has served on the editorial boards of over 10 other journals and as a member of numerous scientific and advisory boards. Dr. Noori has been a Distinguished Visiting Professor at several highly ranked global universities and directed the Sensors Program at the National Science Foundation in 2014. Dr. Noori has been a founding director or co-founder of three industry–university research centers and held Chair Professorships at two major universities. He served as the Dean of Engineering at Cal Poly for five years, has also served as the Chair of the National Committee of Mechanical Engineering department heads, and was one of seven co-founders of the National Institute of Aerospace, in partnership with NASA Langley Research Center. Dr. Noori also serves as the Chief Technical Advisor for several scientific organizations and industries.

**Dr. Paolo Gardoni** is a Professor and Excellence Faculty Scholar in the Department of Civil and Environmental Engineering at the University of Illinois at Urbana-Champaign. He is the Director of the MAE Center that focuses on creating a Multi-hazard Approach to Engineering, and the Associate Director of the NIST-funded Center of Excellence for Risk-based Community Resilience Planning. Dr. Gardoni is the founder and Editor-in-Chief of the international journal *Sustainable and Resilient Infrastructure* published by Taylor & Francis Group. He is a member of the Board of Directors of the International Civil Engineering Risk and Reliability Association (CERRA), of the Advisory Council of the International Forum on Engineering Decision Making (IFED), and of a number of national and international committees and associations that focus on risk and reliability analysis. Dr. Gardoni's research interests include sustainable and resilient infrastructure; reliability, risk, and life cycle analysis; decision-making under uncertainty; performance assessment of deteriorating systems; ethical, social, and legal dimensions of risk; and policies for natural hazard mitigation and disaster recovery. He is the author of more than 150 refereed journal papers, 27 book chapters, and 7 edited volumes. He has received more than $50 million in research funding from multiple national and international agencies including the National Science Foundation (NSF), the Qatar National Research Funds (QNRF), the National Institute of Standards and Technology (NIST), the Nuclear Regulatory Commission (NRC), and the United States Army Corps of Engineers. Dr. Gardoni has given more than 40 invited and keynote lectures around the world.

**Dr. Izuru Takewaki** is a Professor at Kyoto University, Kyoto, Japan. He also graduated in Architectural Engineering from Kyoto University. After he earned his PhD in 1991 on structural optimization and inverse vibration problems at Kyoto University, he focused on critical excitation method and earthquake resilience at Kyoto University, where he is currently a professor. He is the 56th President of Architectural Institute of Japan (AIJ) since 2019. He is the Field Chief Editor of *Frontiers in Built Environment* (Switzerland) and also the Specialty Chief Editor of "Earthquake Engineering" section in the same journal. Dr. Takewaki is serving as an editorial board member on several world-leading international journals including *Soil Dynamics and Earthquake Engineering* (Elsevier). He published more than 200 papers in international journals and his h-index is 32. His research interests include resilience-based design of structures, seismic-resistant design of building structures, structural control, base isolation, inverse problem in vibration, structural optimization, soil-structure interaction, random vibration theory, critical excitation method, structural health monitoring, and system identification. Dr. Takewaki was awarded numerous prizes, including the Research Prize of AIJ in 2004, the 2008 Paper of the Year in Journal of *The Structural Design of Tall and Special Buildings* (Wiley), the Prize of AIJ for Book in 2014.

**Dr. Humberto Varum** is a Professor and Director of the PhD Program in Civil Engineering at the Faculty of Engineering of the University of Porto (FEUP), Portugal. He is an Honorary Lecturer at the Department of Civil, Environmental and Geomatic Engineering, University College of London (UCL), United Kingdom since 2010. He is Visiting Professor at the College of Civil Engineering, Fuzhou University, Fujian, China since July 2016. Dr. Varum has been Seconded National Expert to the ELSA laboratory, Joint Research Centre, European Commission, Italy, in the period July 2009 to August 2010. He is an Integrated member and Vice-Coordinator of CONSTRUCT research unit: Institute of R&D in Structures and Construction (FEUP). Since May 2015, he is a member of the Directorate Body of the Construction Institute of the University of Porto, and President since May 2019. Dr. Varum is a member of the National Committee of the International Council on Monuments and Sites (ICOMOS), since 2009, and expert member of the ICOMOS's International Scientific Committee of Earthen Architectural Heritage (ISCEAH). He is a member of the Project Team 2 for the development of the second generation of EN Eurocodes. He is a member of the Mexican Academy of Engineering, since 2018. Since June 2019, he is External Researcher of Grupo de Gestión de Riesgos de Desastres en Infraestructura Social y Vivienda de Bajo Costo (GERDIS), Pontificia Universidad Católica del Perú (PUCP), Peru. Dr. Varum is a member of the International Scientific Board (ISB) of AdMaS (Advanced Materials, Structures and Technologies) center, Faculty of Civil Engineering, Brno University of Technology, Czech Republic, since 2013. He has participated in post-earthquake field reconnaissance missions, in particular to L'Aquila (Italy, 2009), Lorca (Spain, 2011), Emilia-Romagna (Italy, 2012), Gorkha (Nepal, 2015), and Puebla (Mexico, 2017). His main research interests include assessment, strengthening and repair of structures, earthquake engineering, historic constructions conservation, and strengthening.

**Dr. Aleksandra Bogdanovic** is an Associate Professor in the subject design and analysis of structures with seismic isolation and passive system for energy dissipation, at the International Master and Doctoral Studies at Institute of Earthquake Engineering and Engineering Seismology IZIIS, University "Ss. Cyril and Methodius." Her main professional activities are related to research and education in civil engineering, earthquake engineering, structural control and health monitoring, experimental testing, and analysis of the structures. Her doctoral studies were performed within the framework of the international SEEFORM/DAAD Program under which she completed several months of study at RWTH Aachen in 2009. She has been a participant for IZIIS in FP7 SERIES Project (2009-2013, Grant No 227887), FP7 UREDITEME Project (2009-2012, Grant No 230099), as well as deputy assistant at WP 14 in H2020 SERA Project (2017-2020, Grant No 730900). Dr. Bogdanovic has published more than 40 scientific papers in national and international conferences, more than 60 professional reports, and more than 20 publications in journals. She has received an award for the notable achievement in the civil engineering practice granted by the Macedonian Association of Structural Engineers in 2015. She is a member of the Macedonian Association of Earthquake Engineering (MAEE) and European Association of Earthquake Engineering (EAEE).

# Contributors

**Naida Ademović**
University of Sarajevo
Bosnia and Herzegovina

**Donatello Cardone**
University of Basilicata
Italy

**Paolo Castaldo**
Geotechnical and Building Engineering
 (DISEG), Politecnico di Torino,
Turin, Italy

**Ji Dang**
Saitama University
Saitama, Japan

**Moshe Danieli**
Ariel University
Ariel, Israel

**Jian Deng**
Lakehead University,
Thunder Bay, ON, Canada

**Anjan Dutta**
Indian Institute of Technology Guwahati
Guwahati, Assam, India

**Christopher D. Eamon**
Wayne State University
Detroit, Michigan, USA

**Ehsan Noroozinejad Farsangi**
Graduate University of Advanced Technology
Kerman, Iran

**Goutam Ghosh**
Motilal Nehru National Institute of Technology
Allahabad, Uttar Pradesh, India

**Vasudha A. Gokhale**
S. P. Pune University
Pune, India

**Çağlayan Hızal**
Izmir Institute of Technology
Urla, Izmir, Turkey

**Ke Huang**
Changsha University of Science and
 Technology
Changsha, China

**Alexander Humer**
Johannes Kepler University
Linz, Austria

**Hui Jin**
Southeast University
Nanjing, China

**Jichao Li**
Institute of Engineering Mechanics, China
 Earthquake Administration
Harbin, China

**Pei-Pei Li**
Kanagawa University
Yokohama, Japan

**Zhao-Hui Lu**
Beijing University of Technology
Beijing, China

**Luís Martins**
Global Earthquake Model Foundation
Pavia, Italy

**Tiago P. Ribeiro**
Tal Projecto, Lda
Lisboa, Portugal

**T.V. Santhosh**
University of Liverpool
Liverpool, UK

**Yuanfeng Shi**
Sichuan University
Chengdu, China

**Maria João Falcão Silva**
Laboratório Nacional de Engenharia Civil
    (LNEC) Lisboa, Portugal

**Izuru Takewaki**
Kyoto University
Kyoto, Japan

**Selcuk Toprak**
Gebze Technical University
Kocaeli, Turkey

**Marcos A. Valdebenito**
Universidad Adolfo Ibáñez
Chile

**Cao Wang**
University of Wollongong
Wollongong, NSW, Australia

**T.Y. Yang**
The University of British Columbia
Vancouver, Canada

**Tayyab Zafar**
University of Electronic Science and
    Technology of China
Chengdu, China
National University of Sciences and
    Technology
Islamabad, Pakistan

# 1 Reliability-Based Design Optimization

*Christopher D. Eamon*

Department of Civil and Environmental Engineering,
Wayne State University, Detroit, Michigan, USA

## CONTENTS

## 1.1 INTRODUCTION

Reliability-based design optimization (RBDO) commonly uses probability theory to account for uncertainties in load and resistance parameters in the optimization process. Although various ways of formulating an RBDO problem exist, in general, an optimal vector of design variables (DVs) $Y = \{Y_1, Y_2, ..., Y_{NDV}\}^T$ is desired that would minimize an objective function of interest while subjected to a series of deterministic and probabilistic constraints. A typical problem can be written as:

$$\min f(X,Y)$$

$$\text{s.t. } P_{fi} = P[G_i(X,Y) \le 0] \le P_{ai}; i = 1, N_p$$

$$q_j(Y) \le 0; j = 1, N_q \tag{1.1}$$

$$Y_k^l \le Y_K \le Y_k^u; k = 1, NDV$$

$$x_m^l \le x_m \le x_m^u; m = 1, NRV$$

where $X = \{X_1, X_2, ..., X_{NRV}\}^T$ is a vector of random variables (RVs) used to characterize uncertainty; $f(Y,X)$ is the objective function; $P_{ai}$ is the allowable bound on the $i$th failure probability, $P_{fi}$, corresponding to limit state function $G_i(X,Y)$ for one of $N_p$ total probabilistic constraints; $q_j$ is the $j$th deterministic inequality constraint of $N_q$ total deterministic constraints; $Y_k$ is the $k$th DV among number of decision variables *(NDV)* total DVs with lower and upper bound (side) constraints $Y_k^l$ and $Y_k^u$; and $x_m$ is the $m$th realized or design point value of normal random variable *(NRV)* total RVs with side constraints $x_m^l$ and $x_m^u$. Alternatively, utilizing reliability index (β) directly rather than failure

DOI: 10.1201/9781003194613-1

probability, as commonly done in structural reliability analysis, the probabilistic constraint $P_{fi} \leq P_{ai}$ can be written as: $\beta_i \geq \beta_{min}$; $i = 1, N_p$, where $\beta_i$ is the $i$th reliability index constraint and $\beta_{min}$ is the minimum acceptable reliability index. Although $X$ and $Y$ in Eq. (1.1) are given as two different vectors, elements within them often overlap, as will be discussed below.

## 1.2   CHOICE OF OBJECTIVE FUNCTION, LIMIT STATE FUNCTION, AND CONSTRAINTS

One of the most common objectives is to minimize weight, which is often a surrogate for the cost (Rais-Rohani et al. 2010; Thompson et al. 2006). However, there is no limitation in this regard; any objective function can be used, as long as it can be mathematically expressed in terms of the DVs considered in the problem. In RBDO, the objective itself may even involve a probabilistic quantity, such as to maximize reliability index or minimize failure probability.

The limit state, or performance function(s) $G(X,Y)$, is used in the reliability analysis to evaluate the probabilistic constraints in the RBDO. As such, it is a separate concept from the objective function $f(X,Y)$ subjected to minimization. As with the choice of the objective function, there is no limitation to the choice of limit state function, provided that it can be expressed in terms of the RVs considered in the problem. Most limit states considered in reliability analysis are strength-based. However, serviceability-based states are also possible, such as a limit on deflection, cracking, or vibration frequency. Such types of limit states may serve in place of direct strength-based limits as well, which may greatly simplify the reliability analysis in some cases. For example, conducting the reliability analysis of a complex structural system in terms of failure path analysis through multiple components may be difficult to implement. In lieu of such a formulation, a global metric for system failure might be used such as a deflection or buckling (Thompson et al. 2006; Eamon and Rais-Rohani 2009).

Probabilistic limits set in the RBDO are often taken as those corresponding to an existing reliability-based code calibration (Siavashi and Eamon 2019; Kamjoo and Eamon 2018). If this is the case, any deviation in the reliability model from that used in the calibration must be carefully considered as, in general, such deviations will render the reliability level calculated in the RBDO invalid for comparison to that established for the standard. In certain circumstances, however, deviation from the code reliability model is warranted, such as when the proposed change is not simply due to modeling preference but represents a data set unique to the structure.

Appropriate side constraints are often needed to develop an allowable range of DV values as well as realized RV values to ensure physically possible and acceptable outcomes. When side constraints are placed on RVs, this results in the use of truncated RV distributions in many cases. Using analytical, reliability index-based methods (such as the first-order reliability method [FORM]) with truncated RV, distributions can become complex as well as numerically difficult in some cases. If needed, a truncated RV of any distribution type can be readily achieved if a direct simulation reliability approach is used, by simply placing limits on the outcome, and discarding and re-running simulations wherein any RV exceeds these limits. Before the analysis is potentially complicated by instituting RV bounds, however, it is advisable to check whether this consideration is practically necessary. Unless RV variance is high and the mean close to an infeasible outcome, the probability of generating an unrealistic value may be, and often is, practically negligible.

## 1.3   RANDOM AND DESIGN VARIABLES

For typical reliability problems, failure probability is usually most sensitive to load RVs, primarily because these are generally associated with the greatest uncertainty (Nowak and Eamon 2008; Eamon and Nowak 2005). An issue of concern is the action of multiple load RVs on the structure. Properly combining multiple time-dependent loads can be a complex issue to resolve, but commonly defaults to a form of Turkstra's rule in practice. This approach implies that the probability of multiple extreme (i.e., once in a design lifetime) loads occurring simultaneously is so small that it can be practically disregarded. Thus, complex load combinations often default to the consideration of a single load at its transient or extreme (design lifetime) value and the remaining loads at their sustained or arbitrary-point-in-time values (Naess and Royset 2000).

For resistance, uncertainty in material strength is usually critical, followed by stiffness properties and geometric dimensions, depending on the components considered. These different uncertainties are traditionally divided into those of the material ($M$) such as strength and stiffness, and those due to fabrication ($F$) such as dimensions. For many factory-fabricated components, dimensional variations are often so small as to be practically insignificant in the reliability analysis. This is not so in all cases, however, such as when reinforced concrete components are considered, which have shown to have a significant variation in dimensions as well as the location of reinforcing steel (Nowak and Szerszen 2003).

Resistance parameters are sometimes available in the literature for complete structural components as well, often given in terms of bias factor ($\lambda$), the ratio of the mean value to the nominal value, and coefficient of variation (COV), the ratio of standard deviation to mean value. Such values can be determined by simple statistical analysis of a sample of model results developed from a Monte Carlo simulation (MCS) of the fundamental RVs that affect the response of interest, such as the moment capacity of a beam. Although these gross component statistics are commonly used for code calibration efforts, they may have limited usefulness for detailed RBDO procedures. This is because component statistics ($\lambda$ and COV) may in fact vary as section geometry changes, such as what may occur during the optimization, sometimes significantly (Chehab and Eamon 2019).

An additional parameter important to consider when comparing to code reliability levels is the analysis, or "professional" factor ($P$). This RV characterizes the uncertainty in the modeling approach itself, and accounts for discrepancies that result from the analysis prediction as compared to the available experimental results. Such a parameter is often unknown, but has been approximated for some cases (Chehab and Eamon 2019).

It is not uncommon that overlap will occur between parameters taken as DVs and RVs; i.e., vectors $X$ and $Y$ within Eq. (1.1) will share some (or all) elements. For example, perhaps the thickness of a beam flange is taken as a DV, but the fabricated uncertainty in this width is also considered, and is thus taken as an RV as well. In this scenario, a rational approach is to take the mean value of the RV equal to the current value of the DV. Thus, the mean value of the RV will change throughout the optimization as the DV value changes. Although some exceptions exist, in most cases, the standard deviation of the RV should similarly change, as COV is typically taken as a constant rather than standard deviation.

## 1.4 SOLUTION PROCEDURE

In general, the solution to the RBDO problem in Eq. (1.1) involves two major parts: the search for the optimum design point in DV space, and the evaluation of component failure probability (or reliability index) in RV space at every updated design point. A simple RBDO approach, a "double-loop" procedure, typically requires two iterations; one iteration, the primary outer loop, involves the optimizer, while the inner, nested loop concerns the reliability algorithm (with the assumption that iteration, in the form of a most probable point (MPP) of failure search, or a repeated random simulation, is required for calculation of reliability). A typical double-loop RBDO process, assuming a combination of deterministic and probabilistic constraints exists, is as follows. In the steps below, the outer loop corresponds to steps 1–8, while the inner loop is given by steps 4–6. Here it is assumed that finite element analysis (FEA) is used to evaluate structural response, though any analysis method could be used.

1. The optimizer perturbs the DV values.
2. The FEA model is updated with the current DV values and the FEA is performed (for the optimizer).

3. Results required by the optimizer are extracted and the deterministic constraints are evaluated.
4. The reliability algorithm perturbs (or simulates, in accordance with the method) RV values. For RVs that are also DVs, the mean values of the RVs are first updated to match the current values of the corresponding DVs. Affected RV standard deviations ($\sigma$) are also recomputed (as a product of mean value and COV) prior to the perturbation.
5. The FEA model (with DV values from step 2) is updated with the current simulated or design point RV values and an FEA is performed (for the reliability algorithm).
6. The result needed by the reliability algorithm is extracted and the limit state function evaluated.
7. Steps 4–6 are repeated as needed to compute reliability.
8. Probabilistic constraints are evaluated.
9. Steps 1–8 are repeated until convergence of the optimizer.

This is illustrated in Figure 1.1. The above steps are generic, and are direct when using simulation-based methods of optimization and reliability analysis. When gradient-based optimization or reliability analysis approaches are implemented, numerical evaluation of derivatives is required. For example, say FORM is chosen as the reliability algorithm, which generally requires direction cosines to update the RV values (i.e., the "design point") for the iterative search for the MPP of failure. Here, the RV perturbation in step 4 would refer to an increment needed to compute the numerical derivatives $\partial G_i / \partial X_m$ to locate the new search direction. Step 5 is unchanged, but step 6 would be composed of three substeps: 6a) extract the response $g_i$ and evaluate the derivative $\partial g_i / \partial x_m \approx \Delta g_i / \Delta x_m$; 6b) repeat steps 4–6a for all RVs; 6c) update the new design point, as a function of derivative information. As per step 7, steps 4–6 would be repeated until convergence of the design point (and reliability index). When numerical derivatives are estimated, it is suggested that a sensitivity study is conducted with regard to the size of the increment $\Delta$, as changing this increment may alter results.

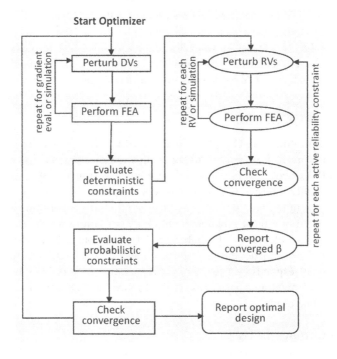

**FIGURE 1.1** Typical RBDO flowchart.

## 1.5 SELECTION OF OPTIMIZER AND RELIABILITY METHOD

Numerous optimization methods have been proposed in the last several decades. Broadly, these might be grouped into two categories: gradient-based formulations and simulation-based formulations. When DVs have continuous rather than discrete values, the solution to Eq. (1.1) can be found using a gradient-based solver such as sequential quadratic programming or the modified method of feasible directions (Vanderplaats 1999), among numerous other available approaches. In these methods, gradients of the objective function with respect to the DVs are taken each cycle to determine the new search direction to locate the optimal DV set. This procedure is repeated until the solution convergence. Although generally efficient, the use of gradients has potential drawbacks. For objective function responses that are complex, such as those which are discontinuous, contain multiple local minimums, or are non-convex, the search direction may be inaccurately computed, or a local rather than the global minimum might be found. One way to minimize this latter problem is to re-run the RDBO, considering multiple initial design points.

A different approach to optimization is represented by heuristic methods, which often use a form of probabilistic simulation in lieu of computing numerical derivatives. Some of these methods include simulated annealing, insect colony optimization, genetic algorithms, and particle swarm optimization. These methods have the potential to address RBDO problems of any complexity, but they are often significantly more computationally costly than gradient-based approaches.

As with optimization algorithms, numerous reliability methods are potentially suitable for RBDO, and the most popular of these can be similarly grouped into the same two divisions noted above: simulation methods such as MCS and its many variants; and gradient-based approaches that compute reliability index as a surrogate for direct evaluation of failure probability. This latter group of methods includes FORM. Although not as frequently used, second-order reliability methods (SORMs) have also been proposed, as well as other algorithms. Such methods attempt to locate the MPP, the peak of the joint probability density function on the failure boundary of the limit state function in standard normal space. The MPP search is itself an optimization problem, where the minimum distance from the MPP to the origin is identified; hence, the inner loop for reliability analysis in the RBDO. Once this optimal (minimum) distance is found, the reliability index can then be calculated as the distance from the MPP to the origin, from which a simple transformation to failure probability can be obtained if desired ($P_f = \Phi\,(-\beta)$).

Although computationally efficient, generally offering vast reductions in computational effort compared to simulation approaches for moderate- to high-reliability levels, such gradient-based approaches used for reliability encounter the same problems when used for deterministic optimization. These methods cannot guarantee convergence to the true solution, unlike MCS if the sample size is sufficiently increased. For complex responses, search algorithms may inaccurately identify, or be unable to identify, the MPP (Eamon and Charumas 2011). Moreover, since reliability index methods rely upon an approximation of the limit state boundary at the MPP (a linear approximation in the case of FORM), nonlinearities in standard normal space, either from inherent nonlinearities in the structural response used for the limit state function or when non-normal RVs are introduced, will result in some degree of error. Although this error is often small, in some cases, particularly for complex nonlinear limit state functions, it can be unacceptably large (Eamon and Charumas 2011).

In contrast, although direct simulation such as MCS will approach the true solution as the sample size is increased, as noted, the drawback is the large computational demand. To increase efficiency, numerous variance reduction methods were developed, such as stratified sampling, importance sampling, directional simulation, subset simulation, dimensional reduction and integration, as well as many others. As with any approach, each of these methods has particular advantages and disadvantages that must be carefully considered.

## 1.6  PROBLEM SIMPLIFICATION

For gradient-based optimization and reliability methods, one way to lower RBDO problem cost is to reduce the number of variables and constraints. A sensitivity analysis can provide quantitative guidance in this regard. For DVs, the sensitivity analysis can be conducted by computing changes in the objective function ($f$) for a given DV $Y_k$:

$$\propto_j = \frac{\partial f}{\partial Y_k}\left(\frac{Y_{0k}}{f(Y_{0k})}\right) \tag{1.2}$$

A similar analysis should be conducted with regard to deterministic constraints ($q_j$), where ($f$) is simply replaced with $q_j$ in Eq. (1.2). The term in brackets is used to non-dimensionalize the sensitivities for consistent comparison, where here the pre-perturbed value of the DV, $Y_{0k}$, is used directly as well as to evaluate the objective and constraint function of interest. For RVs, although various ways are possible to estimate importance, one such estimate can be obtained by:

$$\propto_j = \frac{\partial \beta_i}{\partial \sigma_m}\left(\frac{\sigma_{0m}}{\beta_{0i}}\right) \tag{1.3}$$

The above expression considers the change in reliability index $\beta$ associated with reliability constraint $i$, with respect to the standard deviation $\sigma$ of particular RV $m$. It is also possible to consider failure probability directly rather than $\beta$, if appropriate. However, potential numerical difficulties may arise since $P_f$ values can become very small when reliability levels increase. An alternate way to determine RV significance is simply by elimination. That is, the RV can be reduced to a deterministic parameter to quantify its effect on reliability.

Assuming that the response evaluation is the expensive part of the RBDO, where a call to an FEA code or other time-consuming procedure is required, it should be kept in mind that the effort needed for simulation-based approaches is not significantly affected by the number of RVs in the problem. Here, relative to calling an FEA code, generating an additional random number and its associated transformation to a basic variable is computationally inconsequential. Rather, savings for simulation methods are a function of the number of simulations (i.e., calls to the FEA code required). In contrast, FORM-based procedures will directly benefit by reducing the number of RVs, since each additional RV requires an additional evaluation of the response. Although RV (and DV) reduction can provide some potential savings, these savings will in general be insignificant when compared to that which could result by switching from a typical simulation-based to a reliability index-based approach.

Since each active reliability-based constraint requires probabilistic evaluation, the computational expense can be lowered by reducing the number of such constraints, regardless of the reliability method utilized. Alternatively, reduction of a probabilistic constraint to an equivalent deterministic constraint, where possible, may be effective as well. Even for deterministic constraints, for gradient-based optimization methods, the resolution of each active inequality constraint requires a response call for each DV that the constraint is a function of, to evaluate gradients. Thus, whether deterministic or probabilistic, it may be beneficial to reduce the constraint set to those that are critical for successful problem-solving.

Another issue to consider is convergence criteria. When reliability methods are used that search for the MPP, convergence is often specified in terms of the gradients or direction cosines for each RV, RV values at the MPP, as well as reliability index. Specifying very tight tolerances for convergence may dramatically increase the number of response evaluations required but produce no practical change in $\beta$. Thus, an informal sensitivity study might be performed on the reliability problem prior to the RBDO to assess the stringency of the convergence criteria that are actually needed.

In addition to the reliability method, the limit state function itself can be simplified, such as by using a metamodel or response surface, where a computationally demanding response is replaced with a simpler, analytical surrogate function. The response surface, now in analytical form, can then be solved with various traditional reliability analysis methods such as FORM to provide fast solutions. Alternative surrogate models include those developed from polynomial chaos expansion, Kriging, genetic algorithms, and artificial neural networks (ANNs), among others (Gomes 2019; Guo et al. 2020). Note that more generally, a response surface can be substituted not only for the limit state function but to evaluate the objective function and deterministic constraints as well.

In the next sections, three example RBDO problems and solution strategies are briefly discussed.

## 1.7 EXAMPLE 1: COMPLEX MARINE STRUCTURE

This problem concerned the weight minimization of a large composite structure (a submarine sail; see Eamon and Rais-Rohani 2009), which was composed of an outer shell supported by stiffeners, as shown in Figure 1.2. The shell and its stiffeners were made of laminated composites, while the top portion was made of steel. The critical design load was caused by an ocean wave striking the sail. The FE model had approximately 100,000 degree-of-freedom (DOF). The problem had 43 DVs, all of which were taken as material thickness values. These were the composite shell, divided into three thickness regions; a steel crown; and individual stiffener web and flange layer thicknesses of bidirectional (FGRP[Fiber Glass Reinforced Plastic]) and unidirectional (UGRP[Unidirectional Glass Reinforced Plastic]) material.

Constraints were required such that the reliability of all failure modes of each component were to be no less than the global buckling reliability of structure, found from a preliminary reliability analysis ($\beta = 2.51$). Failure was specified as any composite material layer reaching an ultimate strain limit in terms of axial tension or compression in the fiber or transverse directions, or in-plane shear strain. For the steel crown, failure was based on von Mises criterion. Specifying a different limit state function for each structural component and possible failure mode resulted in 116 potential probabilistic constraints. Initial RVs were the material elastic moduli in the principal material directions; shear modulus; Poisson's ratio; the ultimate tensile, compressive, and in-plane shear strains; and yield strain. A load RV was also specified corresponding to a 30-year maximum wave-slap load. Assuming component material properties were uncorrelated resulted in 205 total RVs.

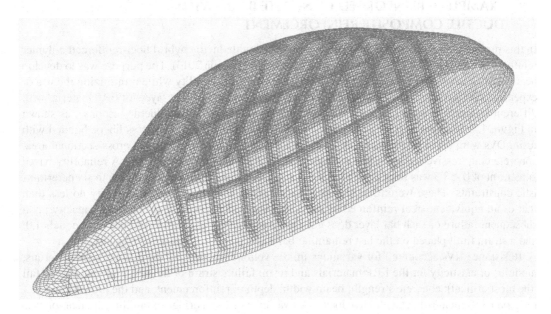

**FIGURE 1.2** Composite sail structure.

This initial number of RVs and probabilistic constraints represented an unfeasibly large computational effort. For example, allowing only five reliability iterations and 30 seconds per FEA solution would require nearly 1000 hours to evaluate all probabilistic constraints for a single optimization cycle. Thus, reduction of computational burden was paramount. This was accomplished in several ways. First, the number of optimization cycles was reduced by choosing a starting point for DV values close to the expected optimum. This was determined by conducting a deterministic optimization using constraints set to allowable strain levels that would provide approximately the same level of component reliability as specified with the reliability constraint. Second, a FORM-based reliability method was used that could locate the MPP with a small number of iterations such as that assumed in the example calculation above. Third, to reduce the number of expensive probabilistic constraints, an initial reliability analysis was conducted to determine the most critical limits. In particular, three failure modes resulted in substantially lower reliability indices than the others: shear strain in the FGRP layers, yield of the steel crown, and axial tension of the UGRP layers. Thus, rather than evaluate probabilistic limit states for every component failure individually, only three probabilistic constraint functions were formed, based on the highest shear strains within any FGRP layer, the highest axial tension strains within any UGRP layer, and von Mises strain in the crown. The remaining limit states throughout the components were reduced to deterministic limits. Note that using the highest strain result from any component would not be possible if multiple component failures were used to define structural system failure when calculating reliability. Finally, the initial reliability analysis was also used as a sensitivity study to determine critical RVs, where it was found that only four RVs significantly affected reliability (live load, ultimate FGRP shear strain, ultimate UGRP tensile strain, and steel yield strain) and thus only these were included in the RBDO.

The greatly simplified, constrained optimization problem was solved using a gradient-based method, using an ad hoc linking program to coordinate the optimizer, reliability algorithm, and FEA code. The RBDO problem converged in seven cycles using 170 CPU (central processing unit) hours to an optimal design with 5% greater weight than the initial structure. This increase in weight was due to the imposition of the minimum reliability index constraint of 2.51, which was higher than the level of reliability in some components in the initial deterministic design.

## 1.8 EXAMPLE 2: REINFORCED CONCRETE BEAM WITH DUCTILE COMPOSITE REINFORCEMENT

In this problem, the cost of concrete beams reinforced with ductile hybrid fiber-reinforced polymer reinforcing bars (DHFRP) was minimized (Behnam and Eamon 2013). The purpose was to develop beams with corrosion less reinforcement that have steel-like ductility while minimizing the use of expensive composite materials. The DHFRP bar was composed of four layers of FRP material with different failure strains, such that successive layer failures produced a ductile response, as shown in Figure 1.3. Bar materials were different grades of carbon, aramid, and glass fibers, bonded with resin. DVs were taken as the material fractions for the bar materials, total bar cross-sectional area, concrete compressive strength, beam width, and DHFRP reinforcement depth. A reliability-based constraint of $\beta \geq 3.5$ was imposed on the moment capacity of the beam, in addition to six deterministic constraints. These were: the minimum code-specified design capacity; ductility no less than that of an equivalent steel-reinforced beam; a maximum deflection; beam moment capacity upon subsequent failure of each bar layer does not decrease; the sum of bar material fractions equals 1.0; and a strain limit placed on the last remaining bar material at beam ultimate capacity.

Resistance RVs accounted for variations in: the volume fractions of the different bar materials; modulus of elasticity for the FRP materials and resin; failure strain of the first FRP material to fail (the most critical); concrete strength; beam width; depth of reinforcement; and the professional factor. Load RVs included dead and live loads, and varied whether a bridge or building application was considered. The total number of RVs varied from 17 to 19.

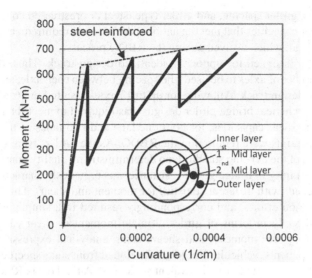

**FIGURE 1.3**  Example DHFRP bar and beam moment-curvature diagram.

In this problem, evaluation of the moment capacity limit state function could be accurately accomplished with analytically formulated, closed-form algebraic expressions. Thus, the computational effort for a single response call was trivial. As such, direct MCS was used to determine failure probability. The RBDO was conducted with an iterative procedure that systematically incremented through feasible sets of DV values to find the minimum cost solution. The process was implemented in two stages, where first a set of feasible bar configurations was developed considering the volume fraction DVs and initial constraints on bar performance requirements. Next, a set of beams containing the bars was developed by incrementing through combinations of the remaining DVs (bar area and concrete beam properties) which included a final evaluation of all constraints. Of this set of feasible beams, the minimum cost solutions were selected.

This incremental approach, particularly when combined with MCS, required a very large number of iterations compared to gradient-based solution procedures. This was only feasible due to the minimal effort required for the evaluation of the response. The advantage of this approach was its ability to directly develop a set of discrete optimal DV values that were required for practical beam designs. Although the DV values resulting from a gradient-based optimization process could be rounded off to discrete values, this approach was found to produce less-than-optimal solutions for this particular problem. The final, optimized designs were from 10% to 30% less expensive than the initial feasible designs considered.

## 1.9  EXAMPLE 3: DEVELOPMENT OF OPTIMAL LIVE LOAD MODEL

This example was an atypical application of RBDO (Kamjoo and Eamon 2018), where the objective was to minimize the variation in the reliability of different bridge girders designed to a national standard. Specifically, when designing bridges in the State of Michigan, the traffic live load specified in the standard produced girder designs with significant variations in reliability. This was caused by traffic loads allowed within the state that differed substantially from those used to develop the national design standard. Correspondingly, the goal was to develop a state-specific live load model to minimize the variation in reliability found.

Used as a metric for variance, the objective function was taken as the sum of the squares of the differences between the reliability index of each bridge girder considered and the minimum acceptable index ($\beta = 3.5$). Approximately, 40 representative bridge designs were considered, with different span lengths, girder spacing, and girder type (steel or prestressed concrete). These girders had moment and shear capacities that met the national design code requirements, with the exception of the specified live load, which evolved during the RBDO process.

Bridges are usually designed to support an idealized design truck. Thus, the load model DVs were taken as: the number of axles to be used; the spacing between each axle; and the weight of each axle for the potential design truck. Allowing for up to 13 axles resulted in 26 possible DVs.

Each of the 40 hypothetical bridge girder designs was required to meet the minimum reliability index for moment and shear capacities, for both one lane and two lanes of traffic load. Thus, 160 reliability-based constraints were required in the RBDO. As the girders were specified to meet the strength requirements of the existing standard, no deterministic inequality constraints were needed.

Since overall uncertainty in bridge girder resistance was previously characterized, only a single strength RV was needed (with different statistics for moment and shear). This was consistent with previous code calibration efforts, and correspondingly resulted in a simple, analytical limit state function expressed directly in terms of girder ultimate moment or shear capacity, with the traffic loads converted to girder moments and shears using analytical expressions. Load RVs were bridge dead load components, vehicular live loads developed from state-specific data, and vehicular dynamic load, the latter converted to an equivalent static load. A total of six RVs were used to evaluate each limit state.

Due to the use of a discrete DV (number of axles), rather than a gradient-based optimization scheme, a simulation-based approach, a genetic algorithm, was used. This required a large number of response calls, approximately 500,000 per cycle, and a large number of cycles (approximately 1000) for convergence.

To keep computational costs feasible, a modified version of the first-order second-moment method (i.e., direct solution of $\beta$ without iteration) was used (Eamon et al. 2016) that accounted for non-normal RVs and provided acceptable accuracy. Two different design vehicles were generated, one each for moment and shear effects.

At the conclusion of the RBDO, a five-axle optimal design vehicle for moment and a six-axle optimal vehicle for shear were developed. As illustrated in Figure 1.4, consistency in girder reliability was greatly improved using the optimized load model.

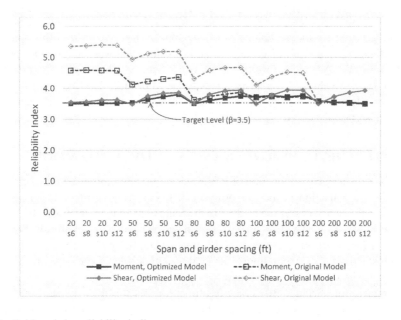

**FIGURE 1.4** Bridge girder reliability indices.

## REFERENCES

Behnam, B., and Eamon, C. "Reliability-Based Design Optimization of Concrete Flexural Members Reinforced with Ductile FRP Bars." Construction and Building Materials, Vol. 47, pp. 942–950, 2013.

Chehab, A., and Eamon, C. "Reliability-Based Shear Rating of Prestressed Concrete Bridge Girders Considering Capacity Adjustment Factor." ASCE Journal of Risk and Uncertainty in Engineering Systems, Vol. 5, No. 2, 2019.

Eamon, C., and Charumas, B. "Reliability Estimation of Complex Numerical Problems Using Modified Conditional Expectation Method." Computers and Structures, No. 89, pp. 181–188, 2011.

Eamon, C., Kamjoo, V., and Shinki, K. "Design Live Load Factor Calibration for Michigan Highway Bridges." ASCE Journal of Bridge Engineering, Vol. 21, No. 6, 2016.

Eamon, C., and Nowak, A.S. "Effect of Edge-Stiffening and Diaphragms on the Reliability of Bridge Girders." ASCE Journal of Bridge Engineering, Vol. 10, No. 2, pp. 206–214, 2005.

Eamon, C., and Rais-Rohani, M. "Integrated Reliability and Sizing Optimization of a Large Composite Structure." Marine Structures, Vol. 22, No. 2, pp. 315–334, 2009.

Gomes, W. "Structural Reliability Analysis Using Adaptive Artificial Neural Networks." Journal of Risk and Uncertainty in Engineering Systems Part B: Mechanical Engineering, Vol. 5, pp. 1–8, 2019.

Guo, Q., Liu, Y., Chen, B., and Zhao, Y. "An Active Learning Kriging Model Combined with Directional Importance Sampling Method for Efficient Reliability Analysis." Probabilistic Engineering Mechanics, No. 60, 2020.

Kamjoo, V., and Eamon, C. "Reliability-Based Design Optimization of a Vehicular Live Load Model." Engineering Structures 168, pp. 799–808, 2018.

Naess, A., and Royset, J. "Extensions of Turkstra's Rule and Their Application to Combination of Dependent Load Effects." Structural Safety, Vol. 22, No. 2, pp. 129–143, 2000.

Nowak, A., and Eamon, C. "Reliability Analysis of Plank Decks." ASCE Journal of Bridge Engineering, Vol. 13, No. 5, pp. 540–546, 2008.

Nowak, A., and Szerszen, M. "Calibration of Design Code for Buildings (ACI 318): Part 1-Statistical Models for Resistance." ACI Structural Journal, Vol. 100, No. 3, pp. 377–382, 2003.

Rais-Rohani, M., Solanki, K., Acar, E., and Eamon, C. "Shape and Sizing Optimization of Automotive Structures with Deterministic and Probabilistic Design Constraints." International Journal of Vehicle Design, Vol. 54, No. 4, pp. 309–338, 2010.

Siavashi, S., and Eamon, C. "Development of Traffic Live Load Models for Bridge Superstructure Rating with RBDO and Best Selection Approach." ASCE Journal of Bridge Engineering, Vol. 24, No. 8, 2019.

Thompson, M., Eamon, C., and Rais-Rohani, M. "Reliability-Based Optimization of Fiber-Reinforced Polymer Composite Bridge Deck Panels." ASCE Journal of Structural Engineering, Vol. 132, No. 12, pp. 1898–1906, 2006.

Vanderplaats, G. N. "Numerical optimization techniques for engineering design, Research & Development." Vanderplaats Research and Development, Inc., Colorado Springs, CO, 1999.

# 2 Reliability-Based Structural Vibration Control

*Yuanfeng Shi*
Sichuan University, Chengdu, China

*Ka-Veng Yuen*
University of Macau, Macau, China

## CONTENTS

## 2.1 INTRODUCTION

### 2.1.1 RELIABILITY-BASED DESIGN UNDER UNCERTAINTIES

Due to the existence of various sources of uncertainty inherent in engineering design, accurate and efficient techniques are necessary to analyze and assess uncertainties from external loadings, materials, geometry, construction processes, and operational environments, etc., and their influence on the performance of structures. Uncertainty analysis can facilitate robust design, providing the designer with a satisfactory tool for design under a given amount of uncertainty [1]. Treatment of uncertainties can be performed in a probabilistic or a non-probabilistic framework. When the uncertainty is recognizable and quantifiable, the probability-based approach is suggested because it provides a rational and consistent framework for uncertainty quantification and analysis in engineering design [2].

DOI: 10.1201/9781003194613-2

Reliability is an important notion in almost all applications of civil engineering and is receiving continual concerns in the design process of engineering structures under uncertainties [3–5]. Qualitatively, structural reliability is the ability of the structure to behave in its designed function properly under various prescribed conditions during its design life. Quantitatively, structural reliability deals with the determination and prediction of the probability of the failure event at any stage during the design life of a structure. Various techniques on theory, calculation, analytical approximation, stochastic simulation, etc. for reliability have been developed. Reliability-based analysis and design nowadays is a common accepted design practice in almost all fields of engineering [4, 5]. For a robust and cost-effective design of engineering systems, reliability-based design optimization (RBDO) has thus emerged as one of the standard tools to deal with uncertainties. Two types of objective function in RBDO are commonly considered: (1) life cycle cost function as the objective function along with target reliability as the constraint [6]; (2) reliability measure as the objective function [7].

## 2.1.2　Structural Vibration Control under Uncertainties

Modern control techniques have been extensively used in the vibration control of civil structures under dynamic actions such as extreme earthquakes and strong winds [8, 9]. Various vibration control systems for structures categorized as passive, active, semi-active, and hybrid control systems have been developed and applied. Often called structural protective systems, these systems can improve the structural performance by modifying the dynamical properties of a structure in a prescribed manner. Hence, all vibration control applications for structures seek to mitigate the response of a structure to dynamic loadings. The development of effective and robust control algorithms to command the control devices to achieve the optimal control performance becomes crucial for a successful realization of the structural control technology [10].

Since the full information of a dynamical system is never available, it may fail to provide satisfactory control performance for those classical control approaches developed from a single nominal (deterministic) model of the system. Thus, robustness of a controller is often sought to enhance the control performance when uncertainties are encountered [11]. Robust control techniques, such as $\mathcal{H}_2$, $\mathcal{H}_\infty$, and μ-synthesis, etc., were therefore developed so as to achieve robust performance and stability of the controlled system considering bounded modeling errors for a set of "possible" models of the system.

However, existing $\mathcal{H}_\infty$/μ-synthesis robust control methodologies give only the guaranteed (i.e., "worst-case") system performance [12] and do not consider the information about the system's probable performance, which can be of interest to civil engineers. The "worst" model leading to worst-case performance may be quite improbable in practice. On the other hand, engineering knowledge from considerable information of engineering applications gives the relative likelihood of the parameter values considered. Thus, studies based on a probabilistic interpretation were developed in dealing with the robustness of the controlled systems. For example, reliability-based measures for assessing the robustness of the controlled structural systems under uncertainties were studied in reference [12], and robust stabilization was addressed in reference [13] for uncertain dynamical systems using a probabilistic approach. The Monte Carlo method and first-order reliability method/second-order reliability method (FORM/SORM) were adopted, respectively, to evaluate the reliability in these studies. The performance comparison of different control laws is also feasible by reliability-based assessment.

Since the information about a system and its surrounding environment is never completely available, model parameters of the system and its excitation cannot be known exactly. Probabilistic description for uncertainties gives a logical measure of the plausibility of the uncertain parameters. It thus becomes relevant that optimal vibration control of civil structures should achieve to maximize their reliability or minimize the probability of failure for the defined event in the presence of uncertainties. The concept of robust reliability [14] introduced for the circumstance of incomplete information about the system in nature can serve as a reasonable metric for a controller to be assessed. Furthermore, reliability-based design can be updated using updated probabilistic information of the model uncertainties from tested or monitoring data using the Bayesian system identification techniques [15, 16].

A reliability-based design approach to structural vibration control is described in this chapter. This reliability problem here is the classic first-passage problem [17], which aims to determine the probability of the response trajectory of the structure that out-crosses the bounds of a safe domain within a time duration. The design variables related to the control devices or their control laws are sought for an optimal-reliability control performance. The methodology can be directly applied for the design of optimal passive and active controllers, while for the semi-active systems, the reliability-based design concept can be achieved by clipping the reliability-based active controller.

## 2.2 RELIABILITY-BASED ROBUST CONTROL PROBLEM

The controlled dynamical structure can be treated generally using a state-space representation with a model class $\mathcal{M}(\theta)$ as

$$\begin{cases} \dot{\mathbf{x}} = \mathbf{f}\left(\mathbf{x}, \mathbf{w}, t | \varphi, \theta\right) \\ \mathbf{y} = \mathbf{g}\left(\mathbf{x}, \mathbf{w}, t | \varphi, \theta\right) \\ \mathbf{z} = \mathbf{h}\left(\mathbf{x}, \mathbf{w}, t | \varphi, \theta\right) \end{cases} \tag{2.1}$$

where $\mathbf{x}$ is the system state vector containing the structural states and those ancillary states that model the dynamics of actuators and sensors, spectral characteristics of the excitation, and so on; $\mathbf{w}$ is the white noise vector modeling both the structural unknown excitation and output noise; the feedback vector $\mathbf{y}$ is relating to the design variables of the control devices or the feedback control law; $\mathbf{z}$ is the control performance vector used for the criterion of design for the control devices or controllers, including displacements, interstory drifts, control forces, etc.; $\theta \in \Theta$ is the vector of the model parameters with the allowable parameter space $\Theta$; $\varphi$ is the vector of the control design variables such as the design parameters of the control devices and/or the control gains of a control algorithm.

Under the influence of stochastic excitation and model uncertainty, the robustness of the performance of the controlled system can be well described in a reliability perspective. The optimal controller is the one that achieves the maximum robust reliability in the presence of uncertainties. It is equivalent to the one with the least robust failure probability as:

$$\varphi^* = \underset{\varphi \in \Omega}{\operatorname{argmin}} P\left(F | \varphi, \Theta\right) \tag{2.2}$$

where $\varphi^*$ is the optimal design; $\Omega$ is the space of the design variables; and $P\left(F | \varphi, \Theta\right)$ is the robust failure probability of the considered failure event $F$. Determination of the value $P\left(F | \varphi, \Theta\right)$ depends on the prior knowledge or the assumptions of the model in Eq. (2.1) for a given problem. Considering the uncertainties from both the external excitation and model parameters, the robust failure probability can be written as [14]:

$$P\left(F | \varphi, \Theta\right) = \int_\Theta P\left(F | \varphi, \theta\right) p(\theta | \Theta) \, \mathrm{d}\theta \tag{2.3}$$

where $P\left(F | \varphi, \theta\right)$ is the conditional failure probability under stochastic excitation on a set of specified model parameters for a considered failure event; $p(\theta | \Theta)$ is the probabilistic density function describing the plausibility of the model parameters. In particular, if the model parameters $\theta$ are known to be $\theta_0$, which is the deterministic case as $p(\theta|\Theta) = \delta(\theta - \theta_0)$, then the uncertainty source for the problem is only from the stochastic input $\mathbf{w}$. In general, evaluation of the integral of Eq. (2.3) cannot be performed analytically and is often resorted numerically although it is also nontrivial when the dimension of $\theta$ is not small (say, more than 3) [7].

### 2.2.1 STATE-SPACE REPRESENTATION OF LINEAR FEEDBACK CONTROL

Since in stochastic analysis of general dynamical systems, the calculation of the reliability is often a challenging problem, most reliability-based robust design is performed with linear models [7]. With linearity and further stationarity assumptions for the stochastic system, it allows for the use of an analytical approximation to the reliability calculation of the control performance measures, thereby realizing efficient optimization.

When a linear time-invariant system is considered, Eq. (2.1) can be simply expressed in continuous time as:

$$\begin{cases} \dot{\mathbf{x}} = \mathbf{A}(\varphi,\theta)\mathbf{x} + \mathbf{B}(\varphi,\theta)\mathbf{w} \\ \mathbf{y} = \mathbf{L}_1\mathbf{x} + \mathbf{L}_2\mathbf{w} \\ \mathbf{z} = \mathbf{C}(\varphi,\theta)\mathbf{x} + \mathbf{D}(\varphi,\theta)\mathbf{w} \end{cases} \qquad (2.4)$$

or in discrete time as:

$$\begin{cases} \mathbf{x}_{k+1} = \mathbf{A}_d(\varphi,\theta)\mathbf{x}_k + \mathbf{B}_d(\varphi,\theta)\mathbf{w}_k \\ \mathbf{y}_k = \mathbf{L}_1\mathbf{x}_k + \mathbf{L}_2\mathbf{w}_k \\ \mathbf{z}_k = \mathbf{C}(\varphi,\theta)\mathbf{x}_k + \mathbf{D}(\varphi,\theta)\mathbf{w}_k \end{cases} \qquad (2.5)$$

where the model matrices $\{\mathbf{A}, \mathbf{B}, \mathbf{L}_1, \mathbf{L}_2, \mathbf{C}, \mathbf{D}\}$ or $\{\mathbf{A}_d, \mathbf{B}_d, \mathbf{L}_1, \mathbf{L}_2, \mathbf{C}, \mathbf{D}\}$ are functions of $\varphi$ and $\theta$ ; the feedback vector is linearly related to the state vector and stochastic input by the deterministic matrix $\mathbf{L}_1$ and $\mathbf{L}_2$; the performance variable vector for reliability measures $\mathbf{z}$ or $\mathbf{z}_k$ is often taken to be a linear combination of the system state vector.

If linear feedback control is considered, the control force $\mathbf{f}_c$ can be simply written as:

$$\mathbf{f}_c = \varphi\mathbf{y} \qquad (2.6)$$

The above form of controller establishes an instantaneous relationship between the feedback and control force vectors. Note that Eq. (2.6) implicitly allows for the inclusion of dynamics of the controllers in the feedback as well when the state vector is composed of a desired set of response states [18].

## 2.3 THE FIRST-PASSAGE PROBLEM OF DYNAMICAL SYSTEM

The problem of structural control is to mitigate the vibrational response of structures under dynamical loadings such as earthquakes and winds using control devices so as to improve or maintain the reliability. In the reliability-based design, a limit state or failure event for selected control performance quantities should be predefined. The optimal design variables thus also depend on the selected failure event.

The first-passage problem is often considered as the control performance for RBDO for civil applications [18–24]. Satisfactory performance is that the selected important response quantities remain in the safe domain for a time duration. Structural failure is the occurrence of the event that at least one of the response quantities exits the safe domain for the first time.

The safety domain is defined as:

$$S_i := \{|z_i(t)| < \beta_i, t \in [0, T]\}, i = 1, 2, \cdots, n_z$$

$$S := \bigcap_{i=1}^{n_z} S_i \qquad (2.7)$$

where $\beta_i$ is the threshold of the control performance quantity $z_i$; $n_z$ is the number of performance quantities considered.

The complementary failure domain is:

$$F_i := \{|z_i(t)| \geq \beta_i, t \in [0, T]\}, i = 1, 2, \cdots, n_z$$

$$F := \bigcup_{i=1}^{n_z} F_i \qquad (2.8)$$

There is no known exact solution for the calculation of the failure probability for the classic first-passage problem for a system under random excitation [17]. It is often tackled approximately based on threshold-crossing theory by estimating the "out-crossing rate," which is the mean rate that a random process crosses a specified boundary in the outward direction. Moreover, the failures are assumed as independent arrivals of a Poisson process when calculating the failure probability of the first-passage problem from the out-crossing rate. By this, the probability of each failure event $F_i$, assuming an unfailed system initially, can be approximated as:

$$P(F_i|\varphi, \theta) \approx 1 - \exp\left[-\int_0^T v_{\beta_i}(\varphi, \theta, t)dt\right] \qquad (2.9)$$

where $P(F_i|\varphi, \theta)$ is also called the conditional probability of failure under specified model parameter $\theta$ and controller variable $\varphi$, $v_{\beta_i}(\varphi, \theta, t)$ is the mean out-crossing rate of the response that out-crosses the threshold level $\beta_i$ at time $t$.

Under the assumption of stationarity for the response of linear dynamical systems, the mean out-crossing rate becomes independent of time, that is, $v_{\beta_i}(\varphi, \theta, t) = v_{\beta_i}(\varphi, \theta)$. Thus, the conditional failure probability of Eq. (2.9) becomes [17]:

$$P(F_i|\varphi, \theta) \approx 1 - \exp\left[-v_{\beta_i}(u, \theta)T\right] \qquad (2.10)$$

where $v_{\beta_i}(\varphi, \theta)$ is in the form as

$$v_{\beta_i}(\varphi, \theta) = \frac{\sigma_{\dot{z}_i}}{\pi \sigma_{z_i}} \exp\left(-\frac{\beta_i^2}{2\sigma_{z_i}^2}\right) \qquad (2.11)$$

where the standard deviations of the stationary response quantities can be obtained using the Lyapunov equation for the system in Eq. (2.4) or (2.5).

For vector process, the conditional probability of failure for the whole failure domain can be approximated as [25]:

$$P(F|\varphi, \theta) \approx 1 - \exp\left[-\sum_i^{n_z} v_{\beta_i}(\varphi, \theta)T\right] \qquad (2.12)$$

The approximation by Eq. (2.12) is only justified if the failures encountered are unlikely or not highly correlated. For the case of high correlation between the failures, Taflanidis and Beck [26] provide more details on the analytical approximations and computational aspects for reliability of stationary linear stochastic systems with improved accuracy comparing the result of Eq. (2.10).

When the information about the dynamical system is not complete, the robust failure reliability of the failure event $F$ formulated by Eq. (2.3) should be evaluated.

## 2.4   COMPUTATIONAL ASPECTS

Note that due to the complex coupling effect among system modeling, stochastic analysis, and parameter optimization, the computational burden considering uncertainties has restricted the range of reliability-based design applications although the theoretical reliability concept was introduced many decades ago. Its applicability is often restricted by the ability to perform the design optimization and the availability of computational resources. For complex systems, this issue has been often resorted by: (1) the use of surrogate modes that keep only necessary but not full characteristics of the true system behavior, or (2) the use of approximate techniques to evaluate their probabilistic measures. Recent advances in computing technology with respect to both software and hardware have been made to overcome many of these barriers.

1. Asymptotic approximation

   The optimization problem for solving the design variable $\varphi$ requires repeated evaluations of Eq. (2.3) in any iterative optimization algorithm. First, the exact analytical expression of the integral for Eq. (2.3) is not available. On the other hand, numerical integration is very costly and usually unaffordable for high-dimensional space. To make the optimal design approach feasible, one needs to resort to approximate, but computationally, economical methods for evaluating the integral of Eq. (2.3).

   An asymptotic expansion in which a Gaussian distribution is fitted to the integrand of Eq. (2.3) at its global maximum derived in reference [27] can be used:

   $$P(F|\Theta) \approx (2\pi)^{N_\theta/2} \frac{P(F|\theta^*) p(\theta^*|\Theta)}{\sqrt{\det \mathbf{L}(\theta^*)}} \tag{2.13}$$

   where $\theta^*$ is called the design point that globally maximizes $l(\theta) = \ln\left[P(F|\theta)\right] + \ln\left[p(\theta|\Theta)\right]$, $\mathbf{L}(\theta^*)$ represents the Hessian matrix of $-l(\theta)$ evaluated at $\theta^*$. Due to the high nonlinearity of $l(\theta)$, there may exist multiple local maxima, that is, multiple design points. For this case, these design points can be included to achieve improved results, as treated in reference [28].

2. Stochastic simulation techniques

   Stochastic simulation techniques, such as importance sampling [29] and subset simulation [30, 31], have been used to evaluate the integral like Eq. (2.3) when at least one design point is available. With the feature of a predefined level of accuracy to estimate system reliability, these stochastic simulation techniques can be readily applied for reliability evaluation concerning uncertainties from both the model parameters and the stochastic excitation, and can be also extended to complex issues of stochastic dynamical system such as system nonlinearity and nonstationary characteristics [7].

   The two computational categories above, for the evaluation of the robust reliability, can be used to efficiently perform the reliability-based design optimization. Note that both methods require the information of the design points beforehand. For the asymptotic approximation to the reliability integral, computational effort required to identify all design points and accurately evaluate the Hessian information increases with the dimension of the uncertain parameters and the output of the system [26]. This computational cost can be significant and the approximation error can be large for high-dimensional space. On the other hand, although the computational expense is larger when using importance sampling for the reliability integral, this technique was proven to give a robust estimation regardless of the existence of multiple design points, level of uncertainty, extent of failure probability, and dimension of the uncertain parameter space.

Furthermore, it was mentioned in reference [26] that, in most applications, the approximation error introduced by Eq. (2.10) does not affect the efficiency of the reliability evaluation because consistent estimation errors are produced in the calculated failure probabilities. This property is quite important for RBDO applications. That is to say, the comparison between the reliability objective $P(F|\varphi^1, \Theta)$ and $P(F|\varphi^2, \Theta)$ corresponding to two design choices $\varphi^1$ and $\varphi^2$, respectively, will be correct as long as the estimation error is consistent.

## 2.5 RELIABILITY-BASED ROBUST CONTROL APPLICATIONS

Control applications of the reliability-based robust structural control methodology to passive, active, and semi-active control systems are summarized in this section.

### 2.5.1 PASSIVE CONTROL SYSTEM

Reliability-based passive tuned mass damper (TMD) design under structural uncertainties was studied in reference [19]. The linear TMD parameters were sought in an optimal way that achieves the least robust failure probability by Eq. (2.2). Asymptotic expansion was used to calculate the robust reliability. The robustness of the TMD system designed by the reliability-based approach was validated. The reliability-based design was compared to the design using a mean-square response (MSR) index as:

$$\varphi^* = \underset{\varphi \in \Omega}{argmin} \int_\Theta \sigma_z^2(\varphi, \theta) p(\theta|\Theta) d\theta \qquad (2.14)$$

where $\sigma_z^2(\varphi, \theta)$ is the MSR given the model and TMD parameters assuming stationary response of the structure under stochastic excitation.

Design of liquid column mass dampers using the robust reliability-based approach was studied in reference [20]. The nonlinear characteristics of damper were approximated by a statistically equivalent linearization so that the practically applicable reliability-based design approach for linear systems can be used. An improved analytical approximation to the reliability for stationary responses was developed, allowing for a computationally efficient gradient-based design optimization.

### 2.5.2 ACTIVE CONTROL SYSTEM

Applications of the robust reliability-based design approach to active control systems were also studied in references [21–23]. In reference [21], the active controller was proposed to be the full-state feedback at the current time only, where state estimation was also required for the practical case of incomplete response measures. Considering the fact that past output measurements contain additional information, improved control performance was observed in reference [22] that developed a control law by the feedback of incomplete response measurements at previous time steps. Considering both the current and the previous output measurements, the control force can be written as [22]:

$$\mathbf{f}_c[k] = \sum_{p=0}^{N_p} \mathbf{G}_p \mathbf{z}_f[k-p] \qquad (2.15)$$

where $\mathbf{G}_p$, $p = 0, 1, \cdots, N_p$ are the control gains to be determined, $\mathbf{z}_f[k-p]$ is the filtered measurement vector at time $(k-p)\Delta t$. The $\mathbf{z}_f$ is introduced to allow more options of output types to be fed back, but it can simply be $z[k-p]$, which is the output measurement vector.

### 2.5.3 SEMI-ACTIVE SYSTEM

Strong efforts have been put toward the applications of semi-active control systems in the past several decades [32] as these systems have promising advantages over the passive and active control systems. Semi-active systems can maintain the stability feature of passive systems but also having the feature of active control systems with adjustable parameters. Similar to the passive systems, the control forces developed with are opposite to the motion of the structure, and like the active systems, controllers are required to monitor feedbacks and generate appropriate commands to the semi-active instruments. Another feature of the semi-active systems is that a small power source is required to operate them.

Due to the inherent nonlinear behavior of semi-active devices, it is always difficult to develop the optimal control law. Control strategy for semi-active devices is often made to mimic the performance of a referenced active controller. The command signal to the semi-active device is thus tuned in a suboptimal way.

A reliability-based design methodology was developed for a semi-active system using magnetorheological (MR) dampers by adopting the clipped-optimal control algorithm to command the voltage of the semi-active device in reference [24]. The purpose of the clipped control algorithm is to mimic the optimal control of the referenced active controller [33]. The guidelines for command, the voltage to MR damper is assumed as: (1) the voltage to each MR damper restricted to the range of $v \in [0, V_{\max}]$; (2) the magnitude of the damper force $|f_{mr}|$ increases or decreases when $v$ increases or decreases for a fixed set of states.

The clipped controller develops feedback loops of control forces to command each MR damper to approximately produce a referenced active control force, attempting to mimic the performance of the active control system. The commanded voltage at time $k\Delta t$ for the MR damper is:

$$v[k] = V_{\max} H\left[\left(f_{\text{ref}}[k] - f_{mr}[k]\right) f_{mr}[k]\right] \tag{2.16}$$

where $H[\cdot]$ is the Heaviside unit step function; $f_{mr}[k]$ is the actual MR damper force measured at time $k\Delta t$; and $f_{\text{ref}}[k]$ is the control force from a referenced active controller.

When large and fast changes in the control forces applied to the structure result in high local acceleration values for the structure with low dominant frequencies, a modified clipped-optimal control law by tuning the control voltage between 0 and $V_{\max}$ [34] can be used

$$v[k] = V[k] H\left[\left(f_{\text{ref}}[k] - f_{mr}[k]\right) f_{mr}[k]\right] \tag{2.17}$$

and

$$V[k] = \begin{cases} \dfrac{|f_{\text{ref}}[k]|}{f_{\max}} V_{\max} & \text{for } |f_{\text{ref}}[k]| \leq f_{\max} \\ V_{\max} & \text{for } |f_{\text{ref}}[k]| > f_{\max} \end{cases} \tag{2.18}$$

where $f_{\max}$ is the force capacity of the MR damper.

The clipped robust reliability-based controller is the one that determines the reference control forces using a robust reliability-based active control approach such as discussed in references [21, 22]. The purpose is to achieve a robust controller for the semi-active control devices under stochastic excitation and model uncertainties from a reliability-based perspective.

## 2.6 ILLUSTRATIVE EXAMPLE

Control application of the reliability-based design approach to a semi-active system using MR dampers is demonstrated.

### 2.6.1 THE CONTROLLED STRUCTURE WITH MR DAMPERS

Consider an $N_d$-degree-of-freedom (DOF) linear structure controlled by $N_{mr}$ MR dampers under stochastic ground excitation, the equation of motion of which can be written as:

$$\mathbf{M}(\boldsymbol{\theta}_s)\ddot{\mathbf{x}}_s + \mathbf{C}(\boldsymbol{\theta}_s)\dot{\mathbf{x}}_s + \mathbf{K}(\boldsymbol{\theta}_s)\mathbf{x}_s = -\mathbf{M}(\boldsymbol{\theta}_s)\boldsymbol{\Gamma}\ddot{x}_g + \mathbf{T}\mathbf{f}_{mr} \qquad (2.19)$$

where $\mathbf{M}(\boldsymbol{\theta}_s)$, $\mathbf{C}(\boldsymbol{\theta}_s)$, and $\mathbf{K}(\boldsymbol{\theta}_s)$ are the matrices of the mass, damping, and stiffness information, respectively, and are parameterized with the overall structural parameter vector $\boldsymbol{\theta}_s$; $\mathbf{x}_s$ is the structural relative displacement vector to the ground; $\mathbf{f}_{mr}$ is the control force vector produced from MR dampers; $\ddot{x}_g$ is the ground acceleration; the matrices $\boldsymbol{\Gamma}$ and $\mathbf{T}$ indicate the distribution of the ground excitation and control forces to the structure. A control strategy is to be determined for $\mathbf{f}_{mr}$ by tuning the applied voltage of the MR dampers through feedback of the measured responses.

A 10-story shear-beam building installed with two identical MR dampers at the lowest two stories is demonstrated in Figure 2.1. It is assumed that the structure has identical nominal values of all floor

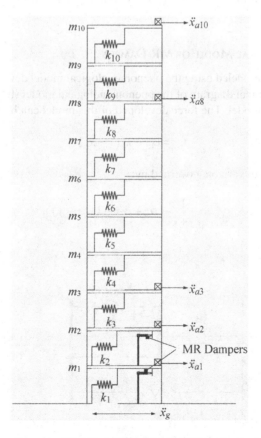

**FIGURE 2.1**  Ten-story shear-beam building installed with two MR dampers.

masses and story stiffnesses. The nominal values are taken as $m_i = 0.5$ N/(cm/s²) = 50 kg and $k_i = 883.65$ N/cm for the floor mass and story stiffness, respectively, giving $\mathbf{M}(\boldsymbol{\theta}_s) = \mathrm{diag}(m_1, m_2, \cdots, m_{10})$ and

$$\mathbf{K}(\boldsymbol{\theta}_s) = \begin{bmatrix} k_1 + k_2 & -k_2 & \mathbf{0} & 0 & 0 \\ -k_2 & k_2 + k_3 & \mathbf{0} & 0 & 0 \\ \mathbf{0} & \mathbf{0} & \ddots & \mathbf{0} & \mathbf{0} \\ 0 & 0 & \mathbf{0} & k_9 + k_{10} & -k_{10} \\ 0 & 0 & \mathbf{0} & -k_{10} & k_{10} \end{bmatrix} \tag{2.20}$$

This nominal model has the first natural frequency of 1.0 Hz. Rayleigh damping is further considered for the nominal model by $\mathbf{C}(\boldsymbol{\theta}_s) = \alpha_1 \mathbf{M} + \alpha_2 \mathbf{K}$ with $\alpha_1 = 0.094$ s⁻¹ and $\alpha_2 = 8.0 \times 10^{-4}$ s to let the damping ratios for the first two modes to be 1%.

In order to take into account the model uncertainties during the controller design, Gaussian distribution for the parameters of the stiffness and the two damping coefficients is considered by taking their nominal values as the means, and coefficients of variation to be 5.0% (stiffness) and 20.0% (damping). It is worth pointing out that Bayesian system identification techniques could be used to obtain the updated probability distribution of the structural parameters using the measured dynamic data [15, 16]. To be more realistic for control performance demonstration, the structure actually controlled is the one with a set of sampled model parameters from the prescribed distributions rather than the one with the nominal values. Here, the set of model parameters of the "actual" structure are {948.70, 836.99, 886.11, 889.33, 925.77, 881.83, 833.79, 824.03, 872.11, and 829.86} N/cm for the story stiffnesses from the bottom to the top, and {0.1 s⁻¹, 7.36 × 10⁻⁴ s} for the Rayleigh damping coefficients $\alpha_1$ and $\alpha_2$.

### 2.6.2 PHENOMENOLOGICAL MODEL OF MR DAMPER

A prototype MR damper modeled using the phenomenological model developed in reference [35] is adopted here. The mechanical diagram of the phenomenological model shown in Figure 2.2 utilizes the Bouc-Wen hysteresis model. The force developed of this model can be written as:

$$f_{mr} = c_1 \dot{y} + k_1 (x - x_0) \tag{2.21}$$

where $y$ is the internal state variable governed by:

$$\dot{y} = \frac{1}{c_0 + c_1} \left[ \alpha z + c_0 \dot{x} + k_0 (x - y) \right] \tag{2.22}$$

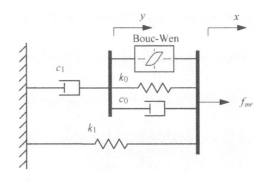

**FIGURE 2.2** Mechanical diagram of the prototype MR damper.

**TABLE 2.1**

**Model Parameters of the Prototype MR Damper**

| Parameter | Value | Parameter | Value |
|---|---|---|---|
| $c_{0a}$ | 21.0 N s/cm | $\alpha_a$ | 140 N/cm |
| $c_{0b}$ | 3.50 N s/cm V | $\alpha_b$ | 695 N/cm V |
| $k_0$ | 46.9 N/cm | $\gamma$ | 363 cm$^{-2}$ |
| $c_{1a}$ | 283 N s/cm | $\beta$ | 363 cm$^{-2}$ |
| $c_{1b}$ | 2.95 N s/cm V | $A$ | 301 |
| $k_1$ | 5.00 N/cm | $n$ | 2 |
| $x_0$ | 14.3 cm | $\eta$ | 190 s$^{-1}$ |
| $V_{max}$ | 2.25 V | | |

in which $z$ is the evolutionary variable governed by the Bouc-Wen model as:

$$\dot{z} = -\gamma|\dot{x} - \dot{y}|z|z|^{n-1} - \beta(\dot{x} - \dot{y})|z|^n + A(\dot{x} - \dot{y}) \tag{2.23}$$

Here, $k_1$ denotes the stiffness of the accumulator; $k_0$ is introduced to model the stiffness when the velocity of the damper is high; and $c_0$ and $c_1$ are used to model the viscous effect at high and low velocities, respectively; $\alpha$ is used to model the evolutionary behavior; and parameters $\gamma$, $\beta$, $n$, and $A$ are used to model the hysteresis loops developed in the prototype MR damper.

In order to model the dynamic behavior from the applied voltage $v$ to the MR damper, the model parameters related are linearly expressed as:

$$\alpha = \alpha_a + \alpha_b u; \ c_0 = c_{0a} + c_{0b}u; \ c_1 = c_{1a} + c_{1b}u; \ \dot{u} = -\eta(u - v) \tag{2.24}$$

where the dynamics for the MR damper to reach rheological equilibrium and drive the electromagnet is modeled by a first-order filer; $\eta$ is used to reflect the respond rate of the damper to the applied voltage.

Table 2.1 lists the set of model parameters for the prototype MR damper, which are adopted from reference [35].

### 2.6.3 REFERENCE ROBUST ACTIVE CONTROLLER DESIGN

The reference robust active controller is designed using the reliability-based design methodology for the linear structure. Here, discrete Gaussian white noise with a spectral intensity $S_0 = 100$ cm$^2$s$^{-3}$ is simply considered as the stationary ground excitation. However, non-white case can also be treated through filtering the white noise and then augmenting the system state. A time duration of 50 s is considered in the first-passage problem. The chosen thresholds for the selected response quantities are 11.0 cm, 2.0 cm for the top floor displacement, each interstory drift, respectively. The threshold level for the control force is chosen to be 2000 N in order to match the capacity of the MR dampers approximately. The event of failure corresponds to the exceedance of any response quantities over their threshold levels at first time during the considered time duration. Five sensors used to measure the absolute accelerations are installed at the 1st, 2nd, 3rd, 8th, and 10th floors with a sampling frequency 1000 Hz. This sensor configuration tries to extract as much response information if there is only a limited number of sensors. Also, discrete Gaussian white noise with standard deviation 10 cm/s$^2$ is added as the measurement noise.

The filter with transfer function $39.5s/(39.5s^2 + 8.89s + 1)$ is used to process the measured acceleration responses to get the pseudo-velocity since it was shown that the feedback of the velocity responses gave better control performance than the feedback of acceleration responses directly for

**TABLE 2.2**

**Optimal Control Gains and Robust Failure Probability of Active Controllers 1 and 2**

| Variable | Controller 1 | | Controller 2 | |
|---|---|---|---|---|
| | **1st Story** | **2nd Story** | **1st Story** | **2nd Story** |
| $G_0(1)$ | −119.54 | −111.01 | −612.07 | −164.33 |
| $G_0(2)$ | 113.87 | 98.39 | 54.24 | −41.37 |
| $G_0(3)$ | −11.61 | −7.42 | 66.86 | −33.06 |
| $G_0(8)$ | 17.44 | 19.68 | 64.27 | −58.76 |
| $G_0(10)$ | −0.27 | 1.15 | −103.24 | −73.65 |
| $G_1(1)$ | — | — | 516.78 | 69.86 |
| $G_1(2)$ | — | — | 17.45 | 108.15 |
| $G_1(3)$ | — | — | −60.29 | 40.86 |
| $G_1(8)$ | — | — | −45.46 | 79.39 |
| $G_1(10)$ | — | — | 101.43 | 72.69 |
| $P(F|\Theta)$ | $3.21 \times 10^{-3}$ | | $6.33 \times 10^{-4}$ | |

this type of structure as studied in reference [36]. To compare the control performance of different feedback strategies, two active controllers using the reliability-based approach are designed and summarized below:

- Active Controller 1: Feedback of the filtered acceleration responses only measured at the current time step; namely, the control design variables are the elements of $\mathbf{G}_0$ in Eq. (2.15).
- Active Controller 2: Feedback of the filtered acceleration responses measured at both the current and the previous time steps; namely, the control design variables are the elements of $\mathbf{G}_0$ and $\mathbf{G}_1$ in Eq. (2.15).

The asymptotic approximation Eq. (2.12) is used to calculate the robust failure probability during the design optimization for the optimal control gains. The optimal gains for the two robust active controllers along with their robust failure probability are listed in Table 2.2. A much smaller failure probability can be observed for Controller 2 over Controller 1, indicating that the inclusion of previous responses to be fed back provides more dynamic information of the structure to Controller 2.

### 2.6.4 SEMI-ACTIVE CONTROL PERFORMANCE

The active control forces determined by the two robust active controllers are used as the reference control forces for the clipped semi-active control law to command the MR dampers. It is nontrivial to analytically determine the robust failure probability of semi-active controlled structures with MR dampers due to their nonlinear characteristics. To have a sense of the probabilistic control performance of the semi-active system, the conditional failure probability is approximated by Eq. (2.12) with the stationary statistics of the required quantities determined through stochastic simulation for a time duration of 1000 s. The active LGQ controller is also considered for a comparison to the reliability-based controllers. The clipped semi-active controllers are denoted as Clip-Con1, Clip-Con2, and Clip-LQG (linear quadratic Gaussian) for the reference active Controller 1, Controller 2, and LGQ, respectively. The approximated conditional failure probabilities for different controlled

**TABLE 2.3**

**Conditional Failure Probability of the Structure with MR Dampers**

|  | Passive-On | Clip-Con1 | Clip-Con2 | Clip-LQG |
|---|---|---|---|---|
| $P(F|\theta)$ | 0.129 | 0.022 | 0.013 | 0.025 |

cases are given in Table 2.3. It can be seen that Clip-Con1 and Clip-Con2 have much larger failure probabilities than the corresponding active controllers in Table 2.2. The reason is that the reference active control forces cannot always be reached by the MR damper forces. The failure probability of the Clip-LGQ controller is slightly larger than the optimal-reliability ones. However, all the semi-active controllers perform better probabilistically than the passive-on one in which the voltage is held to the maximum. For the two optimal-reliability controllers, Clip-Con2 is slightly better than Clip-Con1 in terms of failure probability as it feedbacks more previous states of the structure.

### 2.6.4.1 Stationary Ground Excitation

Control performance of the different controlled cases is examined for the structure under stochastic excitation modeled by Gaussian white noise with spectral density $S_0 = 100$ cm$^2$s$^{-3}$ for 1000 s. Performance statistics in terms of the root-mean-square (RMS) and peak values of the response quantities including displacements, interstory drifts, accelerations, and control forces under different controlled cases are summarized in Table 2.4. The extents of reduction in the response quantities are represented by the percentages in parentheses when comparing the results of the uncontrolled case. The values of the controller with the best control performance are in bold.

For the passive-on system, it does reduce the response quantities significantly over the uncontrolled case, where the reductions of the maximum of the RMS and the peaks are 71% and 67% for the displacements, 67% and 64% for the interstory drift, and 36% and 28% for the accelerations, respectively. On the other hand, Table 2.4 shows that the control forces produced from the semi-actively controlled cases are much smaller than those of the passive-on one by looking at the maximum RMS values. However, the semi-actively controlled structure has better performance than the passive-on case despite the fact that the control forces of the semi-active cases are smaller. This observation indicates that by passively setting the MR dampers at their maximum voltage level to achieve their highest damping capacity may not be always the best choice to mitigate the structural response. For the two reliability-based controlled cases, Table 2.4 shows that Clip-Con2 performs

**TABLE 2.4**

**Statistics of Control Performance under Stationary Stochastic Excitation**

| Performance Quantity | Uncontrolled | Passive-On | Clip-Con1 | Clip-Con2 |
|---|---|---|---|---|
| $\max_i \sigma_{x_i}$ (cm) | 9.79 | 2.84 (−71%) | 2.64 (−73%) | 2.56 (−74%) |
| $\max_{i,t} |x_i(t)|$ (cm) | 33.48 | 11.18 (−67%) | 9.88 (−70%) | 9.79 (−71%) |
| $\max_i \sigma_{d_i}$ (cm) | 1.52 | 0.50 (−67%) | 0.43 (−72%) | 0.41 (−73%) |
| $\max_{i,t} |d_i(t)|$ (cm) | 5.63 | 2.02 (−64%) | 1.73 (−69%) | 1.62 (−71%) |
| $\max_i \sigma_{\ddot{x}_{ai}}$ (g) | 0.50 | 0.32 (−36%) | 0.21 (−57%) | 0.23 (−54%) |
| $\max_{i,t} |\ddot{x}_{ai}(t)|$ (g) | 2.00 | 1.44 (−28%) | 1.51 (−24%) | 1.46 (−27%) |
| $\max_i \sigma_{f_{mri}}$ (N) | — | 461.43 | 341.97 | 331.52 |
| $\max_{i,t} |f_{mri}(t)|$ (N) | — | 1354.12 | 1294.56 | 1312.38 |

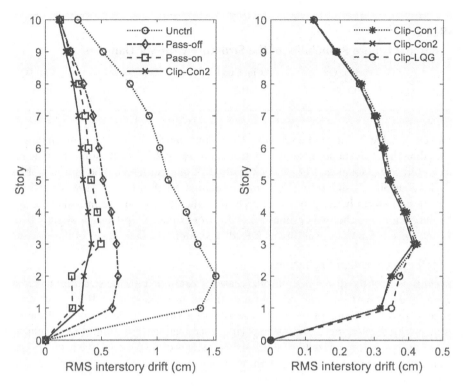

**FIGURE 2.3**   RMS of interstory drift under stationary stochastic excitation.

better than Clip-Con1 in terms of lower values of maximum RMS and peaks for the displacement and the interstory drift. This phenomenon of control performance comparison for the two reliability-based controllers is consistent with that of the results in Table 2.3, in which Clip-Con2 gives the smallest failure probability.

For a further visual illustration of the control performance of different controlled cases, the RMS values of all the interstory drifts of the structure are drawn in Figure 2.3. As can be seen, the maximum RMS value of the interstory drift occurs at the 2nd story for the uncontrolled case and the passive-off case for which no voltage is applied; however, it occurs at the 3rd story for the passive-on and semi-active cases. This indicates that the MR dampers are strong enough in reducing the responses of the stories where they were installed. As seen from Figure 2.3a, Clip-Con2 performs better in response mitigation than the passive-on case because it tries to tune the interstory drifts at the lower stories more uniformly. Figure 2.3b further compares the three clipped semi-active controllers, showing better control performance of the controllers designed with the reliability-based approach.

### 2.6.4.2   El Centro Earthquake Excitation

In order to further study the control performance of the reliability-based semi-active system, the structure excited by the 1940 El Centro earthquake is tested. Table 2.5 lists the performance quantities of the structure, showing that significant reductions of the response quantities are achieved under all controlled cases. However, the semi-active systems are again shown to be more effective than the passive-on system from the percentage reductions of most responses. On the other hand, the reliability-based controllers are superior in the reductions of all performance quantities except the maximum peak absolute acceleration and Clip-Con2 is slightly better than Clip-Con1.

**TABLE 2.5**

**Statistics of Control Performance under El Centro Earthquake**

| Performance Quantity | Uncontrolled | Passive-On | Clip-Con1 | Clip-Con2 |
|---|---|---|---|---|
| $\max_i \sigma_{x_i}$ (cm) | 5.53 | 2.12 (−62%) | 1.73 (−69%) | 1.69 (−69%) |
| $\max_{i,t} \lvert x_i(t) \rvert$ (cm) | 23.66 | 10.66 (−55%) | 9.29 (−61%) | 8.98 (−62%) |
| $\max_i \sigma_{d_i}$ (cm) | 0.85 | 0.36 (−57%) | 0.28 (−67%) | 0.27 (−68%) |
| $\max_{i,t} \lvert d_i(t) \rvert$ (cm) | 3.63 | 1.78 (−51%) | 1.35 (−63%) | 1.35 (−63%) |
| $\max_i \sigma_{\ddot{x}_{ai}}$ (g) | 0.24 | 0.14 (−44%) | 0.11 (−53%) | 0.11 (−54%) |
| $\max_{i,t} \lvert \ddot{x}_{ai}(t) \rvert$ (g) | 1.13 | 0.83 (−**27%**) | 0.85 (−25%) | 0.88 (−22%) |
| $\max_i \sigma_{f_{mri}}$ (N) | — | 295.77 | 236.07 | 226.37 |
| $\max_{i,t} \lvert f_{mri}(t) \rvert$ (N) | — | 1208.22 | 1243.40 | 1254.53 |

The peak drifts of all stories of the structure subjected to El Centro earthquake under different controlled cases are shown in Figure 2.4. The control performance of the passive-off case is worst among all the controlled cases. The passive-on system is the best in controlling the interstory drift of the bottom two stories, where MR dampers are set with the maximum energy dissipation. However, the peak drifts at other stories without MR dampers of the passive-on system are larger

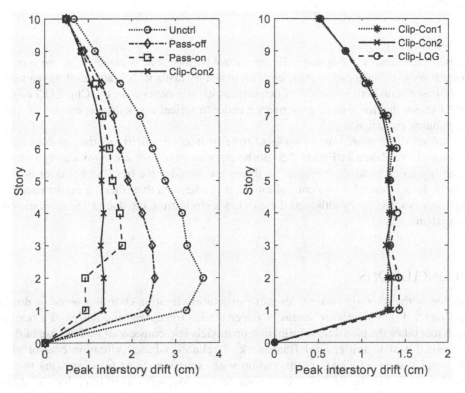

**FIGURE 2.4** Peak interstory drift under El Centro earthquake.

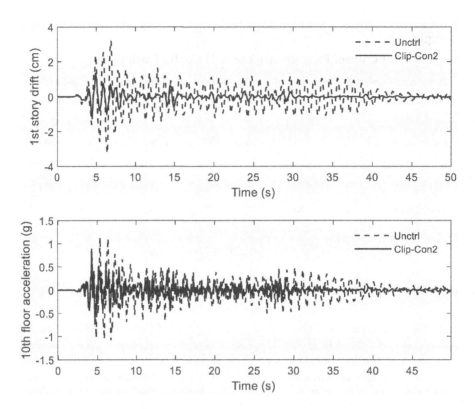

**FIGURE 2.5** The 1st interstory drift and 10th-floor acceleration of the uncontrolled and the controlled with Controller 2 under El Centro earthquake.

than those of the semi-active systems. Better control performance is achieved by the semi-active systems that try to balance the magnitudes of all interstory drifts with a more uniform pattern. The reliability-based semi-active controllers outperform slightly better than the Clip-LQG controller. Figure 2.4 shows that the control performance under historical earthquake is consistent with that under stationary excitation.

As a further illustration, time histories of the first interstory drift and the top floor acceleration responses are shown in Figure 2.5. Both response quantities are significantly reduced by the reliability-based controller. Figure 2.6 draws the actual force tracking behavior of the MR dampers to the referenced active control forces. It can be seen that tracking performance of the MR dampers is satisfactory although there exists a slight time lag due to the dynamics of the control system.

## 2.7  CONCLUSIONS

Application of the reliability-based design to structural vibration control problem is described in this chapter. Probabilistic treatment of uncertainties in engineering design is a reasonable approach that takes the plausible information on models into consideration. Because of the high computational cost in the general framework of reliability-based vibration control problem of dynamical system, efficient approximation with consistent error for evaluating the robust reliability is critical in the RBDO of controller design. Assumptions of linearity and stationarity allow for an analytical approximation to be used to calculate the reliability performance

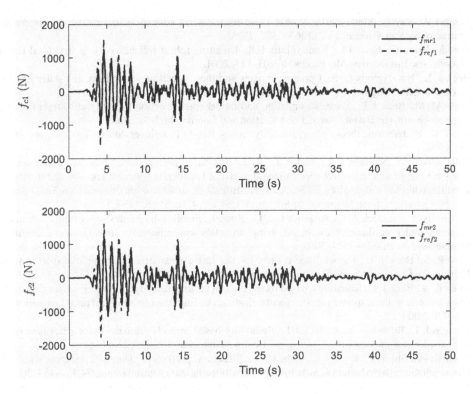

**FIGURE 2.6** The control force responses of the MR dampers and the referenced active Controller 2 under El Centro earthquake.

measures and further for efficient design optimization. The method of reliability-based structural vibration control is demonstrated on a semi-active controller design for a building installed with MR dampers.

## REFERENCES

[1] Ayyub B. M., Klir G. J., Uncertainty modeling and analysis in engineering and the sciences, Chapman & Hall/CRC, Taylor & Francis Group, 2006.

[2] Jaynes E. T., Probability theory: The logic of science, Cambridge University Press, 2003.

[3] Madsen H. O., Krenk S., Lind N. C., Methods of structural safety, Prentice-Hall, 1986.

[4] Choi S. -K., Grandhi R. V., Canfield R. A., Reliability-based structural design, Springer, 2007.

[5] Melchers R. E., Beck A. T., Structural reliability analysis and prediction, Third Ed., Wiley, 2018.

[6] Jensen H. A., Reliability-based optimization of uncertain systems in structural dynamics, AIAA Journal, 40(4):731–738, 2002.

[7] Taflanidis A. A., Reliability-based optimal design of linear dynamical systems under stochastic stationary excitation and model uncertainty, Engineering Structures, 32:1446–1458, 2010.

[8] Housner G. W., Bergman L. A., Caughey T. K., Chassiakos A. G., Claus R. O., Masri S. F., Skelton R. E., Soong T. T., Spencer B. F., Yao J. T. P., Special issue on structural control: Past, present, and future. Journal of Engineering Mechanics, 123(9):897–971, 1997.

[9] Spencer B.F., Nagarajaiah S., State of the art in structural control, Journal of Structural Engineering, 129(7):845–856, 2003.

[10] El-Khoury O., Adeli H., Recent advances on vibration control of structures under dynamic loading, Archives of Computational Methods in Engineering, 20:353–360, 2013.

[11] Liu K. Z., Yao Y., Robust control: Theory and applications, Wiley, 2016.

[12] Spencer B. F., Sain M. K., Won C. -H., Kaspari D. C., Sain P. M., Reliability-based measures of structural control robustness, Structural Safety, 15:111–129, 1994.

[13] Stengel R., Ray L., Marrison C., Probabilistic evaluation of control-system robustness, International Journal of System Science, 26(7):1363–1382, 1995.

[14] Papadimitriou C., Beck J.L., Katafygiotis L.S., Updating robust reliability using structural test data, Probabilistic Engineering Mechanics, 16:103–113, 2001.

[15] Beck J. L., Katafygiotis L. S., Updating models and their uncertainties. I: Bayesian statistical framework, Journal of Engineering Mechanics, 124(4):455–461, 1998.

[16] Muto M. M., Beck J. L., Bayesian updating and model class selection for hysteretic structural models using stochastic simulation, Journal of Vibration and Control, 14:7–34, 2008.

[17] Lin Y. K., Probabilistic theory of structural dynamics, Robert E. Krieger Publishing Company, Malabar, FL, 1976.

[18] Taflanidis A. A., Scruggs J. T., Beck J. L., Reliability-based performance objectives and probabilistic robustness in structural control applications, Journal of Engineering Mechanics, 134(4):291–301, 2008.

[19] Papadimitriou C., Katafygiotis L. S., Au S. K., Effects of structural uncertainties on TMD design: A reliability-based approach, Journal of Structural Control, 4(1):65–88, 1997.

[20] Taflanidis A. A., Beck J. L., Angelides D. C., Robust reliability-based design of liquid column mass dampers under earthquake excitation using an analytical reliability approximation, Engineering Structures, 29(12):3525–3529, 2007.

[21] May B. S., Beck J. L., Probabilistic control for the active mass driver benchmark structural model, Earthquake Engineering and Structural Dynamics, 27(11):1331–1346, 1998.

[22] Yuen K. V., Beck J. L., Reliability-based robust control for uncertain dynamical systems using feedback of incomplete noisy response measurements, Earthquake Engineering and Structural Dynamics, 32(5): 751–770, 2003.

[23] Scruggs J. T., Taflanidis A. A., Beck J. L., Reliability-based control optimization for active base isolation systems, Journal of Structural Control and Health Monitoring, 13:705–723, 2006.

[24] Yuen K. -V., Shi Y. F., Beck J. L., Lam H.- F., Structural protection using MR dampers with clipped robust reliability-based control, Structural and Multidisciplinary Optimization, 34:431–433, 2007.

[25] Veneziano D., Grigoriu M., Cornell C. A., Vector-process models for system reliability, Journal of Engineering Mechanics, 103(3):441–460, 1977.

[26] Taflanidis A. A., Beck J. L., Analytical approximation for stationary reliability of certain and uncertain linear dynamic systems with higher-dimensional output, Earthquake Engineering and Structural Dynamics, 35:1247–1267, 2006.

[27] Papadimitriou C., Beck J. L., Katafygiotis L. S., Asymptotic expansions for reliability and moments of uncertain systems, Journal of Structural Engineering, 123(12):1219–1229, 1997.

[28] Au S. K., Papadimitriou C., Beck J.L., Reliability of uncertain dynamical systems with multiple design points, Structural Safety, 21:113–133, 1999.

[29] Au S. K., Beck J. L., First-excursion probabilities for linear systems by very efficient importance sampling, Probabilistic Engineering Mechanics, 16:193–207, 2001.

[30] Au S. K., Beck J. L., Subset simulation and its applications to seismic risk based on dynamic analysis, Journal of Engineering Mechanics, 129(8):901–917, 2003.

[31] Taflanidis A. A., Beck J. L., Stochastic subset optimization for reliability optimization and sensitivity analysis in system design, Computers and Structures, 87:318–331, 2009.

[32] Symans M. D., Constantinou M. C., Semi-active control systems for seismic protection of structures: A state-of-the-art review, Engineering Structures, 21(6):469–487, 1999.

[33] Dyke S. J., Spencer B. F., Sain M. K., Carlson J. D., Modeling and control of magnetorheological dampers for seismic response reduction, Smart Materials and Structures, 5(5):565–575, 1996.

[34] Yoshida O., Dyke S.J., Seismic control of a nonlinear benchmark building using smart dampers, Journal of Engineering Mechanics, 130(4):386–392, 2004.

[35] Spencer B. F., Dyke S. J., Sain M. K., Phenomenological model of magnetorheological dampers, Journal of Engineering Mechanics, 123(3):230–238, 1997.

[36] Spencer B. F., Dyke S. J., Deoskar H. S., Benchmark problems in structural control: Part I – active mass driver system, Earthquake Engineering & Structural Dynamics, 27(11):1127–1139, 1998.

# 3 Seismic Reliability-Based Design of Elastic and Inelastic Structural Systems Equipped with Seismic Devices

*Paolo Castaldo and Guglielmo Amendola*
Geotechnical and Building Engineering (DISEG),
Politecnico di Torino, Turin, Italy

## CONTENTS

## 3.1 INTRODUCTION

Over the years, friction pendulum system (FPS) has experienced a great increase becoming a very effective seismic isolation technique (Mokha, Constantinou & Reinhorn, 1990; Constantinou, Whittaker, Kalpakidis, Fenz & Warn, 2007; Jangid, 2005). Structural reliability methods and reliability-based studies, including reliability-based optimization of base-isolated structures, have been presented by Alhan and Gavin (2005), and Castaldo, Palazzo and Della Vecchia (2015). The seismic performance of base-isolated buildings has been analyzed as a function of the FPS isolator characteristics by adopting a two-degree-of-freedom (2dof) model able to simulate the superstructure flexibility with a velocity-dependent model (Mokha, Constantinou & Reinhorn, 1990) for the FPS behavior. Particularly, nondimensional motion equations describing the seismic response of the 2dof system have been proposed by Castaldo and Tubaldi (2015) along with the description of the nondimensional statistics of the relevant response parameters as a function of both isolator and system properties. The abovementioned results can be useful to derive and assess fragility and seismic risk curves.

The present chapter assesses focuses on the seismic reliability of both elastic and inelastic superstructures isolated by FPS devices with the objectives to propose seismic reliability-based design (SRBD) results. As for the FPS isolators, SRBD expressions are presented to design these devices considering either an elastic or inelastic response of the superstructure. As for yielding superstructures, designed in line with Structural Engineering Institute (2010), European Committee

DOI: 10.1201/9781003194613-3

for Standardization, Eurocode 8 (2004), Italian Building Codes, NTC2018 (2018), and Japanese Ministry of Land, Infrastructure and Transport (2000) and assuming a perfectly elastoplastic law or a hardening and softening post-yielding rule, relationships between the ductility-dependent strength reduction factors and the displacement ductility demand are proposed. All the proposed results are achieved for several inelastic and elastic building properties, increasing seismic intensities and selecting the value of 50 years as reference life and L'Aquila (Italy) as reference site (geographic coordinates: 42°38'49''N, 13°42'25''E). In addition, aleatory uncertainties are explicitly considered on both the friction coefficient and the seismic input, so as to treat that as the main aleatory uncertainties. As far as the uncertainty in the seismic inputs is concerned, either artificial records defined through Monte Carlo simulations in line with the power spectral density (PSD) method (Shinozuka & Deodatis, 1991) or natural records have been adopted. As for the sliding friction coefficient, the value at large velocity is modeled as a random variable through a Gaussian probability density function (PDF) and the corresponding sampled values are defined in accordance with the Latin hypercube sampling (LHS) method (Celarec & Dolšek, 2013). Afterward, incremental dynamic analyses (IDAs) (Vamvatsikos & Cornell, 2002) are developed to assess the superstructure and bearing response of the elastic and inelastic systems. These results are processed for computing seismic fragility curves of both the inelastic and elastic superstructure and the FP bearings, selecting specific values of the corresponding limit states (*LSs*). Successively, the fragility curves are integrated with the site-related seismic hazard ones to derive the seismic reliability curves of the investigated structures. Finally, SRBD regressions where the ductility-dependent strength reduction factors are related to the displacement ductility demand together with SRBD abacuses for the single-concave FP devices are proposed to ensure a reliable design of structures with FPS isolators.

## 3.2  EQUATIONS OF MOTION FOR LINEAR AND NONLINEAR STRUCTURES WITH FPS DEVICES

The 2dof system, adopted by Naeim and Kelly (1999) and correctly modified in relation to the inelastic behavior of both the superstructure and the device as depicted in Figure 3.1, is used. Under the assumption of neglecting the vertical component (i.e., large radii of curvature $R$ of the FP devices),

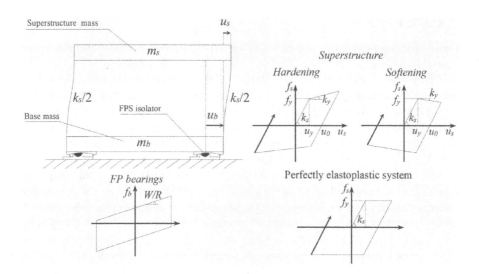

**FIGURE 3.1**  2dof model of a perfectly elastoplastic, hardening, or softening building isolated with FPS devices. (Modified from Castaldo, Amendola & Palazzo, 2017; Castaldo, Palazzo & Ferrentino, 2017; Castaldo, Palazzo, Alfano & Palumbo, 2018, with permissions.)

the device displacement can be computed only in accordance with the horizontal component and so the isolator restoring force applies (Zayas, Low & Mahin, 1990):

$$f_b = \frac{W}{R} u_b + \mu_d W \, \text{sgn}(\dot{u}_b)$$ (3.1)

where $W = (m_b + m_s)g$ denotes the weight on the isolator, $g$ the gravity constant, $u_b$ the bearing displacement with respect to the ground, $\mu_d$ the sliding friction coefficient, sgn the signum function applied to the sliding velocity, $\dot{u}_b$. The nonlinear relationship between the friction coefficient and the sliding velocity $\dot{u}_b$ applies (Mokha, Constantinou & Reinhorn, 1990; Constantinou, Whittaker, Kalpakidis, Fenz & Warn, 2007):

$$\mu_d = f_{\max} - (f_{\max} - f_{\min}) \exp(-\alpha \dot{u}_b)$$ (3.2)

where $f_{\min}$ and $f_{\max}$ denote, respectively, the friction coefficients at very low and high sliding velocities, $\alpha$ and the ratio $f_{\max}/f_{\min}$ are herein assumed as deterministic parameters equal, respectively, to 30 and 3 (Castaldo & Tubaldi, 2015; Castaldo, Palazzo & Della Vecchia, 2015; Castaldo, Amendola & Palazzo, 2017). In this way, the equations describing the nonlinear response of the 2dof system with FP devices, to the seismic input $\ddot{u}_g(t)$, apply:

$$(m_b + m_s)\ddot{u}_b + m_s\ddot{u}_s + c_b\dot{u}_b + \frac{W}{R}u_b + \mu_d W \, \text{sgn} \, \dot{u}_b = -(m_b + m_s)\ddot{u}_g$$

$$m_s\ddot{u}_b + m_s\ddot{u}_s + c_s\dot{u}_s + f_s(u_s) = -m_s\ddot{u}_g$$ (3.3a,b)

where $m_b$ and $m_s$ denote, respectively, the mass of the device and of the superstructure, $c_b$ and $c_s$, respectively, the viscous damping factor of the isolator and of the superstructure. Dividing Eq. (3.3a) by $m_b + m_s$ as well as Eq. (3.3b) by $m_s$, and introducing the mass ratio $\gamma = m_s/(m_s + m_b)$ (Naeim & Kelly, 1999), the structural $\omega_s = \sqrt{k_s/m_s}$ and isolation $\omega_b = \sqrt{k_b/(m_s + m_b)} = \sqrt{g/R}$ circular frequency, the structural $\xi_s = c_s/2m_s\omega_s$ and isolation $\xi_b = c_b/2(m_b + m_s)\omega_b$ inherent damping ratio, the equations in nondimensional form derive:

$$\ddot{u}_b + \gamma\ddot{u}_s + 2\xi_b\omega_b\dot{u}_b + \frac{g}{R}u_b + \mu_d g \, \text{sgn} \, \dot{u}_b = -\ddot{u}_g$$

$$\ddot{u}_b + \ddot{u}_s + 2\xi_s\omega_s\dot{u}_s + a_s(u_s) = -\ddot{u}_g$$ (3.4a,b)

where $a_s(u_s) = f_s(u_s)/m_s$ is the nondimensional superstructure force, which is a function, respectively, of the stiffness $k_s$ in the elastic response and of the yielding condition in the plastic one. Specifically, the superstructure responses in the elastic phase if Eq. (3.5) is satisfied and Eq. (3.6) is useful to assess the corresponding restoring force.

$$|u_{s,i} - u_{0,i-1}| < y(u_{s,i})$$ (3.5)

$$f_{s,i}(u_{s,i}) = k_s(u_{s,i} - u_{0,i-1})$$ (3.6)

In Eqs. (3.5) and (3.6), $f_{s,i}$ represents the superstructure restoring force at time instant $i$, $u_{s,i}$ the superstructure displacement with respect to the base at time instant $i$, $u_{0,i-1}$ the peak plastic excursion at time instant $(i-1)$, $k_s$ the superstructure elastic stiffness. The function $y(u_{s,i})$ represents the yielding condition that is not univocally determined due to the translation of the elastic domain, so the yielding limits depend on the displacement direction. In detail, the yielding condition

(Hong & Liu, 1999) is a function of the yield displacement $u_y$, being $f_y$ the yield force, and of the hardening or softening post-yield divided by the elastic stiffness (Hatzigeorgiou, Papagiannopoulos & Beskos, 2011). This latter ratio is denoted as $H$ or $S$ and expressed as follows:

$$H = S = k_y / k_s \tag{3.7}$$

The term ($H$ or $S$) distinguishes the hardening to the softening response. The superstructure responses inelastically if Eq. (3.8) is respected and the restoring force applies (Eq. 3.9):

$$\left| u_{s,i} - u_{0,i-1} \right| \geq y(u_{s,i}) \tag{3.8}$$

$$f_{s,i}(u_s) = k_s(u_{s,i} - y(u_{s,i})) \operatorname{sgn}(u_{s,i} - u_{0,i-1}) \tag{3.9}$$

From Eq. (3.4a), it derives that the isolation period depends only on the curvature radius of the FP isolators. In addition, the isolation period $T_b = 2\pi/\omega_b$ divided by the structural one $T_s = 2\pi/\omega_s$ provides the seismic isolation degree $I_d = T_b/T_s$ (Palazzo, 1991).

### 3.2.1 Inelastic Properties of the Nonlinear Superstructures

When it comes to the inelastic superstructure response (Figure 3.1), identified by a perfectly elastoplastic or hardening/softening behavior representative of the global response of multistory building frames (Fanaiea & Asfar Dizaj, 2014; Abdollahzadeh & Banihashemi, 2013), the strength reduction factor or behavior factor, $q$, associated only to the ductility-dependent component (Fanaiea & Asfar Dizaj, 2014; Abdollahzadeh & Banihashemi, 2013), is defined as:

$$q = \frac{f_{s,el}}{f_y} = \frac{u_{s,el}}{u_y} \tag{3.10}$$

where $u_{s,el}$ and $f_{s,el}$ denote, respectively, the yield deformation and minimum yield strength necessary for the superstructure to behave elastically.

The displacement ductility, $\mu$, of the superstructure applies:

$$\mu = \frac{u_{s,\max}}{u_y} \tag{3.11}$$

where $u_{s,\max} = \left| u_s(t) \right|_{\max}$ means the peak response during the inelastic behavior.

Note that referring to perfectly elastoplastic law, the abovementioned behavior factor $q$ is different from the one codified by Structural Engineering Institute (2010), European Committee for Standardization, Eurocode 8 (2004), Italian Building Codes, NTC2018 (2018), Japanese Ministry of Land, Infrastructure and Transport (2000), and FEMA P695 (2009) because the overstrength capacities are not taken into account explicitly. Contrarily, as for a hardening response, the behavior factor is in accordance with the code provisions. In addition, as for softening systems due to the $P - \Delta$ effects, $q$ is also consistent with the codes for the absence of the overstrength capacities.

## 3.3  ALEATORY UNCERTAINTIES IN THE SEISMIC RELIABILITY EVALUATION

Seismic reliability definition of a structure, in accordance with the structural performance (SP) method (Collins & Stojadinovic, 2000; Bertero & Bertero, 2002), is developed by coupling SP levels (SEAOC Vision 2000 Committee, 1995) and related exceeding probabilities within its reference life (CEN – European Committee for Standardization, Eurocode, 2006; Saito, Kanda & Kani, 1998). In accordance with the PEER (Pacific Earthquake Engineering Research Center)like modular approach (Cornell & Krawinkler, 2000) and performance-based earthquake engineering

(PBEE) approach (Aslani & Miranda, 2005), let us introduce an intensity measure (*IM*) to separate the uncertainties associated with the seismic input intensity from the ones related to the record characteristics. Particularly, the randomness on the seismic intensity can be represented by a hazard curve specific for the site, whereas the record characteristics uncertainty for a specific intensity level can be represented through a group of ground motions selected with different characteristics (e.g., duration and frequency content) and scaled to the same *IM* value. It is important to put in evidence that, as suggested by Shome, Cornell, Bazzurro and Carballo (1998), Luco and Cornell (2007), and Pinto, Giannini and Franchin (2003), the *IM's* choice should respect the hazard computability and criteria of both efficiency and sufficiency. In this study, the *IM* coincides with the spectral displacement, $S_D\left(\xi_b, T_b\right)$, at the period of the isolated system, $T_b = 2\pi/\omega_b$ and for the damping ratio $\xi_b$ taken equal to zero (Castaldo & Tubaldi, 2015; Ryan & Chopra, 2004). Either artificial records, achieved in line with the PSD method (Shinozuka & Deodatis, 1991; Castaldo, Amendola & Palazzo, 2017) or natural records (Castaldo, Palazzo & Ferrentino, 2017; Castaldo, Palazzo, Alfano & Palumbo, 2018) are considered to consider the record-to-record variability in the responses.

As for the aleatory randomness on the sliding friction coefficient at high velocity of the FPS isolators (Mokha, Constantinou & Reinhorn, 1990; Constantinou, Whittaker, Kalpakidis, Fenz & Warn, 2007), a specific Gaussian PDF is utilized and, successively, sampled through the LHS method. An in-depth discussion on the PDF and sampled values may be acknowledged in Castaldo, Amendola and Palazzo (2017), Castaldo, Palazzo and Ferrentino (2017), and Castaldo, Palazzo, Alfano and Palumbo (2018).

## 3.4  INCREMENTAL DYNAMIC ANALYSES (IDAs) RESULTS AND CURVES

Since the main goal of the present study concerns developing IDAs (Vamvatsikos & Cornell, 2002), the system responses for increasing *IM* levels and for different superstructure and isolator properties have been computed, in accordance with Eq. (3.4). These deterministic values are also combined with the sampled ones of the friction coefficient (Castaldo, Amendola & Palazzo, 2017; Castaldo, Palazzo & Ferrentino, 2017; Castaldo, Palazzo, Alfano & Palumbo, 2018) and with the artificial or natural ground motions scaled to the increasing levels of the $IM = S_D\left(T_b\right)$. In detail, the deterministic parameters (Castaldo, Amendola & Palazzo, 2017; Castaldo, Palazzo & Ferrentino, 2017; Castaldo, Palazzo, Alfano & Palumbo, 2018) are: the seismic isolation degree $I_d$, the isolation period of vibration $T_b$, the mass ratio $\gamma$, and the (ductility-dependent) strength reduction factor $q$ varying from 1.1 to 2 (Structural Engineering Institute, 2010; European Committee for Standardization, Eurocode 8, 2004; Italian Building Codes, NTC2018, 2018; Japanese Ministry of Land, Infrastructure and Transport, 2000). In addition, different cases of post-yield stiffness are considered (both hardening and softening behaviors): 0.03 and 0.06 for *H* and *S* (Hatzigeorgiou, Papagiannopoulos & Beskos, 2011). It is important to highlight that the yielding characteristics of the inelastic systems have been designed in line with Structural Engineering Institute (2010), European Committee for Standardization, Eurocode 8 (2004), Italian Building Codes, NTC2018 (2018), and Japanese Ministry of Land, Infrastructure and Transport (2000). Particularly, scaling the records to the $S_D\left(T_b\right)$ value specific for the life safety limit state and L'Aquila site, the average yield displacement $u_{y,average}$ and strength $f_{y,average}$ of the superstructures have been calculated in Matlab-Simulink (Math Works Inc., 1997) for each value of $q$. More details may be found in Castaldo, Palazzo and Ferrentino (2017) and Castaldo, Palazzo, Alfano and Palumbo (2018).

The different structural systems, designed as described above, are then subjected to the seismic records, scaled to several intensity levels, consistent with the IDAs procedure. Moreover, different friction coefficient values have been sampled and the simulations have been performed in Matlab-Simulink (Math Works Inc., 1997) to solve Eq. (3.4). Observe that we have not adopted any limits on the response parameters in order to numerically calculate the statistics. The results of the IDAs are computed as displacement relative to the base or ductility demand $\mu$, coherently with the superstructure behavior (e.g., linear or nonlinear), and peak displacement of the bearings with respect to the ground $u_{b,\max} = |u_b(t)|_{\max}$. These response parameters $u_{s,\max}$, $\mu$, and $u_{b,\max}$ represent the engineering demand parameters (EDPs) for the investigated systems and the corresponding peak values have been processed through lognormal PDFs (Castaldo & Tubaldi, 2015; Castaldo, Palazzo & Della Vecchia, 2015; Castaldo, Amendola & Palazzo, 2017; Cornell & Krawinkler, 2000; Ryan & Chopra, 2004) by computing the statistics by means of the maximum likelihood estimation technique (MLET), for each parameter combination and *IM* level. Therefore, it has also been suitable to calculate the 50th, 84th, and 16th percentile for each lognormal PDF (Castaldo & Tubaldi, 2015). Observe that, for softening structural systems, the maximum available ductility capacity coincides with the numerical failure condition that occurs if the softening strength is totally annulled.

Figure 3.2 shows the IDA curves in relation to the EDP $u_{b,\max}$ for elastic superstructure (Figure 3.2a) perfectly elastoplastic (Figure 3.2b) system and both hardening (Figure 3.2c) and softening superstructure (Figure 3.2d) in order to outline the main differences among the above-mentioned systems and together with the influence of both hardening and softening post-yield stiffness. Particularly, when an inelastic (hardening or softening) system is considered, higher response values are in general required for the FP bearings differently from the elastic or perfectly elastoplastic case. (Figure 3.2a) illustrates the IDA results for the FPS response $u_{b,\max}$ considering an isolation period $T_b \cong 3s$ with the elastic superstructure period ranging from $T_s = 0.3s$ to $T_s = 1.5s$, whereby the isolation degree $I_d$ is considered ranging between 2 and 10. The lognormal mean values of $u_b$ slightly increase for decreasing $T_s$. As for the inelastic system, an isolation degree $I_d = 2$ and an isolation period of vibration $T_b = 3s$ have been assumed in Figures 3.2b–d. As is evidenced in Figure 3.2, the increase of $q$ has a little influence on the isolator response showing a slight decrease of $u_{b,\max}$. Finally, Figure 3.2b shows how $\gamma$ affects the statistics: the isolation displacement goes down

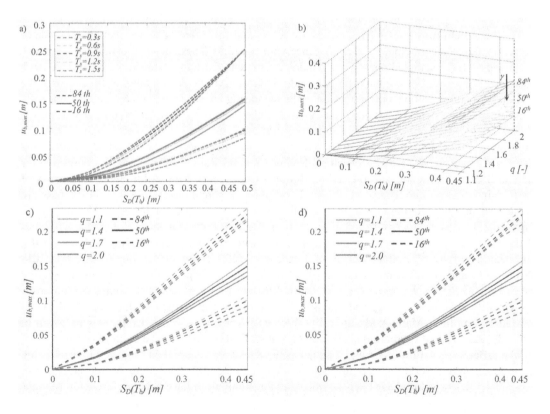

**FIGURE 3.2** IDA curves of the isolation level related to elastic (a), perfectly elastoplastic (b), nonlinear hardening (c), and softening (d) behavior, with $H = 0.03$, $S = 0.03$. (Modified from Castaldo, Amendola & Palazzo, 2017; Castaldo, Palazzo & Ferrentino, 2017; Castaldo, Palazzo, Alfano & Palumbo, 2018, with permissions.)

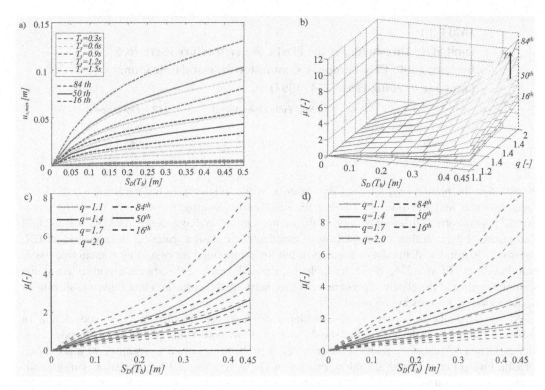

**FIGURE 3.3** IDA curves of the superstructure related to elastic (a), perfectly elastoplastic (b), nonlinear hardening (c), and softening (d) behavior, with $H = 0.03$, $S = 0.03$. (Modified from Castaldo, Amendola & Palazzo, 2017; Castaldo, Palazzo & Ferrentino, 2017; Castaldo, Palazzo, Alfano & Palumbo, 2018, with permissions.)

for increasing $\gamma$. Note that Figures 3.2c and 3.2d are considered with $\gamma = 0.6$; a value of $\gamma = 0.7$ is assumed for the elastic superstructure in Figure 3.2a.

Figure 3.3 depicts the IDA curves for the superstructure response ($T_b \cong 3s$, $\gamma = 0.7$) in terms of $u_s$ valid for elastic superstructure (Figure 3.3a) and of the EDP $\mu$ (Figures 3.3b–d). As depicted in Figures 3.3b–d, the statistics of the EDP $\mu$ are highly influenced by $q$ since its increase leads to a noticeable displacement ductility demand $\mu$, particularly for the softening behavior (Figure 3.3d). Figure 3.3b puts in evidence an increase of the structural damage as the mass ratio increases. The parameter $H$ (Figure 3.3c) allows to achieve a strong reduction of the statistics of the inelastic super-structure response because the displacement ductility demand is reduced for all structural parameter combinations, differently from the perfectly elastoplastic systems. Note that an isolation degree $I_d = 2$ and an isolation period of vibration $T_b = 3s$ have been assumed in Figures 3.3b and c, whereas $I_d = 8$ and $T_b = 3s$ for the softening behavior (Figure 3.3d). In Figure 3.3c and d, the mass ratio is $\gamma = 0.6$. Regarding the elastic system (Figure 3.3a), the lognormal mean decreases for lower values of $T_s$.

An in-depth description of the IDAs results is presented in Castaldo, Amendola and Palazzo (2017), Castaldo, Palazzo and Ferrentino (2017), and Castaldo, Palazzo, Alfano and Palumbo (2018).

## 3.5 SEISMIC FRAGILITY OF ELASTIC AND INELASTIC BASE-ISOLATED STRUCTURES WITH FPS BEARINGS

In the present section, it is described the computation of the seismic fragility representative of the probabilities $P_f$ exceeding different $LS$ thresholds at each level of the $IM$, related to both the FPS bearings and the elastic as well as inelastic superstructure. The $LS$ thresholds are herein expressed

**TABLE 3.1**

**Limit State Thresholds for the Elastic Superstructure (Bertero & Bertero, 2000; CEN-European Committee for standardization, Eurocode 0, 2006; FEMA 274, 1997)**

| | LS1 – Fully Operational | LS2 – Operational |
|---|---|---|
| *Interstory drift (ISD) index* | 0.1% | 0.2% |
| $p_f$ *(50 years)* | $5.0 \cdot 10^{-1}$ | $1.6 \cdot 10^{-1}$ |

as radius in the plan, *r [m]*, for the FPS isolators, while, referring to the performance of elastic superstructure, four *LS*s (i.e., *LS1*, *LS2*, *LS3*, *LS4*), corresponding, respectively, to "fully operational," "operational," "life safety," and "collapse prevention" are considered (SEAOC Vision 2000 Committee, 1995). Within the displacement-based seismic design, each *LS* is expressed as IDI, interstory drift index, defined as a fraction of the limits provided for designing comparable fixed-base buildings (FEMA 274, 1997). In Table 3.1, the *LS1* and *LS2* thresholds useful to assess the seismic fragility of the elastic superstructures together with the corresponding failure probabilities in the period of 50 years are reported.

As for the inelastic superstructure, available displacement ductility $\mu$ *[-]* is considered. In Tables 3.2 and 3.3, different *LS* thresholds in relation to both *r* and $\mu$ are listed. Besides this, Tables 3.2 and 3.3 report the related reference failure probabilities in a time frame of 50 years (Aoki, Ohashi, Fujitani et al., 2000; Castaldo, Palazzo & Della Vecchia, 2015; European Committee for Standardization, Eurocode 8, 2004): this latter corresponds to the collapse *LS* for the FP isolators and to the life safety *LS* for the superstructure (also in line with the design).

The exceedance probabilities $P_f$, at each *IM* level and for each parameter combination, are numerically calculated using complementary cumulative distribution functions (CCDFs) and then fitted through lognormal PDFs (Castaldo, Amendola & Palazzo, 2017). Particularly, for softening behavior, within the computation of the exceeding probabilities $P_f$, the numerical results deriving from both the collapse and not-collapse cases have been accounted for through the total probability theorem (Shome, Cornell, Bazzurro & Carballo, 1998), as follows:

$$P_{SL}(IM = im) = (1 - F_{EDP|IM=im}(LS_{EDP})) \cdot \frac{N_{not-collapse}}{N} + 1 \cdot \left(1 - \frac{N_{not-collapse}}{N}\right) \qquad (3.12)$$

**TABLE 3.2**

**LS Thresholds for the FP Isolators with the Corresponding Reference Exceedance Probability**

| | $LS_{b,1}$ | $LS_{b,2}$ | $LS_{b,3}$ | $LS_{b,4}$ | $LS_{b,5}$ | $LS_{b,6}$ | $LS_{b,7}$ | $LS_{b,8}$ | $LS_{b,9}$ | $LS_{b,10}$ |
|---|---|---|---|---|---|---|---|---|---|---|
| *r [m]* | 0.05 | 0.1 | 0.15 | 0.2 | 0.25 | 0.3 | 0.35 | 0.4 | 0.45 | 0.5 |

$$p_f\,(50\ years) = 1.5 \cdot 10^{-3}$$

**TABLE 3.3**

**LS Thresholds for the Inelastic Superstructure with the Corresponding Reference Exceedance Probability**

| | $LS_{\mu,1}$ | $LS_{\mu,2}$ | $LS_{\mu,3}$ | $LS_{\mu,4}$ | $LS_{\mu,5}$ | $LS_{\mu,6}$ | $LS_{\mu,7}$ | $LS_{\mu,8}$ | $LS_{\mu,9}$ | $LS_{\mu,10}$ |
|---|---|---|---|---|---|---|---|---|---|---|
| $\mu$ *[-]* | 1 | 2 | 3 | 4 | 5 | 6 | 7 | 8 | 9 | 10 |

$$p_f\,(50\ years) = 2.2 \cdot 10^{-2}$$

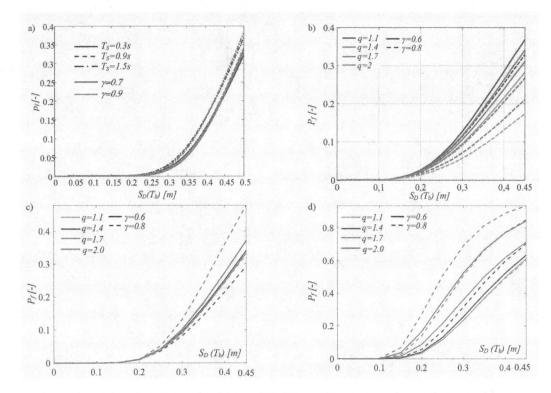

**FIGURE 3.4** Seismic fragility curves of the isolation level related to elastic (a), perfectly elastoplastic (b), nonlinear hardening (c), and softening (d) behavior, with $H = 0.03$, $S = 0.03$. (Modified from Castaldo, Amendola & Palazzo, 2017; Castaldo, Palazzo & Ferrentino, 2017; Castaldo, Palazzo, Alfano & Palumbo, 2018, with permissions.)

where $N$ denotes the total number of numerical analyses for the specific structural system at the specific $IM$ level, and $N_{not-collapse}$ represents the number of the successfully performed simulations (i.e., without any collapse).

The fragility curves are illustrated in Figures 3.4 and 3.5 with regard, respectively, to the isolators and the superstructure. In each plot, different curves in relation to the different values of $\gamma$, $T_s$, and $q$ are represented. Furthermore, only some results associated with few $LS$ thresholds are reported because of space constraints, more details are available in Castaldo, Amendola and Palazzo (2017), Castaldo, Palazzo and Ferrentino (2017), and Castaldo, Palazzo, Alfano and Palumbo (2018). Generally speaking, the seismic fragility of both the superstructure and FPS devices increases for decreasing $LS$ thresholds and reaches peak values when the inelastic system presents a post-yield softening behavior.

Figure 3.4 depicts the fragility curves of the FPS for the elastic system (Figure 3.4a), the perfectly elastoplastic behavior (Figure 3.4b) along with hardening and softening systems, respectively (Figure 3.4c and d). For the inelastic systems, the figures are related to $T_b = 3s$ and $I_d = 8$. Figure 3.4a illustrates that for high values of $R$ (i.e., $T_b \cong 3s$), increasing the $T_s$ values lead to higher failure probabilities. This is due to both the sampled friction coefficient values and higher displacement demand. Figure 3.4b shows that higher values of both $\gamma$ and $q$ cause a little decrease of the exceeding probabilities; in stark contrast, Figure 3.4d depicts that high $q$ corresponds to increasing exceeding probabilities.

Figure 3.5 depicts the fragility curves for the superstructure. Similarly to the previous results, the seismic fragility gets higher for decreasing $LS$ thresholds. For the inelastic system ($T_b = 3s$ and $I_d = 8$), higher values of $\gamma$ and of $q$ cause a high increase of the exceeding probabilities. For the inelastic

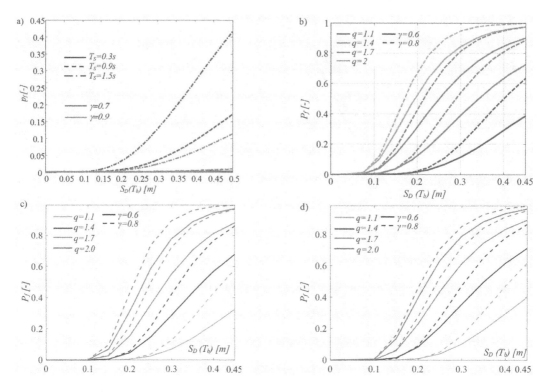

**FIGURE 3.5**  Seismic fragility curves of the superstructure related to elastic (a), perfectly elastoplastic (b), nonlinear hardening (c), and softening (d) behavior, with $H = 0.03$, $S = 0.03$. (Modified from Castaldo, Amendola & Palazzo, 2017; Castaldo, Palazzo & Ferrentino, 2017; Castaldo, Palazzo, Alfano & Palumbo, 2018, with permissions.)

limit states ($\mu > 1$), it is worthy highlighting how the superstructure seismic fragility decreases for higher $T_b$ with a fixed value of the isolation degree since $T_s$ and the corresponding yielding displacement increase. Differently, the superstructure seismic fragility increases for higher seismic isolation degree $I_d$ with a fixed value of $T_b$ since $T_s$ and the corresponding yielding displacement decrease causing resonance effects. This dynamic amplification is reduced if the systems presents a post-yield hardening behavior (Figure 3.5c), whereas is present in the case of the post-yield softening behavior (Figure 3.5d). More details are available in Castaldo, Palazzo and Ferrentino (2017) and Castaldo, Palazzo, Alfano and Palumbo (2018). In relation to the elastic superstructure ($T_b \cong 3s$), the failure probabilities increase as higher values of $T_s$ are considered because of the superstructure flexibility.

## 3.6  SEISMIC RELIABILITY-BASED DESIGN OF BASE-ISOLATED STRUCTURES WITH FP DEVICES

After the seismic fragility is assessed, the abovementioned CCDFs can be integrated with the site-related seismic hazard curves, defined in accordance with the same specific *IM*, $S_D(T_b)$, to compute the mean annual rates $\lambda_{LS}$ (Eq. 3.13) exceeding each *LS* threshold in relation to both the FP isolator and superstructure within each parameter combination.

$$\lambda_{LS}(ls) = \int_0^\infty G_{LS|IM}(ls|im) \left| \frac{d\lambda_{IM}(im)}{d(im)} \right| d(im) \tag{3.13}$$

Afterward, by using homogeneous Poisson distribution, $\lambda_{LS}$ can be used to compute the exceedance probabilities (Eq. 3.14) within the reference life of 50 years and, so, the seismic reliability in 50 years of all the investigated structures.

$$P_f\left(50\ years\right)=1-e^{-\lambda_{LS}\cdot 50\,years}$$

(3.14)

### 3.6.1 SEISMIC RELIABILITY-BASED DESIGN OF THE FPS ISOLATORS

The seismic reliability computation on the FPS devices leads to the definition of the SRBD abacuses useful to select the corresponding radius in plan $r$. As for the elastic superstructure, Figure 3.6 puts in evidence that the seismic reliability of the FPS isolators increases as $R$ decreases, as demonstrated in Castaldo, Amendola and Palazzo (2017). Regarding the inelastic superstructure, as discussed and commented in Castaldo, Palazzo and Ferrentino (2017) and Castaldo, Palazzo, Alfano and Palumbo (2018), the seismic reliability of the FPS devices increases as $T_b$ and $I_d$ decrease and slightly depends on γ. The increase of $q$, specifically for low isolation degrees along with low isolation period $T_b$ and for a perfectly elastoplastic behavior, leads to a reduction of the seismic reliability as illustrated in Figure 3.7a, whereas, for high isolation degrees or high isolation periods (Figure 3.7b), as $q$ decreases, the seismic reliability slightly decreases; Figures 3.7c and d show the results in terms of seismic reliability, respectively, for hardening and softening post-yield stiffness, selecting an isolation degree $I_d = 2$, $T_b = 3$, for $H = 0.03$ and $S = 0.03$. What is worth outlining is that, differently from what was observed for the perfectly elastoplastic case, the increase of $q$ generally reduces the seismic reliability, independently on the isolation degree value, and its influence for high $T_b$ with hardening superstructure is lower. In stark contrast, for softening superstructures, the parameter $S$ strongly affects the displacement demand to the devices. Indeed, the respect of the (collapse)

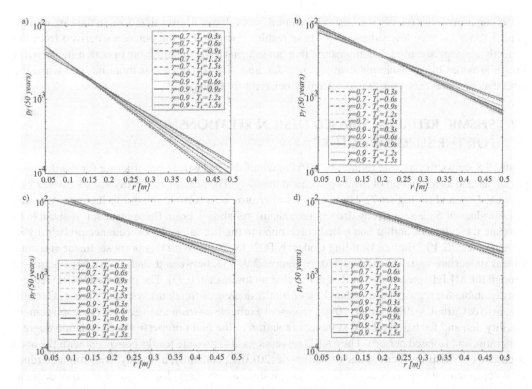

**FIGURE 3.6** SRBD curves of the isolation level in the case of elastic superstructure for $R = 1$ m (a), $R = 2$ m (b), $R = 3$ m (c), and $R = 4$ m (d). (Modified from Castaldo, Amendola & Palazzo, 2017, with permission.)

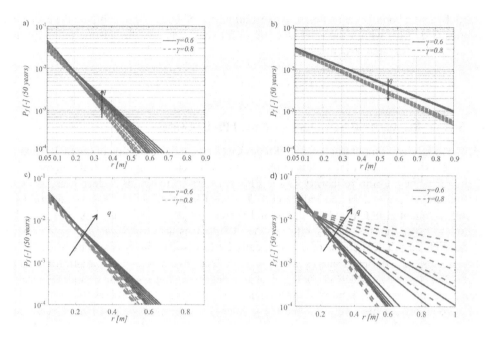

**FIGURE 3.7**  SRBD curves of the isolation level in the case of perfectly elastoplastic system with $I_d = 2$, for $T_b = 3$ s (a), perfectly elastoplastic system with $I_d = 2$, for $T_b = 6$ s (b), nonlinear hardening system with $I_d = 2$, for $T_b = 3$ s, $H = 0.03$ (c), and softening behavior with $I_d = 2$, for $T_b = 3$ s, $S = 0.03$ (d). (Modified from Castaldo, Palazzo & Ferrentino, 2017; Castaldo, Palazzo, Alfano & Palumbo, 2018, with permissions.)

exceeding probability (i.e., $P_f = 1.5 \cdot 10^{-3}$) is ensured by a value of $r$ lower than 1 m, in the case of high $\gamma$ and $I_d$, only for very low values of $q$. These results, compared to the outcomes derived from the perfectly elastoplastic model, demonstrate that considering either a hardening or softening behavior always provides higher isolation displacement demand. This difference is more marked when the superstructure presents a post-yield softening behavior for high $q$.

## 3.7  SEISMIC RELIABILITY-BASED DESIGN RELATIONSHIPS FOR THE SUPERSTRUCTURE

Figure 3.8 shows the seismic reliability (SP) curves of the inelastic superstructure for the different properties and as a function of the displacement ductility. The seismic reliability rises for lower values of $\gamma$, $I_d$, $q$, and for higher $T_b$. Higher values of $H$ provide a seismic reliability increase, whereas higher values of $S$ cause a strong drop in the seismic reliability. From these results, it is suitable to correlate the values of both $\mu$ and $q$ that correspond to the life safety $LS$ exceedance probability in 50 years (i.e., $2.2 \cdot 10^{-2}$; Italian Building Codes, NTC2018, 2018). With this purpose, linear and multilinear univariate regressions, depicted in Figures 3.9–3.11, between $\mu$ and $q$ have been achieved through the MLET (presenting an R-square value no lower than 0.93). The proposed SRBD regressions, suitable for regular base-isolated mdof (multi-degree-of-freedom) systems (Shome, Cornell, Bazzurro & Carballo, 1998; Fajfar, 2000), represent a reliable instrument to go over the displacement ductility demand for base-isolated systems in relation to the main properties (i.e., isolation degree, mass ratio, and isolated period). The SRBD regressions also provide results consistent with the ones discussed by Vassiliou, Tsiavos and Stojadinović (2013). Specifically, a slightly overestimated (ductility-dependent) strength reduction factor may produce a very large displacement ductility demand with a consequential collapse. It is also worthy picking out that, for perfectly elastoplastic systems (Figure 3.9), lower values of $I_d$, for fixed $T_b$, produce a drop of the displacement ductility demand.

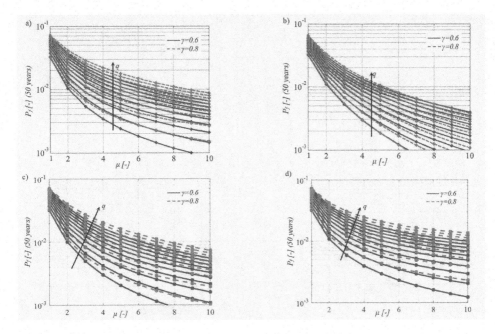

**FIGURE 3.8** SRBD curves of the superstructure in the case of perfectly elastoplastic system with $I_d = 2$, for $T_b = 3$ s (a), perfectly elastoplastic system with $I_d = 2$, for $T_b = 6$ s (b), nonlinear hardening system with $I_d = 2$, for $T_b = 3$ s, $H = 0.03$ (c), and softening behavior with $I_d = 2$, for $T_b = 3$ s, $S = 0.03$ (d). (Modified from Castaldo, Palazzo & Ferrentino, 2017; Castaldo, Palazzo, Alfano & Palumbo, 2018, with permissions.)

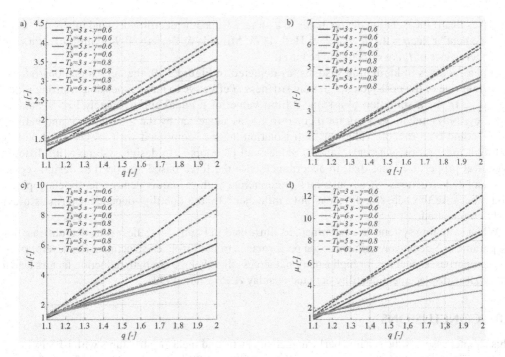

**FIGURE 3.9** Relationships between the displacement ductility demand and (ductility-dependent) strength reduction factors in the case of perfectly elastoplastic system for $I_d = 2$ (a), $I_d = 4$ (b), $I_d = 6$ (c), and $I_d = 8$ (d). (Modified from Castaldo, Palazzo & Ferrentino, 2017, with permission.)

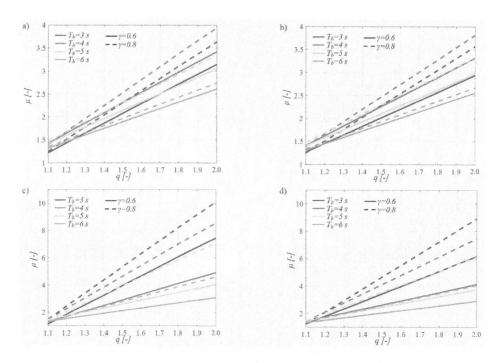

**FIGURE 3.10** Relationships between the displacement ductility demand and (ductility-dependent) strength reduction factors for $I_d = 2$, $H = 0.03$ (a), $I_d = 2$, $H = 0.06$ (b), $I_d = 8$, $H = 0.03$(c), and $I_d = 8$, $H = 0.06$ (d). (Modified from Castaldo, Palazzo, Alfano & Palumbo, 2018, with permission.)

It derives that if the isolation degree tends to a unitary value, the ductility demand meets the "equal displacement" rule, $q = \mu$ (Newmark & Hall, 1973; Miranda & Bertero, 1994). Similarly, for fixed $T_s$, higher values of $I_d$ are effective to reduce $\mu$.

When it comes to hardening systems, as depicted in Figure 3.10, the curves demonstrate the important role of the post-yield hardening stiffness $H$ effective to make it possible a strong reduction of $\mu$. In the assumption to assume a limit value for $\mu$ equal approximately to 5 (Paulay & Priestley, 1992), the upper limit for $q$ equal to 1.5, as suggested by the international and national codes, could be a reliable value for high isolation degrees combined with low isolated periods and high mass ratios; whereas higher $q$ values can be adopted to design systems with different structural properties. Note that, in accordance with the force-based approach to design regular base-isolated structures, these outcomes are coherent with the equivalent perfectly elastoplastic model because the behavior factor is only influenced by the ductility-dependent term as previously commented.

When softening systems are examined, as illustrated in Figure 3.11, the results demonstrate how the parameter $S$ negatively influences the performance of base-isolated structures. For several structural properties, especially for high isolation degrees, slightly high strength reduction factors lead to a very large displacement ductility demand (Paulay & Priestley, 1992).

## 3.8  CONCLUSIONS

This chapter assesses the seismic performance of elastic and inelastic structures with FPS isolators, so as to illustrate useful design recommendations for the FPS isolators as well as for the inelastic superstructure. With this purpose, the study describes the capabilities of the SRBD approach proposing SRBD relationships for the FPS isolators and the isolated superstructures.

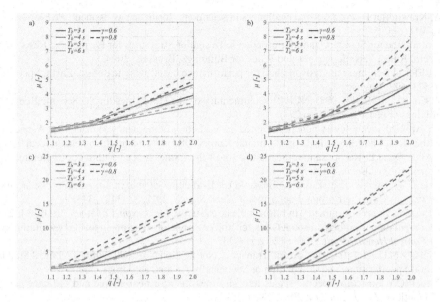

**FIGURE 3.11**   Relationships between the displacement ductility demand and (ductility-dependent) strength reduction factors for $I_d = 2$, $S = 0.03$ (a), $I_d = 2$, $S = 0.06$ (b), $I_d = 8$, $S = 0.03$(c), and $I_d = 8$, $S = 0.06$ (d). (Modified from Castaldo, Palazzo, Alfano & Palumbo, 2018, with permission.)

## REFERENCES

Abdollahzadeh GR, Banihashemi M. Response modification factor of dual moment-resistant frame with buckling restrained brace (BRB). *Steel Compos. Struct.*, 2013; 14(6): 621–636.

Alhan C, Gavin HP. Reliability of base isolation for the protection of critical equipment from earthquake hazards. *Eng. Struct.*, 2005; 27: 1435–1449.

Aoki Y, Ohashi Y, Fujitani H, Saito T, Kanda J, Emoto T, Kohno M. Target seismic performance levels in structural design for buildings. 12WCEE, 2000.

Aslani H, Miranda E. Probability-based seismic response analysis. *Eng. Struct.*, 2005; 27(8): 1151–1163.

Bertero RD, Bertero VV. Performance-based seismic engineering: the need for a reliable conceptual comprehensive approach. *Earth Eng. Struct. Dyn.*, 2002; 31: 627–652.

Building Seismic Safety Council. NEHRP Commentary on the Guidelines for the Seismic Rehabilitation of Buildings. Provisions (FEMA 274). Washington, DC, 1997.

Castaldo P, Amendola G, Palazzo B. Seismic fragility and reliability of structures isolated by friction pendulum devices: seismic reliability-based design (SRBD). *Earth. Eng. Struct. Dyn.*, 2017; 46(3): 425–446, DOI: 10.1002/eqe.2798.

Castaldo P, Palazzo B, Alfano G, Palumbo MF. Seismic reliability-based ductility demand for hardening and softening structures isolated by friction pendulum bearings. *Struct. Control Health Monit.*, 2018; 25: e2256.

Castaldo P, Palazzo B, Della Vecchia P. Seismic reliability of base-isolated structures with friction pendulum bearings. *Eng. Struct.*, 2015; 95: 80–93.

Castaldo P, Palazzo B, Ferrentino T. Seismic reliability-based ductility demand evaluation for inelastic base-isolated structures with friction pendulum devices. *Earth. Eng. Struct. Dyn.*, 2017; 46(8): 1245–1266, DOI: 10.1002/eqe.2854.

Castaldo P, Tubaldi E. Influence of fps bearing properties on the seismic performance of base-isolated structures. *Earth. Eng. Struct. Dyn.*, 2015; 44(15): 2817–2836.

Celarec D, Dolšek M. The impact of modelling uncertainties on the seismic performance assessment of reinforced concrete frame buildings. *Eng. Struct.*, 2013; 52: 340–354.

CEN – European Committee for Standardization. Eurocode 0: Basis of Structural Design. Final draft. Brussels, 2006.

Collins KR, Stojadinovic B. Limit states for performance-based design. 12WCEE, 2000.

Constantinou MC, Whittaker AS, Kalpakidis Y, Fenz DM, Warn GP. Performance of Seismic Isolation Hardware under Service and Seismic Loading. Technical Report, 2007.

Cornell CA, Krawinkler H. Progress and challenges in seismic performance assessment. *PEER Center News*, 2000; 4(1): 1–3.

European Committee for Standardization. Eurocode 8: Design of Structures for Earthquake Resistance. Part 1: General Rules, Seismic Actions and Rules for Buildings, Brussels, 2004.

Fajfar P, M.EERI. A nonlinear analysis method for performance based seismic design. Earth. Spectra, 2000; 16(3): 573–592, August.

FEMA 274: Council, B. S. S. (1997). NEHRP Commentary on the Guidelines for the Seismic Rehabilitation of Buildings (FEMA Publication 274).

FEMA P695. Applied Technology Council, & United States. Federal Emergency Management Agency. (2009). Quantification of building seismic performance factors. US Department of Homeland Security, FEMA.

Fanaiea N, Afsar Dizaj E. Response modification factor of the frames braced with reduced yielding segment BRB. *Struct. Eng. Mech.*, 2014; 50(1): 1–17.

Hatzigeorgiou GD, Papagiannopoulos GA, Beskos DE. Evaluation of maximum seismic displacements of SDOF systems form their residual deformation. *Eng. Struct.*, 2011; 33: 3422–3431.

Hong H-K, Liu C-S. Internal symmetry in bilinear elastoplasticity. *Int. J. Non-Lin Mech.*, 1999; 34: 279–288.

Jangid RS. Computational numerical models for seismic response of structures isolated by sliding systems. *Struct Control Health Monit.*, 2005; 12(1): 117–137.

Japanese Ministry of Land, Infrastructure and Transport, Notification No. 2009–2000, Technical Standard for Structural Specifications and Calculation of Seismically Isolated Buildings, 2000.

Luco N, Cornell CA. Structure-specific scalar intensity measures for near-source and ordinary earthquake ground motions. *Earth. Spectra*, 2007; 23(2): 357–92.

Math Works Inc. MATLAB-High Performance Numeric Computation and Visualization Software. User's Guide. Natick, MA, 1997.

Miranda E, Bertero VV. Evaluation of strength reduction factors for earthquake-resistant design. *Earth. Spectra*, 1994; 10: 357–379.

Mokha A, Constantinou MC, Reinhorn AM. Teflon bearings in base isolation. I: testing. *J. Struct. Eng.*, 1990; 116(2): 438–454.

Naeim F, Kelly JM. Design of Seismic Isolated Structures: From Theory to Practice. John Wiley & Sons, Inc., 1999.

Newmark NM, Hall WJ. Seismic design criteria for nuclear reactor facilities, Report 46, Building Practices for Disaster Mitigation, National Bureau of Standards, 1973.

NTC2018: D.M. 14/01/2008: Norme Tecniche per le Costruzioni (NTC 2008), G.U. Serie Generale n. 29 del14/02/2008 - S.O. n.30, Roma, Italia, 2008.

Palazzo B. Seismic behavior of base-isolated buildings. Proc. International Meeting on Earthquake Protection of Buildings, Ancona, 1991.

Paulay T, Priestley MJN. Seismic Design of Reinforced Concrete and Masonry Buildings. John Wiley & Sons, 1992.

Pinto PE, Giannini R, Franchin P. Seismic Reliability Analysis of Structures. IUSS Press, Pavia, Italy, 2003.

Ryan KL, Chopra AK. Estimation of seismic demands on isolators based on nonlinear analysis. *J. Struct. Eng.*, 2004; 130(3): 392–402.

Saito T, Kanda J, Kani N. Seismic reliability estimate of building structures designed according to the current Japanese design code. Proceedings of the Structural Engineers World Congress, 1998.

SEAOC Vision 2000 Committee. Performance-based seismic engineering. Report prepared by Structural Engineers Association of California, Sacramento, CA, 1995.

Shinozuka M, Deodatis G. Simulation of stochastic processes by spectral representation. *App. Mech. Rev.*, 1991; 44(4): 191–203.

Shome N, Cornell CA, Bazzurro P, Carballo JE. Earthquake, records, and nonlinear responses. *Earth. Spectra* 1998; 14(3): 469–500.

Structural Engineering Institute. Minimum design loads for buildings and other structures. *Am. Soc. Civil Eng.*, 2010; 8007(5).

Vamvatsikos D, Cornell CA. Incremental dynamic analysis. *Earth. Eng. Struct. Dyn.*, 2002; 31(3): 491–514.

Vassiliou MF, Tsiavos A, Stojadinović B. Dynamics of inelastic base-isolated structures subjected to analytical pulse ground motions. *Earth. Eng. Struct. Dyn.*, 2013; 42: 2043–2060.

Zayas VA, Low SS, Mahin SA. A simple pendulum technique for achieving seismic isolation. *Earth. Spectra*, 1990; 6: 317–333.

# 4 Why Resilience-Based Design?

*Naida Ademović*
University of Sarajevo, Bosnia and Herzegovina

*Ehsan Noroozinejad Farsangi*
Graduate University of Advanced Technology, Kerman, Iran

## CONTENTS

## 4.1 INTRODUCTION

Structural design is a live matter and it has been changing throughout the years and evolved, and for sure it will be updated and changed in the upcoming years. The first design codes that were used were developed based on the force-based design (FBD) philosophy. For structures to be safe, they need to possess suitable strength and acceptable ductility. In this case, the structure can resist a partial or complete failure. In many seismic codes that are enforced, the FBD approaches are implemented. Priestley (1993), Priestley et al. (2005), and Riva and Belleri (2008) explained the numerous flaws of this philosophy which in some cases may lead even to non-conservative designs. The flaws of this method are manifested through the difficulty of adequate determination of the response modification factor (R), lack of physical explanation for such an analysis, etc. (Ademović and Ibrahimbegović 2020). Based on the experience of past earthquakes and the need to maintain the functionality of the structure and limit the damage, new methods were developed, the so-called displacement-based design (DBD) in the light of performance-based earthquake engineering. An overview of various DBD methods is given in Ademović and Ibrahimbegović (2020), Kumbhar et al. (2020), and Kumbhar et al. (2019). An extension of the performance-based design (PBD) is seen in resilience-based design (RBD). The PBD is limited to a single building/ structure not taking into account the interaction with the other structures and community as a whole. Cimellaro et al. (2010a) indicated the limitation of the PBD after the earthquake that struck Italy in 2009, referring to a single building that suffered only minor damages in the town of Castelnuovo, while the rest of the town was completely wiped out. The main idea of the RBD is for the structure/community to reach again its functionality as quickly as possible and in that sense to be as "Resilient" as possible.

DOI: 10.1201/9781003194613-4

## 4.2 RESILIENCE-BASED DESIGN

Harry Markowitz in 1952 developed the modern portfolio theory (MPT) in the field of finance. This represents the basis for building up the RBD. Cimellaro (2013) in his work defined the MPT as a mathematical formulation of the approach of heterogeneity in investing, intending to select an assembly of investment equities that have mutually a reduced risk than any individual assets. The financial concept is mapped into the engineering field in the sense that buildings are regarded as assets and the fluctuation of the combined loss of building portfolio is identified as the risk of losses at the level of the community. In this way, the building portfolio is demarcated as a weight combination of the performance index of each housing unit (Cimellaro 2013). The four attributes that define resilience are robustness, redundancy, resourcefulness, and rapidity (Bruneau et al. 2003). It is the strength or the ability of the structure or its elements to resist actions to which they are being exposed without enduring damage, failure, or losing function, which is seen as robustness. For the structure to be redundant, it has to have compatible elements or systems that will be activated once the structure is threatened by a certain event, be it a natural or man-made hazard. To achieve certain goals during an event, it is necessary to be able to possess and mobilize human and material resources that are reflected in resourcefulness. Finally, time plays an important factor in this domain which is defined by rapidity in the sense of how fast and in an adequate manner can the priorities be met, enduring acceptable losses, and avoiding future disturbances. Applying different measures to improve resilience, one tries to keep the robustness of the structure/system as much as possible and to recover as fast as possible, so in this sense, rapidity and robustness can be viewed as the anticipated goals, and redundancy and resourcefulness, on the other hand, can be seen as means to achieve these goals (Farsangi et al. 2019).

Additionally, with the application of the RBD, the structure is not seen as an individual item but has to be looked at as a part of the whole community and how it interacts with it. So, in this sense, one moves from an individual structure to a block of structures where this building is located and analyze this interaction (local), which can then be transferred to a broader community. The RBD aims to ensure that not only individual structures but communities stay safe and resilient during and after the extreme events by utilizing different sophisticated technologies, forming hybrid systems, which will be explained later on, and to recover as soon as possible retrieving its functionality rapidly.

In this way, resilience has been analyzed on a global level leading to the development of four classes of resilience: technical, organizational, social, and economical (TOSE) (Bruneau et al. 2003). Technical resilience deals with the functionality of the system in question. Organizational resilience is manifested by the ability of the organization to forecast, be ready, react, and adapt to changes that may be incremental or sudden, all with the aim to continue functioning. The response of society to certain events that cause loss of services is seen through social resilience. In many cases, this social aspect can be critical especially in the case of earthquake activities combined with the COVID-19 pandemic (Zagreb earthquake 2020). Once a system is diverted from its standard operational activities and a blackout occurs due to loss of function, the economy of the system will be affected. The ability of the system to decrease different losses, which may be direct or indirect, is covered by economic resilience (Rose and Liao 2005; Noy and Yonson 2018; Oliva and Lazzeretti 2018; Zhu et al. 2019; Acuti and Bellucci 2020).

## 4.3 EARTHQUAKE-RESILIENCE DESIGN (LOW-DAMAGE DESIGN-LDD)

It is only after devastating events, like major earthquakes, that there is a need for change in the design philosophy and a requirement for upgrading and improvement in the design procedures. Base isolation systems were mainly used as protection means from major earthquakes. However, they have proven not adequate in the aftermath of the New Zealand earthquake in 2011. In that respect, a need for a better understanding of earthquake actions and the development of earthquake-resilient building technologies has been evoked. Kam et al. (2011) observed plentiful localized damage to

**FIGURE 4.1**   (a) Plastic hinge damage; (b) fracture of reinforcement bars in tall buildings (Buchanan et al. 2011).

reinforced concrete (RC) buildings which was concentrated in plastic hinge regions. More than 80% of all structures in Christchurch were up to two-story buildings. The most common type of construction was RC frames and RC walls. The damage in the plastic hinge zones was manifested as the loss of concrete, buckling, and even fracture of the reinforcement bars. The damage was mainly located in the lower story, as shown in Figure 4.1. As illustrated in Figure 4.1a, there is a high detachment and disappearance of concrete cover and serious buckling of the reinforcement in the plastic hinge zone due to an inadequate amount of confinement bars. Figure 4.1b shows the fracture of bars, which was noted on the RC wall with small cracks. This is usually caused by the fact that the built-in concrete is of a higher strength than the designed one, leading to the formation of only one crack in the region of concern, introducing extreme strain requirements on the rebars, which may in some cases lead to their fracture (Buchanan et al. 2011).

The moment-resisting frames exhibited large inelastic deformations, as the consequence of plastic hinges formation in the beams. The deformations that are formed in the plastic hinges (Figure 4.2a) very often lead to a considerable lengthening of the beams, known as "frame elongation." This phenomenon was investigated by numerous researchers (Douglas 1992; Fenwick and Davidson 1993; Peng et al. 2007; Eom and Park 2010). This has caused significant damage to RC floor diaphragms (Figure 4.2b), due to the movement of the columns which resulted due to plastic hinge formation. This kind of damage is very serious as it reduces the rigid diaphragm action whose major task is to carry over the seismic actions to the resisting lateral load systems. Hare (2009) and Fenwick et al. (2010) elaborated on the elongation effect from the formation of the plastic hinges on the diaphragms created from the precast flooring elements. A confirmation has been reported after the 2011 Christchurch earthquake in New Zealand as already presented in Figures 4.2a and 4.2b.

It has been reported that most of these damages needed urgent repair or demolition in the case of too expensive repairs. However, the question that arose was the residual capacity of the damaged structure to further major earthquakes even once repaired. It was concluded that the best scenario would be to construct new buildings.

**FIGURE 4.2**   Two-way plastic hinging in an office tower; extensive damage to the floors (Kam et al. 2011).

In that respect, a new approach should be applied, which is seen in the earthquake-resilient building design. A usual term for the earthquake-resilient building design is "Low-Damage Design" (LDD) or "Damage Avoidance Design" (DAD), as the main idea of the design is that the main structural elements pass without any major damages. Meaning that the structure has an adequate seismic answer with a minimum level of damage, which will enable a fast recovery and reuse of the structure in a short period of time, reducing or even eliminating the expensive downtime (Farsangi et al. 2018; Cruze et al. 2020; Kamaludin et al. 2020).

In the LDD, as per Hare et al. (2012), all four limit states have to be checked, from serviceability limit state (SLS), damage control limit state (DCLS), ultimate limit state (ULS), and collapse limit state (CLS). Most of the codes do not take into account all the abovementioned limit states (Eurocode 8, ATC-58-1, or NZS 1170.5 [SNZ, 2004]). DCLS and CLS are not defined in the current design codes of most standards. Priestley et al. (2007) defined the DCLS as the limit state allowing a repairable level of damage to take place. The repair costs have to be considerably lower in relation to the replacement cost. Regarding CLS, there is no unique definition (Villaverde 2007) and in different codes, it is defined in relation to the engineering demand parameters (EDPs) on various levels (local/global) (Whittaker et al. 2004). Generally, collapse is related to large plastic deformations, usually defined either by displacement or deformations. Chord rotations are used as EDPs in Eurocode 8, while the interstory drift ratio is used in FEMA 356 (2000).

### 4.3.1   OBJECTIVES OF THE LDD

The main idea is to preserve lives following the major earthquake and simultaneously to preserve the load-bearing structure of the building. In order to make the damage as minimal as possible and retain the strength and redundancy of the structure, with adequate ductility, three approaches have to be incorporated. Firstly, the strength and stiffness of the structure need to be increased in order to resist the damage resulted from earthquake activity. This would mean overdesigning the structure which may be acceptable for very important structures like hospitals and schools, however, not for residential buildings as it would be economically unfeasible. So, in this respect, a clear balance has to be found. Secondly, usage of the base isolations, which will isolate the structure from the ground motions, and in this way, the structure will be exposed to lower seismic forces, which will result in a lower magnitude of damage on the buildings. Thereby, one of the objectives would be fulfilled and that is *damage mitigation effectiveness*. One of the prerequisites for this is the application of nonlinear time history analysis, meaning that a more sophisticated design and analysis has to be envisaged. These two strategies have to be combined with the use of special design techniques and the dissipation of earthquake energy by the application of special dampers. Dampers are placed in certain locations of the structure which increases the amount of dissipated energy. In this way, the seismic energy is additionally reduced and the damage is concentrated only in easily replaceable locations. In this case, if energy systems can be unbolted and easily replaced after the earthquake, the second objective would be fulfilled defined as *system reparability*. The chosen systems should have the *self-centering capacity*, meaning that the building after the event returns as close as possible to its original vertical position with slight residual deformations. This objective is mainly covered by base isolation systems. *Nonstructural damage*, as an objective, can be achieved by the application of some kind of dampers to a structural system. For the structure to be *durable* in relation to its service life, adequate maintenance has to be envisaged and performed. Finally, regardless of the selected LDD system, it has to be *affordable* and cost-effective. This is something to be determined by the users and developers, in the case of their involvement in the process of decision-making (Hare et al. 2012; Granello et al. 2020; Voica and Stratan 2020; Dehghani et al. 2021).

## 4.3.2 EXAMPLE OF AN LDD BUILDING

The technology was originally developed in 1990 by Priestley (1991) and Priestley et al. (1999) for precast concrete construction under the US PRESSS and further developed in New Zealand (University of Canterbury) on low-damage PRESSS-technology for buildings and bridges made of both concrete and timber. To fulfill the abovementioned goals in the view of additional damping and moment contribution, dissipaters are located externally which makes the replacement easier in comparison to the first generation of internally located mild steel rebars. In this way, a hybrid system is obtained consisting of unbonded post-tensioning and additional dissipaters. This system was first developed for buildings and then extended for bridges (Hieber et al. 2005; Palermo et al. 2007). The main idea is the formation of the flag-shaped hysteresis response which is obtained by the correct contribution of the two components of the hybrid system (Figure 4.3).

Hysteresis response for recentering components consists of $M_{PT}$ which is the participation to connection strength which is supplied by the axial load attributable to the tendons which are prestressed or bars; $M_N$ is the input provided by the axial load attributable to gravity loads. The energy dissipation component is provided by $M_E$. The recentering ratio is obtained from the equation:

$$\lambda = M_{PT} + \frac{M_N}{M_E} \tag{4.1}$$

Full recentering is obtained once $\lambda$ is higher than 1.15–1.25 in the case of steel-yielding dissipaters with elastoplastic performance (Palermo et al. 2007). Dissipation of energy in these hybrid connections can be achieved by various types of damping, from friction, viscous, viscoelastic to hysteretic damping (White and Palermo 2016). The viscous fluid dampers can either only dissipate energy or trigger displacement if they have a spring that will act against the force. These kinds of dampers are one of the best solutions for damage mitigation due to excessive and rapid ground movement. The drawback of these dampers is their high cost. In this way, the mechanism works like an *internal*

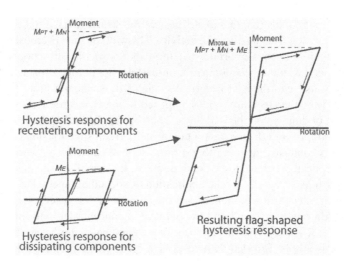

**FIGURE 4.3** Flag-shaped hysteresis response (White and Palermo 2016), with permission from ASCE. This material may be downloaded for personal use only. Any other use requires prior permission of the American Society of Civil Engineers. This material may be found at https://doi.org/10.1061/(ASCE) BE.1943-5592.0000872.

*base isolator*, enough energy will be dissipated with minimum damage to the structure or residual deformation, as the structure is basically kept in the elastic range. So, there is no formation of the plastic hinge, like in the conventional design, which is here replaced by the "controlled rocking mechanism." The plastic hinge would inevitably lead to damage and requirement for its replacement and increase of recovery time.

The first multi-story PRESSS-building was constructed in 2008 in New Zealand (Figure 4.4).

The second structure with the usage of the PRESSS-building system was the Southern Cross Hospital, and the construction of the frame is seen in Figure 4.5a. A detail of the column joint is presented in Figure 4.5b. Details regarding the design, modeling, and construction process can be found in Pampanin et al. (2011). The PRESSS-technology showed excellent behavior during the February 22, 2011 Christchurch earthquake ($M_w = 6.3$), together with the reduced construction time and modest cost (Pampanin et al. 2011). As well, the medical facility was functional immediately after the earthquake event with visible minor damages on the structural elements.

The application of the PRESSS-building system after the 2011 earthquake in New Zealand showed its benefits, first of all, in life safety as the facilities were immediately in function after the major event. In this way, the recovery is very quick as the time to final recovery is very short. There has been limited structural damage and so the need for repair was minor in this respect. More damage was observed on nonstructural elements, which is quite acceptable. In comparison to the base isolation which represents a substantial expense to the cost of a building's foundation, the PRESSS-building system with post-tensioning technology is more affordable. This kind of building's resilience increased the confidence of insurance companies lowering the insurance risk and increasing the confidence of inhabitants.

It should be kept in mind that PRESSS-building technology is not flawless. The issue of beam elongation is present, which may lead to significant damage to the floor system. However, if during the design process expansion in the floor and roof diaphragms is accounted for, such damage will be avoided.

### 4.3.3 EXAMPLE IN BRIDGES

Similar technology is being applied in the bridge industry as well. Bridges as the vain of the transportation network need to maintain their functionality and operations immediately after

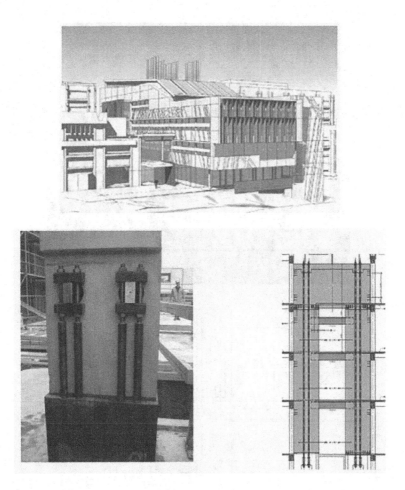

**FIGURE 4.4** PRESSS-building in New Zealand (Structural Engineers: Dunning Thornton Consultants; Cattanach and Pampanin 2008).

moderate to severe earthquake activities. The classical design, as in the case of buildings, foresees the creation of the plastic hinges in the piers which after major earthquakes endure significant damage, that needs to be repaired which is a rather expensive task not only from the repair point of view but as well from the view of traffic disturbance and shut down. In that respect, the same philosophy was applied for bridges in the form of a hybrid system. Palermo et al. (2007) applied the idea of the PRESSS program for buildings for the entire multi-span bridge systems. Experimental campaigns confirmed the high performance of these solutions; however, no widespread application of these techniques is seen in practice. The need for sophisticated numerical finite element modeling, considering the nonlinear material behavior, nonexistence of the guidelines for design, and application of such technologies, as well as the onsite application of the post-tensioning which would delay the construction, had an impact on the application of this techniques. Installation of post-tensioning requires the involvement of experts in this field and prolongs the work on the site as post-tensioning can take place once concrete or grout gains adequate strength. For steel elements, the issue of corrosion is always open. As the post-tensioning bars are located in the pier, it is quite difficult to inspect the state of the post-tensioning bars and their maintenance (Davis et al. 2017). And finally, a significant reduction of the cyclic stresses that the anchorages can bear due to the concentration of stress in the locations where the wedges grip the unbonded strands (Walsh and Kurama 2012).

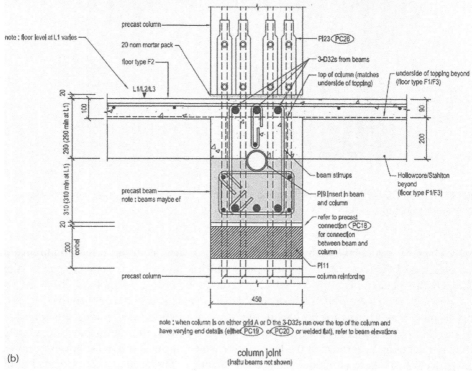

**FIGURE 4.5**   Southern Cross Hospital, with presented details (Structural Engineers: Quoin; Pampanin et al. 2011).

Formation of the new plastic hinge as proposed by Mitoulis and Rodriguez (2016) incorporating the unbonded post-tensioned tendons that permit rocking movement, as shown in Figure 4.6, developed and proposed for the shear walls (Sritharan et al. 2015). In this case, again a mixed system with the flag-shaped hysteretic behavior was achieved, which combines a self-centering capacity with additional damping provided by energy dissipaters (Palermo et al. 2004). The hybrid system

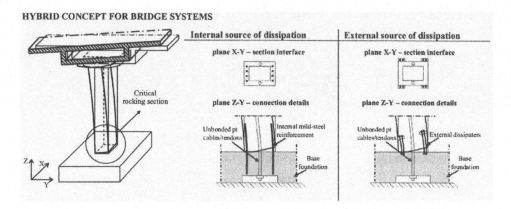

**FIGURE 4.6** Hybrid bridge pier (Palermo et al. 2007), with permission from ASCE. This material may be downloaded for personal use only. Any other use requires prior permission of the American Society of Civil Engineers. This material may be found at https://doi.org/10.1061/(ASCE)0733-9445(2007)133:11(1648).

for the bridge piers is shown in Figure 4.6 (Palermo et al. 2007). During the experimental campaign, no major damage was observed on the structural elements, the cracks were limited to the minor flexural ones, the residual drift was minimized due to the self-centering properties ensured by the unbonded post-tensioned tendons, and high ductility levels were reached with a stable hysteretic behavior. In this case, a very simple lumped plasticity model was used and it showed good agreement with the experimental results.

Davis et al. (2017) applied a similar concept for accelerated bridge construction (ABC in regions prone to seismic activity, as shown in Figure 4.7; however, in this case, pre-tensioning was

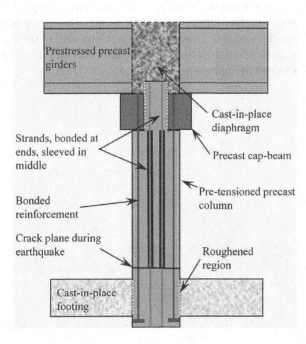

**FIGURE 4.7** Proposed precast pre-tensioned bent system (Davis et al. 2017), with permission from ASCE. This material may be downloaded for personal use only. Any other use requires prior permission of the American Society of Civil Engineers. This material may be found at https://doi.org/10.1061/(ASCE) BE.1943-5592.0000992.

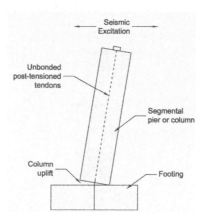

**FIGURE 4.8** New resilient hinge (RH) design (Mitoulis and Rodriguez 2016), with permission from ASCE. This material may be downloaded for personal use only. Any other use requires prior permission of the American Society of Civil Engineers. This material may be found at https://doi.org/10.1061/(ASCE) BE.1943-5592.0000980.

implemented in order to overcome the disadvantages of post-tensioning. In this instance, strand anchors are not required anymore, and the concentration of stress is eliminated, and corrosion is reduced. The construction of the pier is done in the factory under well-controlled conditions so, in this respect, the quality and speed are highly increased in relation to the onsite construction.

Mitoulis and Rodriguez (2016) put forward a novel resilient hinge (RH) that has several functions, from energy dissipation, mitigation of significant effects due to ground movement on bridges, and minimization of the pier drift (Figure 4.8). Due to the existence of numerous bars, it also has a recentering function as some of the steel bars remain in the elastic range. The steel bars are easily changed, which is very important for rapid restoration time. This new solution fulfills the LDD requirements of minimum damage and cost-effectiveness. Compared to the classical RC piers, this new technology reduces the residual drift by 93%. The new resilience hinge proposed by Mitoulis and Rodriguez (2016) is shown in Figure 4.9.

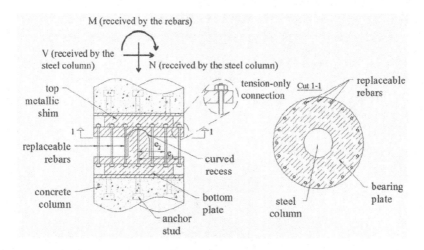

**FIGURE 4.9** New resilient hinge (RH) design (Mitoulis and Rodriguez 2016), with permission from ASCE. This material may be downloaded for personal use only. Any other use requires prior permission of the American Society of Civil Engineers. This material may be found at https://doi.org/10.1061/(ASCE) BE.1943-5592.0000980.

## 4.4 PEOPLES MODEL

The resilience of individual structures is just the first stage in the routine of obtaining a resilient society. Building resilience can be looked at as the ability of the building to endure different extreme events (Ibrahimbegovic and Ademović 2019) or disasters, either coming from nature (earthquakes, floods, etc.) or man-made disasters (explosions, terrorist attacks). It is expected that, after the extreme events, a resilient building would recover as soon as possible and be fully functional. For a community to stay functional and operational after an extreme event resilience of individual structures is not enough. If after an earthquake, only one structure stays intact, not much has been gained by this. It is the resilience of the entire community (preparing for, fighting, absorbing, and speedily recovering from extreme incidents) which is very important. In the community resilience, performance and recovery of individual buildings and/or organizations are seen as an atom and their role and impact on the community as a whole, which can be identified as a molecule. Figure 4.10 shows the management cycle in situations before and after the manifestation of the extreme incidents.

To be able to develop measurable and qualitative models for the continued resilience of communities against extreme incidents or disasters, a conceptual approach widely known as PEOPLES is applied. The acronym PEOPLES accounts for seven dimensions of community resilience: **P**opulation and Demographics, **E**nvironmental/Ecosystem, **O**rganized Governmental Services, **P**hysical Infrastructure, **L**ifestyle and Community Competence, **E**conomic Development, and **S**ocial-Cultural Capital (Renschler et al. 2010). The work suggested by Bruneau et al. (2003) was the starting point for the development of this model. In 2011, Arcidiacono et al. (2011) developed a software platform that was applied in the case of the well-known earthquake in Italy in 2009 for evaluating the resilience of the community to earthquake activity in line with the PEOPLES framework.

Disaster resilience depending on the scale can be grouped into technological units and social systems, the former being connected to the small scale (critical infrastructure), and the latter to the larger scale representing the entire community and all of its components. The PEOPLES approach

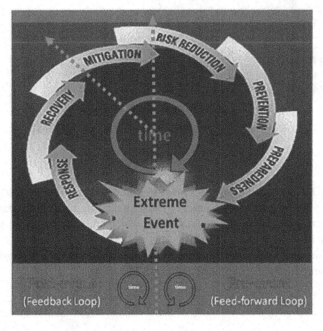

**FIGURE 4.10** Extreme events management cycle (Renschler et al. 2010), with permission from NIST (National Institute of Standards and Technology).

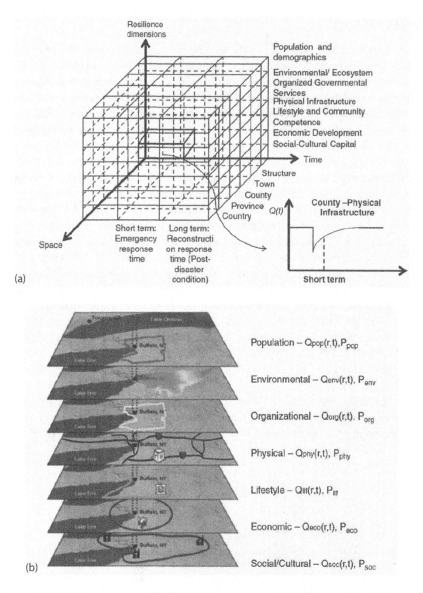

**FIGURE 4.11** (a) PEOPLES approach; (b) PEOPLES layered model (Cimellaro et al. 2015).

can help define RBD in space and time as shown in Figure 4.11a. PEOPLES works on the definitions adopted by the Multidisciplinary Center of Earthquake Engineering Research (MCEER) and the parameters affecting their integrity and resilience which are gathered using a layered model, as seen in Figure 4.11b (Cimellaro et al. 2015).

For the community level, all the parameters defined in the "PEOPLES" framework have to be considered by applying Eq. (4.2) (Reinhorn and Cimellaro 2014).

$$Q_{TOT}(t) = Q_{TOT}(Q_P, Q_{Env}, Q_O, Q_{Ph}, Q_L, Q_{Eco}, Q_s) \qquad (4.2)$$

where each of these aspects takes into consideration their specific elements that impact their functionality. Where $Q_{TOT}(t)$ is the global functionality; and $Q_x(t)$ represents the functionality of each

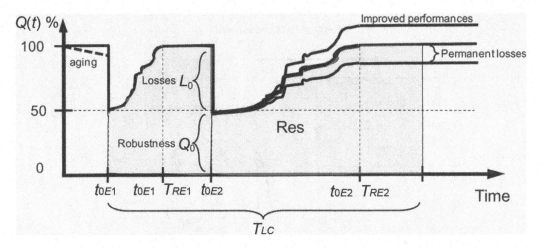

**FIGURE 4.12** Schematic representation of disaster resilience (Kilanitis and Sextos 2018).

of the seven dimensions of the community. As a result, time-dependent functionality maps for a defined region of interest are created. These maps can be transformed into temporal residence contour scaled maps if the temporal scale is defined through the control time of the period of interest $T_{LC}$ (Cimellaro et al. 2010b) required for defining resilience by Eq. (4.3).

$$R(\vec{r}) = \int_{t_{OE}}^{t_{OE} + T_{LC}} \frac{Q_{TOT}(t)}{T_{LC}} dt \qquad (4.3)$$

where:

$t_{OE}$ is the initiation of the extreme event $E$
$T_{RE}$ is the time of recovery from the extreme event $E$
$Q(t)$ is the functionality represented in percentage, which ranges from 0% indicating total loss to 100% referring to the full performance
$R(\vec{r})$ is the position vector defining the position $P$ in a particular region of interest where the resilience index is assessed (Cimellaro et al. 2010b)

Figure 4.12 (Kilanitis and Sextos 2018) shows the reduction of the initial state of the functionality of the system connected to the aging of the transport infrastructure (usually overlooked) which will affect the resilience immediately after the occurrence of a natural disaster.

Cimellaro et al. (2015) proposed an equation for the community resilience index $R_{com}$ that is given by Eq. (4.4) taking into account space and time, as graphically shown in Figure 4.11b.

$$R_{com} = \int_{A_c} \frac{R(\vec{r})}{A_c} dr = \int_{A_c} \int_{t_{OE}}^{t_{OE} + T_{LC}} \frac{Q_{TOT}(t)}{A_c T_{LC}} dt dr \qquad (4.4)$$

Based on this info, a radar graph is produced, and the area will determine the conclusive value of the resilience grade (Figure 4.13) for the assessed region. Information received from these graphs indicates what would be priority activities that have to be taken and additionally it will detect gaps, which all will be used as input data in the decision-making process.

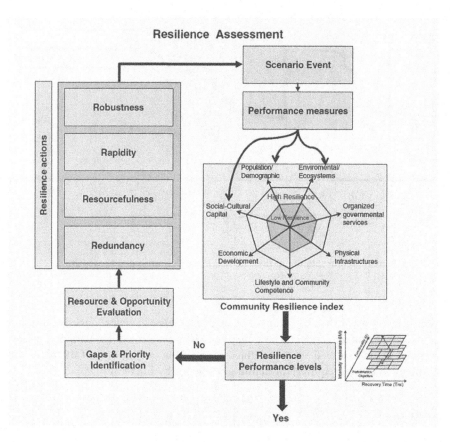

**FIGURE 4.13**   Resilience scoring using PEOPLES methodology (Cimellaro et al. 2015).

The complexity of the PEOPLES framework with all its components and subcomponents is shown in Table 4.1.

## 4.5   CONCLUSION

The need for the change in design philosophy has been triggered by severe damage and economic impact due to downtime of facilities after the earthquake shaking (moderate to strong shaking levels). All this leads to a further step in the design process, to the resilient-earthquake design and LDD. It has been noted that resilience is the key parameter for descriptions of the structures' performance.

For the structures and/or communities to become as resilient as possible, a step forward has to be taken in the domain of design, defined as the RBD and LDD. The main goal is to make not only an individual building but the entire community as resilient as possible, through various advanced technologies that enable the structure/community to recover as quickly as possible and regain its functionality.

The chapter showed a few examples of the advanced technologies that have been carried out in practice in structures; however, it should be noted that these cases are not many. The prerequisites for the application of these techniques are sophisticated numerical finite element modeling taking into account the nonlinear material behavior, on the one hand, and development of the guidelines for design and application of such technologies, on the other.

**TABLE 4.1**

**A Comprehensive List of All Elements of the PEOPLES Framework (Cimellaro 2016)**

**1-Population and Demographics**

a. Distribution/density
  i. Urban
  ii. Suburban
  iii. Rural
  iv. Wildland

b. Composition
  i. Age
  ii. Gender
  iii. Immigrant status
  iv. Race/ethnicity

c. Socio-economic status
  i. Educational attainment
  ii. Income
  iii. Poverty
  iv. Homeownership
  v. Housing vacancies
  vi. Occupation

**2-Environmental/Ecosystem**

a. Water quality/ quantity    b. Air quality    c. Soil quality    d. Biodiversity    e. Biomass (vegetation)    f. Other natural resources

**3-Organized Governmental Services**

a. Executive/administrative
  i. Emergency response
  ii. Health and hygiene
  iii. Schools

b. Judicial

c. Legal/security

**4-Physical Infrastructure**

a. Facilities
  i. Residential
    1. Housing units
  ii. Commercial
    1. Distribution facilities
    2. Hotels accommodation
  iii. Cultural
    1. Entertainment venues
    2. Museums
    3. Religious institutions

  2. Shelters

  3. Manufacturing facilities

  4. Office buildings

  4. Schools
  5. Sports/recreation venues

b. Lifelines
  i. Communication
    1. Internet
    2. Phones
    3. TV
    4. Radio
    5. Postal
  ii. Healthcare
    1. Acute care
    2. Long-term acute care
    3. Psychiatric
    4. Primary care
    5. Specialty
  iii. Food supply
  iv. Utilities
    1. Electrical
    2. Fuel/gas/energy
    3. Waste
    4. Water
  v. Transportation
    1. Aviation
    2. Bridges
    3. Highways
    4. Railways
    5. Transit
    6. Vehicles
    7. Waterways

**5-Lifestyle and Community Competence**

a. Collective action and decision-making
  i. Conflict resolution
  ii. Self-organization

b. Collective efficacy and empowerment

c. Quality of life

*(Continued)*

**TABLE 4.1 (Continued)**

### 6-Economic Development

a. Financial services
   i. Asset base of financial
      institutions
   ii. Checking account balances
       (personal and commercial)
   iii. Consumer price index
   iv. Insurance
   v. Number and average amount
      of loan
   vi. Number of bank and credit
       unions members
   vii. Number of bank and credit
        unions
   viii. Saving account balances
         (personal and commercial)
   ix. Stock market

b. Industry employment services
   i. Agriculture
   ii. Construction
   iii. Education and health services
   iv. Finance, insurance and real estate
   v. Fortune 1000
   vi. Fortune 500
   vii. Information, professional business, others
   viii. Manufacturing
   ix. Leisure and hospitality
   x. Number of corporate headquarters
   xi. Other business services
       1. Employment services
          a. Flexibilities
          b. Opportunities
          c. Placement
       2. Transport and utilities
       3. Wholesales and retails

c. Industry production
   i. Food supply
   ii. Manufacturing

### 7-Social/Cultural Capital

a. Child and elderly
   services
e. Educational services

b. Commercial centers

f. Nonprofit organization

c. Community participation

g. Place attachment

d. Cultural and heritage
   services

## REFERENCES

Acuti, D., Bellucci, M. (2020). "Resilient cities and regions: planning, initiatives, and perspectives", Climate Action, 763–774.

Ademović, N., Ibrahimbegović, A. (2020). "Review of resilience based design", Coupled Syst. Mech. 9(2), 91–110, DOI: 10.12989/csm.2020.9.2.091.

Arcidiacono, V., Cimellaro, G. P., Reinhorn, A. (2011). "A software for measuring disaster community resilience according to the PEOPLES methodology", Proceedings of the 3rd International Conference on Computational Methods in Structural Dynamics and Earthquake Engineering (COMPDYN 2011), 26–28.

ATC-58-1. Seismic Performance Assessment of Buildings – Volume 1 – Methodology, Applied Technology Council, Redwood City, California.

Bruneau, M., Chang, S., Eguchi, R., Lee, G., O'Rourke, T., Reinhorn, A., Shinozuka, M., Tierney, K., Wallace, W., von Winterfeldt, D. (2003). "A framework to quantitatively assess and enhance the seismic resilience of communities", Earthq. Spectra. 19(4), 733–752. https://doi.org/10.1193/1.1623497.

Buchanan, A. H., Bull, D., Dhakal, R., MacRae, G. Palermo, A., Pampanin, S. (2011). "Base Isolation and Damage-Resistant Technologies for Improved Seismic Performance of Buildings A Report Written for the Royal Commission of Inquiry into Building Failure Caused by the Canterbury Earthquakes", Research Report 2011-02, 2011, Final Version DOC ID: ENG.ACA.0010.FINAL.

Cattanach, A., Pampanin, S. (2008). "21st century precast: the detailing and manufacture of NZ's first multi-storey PRESSS-building", NZ Concrete Industry Conference, Rotorua.

Cimellaro, G. P. (2013). "Resilience-Based Design (RBD) Modelling of Civil Infrastructure to Assess Seismic Hazards", Handbook of Seismic Risk Analysis and Management of Civil Infrastructure Systems, 268–303.

Cimellaro, G. P. (2016). Urban Resilience for Emergency Response and Recovery Fundamental Concepts and Applications, Geotechnical, Geological and Earthquake Engineering, 41, Chapter 6, PEOPLES Resilience Framework. Springer International Publishing, Switzerland, DOI: 10.1007/978-3-319-30656-8_6.

Cimellaro, G. P., Christovasilis, I. P., Reinhorn, A. M., De Stefano, A., Kirova, T. (2010a). "L'Aquila earthquake of April 6th, 2009 in Italy: rebuilding a resilient city to multiple hazard", MCEER technical report – MCEER-10-0010, State University of New York at Buffalo (SUNY), Buffalo.

Cimellaro, G. P., Reinhorn, A. M., Bruneau, M. (2010b). "Framework for analytical quantification of disaster resilience", Eng. Struct. 32(11), 3639–3649, DOI: 10.1016/j.engstruct.2010.08.008.

Cimellaro, G. P., Renschler, C., Bruneau, M. (2015). "Introduction to Resilience-Based Design (RBD)", Eds. Cimellaro, G., Nagarajaiah, S., Kunnath, S., Computational Methods, Seismic Protection, Hybrid Testing and Resilience in Earthquake Engineering. Geotechnical, Geological and Earthquake Engineering, 33, Springer, Cham.

Cruze, D., Gladston, H., Farsangi, E. N., Loganathan, S., Dharmaraj, T., Solomon, S. M. (2020). "Development of a multiple coil magneto-rheological smart damper to improve the seismic resilience of building structures", Open Civ. Eng. J. 14(1), 78–93, DOI: 10.2174/1874149502014010078.

Davis, P. M., Janes, T. M., Haraldsson, O. S., Eberhard, M. O., Stanton, J. F. (2017). "Unbonded pretensioned columns for accelerated bridge construction in seismic regions", J. Bridge Eng. 22(5), 04017003, DOI: 10.1061/(asce)be.1943-5592.0000992.

Dehghani, S., Fathizadeh, S. F., Yang, T. Y., Noroozinejad Farsangi, E., Vosoughi, A. R., Hajirasouliha, I., Málaga-Chuquitaype, C., Takewaki, I. (2021). "Performance evaluation of curved damper truss moment frames designed using equivalent energy design procedure", Eng. Struct. 226.

Douglas, K. T. (1992). "Elongation in reinforced concrete frames", Report 526, School of Engineering, Auckland, New Zealand, November.

Eom, T. -S., Park, H. -G. (2010). "Elongation of reinforced concrete members subjected to cyclic loading", J. Struct. Eng. 136(9), 1044–1054, DOI: 10.1061/(asce)st.1943-541x.0000201.

Eurocode 8, EN 1998:1. (2004). Design of structures for earthquake resistance. Part 1: General rules, seismic actions and rules for buildings, European Committee for Standardization, Brussels, Belgium.

Farsangi, E. N., Tasnimi, A. A., Yang, T. Y., Takewaki, I., Mohammadhasani, M. (2018). "Seismic performance of a resilient low-damage base isolation system under combined vertical and horizontal excitations", Smart Struct. Syst. 22(4), 383–397, DOI: 10.12989/sss.2018.22.4.383.

Farsangi, N. E., Takewaki, I., Yang, T. Y., Astaneh-Asl, A., Gardoni, P. (Eds.). (2019). Resilient Structures and Infrastructure. Springer Nature Singapore Pte Ltd. DOI: 10.1007/978-981-13-7446-3.

Federal Emergency Management Agency, FEMA-356. (2000). Prestandard and Commentary for Seismic Rehabilitation of Buildings, Federal Emergency Management Agency, Washington, DC.

Fenwick, R., Bull, D. K., Gardiner, D. (2010). "Assessment of hollow-core floors for seismic performance", Research report 2010-02., Department of Civil and Natural Resources Engineering, University of Canterbury, Christchurch, NZ.

Fenwick, R. C., Davidson, B. J. (1993). "Elongation in ductile seismic resistant reinforced concrete frames", Tom Pauley Symposium, San Diego, USA, September.

Granello, G., Broccardo, M., Palermo, A., Pampanin, S. (2020). "Fragility-based methodology for evaluating the time-dependent seismic performance of post-tensioned timber frames", Earthq. Spectra. 36(1), 322–352.

Hare, J. (2009). "Heritage earthquake prone building strengthening cost study", Prepared for Christchurch City Council. (Internal report). Holmes Consulting Group, Christchurch, NZ.

Hare, J., Oliver, S., Galloway, B. (2012). "Performance objectives for low damage seismic design of buildings", 2012 NZSEE Conference, paper number 35, pp. 1–9.

Hieber, D. G., Wacker, J.M., Stanton, J. F., Eberhard, M. O. (2005). "Precast concrete pier systems for rapid construction of bridges in seismic regions", Washington State Transportation Center (TRAC), Seattle.

Ibrahimbegovic, A., Ademović, N. (2019). Dynamics of Structures Under Extreme Transient Loads, 1st Edition, CRC Press, Taylor & Francis Group, LLC.

Kam, W.Y., Pampanin, S., Elwood, K. (2011). "Seismic performance of reinforced concrete buildings in the 22 February Christchurch (Lyttelton) earthquake", Bull. N.Z. Natl. Soc. Earthq. Eng. 44(4), 239-278., DOI: 10.5459/bnzsee.44.4.239-278.

Kamaludin, P. N. C., Kassem, M. M., Farsangi, E. N., Nazri, F. M., Yamaguchi, E. (2020). "Seismic resilience evaluation of RC-MRFs equipped with passive damping devices", Earthq. Struct. 18(3), 391–405, DOI: 10.12989/eas.2020.18.3.391.

Kilanitis, I., Sextos, A. (2018). "Integrated seismic risk and resilience assessment of roadway networks in earthquake prone areas", Bull. Earthq. Eng. 17(1), 181–210. DOI: 10.1007/s10518-018-0457-y.

Kumbhar, O. G., Kumar, R., Noroozinejad Farsangi, E. (2020). "Investigating the efficiency of DDBD approaches for RC buildings", Structures 27, 1501–1520, DOI: 10.1016/j.istruc.2020.07.015.

Kumbhar, O. G., Kumar, R., Panaiyappanand, P. L., Noroozinejad Farsangi, E. (2019). "Direct displacement based design of reinforced concrete elevated water tanks frame staging", Int. J. Eng. Sci. 32(10), 1395–1406, DOI: 10.5829/IJE.2019.32.10A.09.

Mitoulis, S.A., Rodriguez, J.R. (2016). "Seismic performance of novel resilient hinges for columns and application on irregular bridges", J. Bridge Eng. 22(2), 04016114-1-04016114-12. https://doi.org/10.1061/(ASCE)BE.1943-5592.0000980.

Noy, I., Yonson, R. (2018). "Economic vulnerability and resilience to natural hazards: a survey of concepts and measurements", Sustainability. 10, 2850, 1–16, DOI: 10.3390/su10082850.

NZS 1170.5. (2005). Structural Design Actions – Part 5 Earthquake Actions, Standards New Zealand, Wellington, New Zealand.

Oliva, S., Lazzeretti, L. (2018). "Measuring the economic resilience of natural disasters: an analysis of major earthquakes in Japan City", Cult. Soc. 15, 53–59. DOI: 10.1016/j.ccs.2018.05.005.

Palermo, A., Pampanin, S., Calvi, G. M. (2004). "Use of 'controlled rocking' in the seismic design of bridges", Proc., 13th World Conference on Earthquake Engineering, Vancouver, Canada, Paper 4006.

Palermo, A., Pampanin, S., Marriott, D. (2007). "Design, modeling, and experimental response of seismic resistant bridge piers with posttensioned dissipating connections", J. Struct. Eng. 1648–1661, DOI: 10.1061/(asce)0733-9445(2007)133:11(1648).

Pampanin, S., Kam, W., Haverland, G., Gardiner, S. (2011). "Expectation meets reality: seismic performance of post tensioned precast concrete southern cross endoscopy building during the 22nd Feb 2011 Christchurch earthquake", NZ Concrete Industry Conference, Rotorua, August.

Peng, B. H. H., Dhakal, R. P., Fenwick, R.C., Carr, A.J. (2007). "Analytical model on beam elongation within the reinforced concrete plastic hinges", 2007 NZSEE Conference, paper number 43, pp. 1–8.

Priestley, M. J. N. (1991). "Overview of the PRESSS research programme", PCI J. 36(4), 50–57.

Priestley, M. J. N. (1993). "Myths and fallacies in earthquake engineering – conflicts between design and reality", Bull. N. Z. Natl. Soc. Earthq. Eng. 26(3), 329–341.

Priestley, M. J. N., Calvi, G. M., Kowalsky, M. J. (2007). Displacement Based Seismic Design of Structures, IUSS Press, Italy.

Priestley, M.J.N., Grant, D. N., Blandon, C. A. (2005). "Direct displacement-based seismic design", NZSEE Conference.

Priestley, M. J. N., Sritharan, S., Conley, J. R., Pampanin, S. (1999). "Preliminary results and conclusions from the PRESSS five-story precast concrete test building", PCI J. 44(6), 42–67.

Reinhorn A. M., Cimellaro, G.P. (2014). "Consideration of Resilience of Communities in Structural Design. Chapter 27", Eds. Fischinger, M., Performance-Based Seismic Engineering: Vision for an Earthquake Resilient Society, Geotechnical, Geological and Earthquake Engineering 32, Springer Science + Business Media, Dordrecht, DOI: 10.1007/978-94-017-8875-5__27.

Renschler, C., Frazier, A., Arendt, L., Cimellaro, G. P., Reinhorn, A. M., Bruneau, M. (2010). "Framework for defining and measuring resilience at the community scale: the PEOPLES resilience framework", MCEER Technical Report–MCEER-10-006, University at Buffalo (SUNY), State University of New York, Buffalo, NY, USA.

Riva, P., Belleri, A. (2008). "Direct displacement based design and force based design of precast concrete structures", Life Cycle Civil Eng. 273–278, DOI: 10.1201/9780203885307.ch37.

Rose, A., Liao, S.Y. (2005). "Modeling regional economic resilience to disasters: a computable general equilibrium analysis of water service disruptions", J. Reg. Sci. 45(1), 75–112. https://doi.org/10.1111/j.0022-4146.2005.00365.x.

Sritharan, S., Aaleti, S., Henry, R.S., Liu, K.-Y., Tsai, K.-C. (2015). "Precast concrete wall with end columns (PreWEC) for earthquake resistant design", Earthq. Eng. Struct. D. 44(12), 2075–2092, DOI: 10.1002/eqe.2576.

Villaverde, R. (2007). "Methods to assess the seismic collapse capacity of building structures: state of the art", J. Struct. Eng. 133(1), 57–66. https://doi.org/10.1061/(ASCE)0733-9445(2007)133:1(57).

Voica, T. F., Stratan, A. (2020, March). "Review of damage-tolerant solutions for improved seismic performance of buildings", IOP Conference Series: Materials Science and Engineering (Vol. 789, No. 1, p. 012062). IOP Publishing.

Walsh, K., Kurama, Y. (2012). "Effects of loading conditions on the behavior of unbonded post-tensioning strand-anchorage systems", PCI J. 57(1), 76–96. https://doi.org/10.15554/pcij.01012012.76.96.

White, S., Palermo, A. (2016). "Quasi-static testing of posttensioned nonemulative column-footing connections for bridge piers", J. Bridge Eng. 21(6), 04016025, DOI: 10.1061/(asce)be.1943-5592.0000872.

Whittaker, A., Deierlein, G., Hooper, J., Merovich, A. (2004). "Engineering demand parameters for structural framing systems", ATC 58 Structural Performance Products Team, Redwood.

Zhu, S., Li, D., Feng, H. (2019). "Is smart city resilient? Evidence from China", Sustain. Cities Soc. 50, 101636.

# 5 Resilience Engineering
## *Principles, Methods, and Applications to Critical Infrastructure Systems*

*T.V. Santhosh*
University of Liverpool, Liverpool, UK
Bhabha Atomic Research Centre, Mumbai, India

*Edoardo Patelli*
University of Liverpool, Liverpool, UK
University of Strathclyde, Glasgow, UK

## CONTENTS

## 5.1 INTRODUCTION

Infrastructures are responsible for providing essential services for our society. Modern infrastructures are highly interconnected, and the disruption and failure of some components may trigger the failure of other components. This has the potential to cause the complete failure of the system with potential catastrophic consequences (Qureshi, 2007). Generally, interconnected systems not only provide robustness and flexibility to the entire infrastructure but also make the analysis of such systems more complex and difficult. This requires a new approach to guarantee and assess the safety of such systems.

DOI: 10.1201/9781003194613-5

In addition, the human factor plays a significant role in the safety of such systems, since it represents the ultimate defense measure, yet the weakest link. The importance of the human reliability to the safety of systems has become clear after the accident of the nuclear plant at Three Mile Island (Azadeh, 2016). Recent studies have indicated that more than 90% of the major industrial accidents are related to the human factor (Moura, 2017). In addition, the major accidents in the process industry and aviation industry and failures of critical infrastructures and complex systems have triggered an absolute need for a new and intelligent approach for the safety enhancement of critical infrastructure systems.

The traditional approach to improve the system safety is via redundancy, i.e., the availability of duplicate systems and components that provide the same function or service. However, such an approach increases the complexity and cost of the systems making the infrastructure often unsustainable. In addition, there is no guarantee that a threat cannot cause a cascading failure or triggering the failure of redundant systems. This is particularly true when human intervention is needed (e.g., operators, decision makers).

Instead of focusing on the reliability of component and system and preventing failure at any cost as done in reliability engineering, resilience engineering (RE) moves the attention toward the mitigation and recovery phase from an intentional loss of functionality. As part of RE, human factor engineering (HFE) analyses the performance of operators/decision makers under pressure to successfully mitigate the threat situation and recover the system. However, the flexibility and reconfigurability of complex systems have the potential to create some dynamic instability in the systems, therefore making their control even more challenging.

RE is still in its infancy and not yet adopted in engineering design. This is due to the fact that although several qualitative studies have been conducted, quantitative assessments of real systems are rare. In fact, most of the work on HFE involves qualitative investigations from the data collected through field observations, audit reports, virtual experiments, and expert elicitation (Lay et al., 2015; Musharraf et al., 2018). The situation awareness and mental workload are found to be the key factors in determining the performance of the operator during dynamic threat scenario (Burtscher and Manser, 2012; Azadeh and Salehi, 2014). The conventional human reliability analysis methods fail to address such dynamic behavioral changes of the operator during the unexpected threats. Moreover, the performance shaping factors used to evaluate the operator's performance are strongly dependent on the potential consequences of the threats (Kim et al., 2017; Morais et al., 2020). The situation awareness is especially crucial when dealing with highly dynamic situations with quick changes from a normal operating status to an extremely severe situation. This is also affected by fitness, fatigues, stress, and overconfidence (IAEA, 1998). Although different metrics exist for RE, there is a general consent with resilience principles, namely, anticipation, response, learning, and monitoring (Pillay, 2017). All these definitions and metrics strongly rely on the overlapping concepts of resilience, as detailed in Section 5.2. Although several resilience metrics are currently available, their quantification remains a serious challenge due to many factors involved in such metrics, namely human and organizational factors.

Despite the challenges, several researchers have successfully applied the resilience principles to assess critical systems. Bruneau et al. (2003) proposed a conceptual seismic resilience framework for civil infrastructures based on complementary measures of resilience related to the probability of failure, consequences, and recovery time. Min et al. (2012) suggested a three-stage framework for assessing the resilience of urban infrastructure, with a different strategy at each resilience stage. Hosseini and Barker (2016b) proposed the use of Bayesian networks to understand the causal relationships among variables that affect the infrastructure resilience and applied to a case study of an inland waterway port. Rocchetta et al. (2018) assessed the resilience of repairable power grids by subjecting to weather-induced failures with data deficiency. Santhosh and Patelli (2020) employed dynamic Bayesian networks to model the resilience of a safety-related system of a nuclear reactor addressing the human and organizational factors. Estrada-Lugo et al. (2020) adopted credal networks for estimating the resilience of the primary system of a nuclear reactor in presence of a lack of data and imprecision.

Here, a review of the state-of-the-art research work and a novel framework for RE is presented. This includes RE definitions, principles, methods, quantitative studies, and an application on resilience assessment of a nuclear reactor.

## 5.2 RESILIENCE ENGINEERING

### 5.2.1 DEFINITIONS

There are several definitions to resilience proposed by various researchers from different perspectives (Woods, 2006; Youn et al., 2011; Min and Wang, 2015). One such widely known definition proposed by Cai et al. (2018) defines resilience as the ability of an infrastructure to operate under anticipated or unanticipated threat scenarios after recovering from a sudden disturbance. Though the resilience concept is popular in non-engineering fields, it remains a relatively new subject in the engineering domain and needs further research (Hosseini et al., 2016a). Hollnagel (2006) proposed a resilience framework that considers social, technical, environmental, and economic aspects to deal with complex infrastructure safety. Since then, RE has gradually evolved as a potential alternative method to overcome the limitations of traditional risk assessment techniques (Steen and Aven, 2011). According to the American Society of Mechanical Engineers (2009), resilience is the ability of an infrastructure to continue to perform under external and internal disturbances without discontinuity or to rapidly recover from the discontinuity. Haimes (2009) defined resilience as the ability of an infrastructure to resist and recover from a major disruption within the acceptable time with associated cost and risk. From the initial definition of resilience related to ecological systems (Holling, 1973), the resilience concept has since been widely adopted in many fields including socio-technical systems and economics (Chang and Shinozuka, 2004; Cimellaro et al., 2006). However, the most popular definition is with respect to the resilience of socio-technical systems proposed by Bruneau et al. (2003). They define resilience as the ability of an infrastructure to reduce its probability of failure, limit the consequences, and to recover quickly. The four characteristics suggested by Bruneau et al. (2003) for a resilient system are:

- **Robustness:** the ability of an infrastructure to withstand sudden disturbances and reduce the performance loss by mitigating the consequence of disturbances.
- **Redundancy:** the ability of an infrastructure to continue to deliver its performance even under partial loss of some system's functionality. This can be achieved due to the presence of redundancy feature in some of the system's functions including operator skills.
- **Resourcefulness:** the capability of a system to respond to disturbances ensuring the availability of additional resources such as manpower allocation, emergency mobile equipment, etc., to overcome the disturbances. For example, arrangement of a mobile diesel generator equipment during station blackout event is a characteristic of resourcefulness.
- **Rapidity:** the capability of a system to quickly restore its performance to an acceptable level after a disruption.

Few researchers have proposed analytical relationships among these attributes (Zobel and Khansa, 2014). These characteristics are important attributes of RE but, it is hard to quantify them in real systems.

### 5.2.2 PRINCIPLES

The focus of RE is to ensure that the critical infrastructure delivers resilience performance. Hollnagel (2006) and Hollnagel et al. (2007) have proposed the following six principles related to safety and performance management of resilient systems toward quantitative resilience assessment.

### 5.2.2.1 Top Management Commitment

It is the commitment of the organization management to understand worker's difficulties and problems and provide timely and appropriate solutions to deal with complex situations arising from disruptive events. The resource allocation and safety are the prime factors while responding to the workers' issues.

### 5.2.2.2 Reporting Culture

This property refers to the trustworthiness among organization management and the workers where workers are encouraged to report management the issues related to safety.

### 5.2.2.3 Learning

This principle refers to gaining the knowledge from various incidents and also from routine operations; thus, enabling to design and adopt the best supervisory plans in order to reduce the gap between the managers and workers.

### 5.2.2.4 Awareness

It refers to both management and workers in which the workers should be aware of the system boundaries, their current state, and status of defenses in place within the plant. It is equally important for the management to keep track of the status of the system, quality of the workforce, the current organization state.

### 5.2.2.5 Preparedness

It refers to the ability of the system to anticipate various safety challenges including human performance and preparing to adapt to those challenges.

### 5.2.2.6 Flexibility

It is the capability of the system to respond to variations due to external disturbances by undergoing self-reorganization.

### 5.2.3 RESILIENCE METRICS

The state-of-the-art research in RE shows there are a number of resilience metrics and quantification methods (see, e.g., Cimellaro et al., 2006; Henry and Ramirez-Marquez, 2012; Min et al., 2012; Dessavre et al., 2016). For instance, Dessavre et al. (2016) proposed a resilience evaluation method by adopting visual examination tools to study the infrastructure behavior in a time domain in response to disruptive events. Bruneau et al. (2003) proposed a resilience metric based on the resilience loss concept under a seismic event as in Eq. (5.1):

$$R_l = \int_{t_e}^{t_f} \left[100 - \varphi(t)\right] dt \tag{5.1}$$

where $t_e$ represents the time of occurrence of disruption event, and $t_f$ represents the time when disruption event ends. $\phi(t)$ corresponds to the quality of infrastructure at time $t$. In this model, the resilience is quantified by comparing the initial quality of infrastructure (100) with degraded infrastructure during the recovery period. A large value of $R_l$ corresponds to lower resilience and a small value of $R_l$ corresponds to higher resilience. Henry and Ramirez-Marquez (2012) developed a quantitative resilience metric based on the ratio of performance after recovery at time $t$ to the

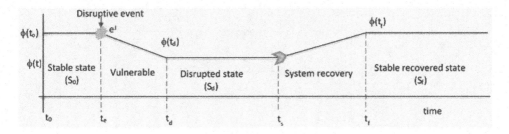

**FIGURE 5.1** Concept of resilience.

performance loss due to disruptive event at time $t_d$. If $R_{es}(t)$ denotes the resilience at time $t$, then resilience of an infrastructure is given by Eq. (5.2):

$$R_{es}(t) = \frac{Recovery(t)}{Loss(t_d)}, t \geq t_d \qquad (5.2)$$

A typical resilient infrastructure system has three performance states, as shown in Figure 5.1, namely the original stable state ($S_0$), a disruptive state ($S_d$), and a stable recovered state ($S_f$) after the disruption. The system is generally considered in a stable state before the disruption event at time $t_e$. After the occurrence of the disruption event, the performance of the system degrades over time due to the failure of some components or due to partial functional loss.

The performance degrades continuously and converges to a maximum disruptive performance state ($S_d$) at time $t_d$. At this stage, restoration measures are implemented to recover the performance that involves repair and recovery actions. The recoverability phase involves identifying the affected system and repair or replacing a failed equipment (Zhang et al., 2018). Upon successful implementation of the recovery, the system starts performing and converging to a new stable state $S_f$.

From Figure 5.1, $\phi(t_0)$ represents the performance of the system corresponding to the initial stable state $S_0$. After the occurrence of disruptive event $e^j$ at time $t_e$, the performance of the system degrades and reaches to a disruptive state $S_d$ at time $t_d$ with a performance value of $\phi(t_d)$ corresponding to disruptive state. During time $t_s - t_d$, the situation of the disruptive state of the system is assessed and necessary recovery actions are initiated at time $t_s$. After successful recovery, the system reaches to a new performance state known as stable recovered state $S_f$ at time $t_f$ with a new performance value of $\phi(t_f)$. Henry and Ramirez-Marquez (2012) proposed a time-dependent relationship to quantify the resilience of the system as defined in Eq. (5.3):

$$R_{es}\left(t_f | e^j\right) = \frac{\phi\left(t_f | e^j\right) - \phi\left(t_d | e^j\right)}{\phi(t_0) - \phi\left(t_d | e^j\right)}, \forall e^j \in D \qquad (5.3)$$

where $D$ is a set of disruptive events.

Chang and Shinozuka (2004) defined another quantification approach to assess the resilience of critical infrastructure subjected to seismic event. The authors use the recovery time and absolute loss as metrics to measure the resilience by comparing the loss in terms of robustness and rapidity under disruptive event with that during normal operation. Cimellaro et al. (2010) proposed a generalized resilience quantification method to overcome the limitations of the method by Chang and Shinozuka (2004) based on the normalized area underneath performance function $\phi(t)$. The authors in this work extend the concept of normalized resilience until the system achieves the complete

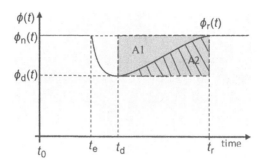

**FIGURE 5.2**  Normalized resilience.

recovery to stable performance implying the successful recovery process, as illustrated in Figure 5.2 and given by Eq. (5.4):

$$R_{es}(t) = \frac{\int_{t_d}^{t_r} \left[ \phi_r(t) - \phi_d(t) \right] dt}{\int_{t_d}^{t_r} \left[ \phi_n(t) - \phi_d(t) \right] dt}, \; t_r \geq t_d \tag{5.4}$$

$R_{es}(t) = 0$, when $\phi_r(t) = \phi_d(t)$, i.e., the system has not recovered from a disruptive event while $R_{es}(t) = 1$, when $\phi_r(t) = \phi_n(t)$, i.e., the system has fully recovered.

Argyroudis et al. (2020) have recently proposed a resilience index quantification approach for critical infrastructure subjected to multiple and diverse hazards based on the classification of various hazard sequences evolving from the disruption event.

Although several resilience metrics and quantitative evaluation methods are currently available, quantifying the resilience of a specific infrastructure system remains a serious challenge due to many factors involved in the calculation of such metrics. Most of the definitions and metrics strongly rely on the overlapping concepts of resilience defined by Filippini and Silva (2014) and Woods (2015). The properties such as adaptability, robustness, redundancy, survivability, and flexibility are determined based on the overall structure of the system, and the time-dependent properties such as rapidity, reparability, resourcefulness, and recoverability are determined based on the maintenance resource (Zobel and Khansa, 2014; Cai et al., 2018). The factors related to external disturbance, threat, and other events are not associated with the resilience metric. Thus, the structure of the system and maintenance resource constitutes a model to measure the resilience. Santhosh and Patelli (2020) have proposed a framework for resilience modeling of critical infrastructure system that models all the phases of system response under a threat. They computed the resilience profiles by adopting the above-discussed resilience properties with respect to the structure of the system and maintenance resource aspects. This novel approach is discussed with a case on a nuclear reactor in Section 5.3.

## 5.3  INTEGRATED FRAMEWORK FOR RESILIENCE ASSESSMENT

The infrastructure resilience can be quantified by a performance measure that can be any quantifiable metric such as reliability, availability, etc. (Cai et al., 2018). Every infrastructure system has its own performance metrics such as reliability, availability, maintainability, and safety. For a continuously operating systems or infrastructure, the widely accepted measure of performance is availability over reliability as the system is usually restored after a disruptive event. Figure 5.3 shows a schematic diagram depicting various phases and factors involved in the resilience modeling of a complex infrastructure system where $\phi(t)$ can be treated as availability of the system which in turn

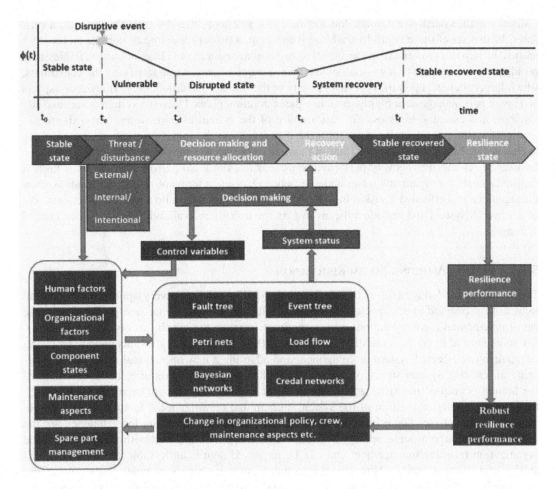

**FIGURE 5.3**   Schematic diagram of resilience modeling.

is a measure of performance. The availability of a system attains a steady-state value after a period from the initial start-up upon implementation of successful operational and maintenance tasks. Upon the occurrence of internal or external disturbance event $e^j$ at time $t_e$, the availability decreases gradually and converges to a maximum disruptive state at time $t_d$ and remains in this state until the restoration takes place. During the time $t_s - t_d$, all the efforts are made in terms of assessing the fault situation, allocating the necessary resources and logistics, identifying the suitable crew for repair actions, etc. Once a successful restoration is implemented at time $t_s$, the availability starts increasing gradually and attains a new stable recovery state corresponding to time $t_f$ and continues to perform in this state until the occurrence of further disruptive events. This process of maintaining an infrastructure system in a regular operational state subjected to several disruptive events accords with the property of resilience (Cai et al., 2018).

As shown in Figure 5.3, the proposed approach models the system resilience taking into account all the resilience principles together with the recovery process under a threat scenario and provides flexibility to achieve the best possible resilience level. The approach presents a complete resilience model of an infrastructure system under any internal or external threat and including the human factors, organizational factors, component operational states, maintenance aspects, and spare part inventory.

A large number of internal factors affect the system restoration; among them, human and organizational factors are found to be the most potential contributors toward achieving an effective resilience of the system (Reinerman-Jones et al., 2019), and hence they are called as the control variables. With a set of these control variables given threat, a proper modeling of the system provides an insight into decision-making toward effective system restoration. These control variables may be adjusted with respect to high- and low-resilience requirements, while satisfying the constraints, which largely depend upon the type and severity of the threat. In addition, the maintenance aspects and spare part management highly influence the restoration plans. Once the system is successfully restored, it is essential to assess the performance of the system after recovery, where the recovered performance is generally expected to be different from the initial original performance due to changes in the failure and repair rates of repaired or replaced components, system structure, etc. In order to achieve the recovery performance as near to as original performance with the highest resilience goals, an organization can undergo policy changes in terms of resource allocation, crew management, and effective decision-making. Thus, it necessitates to build a robust resilience model of the overall system and provide a means to gain the maximum availability factor from a critical system.

### 5.3.1 Factors Affecting System Restoration

The resilience modeling involves the identification of all possible recovery options for the system to be able to respond to unexpected disruptive events and recover the desired performance. The well-implemented recovery actions may sometimes improve the capabilities of the system relative to its normal operating condition (Tran et al., 2017). Such recovery actions include replacing or repairing of affected system or component and adapting a new operational mode or reconfiguring an existing system structure. As most infrastructure systems require the involvement of the human operators and strict adherence to organizational policies for repair or reconfiguration and eventually restoration of the system, human and organizational factors play a key role in the recovery phase. The other important factors that affect the restoration efficiency are the maintenance-related aspects, spare part availability, etc., which are well within the scope of the organization resource management and can be improved significantly subjected to safety, reliability, and cost constraints. However, the human factors are highly complex to determine and may significantly change with respect to the type of disruptive event, and progressive scenarios of the event (Kim et al., 2017, 2018).

### 5.3.2 Modeling Human and Organizational Factors

Performance assessment of human resources in most critical systems is crucial and an important issue for managers, decision makers, and researchers. Among many factors, situation awareness and mental workload have a direct implication on the performance of human operator (Endsley, 1995; Fernandes and Braarud, 2015; Seung et al., 2016). The relationship between key influencing factors for operator performance is shown in Figure 5.4. Situation awareness is related to the perception of the current state of the system, reasoning of the present state, and projecting to an evolving state within the specified time and space (Hwang et al., 2008). It is more likely that an operator with good situation awareness skills is expected to assess the threat situation properly and take efficient corrective actions, which greatly improves the resilience of the system (Salmon et al., 2006). The other important factor is the physical and mental workload to assess the physical and mental needs of an operator during an emergency situation (Yang et al., 2012).

As threats are random and unknown, and operators may not have been trained for all possible threat scenarios, the human factor data related to recovery actions may not be available. Hence, the use of simulator data or expert judgment is recommended in the absence of actual observed data (Musharraf et al., 2014; Kim et al., 2018) better learning from accident and near missing (Moura et al., 2017; Morais et al., 2020). The recovery probabilities for the restoration phase of the system upon a disturbance are generated through the available generic resources taking into consideration

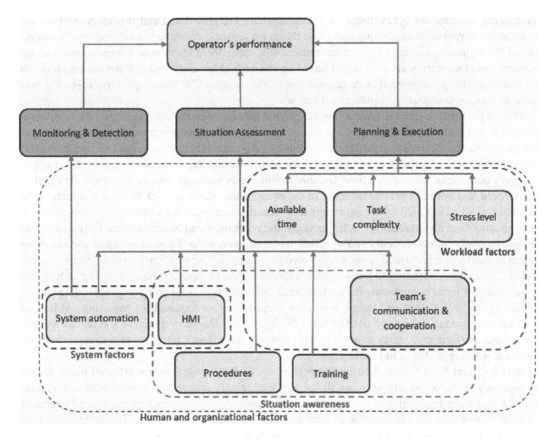

**FIGURE 5.4** Factors influencing the operator's performance.

the highest possible situation awareness and lowest mental workload factors for best resilience requirements.

## 5.4 RESILIENCE ASSESSMENT OF NUCLEAR REACTOR SYSTEM

The proposed resilience framework has been applied to a safety-related reactor regulating system of an advanced thermal nuclear reactor to assess its resilience using Bayesian networks. The Bayesian network has a wide range of applications in reliability, safety, resilience, and decision support systems (Garg et al., 2017; Tolo et al., 2017). The Bayesian network is well suited to applications where strong interdependencies exist among various systems including the human interactions. The main function of the regulation system is to regulate the control rods so as to maintain the reactor under normal operating conditions. The regulating system consists of three different control rods, namely regulation rods, shim rods, and absorber rods (Sinha and Kakodkar, 2006). Regulating rods are used for reactor regulation, shim rods are used for reactor setback, and absorber rods are used for xenon override. The operation of these control rods is based on a 2-out-of-3 sensor logic. The normal operation of a nuclear reactor depends on the availability of regulation function, heat removal function, and neutron moderation function with other associated systems including power supplies. These functions are modeled in a Bayesian network considering a threat to sensor circuit. The performance metric of interest for assessing the resilience of the system under threat is the steady-state availability of the reactor system. The possible threats such as internal, external, or cyber threat on a sensor circuit are postulated and resilience profiles are generated for various threat sequences. Any threat on sensor circuit may result into malfunctioning of the regulating system which, if not mitigated, eventually leads to loss of regulation accident. Several threat scenarios are postulated based on the available safety and restoration measures to demonstrate the possible resilience capabilities of the system. The threat scenario resulting into various resilience sequences is described below:

One of the three sensors is affected by an internal threat. Since the functioning of the regulation system depends on a 2-out-of-3 logic, the failure of one sensor is not going to affect the overall system performance. However, the redundancy in the sensor circuit is lost, as a result of this automatic setback, function is initiated to limit the reactor power within the set limits so as to carry out the recovery operations. During this time, the threat situation is assessed, and the system is expected to be restored to a normal operational state. In the event of automatic setback being not actuated, the manual setback is initiated by the intervention of human operator. In the unlikely scenario of failure of both automatic and manual setback functions, the mitigation and restoration are fully dependent on the available safety systems and eventual human intervention. Depending upon the resulting threat scenario, the resilience sequences evolve accordingly. The resilience of the reactor system may be studied by extending to a case where a threat affects more than one sensor. The failure and repair data for basic components for performance quantification is taken from generic data source IAEA Tecdoc 478 (1988) and the performance shaping factors required for assessing the human performance are taken from NUREG/CR-6949 (2007). The Bayesian model of the reactor system for various performance states under an internal threat on a sensor circuit with automatic setback function working is shown in Figure 5.5.

It is apparent from Figure 5.5 that the performance of the system is not affected much due to redundancy in the sensor circuit when threat is affecting only one sensor. If both auto and manual setback functions fail, which is a very unlikely scenario, the performance of the system degrades to a very low value and prompts for a restoration after a complete shutdown. The system recovers when the threat has been addressed and the component has been repaired or replaced. Then, the system converges to a near-normal performance state. The Bayesian model depicting this recovery

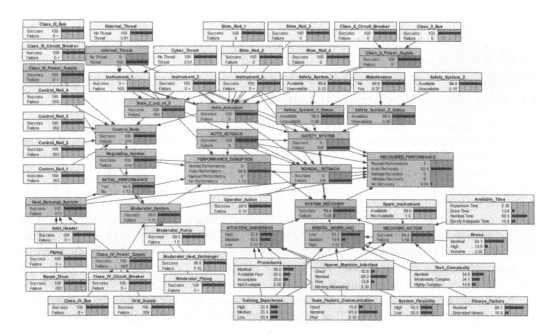

**FIGURE 5.5** Bayesian model with auto setback working.

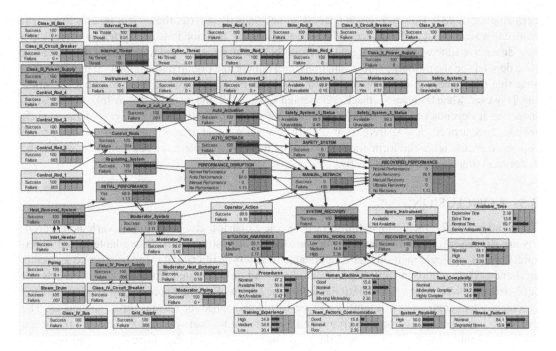

**FIGURE 5.6** Bayesian model with restoration after complete setback failure.

is shown in Figure 5.6. The system restoration with different human and organizational factors is also simulated in the Bayesian model.

The resilience profiles of reactor system under a threat with system recovery are shown in Figure 5.7. It is observed that the performance of the system after restoration reaches a near initial performance level. However, the time required for restoration under each threat sequence is different due to the damage impact. During auto setback working, the performance reduces to 93.4 from an initial

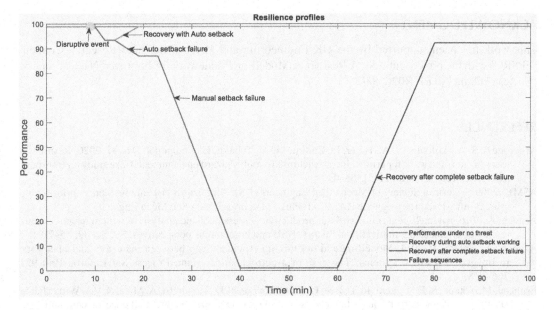

**FIGURE 5.7** Resilience profile for various recovery sequences.

performance of 98.9. However, when restoration becomes successful, the system regains its performance to a near initial performance. When the auto setback function fails, the system converges to a new disruptive state where manual setback from the operator is expected to initiate. The restoration time depends on the available safety functions, and human and organizational factors. It is logical to state that the chances of restoring the system quickly are quite high when the auto setback is working. However, when both auto setback and manual setback fail, the restoration may take significantly longer as it depends on the availability of ultimate safety systems taking into consideration their maintenance aspects, and spare part availability. When a threat affects more than one sensor, the resilience profile takes the form similar to the one during a complete setback failure; however, the restoration time can vary significantly.

## 5.5 CONCLUSIONS

The concepts of RE including definitions, principles, metrics, etc. proposed by various researchers are discussed in this chapter. Despite various quantification methods available in the literature, there is no single method that is applicable to any type of critical infrastructure to assess its resilience. The authors in this chapter have discussed an integrated framework that models the infrastructure system by a computational technique such as Bayesian network by integrating various factors and computing availability as a performance metric for various threat sequences. The approach has been applied to a safety-related system of a nuclear reactor to assess the resilience under a potential internal threat, and the resilience profiles over the entire threat and recovery phase have been generated. The dynamic Bayesian model of the reactor system captures all the dependencies including the human and organizational factors together with the dynamic interaction of the system required for restoration. The system restoration with different human and organizational factors is also simulated in the Bayesian model. The approach discussed in this study is quite flexible for simulating various kinds of threats including extreme events and cyber threats and generating the possible resilience sequences existing within the system with optimal human and organizational factors. The quantitative resilience metrics of the reactor system under a threat provide insights on the significance of various factors to decision-making in terms of the resources within the system, and for probable improvements to build robust resilience into the system.

## ACKNOWLEDGMENT

This work has been supported by the UK Engineering and Physical Sciences Research Council (EPSRC) with the project entitled "A Resilience Modelling Framework for Improved Nuclear Safety (NuRes)," Grant No EP/R020588/2.

## REFERENCES

Argyroudis, S. A., Mitoulis, S. A., Hofer, L., Zanini, M. A., Tubaldi, E., Frangopol, Dan M. 2020. Resilience assessment framework for critical infrastructure in a multi-hazard environment: Case study on transport assets. Sci Total Environ 714: 136854.
ASME. 2009. American Society of Mechanical Engineers (US). All-hazards risk and resilience: prioritizing critical infrastructures using the RAMCAP Plus SM approach. Am Soc Mech Eng.
Azadeh, A. 2016. An intelligent framework for productivity assessment and analysis of human resource from resilience engineering, motivational factors, HSE and ergonomics perspectives. Saf Sci 89: 55–71.
Azadeh, A., Salehi, V. 2014. Modelling and optimizing efficiency gap between managers and operators in integrated resilient systems: The case of a petrochemical plant. Process Saf Environ Prot 92: 766–778.
Bruneau, M., Chang, S. E., Eguchi, R. T., Lee, G. C., O'Rourke, T. D., Reinhorn, A. M., ... & Von Winterfeldt, D. 2003. A framework to quantitatively assess and enhance the seismic resilience of communities. Earthquake Spectra 19(4), 733-752.

Burtscher, M. J., Manser, T. 2012. Team mental models and their potential to improve teamwork and safety: A review and implications for future research in healthcare. Saf Sci 50: 1344–1354.

Cai, B., Xie, M., Liu, Y., Liu, Y., Feng, F. 2018. Availability-based engineering resilience metric and its corresponding evaluation methodology. Reliab Eng Syst Saf 172: 216–224.

Chang, S.E., Shinozuka, M. 2004. Measuring improvements in the disaster resilience of communities. Earthq Spectra 20(3): 739–755.

Cimellaro, G. P., Andrei, M. R., Bruneau, M. 2010. Framework for analytical quantification of disaster resilience. Eng Struct 32: 3639–3649.

Cimellaro, G. P., Reinhorn, A.M., Bruneau, M. 2006. Quantification of seismic resilience. Proceedings of the eighth U.S. national conference on earthquake engineering, 8–22.

Dessavre, D. G., Ramirez-Marquez, J. E., Barker, K. 2016. Multidimensional approach to complex system resilience analysis. Reliab Eng Syst Saf 149: 34–43.

Endsley, M. R. 1995. Toward a theory of situation awareness in dynamic systems. Hum Factors 37(1): 32–64.

Estrada-Lugo, H. D., Santhosh, T. V., Patelli, E. 2020. An approach for resilience assessment of safety critical systems using Credal networks. Proceedings of the 30th ESREL and 15th PSAM, Venice. DOI: 978-981-14-8593-0

Fernandes, A., Braarud, P. Ø. 2015. Exploring measures of workload, situation awareness, and task performance in the Main Control Room. Procedia Manuf 3: 1281–1288.

Filippini, R., Silva, A. 2014. A modelling framework for the resilience analysis of networked systems-of-systems based on functional dependencies. Reliab Eng Syst Saf 125: 82–91.

Garg, V., Santhosh, T. V., Antony, P. D., Gopika, V. 2017. Development of a BN framework for human reliability analysis through virtual simulation. Life Cycle Reliab Saf Eng 6: 223–233.

Haimes, Y. Y. 2009. On the definition of resilience in systems. Risk Anal 29: 4.

Henry, D., Ramirez-Marquez, J. E. 2012. Generic metrics and quantitative approaches for system resilience as a function of time. Reliab Eng Syst Saf 99: 114–122.

Holling, C. S. 1973. Resilience and stability of ecological systems. Annu Rev Ecol Syst 4: 1–23.

Hollnagel, E. 2006. Resilience: The challenge of the unstable. In: Hollnagel, E., Woods, D. D., Leveson, N. C. (Eds.). Resilience engineering: Concepts and precepts, 9–18, CRC Press

Hollnagel, E., Woods, D. D., Leveson, N. 2007. Resilience engineering: Concepts and precepts. Ashgate Publishing Ltd, Farnham, United Kingdom.

Hosseini, S., Barker, K., Ramirez-Marquez, J.E. 2016a. A review of definitions and measures of system resilience. Reliab Eng Syst Saf 145: 47–61.

Hosseini, S., Barker, K. 2016b. A Bayesian network model for resilience-based supplier selection. Int J Prod Econ 180: 68–87.

Hwang, S. -L., Yau, Y. -J., Lin, Y. -T., Chen, J. -H., Huang, T. -H., Yenn, T. -C., Hsu, C. -C. 2008. Predicting work performance in nuclear power plants. Saf Sci 46: 1115–1124.

IAEA Tecdoc 478. 1988. Component reliability data for use in probabilistic safety assessment. IAEA, Vienna.

IAEA. 1998. International Atomic Energy Agency. Developing safety culture in nuclear activities: Practical suggestions to assist progress. Vienna, Safety reports series 11.

Kim, Y., Park, J., Jung, W. 2017. A classification scheme of erroneous behaviors for human error probability estimations based on simulator data. Reliab Eng Syst Saf 163: 1–13.

Kim, Y., Park, J., Jung, W., Choi, S.Y., Kim, S. 2018. Estimating the quantitative relation between PSFs and HEPs from full-scope simulator data. Reliab Eng Syst Saf 173: 12–22.

Lay, E., Branlat, M., Woods, Z. 2015. A practitioner's experiences operationalizing, resilience engineering. Reliab Eng Syst Saf 141: 63–73.

Min, O., Dueñas-Osorio, L., Min, X. 2012. A three-stage resilience analysis framework for urban infrastructure systems. Struct Saf 36–37: 23–31.

Min, O., Wang, Z. 2015. Resilience assessment of interdependent infrastructure systems: With a focus on joint restoration modelling and analysis. Reliab Eng Syst Saf 141: 74–82.

Morais, C., Moura, R., Beer, M., Patelli, E. 2020. Analysis and estimation of human error from report of major accident investigations. ASCE-ASME J Risk Uncertain Eng Syst B 6: 011014.

Moura, R., Beer, M., Patelli, E., Lewis, J., Knoll, F. 2017. Learning from accidents: Interactions between human factors, technology and organisations as a central element to validate risk studies. Saf Sci 99: 196–214.

Musharraf, M., Bradbury-Squires, D., Khan, F., Veitch, B., Mac Kinnon, S., Imtiaz, S. 2014. A virtual experimental technique for data collection for a Bayesian network approach to human reliability analysis. Reliab Eng Syst Saf 132: 1–8.

NUREG/CR-6949. 2007. The Employment of Empirical Data and Bayesian Methods in Human Reliability Analysis: A Feasibility Study. U.S. Nuclear Regulatory Commission.

Pillay, M. 2017. Resilience engineering: An integrative review of fundamental concepts and directions for future research in safety management. Open J Saf Sci Technol 7: 129–160.

Qureshi, Z. H. 2007. A review of accident modelling approaches for complex socio-technical systems. Proceedings of the twelfth Australian workshop on safety critical systems and software and safety-related programmable systems 86: 47–59.

Reinerman-Jones, L. E., Hughesb, N., D'Agostino, A., Matthews, G. 2019. Human performance metrics for the nuclear domain: A tool for evaluating measures of workload, situation awareness and teamwork. Int J Ind Ergon 69: 217–227.

Rocchetta, R., Zio, E., Patelli, E. 2018. A power-flow emulator approach for resilience assessment of repairable power grids subject to weather-induced failures and data deficiency. Appl Energy 210, 339–350.

Salmon, P., Stanton, N., Walker, G., Green, D. 2006. Situation awareness measurement: A review of applicability for C4i environments. Appl Ergon 37: 225–238.

Santhosh, T. V., Patelli, E. 2020. A Bayesian network approach for the quantitative assessment of resilience of critical systems. Proceedings of the 30th ESREL and 15th PSAM, Venice. https://www.rpsonline.com.sg/proceedings/esrel2020/pdf/3699.pdf

Seung, W. L., Kim, A. R., Park, J., Gook Kang, H., Seong, P.H. 2016. Measuring situation awareness of operating team in different main control room environments of nuclear power plants. Nucl Eng Technol 48: 153–163.

Sinha, R. K., Kakodkar, A. 2006. Design and development of the AHWR – the Indian thorium fuelled innovative nuclear reactor. Nucl Eng Des 236: 683–700.

Steen, R., Aven, T. 2011. A risk perspective suitable for resilience engineering. Saf Sci 49(2): 292–297.

Tolo, S., Patelli, E., Beer, M. 2017. Robust vulnerability analysis of nuclear facilities subject to external hazards. Stoch Environ Res Risk Assess 31(10): 2733–2756.

Tran, H. T., Balchanos, M., Domerçant, J. C., Mavris, D.N. 2017. A framework for the quantitative assessment of performance-based system resilience. Reliab Eng Syst Saf 158: 73–84.

Woods, D. D. 2006. Essential characteristics of resilience. In: Hollnagel, E., Woods, D., Leveson, N. (Eds.). Resilience engineering: Concepts and precepts. Burlington, VT: Ashgate Publishing Company.

Woods, D.D. 2015. Four concepts for resilience and the implications for the future of resilience engineering. Reliab Eng Syst Saf 141: 5–9.

Youn, B. D., Hu, C., & Wang, P. 2011. Resilience-driven system design of complex engineered systems. J Mech Desig. 133(10).

Yang, C. -W., Yang, L. -C., Cheng, T. -C., Jou, Y. -T., Chiou, S. -W. 2012. Assessing mental workload and situation awareness in the evaluation of computerized procedures in the main control room. Nucl Eng Des 250: 713–719.

Zhang, X., Mahadevan, S., Sankararaman, S., Goebel, K. 2018. Resilience-based network design under uncertainty. Reliab Eng Syst Saf 169: 364–379.

Zobel, C. W., Khansa, L. 2014. Characterizing multi-event disaster resilience. Comput Oper Res 42: 83–94.

# 6 Resilience-Based Design of Buildings

*S. Prasanth and Goutam Ghosh*
Motilal Nehru National Institute of Technology
Allahabad, Uttar Pradesh, India

## CONTENTS

## 6.1 INTRODUCTION

Resilience is a trait of individual institutions, not of individual buildings or structures. This makes earthquake resilience a good plan for any entity that requires far more than just its physical facilities, a city, a neighborhood, a campus, a town, an industry, a corporation, or a residence. Resilience is about preserving and restoring functionality, and not just about stability. In the framework of building codes and standards, it means that the requirements for resilience design do not only take into account structural or nonstructural features and certain building contents and even certain externalities usually disregarded in the code or specification, such as infrastructure reliability, availability of maintenance contractors, or the efficiency of other facilities. The focal point on functionality also demonstrates that design requirements focused on resilience can differ with the particular usage and occupancy of a structure.

Resilience includes a time function. Unlike the immediate and direct impact of the structural response on earthquake protection and contrary to the nature of emergency structures now intended to be operational immediately, resilience-based design may require the return of any lack of function over hours, days, weeks, or even months. The focus is on the prompt return to normal circumstances, in resilience-based design and not only in the emergency process.

## 6.2 CONCEPT OF SEISMIC RESILIENCE

Resilience has a variety of definitions that are provided by different writers. In general, "resilience is defined as the ability of social units (structure) to mitigate hazards, minimize the effects of disasters when they occur and to carry out the recovery activities in ways that minimize social disruption and mitigate the effects of future earthquakes" (Bruneau et al., 2003). The resilience of the structure will be measured in terms of rapidity, robustness, redundancy, and resourcefulness. Rapidity means the capacity of the structure to fulfill the required objectives in a timely manner in order to contain losses. Robustness is the capacity of the system to withstand the required stress

DOI: 10.1201/9781003194613-6

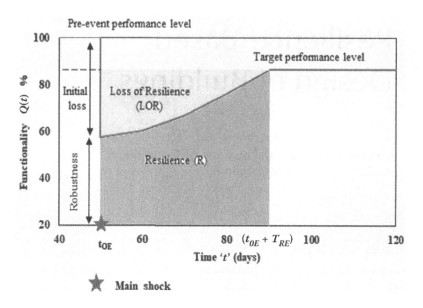

**FIGURE 6.1** Schematic representation of seismic resilience.

level without any degradation. The lack of functionality/degradation has been expressed in terms of redundancy. Resourcefulness is the indicator of mobilizing material and human capital during a crisis. The two main parameters to be analyzed are rapidity and robustness. The rapidity of the structure was measured in terms of "recovery functions" and the robustness was measured in terms of "loss function."

Resilience is about preserving and restoring functionality and is related to building efficiency as determined by the time it takes to restore basic functionality. Resilience has been denoted using the mathematical Eq. (6.1). Graphically, the region below the $Q(t)$ curve function shows the resilience of the system (Figure 6.1). The functionality denotes the target performance level of the structure, which will be represented as a percentage from 0% to 100%. The functionality will be estimated using Eq. (6.2).

$$\text{Resilience}\,(\text{R}) = \int \frac{Q(t)}{T_{LC}}\,dt \tag{6.1}$$

$$\text{Functionality}: Q(t) = 1 - \left\{ L\left(I, T_{RE}\right) \times \left[ H\left(t - t_{OE}\right) - H\left(t - \left(t_{OE} + T_{RE}\right)\right) \right] \times f_{rec}\left(t, t_{OE}, T_{RE}\right) \right\} \tag{6.2}$$

In the above equation, $L(I, T_{RE})$ indicates the loss functions, $T_{RE}$ is the time after the occurrence of the event, $f_{rec}(t, t_{OE}, T_{RE})$ indicates the recovery functions, $t_{OE}$ is the initial time of the event occurred, and $T_{LC}$ is the control period for the estimation of resilience.

## 6.3　VULNERABILITY ASSESSMENT OF THE STRUCTURE

The first phase in the measurement of seismic resilience was a building risk assessment. The susceptibility to fragility curves must be established with HAZUS (Hazards United States) methodology to detect damages and the building's performance level. The nonlinear pushover static analysis (PA) was performed using finite element tools from SAP2000 for the purpose of the vulnerability evaluation. Applied Technology Council (ATC-40) has been responsible for

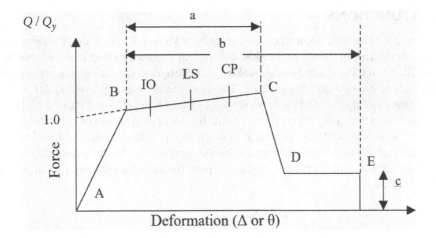

**FIGURE 6.2** Generalized force-deformation curve (ASCE 41-17).

adapting the seismic specifications. For research, different forms of pushover approaches can be used such as Federal Emergency Management Agency (FEMA-440), equivalent linearization approach (EL), displacement modification (DM) method, Advanced Technology Council (ATC-40) approach, etc. In order to test the nonlinear behavior of the structure, plastics are assigned to structural elements such as beams and columns. All the structural members were given the hinge properties based on the ASCE 41-17. The performance levels like Immediate Occupancy (IO), Life Safety (LS), and Collapse Prevention (CP) (Figure 6.2) are the three forms that the structure exhibits. The pushover results showed the spectral displacement, performance, and demand ductility. Based on the HAZUS approach, there are four risk states, such as slight, moderate, extreme, and collapse. Using Eq. (6.3), the probability of exceedance at each damage state can be calculated as follows:

$$P\left(\frac{ds}{S_d}\right) = \phi\left[\frac{1}{\beta_{ds}}\ln\left(\frac{S_d}{S_{d,ds}}\right)\right] \qquad (6.3)$$

where $S_d$ denotes the spectral displacement defining the threshold of a particular damage state, $S_{d,ds}$ denotes the median value of spectral displacement at which the building reaches the threshold of each damage states, $\beta_{ds}$ indicates the standard deviation of the natural logarithm of spectral displacement for damage state ($ds$), and $\phi$ denotes the standard normal cumulative distribution function.

The discrete damage probabilities in each damage state were evaluated by differencing the cumulative damage probabilities as follows:

$$
\begin{aligned}
P[DS = S] &= P[DS = S] - P[DS = M] \\
P[DS = M] &= P[DS = M] - P[DS = E] \\
P[DS = E] &= P[DS = E] - P[DS = C] \\
P[DS = C] &= P[DS = C]
\end{aligned}
$$

S = slight, M = moderate, E = extreme, C = collapse, and DS = damage state

Based on these results, the fragility curve was plotted against the probability of exceedance and the spectral displacements of the building.

## 6.4  LOSS FUNCTIONS

Losses attributable to seismic occurrence are generally unknown. It varies for each seismic scenario and location considered for the study. The loss assessment was the last step in the seismic resilience assessment. Based on these consequences, stakeholders/designers should prepare pre-hazard strategic planning and risk-based decision-making. The loss function was expressed as $L(I, T_{RE})$, a function of the seismic strength ($I$) and the recovery time ($T_{RE}$). Direct ($L_D$) and indirect ($L_{ID}$) economic losses are the two types of losses correlated with the calculation of losses. The direct economic loss applies to structural and non-structural damage to structures. Indirect economic losses are time-dependent, related to relocation expenses, loss of income, etc. As the quantification of indirect economic losses is a complex process, in this chapter, the steps to estimate the direct economic losses are explained.

$$L_D = \sum P_E(DS = K) \times r_K \tag{6.4}$$

Where $K$ is the damage state of the structures, $r_k$ is the damage ratio that corresponds to each damage state taken from HAZUS MR4 technical manual. $P_E(DS = K)$ is the discrete probability of losses to the damage state in the time of the event.

## 6.5  RECOVERY FUNCTIONS

Recovery of the system from a certain amount of damage is a dynamic process, based on time, availability of materials, costs, etc. Several forms of recovery feature models have been proposed by Gian Paolo Cimellaro et al. (2010). Three types of suitable mathematical recovery function models, such as linear, trigonometric, and exponential recovery functions, have been used to minimize complexity. As stated earlier, the rapidity of the structure was dependent on the form of recovery functions used (Figures 6.3–6.5). These recovery functions can be chosen on the basis of the preparedness and the response of the structure. Each recovery function follows various recovery paths and has been given as a mathematical Eqs. (6.5–6.7).

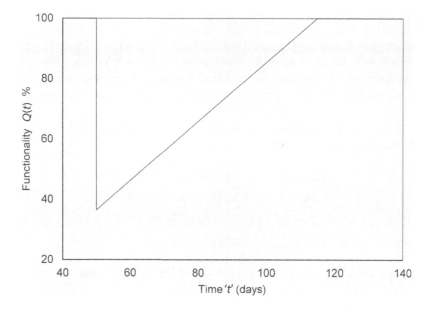

**FIGURE 6.3**  Linear recovery path (Gian Paolo Cimellaro et al., 2010).

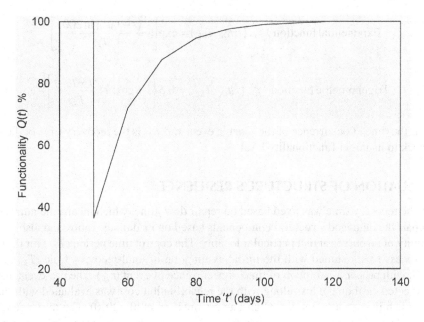

**FIGURE 6.4**   Exponential recovery path (Gian Paolo Cimellaro et al., 2010).

For the real incident, the three possible realistic conditions are followed by each recovery function. Linear functions are continuous in functionality, which means timely availability of services. The exponential functions at the initial path have a higher functionality rate, which implies a high resource inflow at the initial levels, although this is not possible in all cases. At the outset, the trigonometric functions adopted a lower rate of efficiency due to lack of initial capital investment. Depending on the need, one can adopt any of the recovery functions to restore the functionality losses.

$$\text{Linear function}: f_{rec}\left(t, t_{OE}, T_{RE}\right) = \left[1 - \frac{t - t_{OE}}{T_{RE}}\right] \qquad (6.5)$$

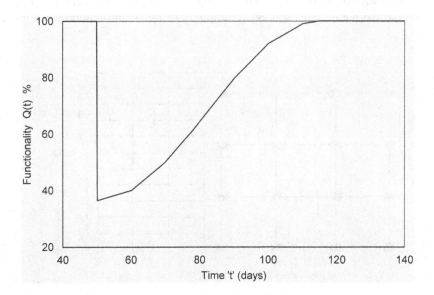

**FIGURE 6.5**   Trigonometric recovery path (Gian Paolo Cimellaro et al., 2010).

$$\text{Exponential function}: f_{rec}\left(t, t_{OE}, T_{RE}\right) = \exp\left[-\frac{\left(t - t_{OE}\right)\left(\ln 200\right)}{T_{RE}}\right] \tag{6.6}$$

$$\text{Trigonometric function}: f_{rec}\left(t, t_{OE}, T_{RE}\right) = 0.5\left\{1 + \cos\left[\Pi\frac{\left(t - t_{OE}\right)}{T_{RE}}\right]\right\} \tag{6.7}$$

Where $t_{OE}$ is the time of occurrence of the seismic event and $T_{RE}$ is the recovery time of the structure to return back to its target functionality level.

## 6.6 EVALUATION OF STRUCTURE'S RESILIENCE

Generally, the recovery time was fixed based on repair downtime, which means the number of days taken to repair the damaged structural components based on its damage ratios and also depends on the availability of resources at that particular locality. The control time period ($T_{LC}$) for the complete recovery process was assumed with the initial assumption of total recovery time ($T_{RE}$) as per the nature/need, with proper assumption of time of occurrence of event ($t_{OE}$). The pre-event assessment with the expected earthquake loss along with the rehabilitation cost was evaluated with the above-mentioned rehabilitation alternatives (Gian Paolo Cimellaro et al., 2010). The target functionality level was fixed based on which the recovery function and time were adopted. The structure's functionality $Q(t)$, which was found using Eq. (6.2), was used to plot the graph functionality $Q(t)$ and the time ($t$). The area beneath the functionality curve gives the loss of resilience, whereas in the case of the exponential and trigonometric recovery functions, using curve fitting the area beneath the curve can be found by integration with respect to $t_{OE}$ and $T_{RE}$.

## 6.7 CASE STUDY G+10 STORY HIGH-RISE BUILDING

A high-rise reinforced concrete building of G+10 story was considered having a plan area of 18 × 18 m and a total height of 44 m (Figure 6.6). The building was subjected to the ground motion of peak ground acceleration (PGA) 0.36 g with significance factors ($I$) 1.5 and response factor ($R$) 5 (IS: 1893–2016) and designed in accordance with Indian requirements (IS: 456-2000). The size of the beam and column was 400 × 600 mm and 600 × 600 mm. The beam tension reinforcing area was 942.477 mm² and the bottom area was 603.186 mm². The maximum reinforcement ratio for the

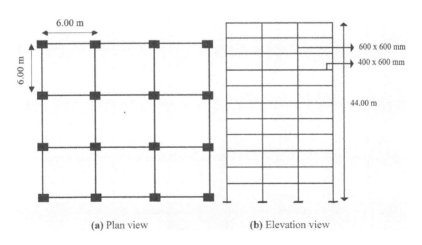

(a) Plan view        (b) Elevation view

**FIGURE 6.6** Plan and elevation view of the building.

**TABLE 6.1**

**Pushover Results**

| S. No. | PGA (g) | Performance Point (kN) | Top Roof Displacement (mm) | Spectral Displacement (mm) | Demand Ductility |
|--------|---------|------------------------|-----------------------------|-----------------------------|------------------|
| 1 | 0.36 | 1799.135 | 75.795 | 61.910 | 1.844 |

**TABLE 6.2**

**Percentage of Cumulative Damage Probability**

| S. No. | PGA (g) | Probability of Exceedance of Damage States (%) | | | |
|--------|---------|------------|----------|---------|----------|
| | | Slight | Moderate | Extreme | Collapse |
| 1 | 0.36 | 38.97 | 20.60 | 1.53 | 0.13 |

column worked was 2.67% (within limit as per IS: 456-2000). The pushover analysis was carried out based on the FEMA-440 EL method. The spectral displacement, demand ductility, and performance level were noted from the pushover result, which is shown in Table 6.1. Using Eq. (6.3), the probability of exceedance corresponding to the spectral displacement was found and shown in Table 6.2. The fragility curve was plotted (Figure 6.7) against spectral displacement that shows the damage percentage of the building that exhibits at each respective spectral displacement. The discrete damage probability, which was found from the cumulative probability of exceedance (Table 6.3), was used to evaluate the direct damage loss ratios (Tables 6.4 and 6.5) using Eq. (6.4).

In this study, the target functionality level was taken as 100%, based on which the recovery time was fixed. The total control time period ($T_{LC}$) of 140 days with recovery time ($T_{RE}$) as 65 days was considered. The time of occurrence of event ($t_{OE}$) was considered to be on the 50th day of the time

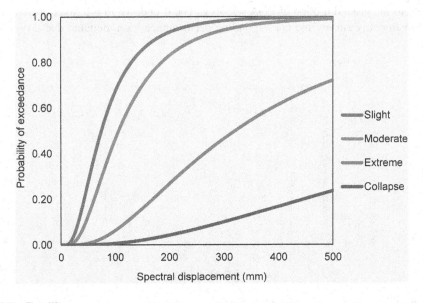

**FIGURE 6.7** Fragility curves.

**TABLE 6.3**
**Discrete Damage Probability**

| S. No. | PGA (g) | Damage States | | | |
|---|---|---|---|---|---|
| | | Slight | Moderate | Extreme | Collapse |
| 1 | 0.36 | 0.184 | 0.191 | 0.014 | 0.001 |

**TABLE 6.4**
**Damage Ratios Based on HAZUS-MR4 Technical Manual for Buildings**

| Damage State (ds) | Damage Ratio (rK) |
|---|---|
| Slight | 0.10 |
| Moderate | 0.40 |
| Extreme | 0.80 |
| Collapse | 1.00 |

**TABLE 6.5**
**Direct Economic Loss Ratios**

| S. No. | PGA (g) | Direct Economic Loss Ratios ($L_D$) |
|---|---|---|
| 1 | 0.36 | 0.107 |

period. The building functionality was evaluated using Eq. (6.2). The functionality curves were plotted using three recovery functions (Figures 6.8–6.10). The building's resilience was calculated with respect to the ground motion of PGA = 0.36 g. The dropdown in the functionality was found from the functionality curves and corresponding performance level, demand ductility is shown in Table 6.6.

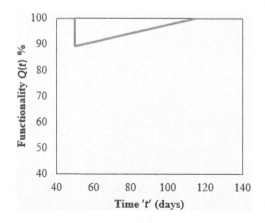

**FIGURE 6.8**　Functionality curves-based linear recovery function.

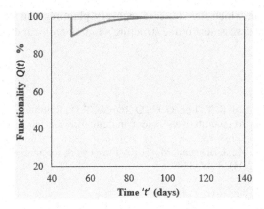

**FIGURE 6.9**   Functionality curves-based exponential recovery function.

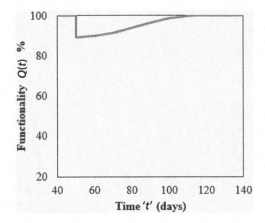

**FIGURE 6.10**   Functionality curves-based trigonometric recovery function.

**TABLE 6.6**
**Seismic Resilience of the Building and Its Performance Level**

| | | Resilience (%) | | | Performance Level |
| S. No. | PGA (g) | Design Horizontal Seismic Coefficient "Ah" (g) | Linear Function | Exponential Function | Trigonometric Function | from Vulnerability Analysis (Figure 6.5) |
|---|---|---|---|---|---|---|
| 1 | 0.36 | 0.157 | 94.65 | 98.13 | 95.03 | IO to LS |

## 6.8   CONCLUSION

One has to make the community/society resilient toward disaster events; hence, there exists the need of resilience-based design. Due to the recent rise in natural disasters, robust society is needed in the aftermath of the catastrophe. The idea of resilience-based design appears in recent times in order to make the system more robust.

The chapter emphasizes the need and implementation of the framework to evaluate the structure's resilience as per their regional conditions. The loss in the structure's functionality and its resilience can be anticipated based on which the strategic recovery planning can be done. It can be used as pre- and post-disaster event assessment of the structure's functionality and can be implemented at the large community level.

## REFERENCES

Bruneau, M., Chang, S. E., Eguchi, R. T., Lee, G. C., O'Rourke, T. D., Reinhorn, A. M., ... & Von Winterfeldt, D. (2003). A framework to quantitatively assess and enhance the seismic resilience of communities. Earthquake Spectra, 19(4), 733–752.

Cimellaro, G. P., Reinhorn, A. M., & Bruneau, M. (2010). Framework for analytical quantification of disaster resilience. Eng Struct, 32(11), 3639–3649.

# 7 Positioning Resilience in Design and Analysis of Structures and Infrastructure

*Vasudha A. Gokhale*
S. P. Pune University, Pune, India

## CONTENTS

## 7.1 INTRODUCTION: RESILIENCE OF STRUCTURE AND INFRASTRUCTURAL SYSTEMS

Resilience represents the capacity of a system, society, or community to sustain and recover from the effects imposed by a disruptive event. Disaster resilience represents a reduced probability of damage to systems, components, critical infrastructures, and reduced adverse socio-economic impacts. It includes the activities for prevention, restoration of basic functions and structures in an effective and timely manner (MCEER 2008). Resilience incorporates the time history of structure and infrastructure performance following significant disturbance and reflects system-performance effects. It is recognized as an intrinsic ability, an inherent quality of the structure, and infrastructural systems comprised of performance-based and temporal properties (Baoping 2017). A resilient engineering system is capable enough to stand against an adverse situation and support a recovery mechanism to minimize the impact of a disaster. The factors affecting structure and infrastructure's resilience include design specifications, damaged components' accessibility, preparedness, ability to recover, and prevailing environmental conditions (Sharma et al. 2018). Engineering resilience is defined as the system's capability to sustain shocks, often evaluated considering the time consumed to return

DOI: 10.1201/9781003194613-7

to its pre-existing equilibrium stage (Cutter et al. 2010). A resilience paradigm is comprised of three capabilities that include absorptive, adaptive, and restorative capabilities. Absorptive capacity refers to the system's capability to respond to the system disturbance-generated impacts and limits undesirable outcomes with minimum efforts. Adaptive capacity is a system's capability to adjust under unforeseen situations by going through reasonable changes, whereas restorative capacity indicates the regain of the normal condition rapidly and reliably (Norris et al. 2008).

## 7.2  QUANTIFICATION OF RESILIENCE

Resilience is a complex phenomenon that possesses organizational, human, temporal, spatial, and technical dimensions. Measuring resilience from an engineering perspective is a challenging task as it calls for a comprehensive understanding of the structural properties and operational aspects of structures and infrastructures. In addition to this, associated uncertainties need to be considered in the context of potential responses and the system's failure against stresses imposed by a disruptive event. For quantification, resilience has been conceptualized considering four interrelated aspects: technical, organizational, social, and economic, referred to with the acronym (TOSE). Resilience has been described as more comprehensively by integrating four resilience measures: robustness, redundancy, rapidity, and resourcefulness (Bruneau et al. 2003). Rapidity represents the ability to achieve goals, satisfy priorities in stipulated time aimed to limit losses, and avoid disruption in the future. It can be expressed as per Eq. (7.1):

$$\theta = \frac{dQ(t)}{dt} \tag{7.1}$$

where $d/dt$ = the differential operator, $Q(t)$ = the functionality function.

Robustness represents the capability of elements, assets, communities to withstand external stresses or demands with no loss of function or degradation. It is a function of geometry and typology, which are unchangeable aspects. Geometry refers to a structure's structural load-bearing element's layout, whereas typology represents the structural configuration corresponding to the site, characterizing exposure expected from extreme imposed loads. Robustness is a variable property of a system that changes with change in design and configuration of the system and can be described by Eq. (7.2):

$$\text{Robustness} = 1 - L^{\sim}(m_L + \alpha.\sigma L) \tag{7.2}$$

where $L^{\sim}$ represents a random variable that expresses the mean function $(m_L)$ and the standard deviation $(\sigma L)$.

Redundancy is the availability of substitutable systems or elements that can satisfy the functional needs in an emergency resulting in degradation or functionality losses. Structural system redundancy depends on various load-carrying components' geometry and structural characteristics. The presence of alternative routes to transfer the lateral forces characterizes structural redundancy (ASCE 2000). Redundancy ensures that if part of the structure fails, the rest of the structure can facilitate the redistribution of the loads and save the whole system's failure (Frangopol and Curley 1987).

However, resourcefulness facilitates problem identification, mobilization of physical, monetary, technological, and workforce to achieve goals based on priorities in the recovery process (Bruneau and Reinhorn 2007). Resilience is a function that represents a structure, lifeline network, bridge, or community's ability to continue performing or functioning to a prescribed level over a defined period. Considering the temporal dimensions, resilience can be described as the function $Q(t)$ representing the system's functionality in a prescribed time. The factor $Q(t)$ is quantified with nonlinear recovery function and loss functions based on two quantitative measures referred to as control time $(T_{LC})$, and time to recover from the consequent disruptions indicates the recovery time. Compared to the original status, the time required to restore the structure's functionality to a similar or improved operation level represents the recovery time ($T_{RE}$). Disaster resilience quantification includes two key components: the event's impact in terms of losses and post-event performance, as expressed in Eq. (7.3) (Cimellaro et al. 2010).

$$R = \int_{t_{OE}}^{t_{OE}+T_{RE}} \frac{Q(t)}{T_{RE}} dt \tag{7.3}$$

Where $t_{0E}$ indicates the time when the event $E$ occurred, $T_{RE}$ indicates the recovery time. $Q(t)$ is system functionality, expressed as a function of time $t$. $Q(t)$ is expressed by Eq. (7.4).

$$Q(t) = 1 - L(I, T_{RE}) \cdot \left[ H(t - t_{0E}) - H(t - (t_{0E} + T_{RE})) \right] \cdot f_{Rec}(t, t_{0E}, T_{RE}) \tag{7.4}$$

Where $L(I, T_{RE})$ = the loss function; $f_{Rec}(t, t_{0E}, T_{RE})$ = the recovery function; $H(t_0)$ = the Heaviside function, $t_{0E}$ = the occurrence time of the event $E$, and $I$ = the intensity. In this equation, the values of the quantities are less than one as the full functionality is equal to 100% = 1. Figure 7.1 describes the concept of resilience.

In Figure 7.1, "$T_{0E}$" shows the time when the disruptive event occurred, and "$T_{RE}$" is the time taken by the system for recovery. The section from point A to B represents a drop in performance due to disruption. The system may follow one of the three performance levels indicated by resilience curves. The phase of disruption and recovery reflects the absorptive capability and restoration capability, followed by a new normal phase. The function mentioned above is suitable for assessing a system's functionality, provided the immediate losses and recovery are identifiable distinctly. It can also analyze individual structures' resilience where the functionality includes human casualties and memory losses (Cimellaro et al. 2010).

For a comprehensive understanding of a disruptive event's impact, resilience refers to resilience functions that include time variation of damage and its relation to the response and recovery

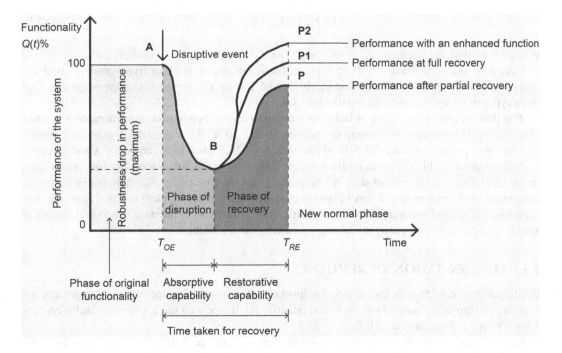

**FIGURE 7.1** Concept of resilience (Adapted from Sarke and Lester 2019).

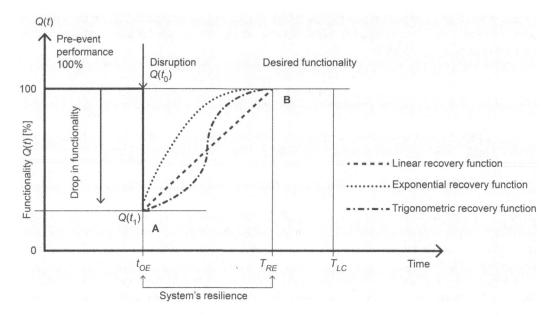

**FIGURE 7.2**   Recovery function (Based on Cimellaro et al. 2006).

mechanisms. From an engineering perspective, resilience can be evaluated based on three key aspects that include time recovery, loss function, and recovery function.

Loss function: In the context of losses due to earthquake occurrence, the loss function describes an earthquake's intensity ($I$) and restoration downtime ($T_{RE}$). Losses due to a disruptive event include direct losses ($L_D$) and indirect losses ($L_I$), where the loss function is expressed as per Eq. (7.5) (Venkittaram 2014):

$$L(I, T_{RE}) = L_D + \alpha_I \cdot L_I \tag{7.5}$$

where $I$ = earthquake intensity, $T_{RE}$ = recovery time, $L_D$ = direct loss, $L_I$ = indirect loss.

Recovery functions: Recovery functions can be linear, exponential, or trigonometric, used for analysis considering the system's characteristics and society's response under investigation. The three types of recovery functions are illustrated in Figure 7.2.

The three recovery functions indicate system performance before and after disruption at time "$T_{OE}$." The performance of the system drops from "$Q(t_0)$" to "$Q(t_1)$" due to the disruption followed by a recovery process to reach the desired functionality. The section between points A and B represents the recovery, which depends on the recovery function type. A linear recovery function is used in the lack of information about society's response to an extreme event. The exponential recovery function is used when an initial flow of resources results in a fast society response and consequently decreased rapidity of recovery; the trigonometric type of recovery – function is adopted when the society's response is slow at the initial stage (Cimellaro et al. 2006).

## 7.3   REPRESENTATION OF RESILIENCE

Resilience metric is an efficient tool to encapsulate infrastructure characteristics that portray an accurate picture of the recovery process and quantify resilience with the appropriate inclusion of a recovery curve (Sharma et al. 2018).

Recovery curve: Recovery curves describe the level of damage at the initial stage and the presence of required resources indicating post-disaster recovery conditions qualitatively. It represents a continuous, discrete, or piecewise non-decreasing function of time. The recovery process is primarily affected by weather conditions, availability of resources, and the disrupting forces, shown as a sudden drop in the recovery curve.

Fragility curves: The prediction of structural damages to structures and infrastructures is of profound importance, which could be estimated and represented with fragility curves. In light of an earthquake's occurrence, fragility curves represent the damage level and the ground motion intensity parameters (Noroozinejad et al. 2014). Fragility curves express the expected damage factor and expected loss, which are functions that indicate the conditional probability in the context of the response of a system subjected to seismic excitations exceeds a given performance limit state. The information is graphically presented, showing the damage distribution pattern indicating the cumulative damage distribution, specifying the continual probability of reaching or exceeding a particular damage level (Shinozuka et al. 2000).

Fragility functions: Structure and infrastructure's vulnerability is often described using fragility functions (FFs) that represent one of the key aspects of seismic risk assessment. They are used to identify the probability to which a system, structure, or a non-structural component will go beyond a particular state of damage corresponding to earthquake intensity level characterized by peak ground velocity (PGV), peak ground acceleration (PGA), or spectral acceleration (Sa) or story drift, floor acceleration representing demand parameters. FFs are estimated with processing data gathered from nonlinear dynamic structural analyses and incremental dynamic analysis (IDA) approach or multiple stripes analysis (Vamvatsikos 2002). The actual data gathered from existing earthquake-damaged structures can be used to prepare a fragility curve, which refers to the function representing the probable response of a structure to seismic excitations exceeds, performance limit states (Koutsourelakis 2010). It has been stated that vulnerability curves estimate resilience more realistically as compared to fragility curves.

Vulnerability curves: Vulnerability curves are developed based on different data sources, including observational data from post-earthquake surveys and analytical simulation. Empirical vulnerability curves represent data gathered from observed damages to the structures and infrastructure subjected to a disruptive event. Hence, they are more realistic as compared to curves based on analytical simulation. However, their use in a general application is not recommended given the limitation of data collected from a specific region that may suffer from non-representative sampling (Noroozinejad 2014).

Resilience triangle: The concept of resilience possesses a temporal dimension that is visualized with a "resilience triangle," which is mostly applied in the context of a technical system to demonstrate the impact of a disaster on resilience (Bocchini et al. 2013). The resilience triangle represents the gradual recovery time, abrupt performance loss, and their impact on the more extensive portfolios and structures. Based on this concept, system resilience quantification considers three factors: the probability of failure, consequences of failure, and repair time. Here, resilience represents a system's ability to survive against a system's declining performance in a specified period, $Q(t)$, as expressed in Eq. (7.6):

$$Q(t) = Q_\infty - (Q_\infty - Q_0)e^{-bt} \tag{7.6}$$

where $Q$ = the structural system's capacity at its full functional level, $Q_0$ = the capacity after the event, $b$ = the recovery process's rapidity, and $t$ = time after the event.

## 7.4  FRAMEWORKS, METHODS, AND TOOLS

Due to the interdependency of infrastructure, their behavior on a disruptive event is complicated. Several methods are in place for analyzing disaster resilience; each has a different concept, epistemological stance, and methodological approach. The next section discusses the most commonly used frameworks, concepts, and methods.

### 7.4.1 The Peoples Framework

Disaster resilience of communities can be analyzed based on the people's framework integrating physical and socio-economic aspects to define the community's disaster resilience at various scales qualitatively. This hierarchical framework consists of seven dimensions that represent Population (P); Environment (E); Organized governmental services (O); Physical infrastructure (P); Lifestyle (L); Economical (E); and Social capital (S), summarized with "PEOPLES" acronym (Kammouh and Cimellaro 2018). It provides a basis to develop qualitative and quantitative models to evaluate communities' resilience against extreme events. Here the specified indicators can be represented in the area's geographic information system (GIS) layer. The global resilience index (Rcom) developed with this approach is illustrated in Eq. (7.7):

$$\mathrm{R_{com}} = \int\limits_{A_c} \int\limits_{t_{0E}}^{t_{0E}+T_{RE}} Q_{TOT\ (r,t)/(A_cT_{LC})}\ dtdr \qquad (7.7)$$

where $A_c$ = the geographical area, $t_{0E}$ = the time when the event occurred, $T_{LC}$ = the time to control corresponding to the specific period of interest, $Q_{TOT}\ (r,t)$ = global functionality-performance function, and $r$ is the spatial vector defining the position (Lu et al. 2020).

Robustness, recoverability, adaptability, and reliability represent the essential aspects of the resilience metric. For assessment of resilience, four stages of disaster are considered that include the stage of disaster prevention, damage propagation, assessment, and recovery. The final stable stage indicates the completion of the recovery process (Ouyang 2014).

### 7.4.2 The Disaster Resilience of Place Model

The disaster resilience of place (DROP) model includes natural and social systems, the built environment that interacts and affects the inherent vulnerability and resilience. The model recognizes interrelation between the natural, built environment, and social system, considering people's and places' role. It includes the associated natural and social systems representing infrastructural, institutional, and ecological components as an integral part of resilience (Bruneau et al. 2003). The DROP model has five indicators: infrastructure, institutional, social, economic resilience, and community capital. However, attributes such as information and communication and community competence are excluded in the DROP model given the associated difficulties in quantification based on data available in the public domain, limiting its applicability to a certain extent (Tiernan 2019).

### 7.4.3 Figure-of-Merit

Resilience is often looked at considering structure and infrastructure systems' delivery capacity during and after a disaster, referred to as figure-of-merit (FOM). The concept of FOM includes the level of performance of a system over time, considering the system's characteristics such as connectivity, reliability, and system's delivery capacity during and after a disaster to represent resilience. FOM metric represents the ratio of recovery to the losses that occurred previously. In this method, the FOM of a system is quantified to estimate the resilience, which includes a single metric; however, multiple FOM is possible in a real system that is not integrated here to find resilience. This approach is not suitable for modeling a complex system with multiple and secondary impacts of a disaster and actions taken for recovery after the event (Tamvakis and Xenidis 2013).

### 7.4.4 Fuzzy Inference Systems

This method considers two concepts: the resilience cycle and the system performance hierarchy. The resilience cycle expresses a system condition flow on its exposure to a disruption based on four components. The system performance hierarchy ranks and defines systems performance based on theory regarding human needs' hierarchy, introduced by Maslow. A fuzzy inference system (FIS) quantifies the values obtained for various variables, including numerical ranges or linguistic terms. The hierarchically structured dependency diagram is prepared to represent a hierarchy of performance levels (Tamvakis and Xenidis 2013).

### 7.4.5 PBEE METHODOLOGY

Performance-based earthquake engineering (PBEE) methodology facilitates the probabilistic assessment of seismic performances of structures and infrastructures, and other facilities. It analyzes the performance objectives that include the site's location and characteristics, the structural systems used in construction, building configuration, occupancy, and non-structural elements. Four logical progression steps calculate the probable damages and consequent losses: analysis of hazard level, structural aspects, damages, and losses (Günay and Mosalam 2013).

### 7.4.6 PROBABILITY RISK ASSESSMENT

The concept of Sliding an Overlaid Multidimensional Bell-curve of Response for Engineering Resilience Operationalization (SOMBRERO) is adopted to quantify engineering seismic resilience using an Orthogonal Limit-space Environment (OLE) (Bruneau and Reinhorn 2007). The probability risk assessment (PRA) approach evaluates the risk to complex engineered technologies such as power plants or spacecraft's life cycle. The damage scenario is created as a chain of events to quantify the severity of consequences. This indicates a link of the initiator to the detrimental consequences at the endpoint. The events' probability and frequencies are modeled with Boolean logic methods using statistical or probabilistic quantification methods. The tools used in the Boolean logic method include event tree analysis (ETA) and fault tree analysis (FTA) (Smith 2019).

Fault tree analysis: FTA is characterized by constructing the fault tree that indicates the likely contributing events to a failure or accidents in the form of braches in a logic diagram multi-casualty concept. The fault tree is prepared with a top-down approach progressing from the "top-event," representing the unwanted event of interest and depicting the various conditions of fault exhibiting the event as a whole in a sequential manner. This method has limitations as the construction of a fault tree suffers from difficulty determining the exact sequence of failure dependencies.

Event tree analysis: ETA is aimed to examine the loss-making event's consequences to facilitate mitigation efforts instead of loss prevention. An event tree includes identifying events and associated controls like alarms, safety mechanisms, etc. The construction of an event tree starts with the event's initiation continuing through the safety function's failures, followed by establishing the sequence of resulting accidents aimed at identifying critical failure that needs immediate attention (Smith 2019).

### 7.4.7 DYNAMIC BAYESIAN NETWORK

Dynamic Bayesian network (DBN) is developed for the assessment of engineered systems considering human and organizational aspects. The DBN-based assessment is probabilistic in nature that incorporates temporal dimensions in adoption and the recovery mechanism to analyze its functionality. Besides, it defines the probability of change in system functionality during and after a disruptive event. The process includes the time-dependent reliability analysis, which facilitates probabilistic modeling where the quantitative analysis of resilience and risk indicates the severity of an extreme event (Yodo 2017).

### 7.4.8 Monte Carlo Analysis

Monte Carlo analysis (MCA) is one of the widely used methods for the analysis of the resilience of complex systems. It is suitable for the quantification of uncertainty as it facilitates probabilistic models. MCA process includes examining a probability distribution aimed to test a hypothesis with a random selection of variables through repetitive sampling to achieve higher accuracy. Monte Carlo simulation often derives the probability distributions from historical data; thus, its adequacy and accuracy in different contexts are questionable. It is rather found suitable for small benchmark cases. Besides, it does not include the impact of the preparedness mechanism, limiting its applicability for evaluating disaster scenarios to a certain extent (Aydin 2018).

### 7.4.9 Performance Assessment Calculation Tool

Federal Emergency Management Agency (FEMA) provides a powerful database with tools such as fragility database and specification, the tool for normative quantity estimation, consequence estimation, and performance assessment calculation tool (PACT) (Alirezaei et al. 2016). PACT facilitates the quantification of earthquake consequences considering indirect and direct economic losses. The tool considers direct economic losses, interruption of business activities, and building downtime, taken into account to obtain indirect economic losses. Data required include basic information about the building, structural and non-structural component's type and quantity, occupants' location, and distribution pattern over time. At each hazard level, the total loss is estimated based on cumulative losses from individual components.

### 7.4.10 Endurance Time Method

Endurance time (ET) method relies on the response history of the structures. In this method, the structures are subjected to dynamic excitations, increasing their intensity gradually to evaluate the response at various levels of excitations corresponding to ground motion's specific intensities. This method is accurate and simple as there is no need for detailed response history analysis that involves large computation. Given this method's ability to incorporate design concepts such as resiliency and life cycle cost, it is widely used for designing advanced structures.

### 7.4.11 Reliability-Based Capability Approach

Society's resilience refers to its ability to sustain the disruption caused by a disaster and move toward normalcy rapidly. The impact of a disruptive event on society can be examined based on the concept of capabilities that define social impact as the change in affected people's well-being, referred to as the reliability-based capability approach (RCA) (Tamvakis and Xenidis 2013).

### 7.4.12 FORM-Based Approach of Resilience

To quantify seismic resilience and analytical, a reliability-based approach is often used, which considers the rapidity of robustness and restoration of structures in a post-disaster situation. FORM provides a framework to develop an algorithm for evaluating resilience in light of the probabilistic distribution of a cluster of structures located at different locations (Bonstrom and Corotis 2016).

## 7.5 MODELING RESILIENCE

Resilience is the prerequisite for adequate functioning of a system as it facilitates anticipating, absorbing any disruption, adapting to the situation, and recovering instantaneously. The existing methods and models developed to analyze infrastructural systems' interdependence can be classified as empirical and predictive based on approaches adopted. The empirical approach-based investigation includes analyzing the empirical data processed based on expert' subjective judgments. In contrast, the predictive approach examines analyzing infrastructural systems through simulation or modeling (Ouyang 2014).

Discrete event simulation (DES) models: DES is an effective tool for analyzing a complex system that provides a simulation model that facilitates analysis of a wide range of observations rapidly, test modifications to improve the system's operation economically. It is widely used to model real-life scenarios such as healthcare facilities, engineering systems, transportation, etc. (Cimellaro 2017).

Human behavior modeling: Infrastructural development is a complex process coupled with a certain degree of interdependence, which increases when subjected to a disturbance due to earthquakes, floods, hurricanes, explosions, terrorism, etc. To contemplate the impact of disasters, simulation is used to predict a system's behavior, accommodating required randomness and details. Human behavior modeling is largely performed with an agent-based approach (Pumpuni 2017).

Agent-based modeling (ABM): ABM is largely used with several agent-based platforms such as NetLogo, Analogic, Repast, Mason, etc., each having a different programming language and features. In ABM, autonomous individuals' interactions and actions referred to as an "agent" are simulated in a particular environment. The behavior and the reciprocal interactions of an agent are performed using logical operations or rules formalized with equations (Pumpuni 2017). Besides, variations in behavior in terms of heterogeneity and stochasticity, which refers to random influences or variations, could be addressed. In this method, several assumptions are made regarding agents' behavior, which may not be justified statistically or theoretically. Given the lack of relevant data and complications associated with modeling agents' behavior, simulation parameters' calibration becomes difficult.

System dynamics-based models: In this, an interdependent lifeline's dynamic behavior is modeled considering causes and effects. The interdependent infrastructures are represented with the help of diagrams indicating causal loop and stock and flow. The diagram showing the causal loop depicts the influence among the identified variables. In contrast, the flow of products and information is described with the diagram indicating stock and flow. The causal loop diagram follows a semi-quantitative approach that relies on a subject matter expert's knowledge; hence this approach needs validation and calibration and needs a large amount of data (Ouyang 2014).

Interdependency of critical infrastructure: Given the interdependency of critical infrastructures, their modeling needs to be based on an in-depth analysis of a system's overall functioning considering the likelihood of failure of a large number of components and the consequent impacts. A probabilistic approach can prove instrumental in modeling critical infrastructure's resilience, including physical aspects and functionality, to estimate recovery mechanisms considering temporal dimensions (Guidotti et al., 2016).

An in-depth understanding of resilience can provide a philosophical and methodological basis to address disaster risk. The resilience-based approach is of profound importance, which provides an avenue for understanding and quantifying a web of complex interconnected networks associated with structures and infrastructural development and their vulnerability for disruption through cascading threat. Preservation of critical functionality, facilitating the fast recovery of complex systems, can be achieved with engineered generic capabilities and robust methods and software tools for rendering structures and infrastructures resilient against a disruptive event in the future.

## REFERENCES

Alirezaei, M., Noori, M., Tatari, O., Mackie, K. R., Elgamal, A. 2016. BIM-based damage estimation of buildings under earthquake loading condition. Procedia Engineering, 145:1051–1058.

ASCE. 2000. Prestandard and Commentary for the Seismic Rehabilitation of Buildings FEMA 356, ASCE, Washington, DC.

Aydin, N. Y., Duzgun, H. S., Heinimann, H. R., Wenzel, F., Gnyawali, K. 2018. Framework for improving the resilience and recovery of transportation networks under geohazard risks. International Journal of Disaster Risk Reduction, 31:832–843.

Baoping, C., Min, X., Yonghong, L., Yiliu, L., Qiang, F. 2017. Availability-based engineering resilience metric and its corresponding evaluation methodology. Reliability Engineering and System Safety, 172:216–224.

Bocchini, P., Frangopol, D. M., Ummenhofer, T., Zinke, T. 2013. Resilience and sustainability of civil infrastructure: Toward a unified approach. Journal of Infrastructure Systems, 20:04014004-1–04014004-16.

Bonstrom, H., Corotis, R.B. 2016. First-order reliability approach to quantify and improve building portfolio resilience. Journal of Structural Engineering, 142(8):C4014001.

Bruneau, M., Chang, S., Eguchi, R., et al. 2003. A framework to quantitatively assess and enhance the seismic resilience of communities. Earthquake Spectra, 19(4):733–752.

Bruneau, M., Reinhorn, A. 2007. Exploring the concept of seismic resilience for acute care facilities. Earthquake Spectra, 23(1):41–62.

Cimellaro, G. P., Malavisi, M., Mahin, S. 2017. Using discrete event simulation models to evaluate resilience of an emergency department. Journal of Earthquake Engineering, 21(2):203–226.

Cimellaro, G.P., Reinhorn, A.M., Bruneau, M. 2006. Quantification of seismic resilience. Proceedings of the 8th US National Conference on Earthquake Engineering, California, USA. https://www.researchgate.net/profile/G_Cimellaro/publication/254603471_QUANTIFICATION_OF_SEISMIC_RESILIENCE/links/54161e3a0cf2fa878ad3fd26/QUANTIFICATION-OF-SEISMIC-RESILIENCE.pdf (accessed March 15, 2020).

Cimellaro, G. P., Reinhorn, A. M., Bruneau, M. 2010. Framework for analytical quantification of disaster resilience. Engineering Structures, 32(11):3639–3649.

Cutter, S. L., Burton, C., Emrich, C. 2010. Disaster resilience indicators for benchmarking baseline conditions. Journal of Homeland Security and Emergency Management, 7:1–22.

Frangopol, D. M., Curley, J. P. 1987. Effects of damage and redundancy on structural reliability. Journal of Structural Engineering, ASCE, New York, 113:1533–1549.

Guidotti, R., Chmielewski, H., Unnikrishnan, V., Gardoni, P., McAllister, T., Van de Lindt, J. 2016. Modeling the resilience of critical infrastructure: The role of network dependencies. Sustainable and Resilient Infrastructure, 1(3–4):153–168. https://doi.org/10.1080/23789689.2016.1254999.

Günay, S., Mosalam, K.M. 2013. PEER performance-based earthquake engineering methodology, Revisited. Journal of Earthquake Engineering, 17(6):829–858.

Kammouh, O., Cimellaro, G. P. 2018. PEOPLES: A tool to measure community resilience. Proceedings of 2018 Structures Congress. Soules, J. G., Ed., ASCE – American Society of Civil Engineering, Texas. April 19–21, 161–171.

Koutsourelakis, P. 2010. Assessing structural vulnerability against earthquakes using multi-dimensional fragility surfaces: A Bayesian framework. Probabilistic Engineering Mechanics, 25(1):49–60.

Lu, X., Liao, W., Fang, D., et al. 2020. Quantification of disaster resilience in civil engineering: A review. Journal of Safety Science and Resilience, 1:19–30. https://www.sciencedirect.com/science/article/pii/S2666449620300086.

MCEER. 2008. Engineering Resilience Solutions, University of Buffalo, USA.

Noroozinejad, E., Hashemi, F., Talebi, M., et al. 2014. Seismic risk analysis of steel-MRFs by means of fragility curves in high seismic zones. Advances in Structural Engineering, 17(9):1227–1240.

Norris, F.H., Stevens, S.P., Pfefferbaum, B., et al. 2008. Community resilience as a metaphor, theory, set of capacities, and strategy for disaster readiness. American Journal of Community Psychology, 41:127–150.

Ouyang, M. 2014. Review on modeling and simulation of interdependent critical infrastructure systems. Reliability Engineering and System Safety, 121:43–60.

Pumpuni, G., Blackburn, T., Garstenauer, A. 2017. Resilience in complex systems: An agent-based approach. System Engineering, 20(2):158–172.

Sarke, P., Lester, H. 2019. Post-disaster recovery associations of power systems dependent critical infrastructures. Journal of Infrastructures, 4(2):30.

Sharma, N., Tabandeh, A., Gardoni, P. 2018. Resilience analysis: A mathematical formulation to model resilience of engineering systems. Sustainable and Resilient Infrastructure, 3(2):49–67. https://doi.org/10.1080/23789689.2017.1345257 (accessed March 18, 2020).

Shinozuka, M., Feng, M., Lee, J., Naganuma, T. 2000. Statistical analysis of fragility curves. Journal of Engineering Mechanics. ASCE, 126(12):1224–1231.

Smith, C., Timothy J. Allensworth. 2019. Quantifying System Resilience Using Probabilistic Risk Assessment Techniques. https://www.semanticscholar.org/paper/Quantifying-System-Resilience-Using-Probabilistic-Smith-3d04a9371ce201acaba8d7325072f7be8?p2df (accessed March 18, 2020).

Tamvakis, P., Xenidis, Y. 2013. Comparative evaluation of resilience quantification methods for infrastructure systems. The 26th IPMA (International Project Management Association), World Congress, 74:339–348.

Tiernan, A., Drennan, L., Nalau, J., et al. 2019. A review of themes in disaster resilience literature and international practice since 2012. Policy Design and Practice, 2(1):53–74.

Vamvatsikos, D. 2002. Seismic Performance, Capacity and Reliability of Structures as Seen through Incremental Dynamic Analysis. Department of Civil and Environmental Engineering, Stanford University: Stanford. http//www.stanford.edu/group/rms (accessed March 15, 2020).

Venkittaraman, A., Banerjee, S. 2014. Enhancing resilience of highway bridges through seismic retrofit. Earthquake Engineering and Structural Dynamics, 43(8):1173–1191.

Yodo, N., Wang, P., Zhou, Z., P. P. 2017. Predictive resilience analysis of complex systems using dynamic Bayesian networks. IEEE Transactions on Reliability, 66(3):761–770.

# 8 Stochastic Finite Element Method

*Kamaljyoti Nath, Anjan Dutta, and Budhaditya Hazra*
Indian Institute of Technology Guwahati, Guwahati, Assam, India

## CONTENTS

## 8.1 INTRODUCTION

Finite element method (FEM) is one of the powerful and most commonly adopted mathematical tools to solve engineering problems. It has been considered in many fields of engineering like structural mechanics, heat conductivity, geotechnical analysis, and electrical circuit analysis. However, the standard finite element model does not consider the randomness in its parameters. Particularly in the case of structural mechanics problems, it is found that the load and material are more likely to be random in nature. In a deterministic model, uncertainties are not considered, whereas in a stochastic model, one or more uncertainties are considered. The effect of random nature of material with and without random loading is studied within stochastic structural mechanics. In stochastic finite element method (SFEM), a probabilistic model is considered for representation of stochastic system and approximate solutions are obtained by combining the standard finite element procedure with the probabilistic model.

Deterministic analyses are generally carried out considering one of the realizations of the input random variables. These types of analyses do not provide complete information about the variability of the output quantities. With the advancement in computational power, there is an increase in demand for a more realistic analysis of engineering problems considering stochasticity in the input parameters. Thus, the random nature of physical parameters is considered in many of the civil engineering problems, like flow through random porous media, modeling soil properties, etc., along with material properties in structural engineering.

DOI: 10.1201/9781003194613-8

The formulation of SFEM is similar to that of its deterministic counterpart, where virtual energy principle, minimization of total potential energy, etc., are considered. The only difference is that the quantities with uncertainties are considered as stochastic instead of treating the same as deterministic. However, in the case of a random field problem, it needs to be discretized before formulation, which is essentially a conversion of continuous random field to a set of random variables [1].

Uncertainty study provides an idea about the nature of the variability of the physical quantities. It also provides the basis for determining the response of physical systems as well as a basis for finding safety factor in engineering design. Many a time it is observed that responses are sensitive to input variability. Moreover, the uncertainty study makes it possible for the assessment of alternative strategy in engineering design [2]. It may, however, be noted that while the uncertainties in a system are inherent, consideration of such uncertainties is not easy due to model complexity, time requirement, lack of data of all the involved physical quantities.

In this chapter, SFEM is discussed in detail with an emphasis on polynomial chaos (PC) based method and some variants are also introduced to address some of the important issues like curse of dimensionality and loss of optimality as prevalent in PC-based approximation. The chapter is organized, with Section 8.1 as introduction, Section 8.2 discusses the modeling of uncertainty in the context of SFEM. It discusses both Gaussian and non-Gaussian representation of civil engineering quantities. The discretization of random field is discussed in Section 8.2.3. Section 8.3 discusses various methods of SFEM. Section 8.4 discusses PC-based method in detail along with the curse of dimensionality. It further discusses the iterative method to address the curse of dimensionality in Section 8.4.3.1. The solution strategy for stochastic structural problem under dynamic loading using time-dependent generalized polynomial chaos (TDgPC) is discussed in Section 8.4.4. Conclusions are discussed in Section 8.5.

## 8.2 UNCERTAINTY MODELING

Input parameters for structural mechanics problems are material properties, geometric properties, boundary conditions, loading, and initial conditions (in case of dynamic problems). Primary material properties like Young's modulus are often considered as random variable or random field, which may be either Gaussian or non-Gaussian. Poisson's ratio ($\nu$) is another independent material parameter in the case of structural mechanics problems. However, the stiffness matrix is a nonlinear function of $\nu$ and hence Young's modulus and Poisson's ratio are often transformed to Lamé's constants as these are linearly related to stiffness matrix. Graham and Deodatis [3] observed that the effect of Poisson's ratio on response variability is lesser compared to that of Young's modulus. Thus, in most of the studies, only Young's modulus is considered as primary random quantity.

### 8.2.1 Gaussian Model

Generally, random fields in civil engineering are assumed to be homogeneous, and the covariance function is of the form $C(x_1, x_2) = f(x_1 - x_2, L_c)$, where $L_c$ is known as correlation length or proportional to correlation length and plays an important role in the accurate representation of simulated random fields. The covariance function approaches to delta correlation process, also known as white noise, when the value of the correlation length is very small. On the other hand, the process becomes a random variable when the correlation length is large compared to the domain under consideration. Huang et al. [4] studied the effect of correlation length on the accuracy of the simulated random field.

One of the commonly considered covariance functions is of exponential type with the absolute value of separation distance between two points as its argument, and for 1D random field, the covariance function is $C(x_1, x_2) = \sigma^2 \exp\left(-\dfrac{|x_1 - x_2|}{L_c}\right)$. $\sigma$ is the standard deviation (SD) of the random field, $x_1, x_2$ are the coordinates within the limit $[-a, a]$, where $a$ is known as length of the process and the value is $L/2$, $L$ being the length of the member. The 2D covariance function is

$$C(x_1, x_2; y_1, y_2) = \sigma^2 \exp\left(\frac{-|x_1 - x_2|}{L_{cx}} + \frac{-|y_1 - y_2|}{L_{cy}}\right),$$ where $L_{cx}$ and $L_{cy}$ are the correlation length in $x$

and $y$ directions, respectively. There are other covariance functions considered by various researchers

like exponential $C(x_1, x_2) = \sigma^2 \exp\left(-(x_1 - x_2)^2/L_c^2\right)$ rational $C(\tau) = \sigma^2 \left[1 - 3\left(\frac{\tau}{L_c}\right)^2\right] / \left[1 + \left(\frac{\tau}{L_c}\right)^2\right]^3$

covariance function, where $\tau = x_1 - x_2$.

## 8.2.2 Non-Gaussian Model

A Gaussian model for physical quantities like Young's modulus, area is suitable for low values of coefficient of variation (COV.), where all the admissible values of physical quantities can be assumed as positive. However, from experimental studies, the material properties are generally observed to be non-Gaussian in nature. Many a time, due to higher values of COV or due to non-Gaussian property of the physical quantity, it may be required to model the random quantity as non-Gaussian or put some restriction on the idealization as Gaussian random field. Yamazaki et al. [5] considered such restriction on the Gaussian random field as $-1 + \delta \leq f(x) \leq 1 - \delta$, $0 < \delta < 1$. The lower and the upper limits ensure that the random quantities always remain positive and the data is symmetric about the mean (i.e., mean value remains unchanged), respectively. However, this leads to change in SD of the random field. Thus, modeling of uncertainty in physical quantities by considering other forms of random field would be a better option, where all the admissible values of the random field are always positive.

The joint probability density function (PDF) required to represent a non-Gaussian random field is often not easy to obtain, and thus marginal PDFs with correlation function are generally considered. However, in many cases, only a few of the lower order moments (mean, variance, skewness, kurtosis) are specified, which was studied by Gurley et al. [6]. The major shortcoming is the possibility of having different admissible realizations that have same lower order moments, but different marginal PDFs.

For a given marginal PDF and correlation structure or spectral function, there are mainly two categories of simulation of non-Gaussian fields (or processes), namely, (i) memoryless transformation to Gaussian random field and (ii) direct simulation of non-Gaussian random field. The basic principle of memoryless transformation is to generate a Gaussian random field of some correlation/spectral function so that the nonlinear transformation can generate the non-Gaussian random field of target correlation/spectral function and prescribed marginal cumulative distribution function (CDF). The process is known as translation process [7] and can be written as $\gamma(x, \theta) = F^{-1}\left(\Phi(\alpha(x, \theta))\right)$, where $\Phi$ is the marginal CDF of Gaussian random field $\alpha(x, \theta)$ and $F$ is the marginal CDF of non-Gaussian random field $\gamma(x, \theta)$.

Yamazaki and Shinozuka [8] proposed an iterative method to generate non-Gaussian random field of spectral density function ($S_{\gamma\gamma}^T(\kappa)$), and marginal CDF. The iteration process is given

as $S_{\alpha\alpha}^{i+1}(\kappa) = \left[\dfrac{S_{\alpha\alpha}^i(\kappa)}{S_{\gamma\gamma}^i(\kappa)}\right] S_{\gamma\gamma}^T(\kappa)$, where $S_{\alpha\alpha}^i(\kappa)$ and $S_{\alpha\alpha}^{i+1}(\kappa)$ are the spectral density function of

the Gaussian random field at $i^{\text{th}}$ and $(i+1)^{\text{th}}$ iterations, respectively and generated using spectral method. $S_{\gamma\gamma}^i(\kappa)$ is the spectral density function of the sample non-Gaussian random field and is generated from the sample random field. The method generates slightly skewed marginal distribution. Deodatis and Micaletti [9] further improved the method by introducing an exponent, $\beta$ to the iteration process so as to simulate highly skewed marginal distribution, thus

$$S_{\alpha\alpha}^{i+1}(\kappa) = S_{\alpha\alpha}^i(\kappa) \left[\frac{S_{\gamma\gamma}^T(\kappa)}{S_{\gamma\gamma}^i(\kappa)}\right]^\beta.$$ Both Gaussian and non-Gaussian random field can be generated

using PC-based translation algorithm, where the Gaussian random field is generated using

Karhunen-Loève (KL) expansion from the covariance function, and PC can be used to generate the non-Gaussian field [10]. Direct simulation of non-Gaussian random field using KL expansion [11] is discussed later. Bocchini and Deodatis [12] have conducted a detailed review on the simulation methods of non-Gaussian random fields.

### 8.2.3 DISCRETIZATION OF RANDOM FIELD

The process of representation of a continuous function random field with a countable number of random variables is known as stochastic discretization or simply discretization [1]. Different approaches of discretization like spatial average (SA) discretization [13], where the random field over an element is represented by the SA of the field over the element, interpolation of nodal points (INP) method [14], where the random field over an element is represented by the interpolation of the values at the nodal points of the element, and midpoint (MP) discretization method, where the random field is approximated using the value of the random field at MP (or centroid) over the element dimension [15], are observed in the literature. In INP, the random field is approximated in each element by the nodal values of random function and associated shape functions [16].

Another category of discretization method is the series expansion method, comprising of KL expansion [17, 18] method and orthogonal series expansion [19]. A zero-mean random function can be expressed as a sum of multiplication of a deterministic orthogonal function of coordinates and a random function. In the case of KL expansion, both the deterministic function and the random variables are orthogonal, which is one of the main advantages over other series expansions [4]. Series expansion methods take advantage of the orthogonal series. The accuracy can be controlled by the inclusion of number of terms in the truncated series as desired.

In optimal linear estimation (OLE) method [20], the random field in the domain is described by the nodal values of the random field by a linear combination. The unknowns are evaluated by minimizing the variance in error of the actual random field and the approximation with the condition that the approximated random field is an unbiased estimate of the random field in mean. Expansion optimal linear estimation (EOLE) method [20] is an expansion and improvement of OLE method, which can be achieved by spectral decomposition of nodal variables.

Brenner and Bucher [21] proposed integration point method, where the random field is represented by the values of the random field at its integration (Gauss points) points rather than MP or node of element. If the random field is considered at Gauss points, it can be directly implemented in finite element during numerical integration using Gauss quadrature method.

In the next section, KL expansion is considered in depth. The method is applicable to discretization of Gaussian random field [22] and can be applied for non-Gaussian random field using an iterative process [11, 23].

#### 8.2.3.1 KL Expansion

KL expansion [17] is considered as one of the popular methods for discretization of the input random field. A zero-mean random field $\alpha(x,\theta)$ with covariance function $C(x_1, x_2)$ and finite variance is a function of position vector $x$ defined over the domain $\mathcal{D}$, with $\theta$ belong to the space of random event $\omega$, and can be expressed using KL expansion as [17],

$$\alpha(x,\theta) = \sum_{n=1}^{\infty} \xi_n(\theta)\sqrt{\lambda_n}\ f_n(x) \tag{8.1}$$

where $\lambda_n$ are the eigenvalues and $f_n(x)$ are the eigenvectors of the covariance function and

$$\xi_n(\theta) = \frac{1}{\sqrt{\lambda_n}} \int_{\mathcal{D}} \alpha(x,\theta) f_n(x) dx \tag{8.2}$$

$\xi_n(\theta)$ are uncorrelated random variables with

$$\mathbb{E}\left[\xi_n(\theta)\right] = 0, \text{ and } \mathbb{E}\left[\xi_n(\theta)\xi_m(\theta)\right] = \delta_{nm} \tag{8.3}$$

where $\mathbb{E}[.]$ is the expectation operator. Specially, the random variables $(\xi_n(\theta))$ in the case of Gaussian random field are standard Gaussian random variables. The covariance function $(C(x_1, x_2) = \mathbb{E}[\alpha(x_1, \theta)\alpha(x_2, \theta)])$ can be evaluated as

$$C(x_1, x_2) = \sum_{n=1}^{\infty}\sum_{m=1}^{\infty}\mathbb{E}\left[\xi_n(\theta)\xi_m(\theta)\right]\sqrt{\lambda_n\lambda_m}\ f_n(x_1)f_m(x_2) = \sum_{n=1}^{\infty}\lambda_n f_n(x_1)f_n(x_2) \tag{8.4}$$

Multiplying Eq. (8.4) by $f_n(x_1)$ and taking integral over the domain give an integral equation as (homogeneous Fredholm integral equation of the second kind)

$$\int_{\mathcal{D}} C(x_1, x_2) f_n(x_1) dx_1 = \lambda_n f_n(x_2) \tag{8.5}$$

and solution of this integral equation gives the eigenvalue and eigenvector of the covariance function. The equation is often solved using numerical solution as analytical solution is limited to only few of the covariance functions. There are numerous methods for the numerical solution, few of these were reviewed by Betz et al. [24]. The accuracy of simulated data is more in case of analytical method of calculation of eigenpairs [4]. The KL expansion can be truncated at an optimal number of terms, $Q$, which can be obtained by considering the expected energy criteria of the covariance function [4].

Iterative KL expansion (direct simulation) can be used to simulate and discretize non-Gaussian random field with a target covariance function and marginal CDF, which can simulate both stationary and nonstationary random fields [11]. In the case of direct simulation of non-Gaussian random field, for a large number of samples, the covariance function always matches with the theoretical covariance irrespective of the iteration number. Thus, it is required to match only the target marginal CDF. The method is suitable particularly with low non-Gaussianity as it is unable to match the tail of the distribution for strong non-Gaussian process. In the iterative method, the random variables, $\xi_n(\theta)$, of KL expansion are updated iteratively, so that the simulated random field matches with the target marginal CDF satisfying Eq. (8.3) at each iteration. It can be observed from Eqs. (8.1) and (8.2) that $\alpha(x, \theta)$ and $\xi_n(\theta)$ are to be generated iteratively to match with the desired properties. The random variables generated, however, may not remain uncorrelated. Phoon et al. [11] considered updated Latin hypercube sampling technique [25] to reduce the correlation among the random variables. The method was further improved by Phoon et al. [23] considering product-moment orthogonalization. The iteration process generally starts with random variables satisfying Eq. (8.3) and that of target marginal distribution of the random field. Though, at each iteration, the simulated covariance matches with theoretical covariance for a large sample size, there is a difference in theoretical and target covariance. This is due to the fact that theoretical covariance depends on the number of eigenvectors considered. It may be observed from Eq. (8.4) that the covariance function always matches with the theoretical covariance, while the updating is required only for the random variables, $\xi_n(\theta)$.

Further, the calculated random variables considered to generate the non-Gaussian random field may not be necessarily independent though uncorrelated. Independent component analysis (ICA) [26, 27] can be performed on the random variables, which replaces the random variables by a linear combination of independent random variables. The uncorrelated random variables, $\xi(\theta)$, represented using ICA as $\xi(\theta) = H_{\text{mixing}}\eta(\theta)$, where $H_{\text{mixing}}$ is an unknown mixing matrix, which mapped the independent random variables $\eta$ to $\xi$. Considering the eigenvalue and eigenvector as one quantity $h_i(x) = \sqrt{\lambda_i} f_i(x)$, the KL expansion (Eq. 8.1) for non-Gaussian random field can be written as

$$\alpha(x, \theta) = \sum_{n=1}^{Q} \xi_n(\eta(\theta)) h_n(x) = \sum_{n=1}^{M} \eta_n(\theta) h_n^{\text{ICA}}(x) \qquad (8.6)$$

where ICA modes $\left(h_n^{\text{ICA}}(x)\right)$ are linear combinations of eigenvector of KL expansion ($\boldsymbol{H}_{\text{mixing}} h(x)$). Nath et al. [28] carried out structural analysis with non-Gaussian random field and considered ICA in order to obtained independent random variables for the random field.

## 8.3 STOCHASTIC FINITE ELEMENT METHOD – AN OVERVIEW

After the discretization of random field, the next step in SFEM is to formulate the SFE equations followed by its solution. The formulation methodologies are same as those of deterministic FEMs, like minimization of total potential energy, virtual work principle, variational principle, etc. The main difference in the formulation is the consideration of quantities with randomness as stochastic. For example, the stiffness matrix for a deterministic case is given by $\boldsymbol{K} = \displaystyle\int_V \boldsymbol{B}^T \tilde{\boldsymbol{D}} \boldsymbol{B} \, dV$, the same would be given by $\boldsymbol{K} = \displaystyle\int_V \boldsymbol{B}^T \tilde{\boldsymbol{D}}_0 \boldsymbol{B} \, dV + \displaystyle\int_V \boldsymbol{B}^T \tilde{\boldsymbol{D}}_0 \boldsymbol{B} f(x, y, z) \, dV$ for a stochastic case, where $\tilde{\boldsymbol{D}} = \tilde{\boldsymbol{D}}_0(1 + f(x, y, z))$ with $\tilde{\boldsymbol{D}}_0$ and $f(x, y, z)$ as mean value of the random field and zero-mean stochastic field, respectively. The formulation of SFE under dynamic loading is similar to the standard procedure used in deterministic FE formulation and is discussed in Section 8.4.4.

One of the most commonly used methods for solution of SFEM is Monte Carlo simulation (MCS) or one of its variants. In MCS, samples for the random quantities are generated based on the assumed distributions of the physical quantities. For a particular realization of the generated random sample, the problem is deterministic and can be solved using an appropriate deterministic analysis method. Once the system is solved for all the realizations of input samples, the ensembles for the responses are obtained and statistical information (e.g., mean, variance, etc.) of responses are evaluated. Thus, MCS is straightforward to apply and requires repetitive analysis using deterministic method. However, typically a large number of samples are needed as convergence is relatively slow. Moreover, with large number of degree of freedom (DOF), the computational burden is too high. Researchers are improving the computational efficiency of MCS by proposing alternative MCS procedure like important sampling [29], quasi-Monte Carlo sampling [30], and directional sampling [31]. However, the computational demand of these methods is higher compared to other approximate methods [32], and thus alternative approximate methods are generally considered for evaluation of statistical responses. MCS methods are generally considered when alternative approaches are not applicable or alternative methods are required to be verified [32].

Perturbation method is one of the methods considered to solve SFEM, where all the input random variables are expanded about their individual mean using Taylor series expansion w.r.t. to stochastic parameter. The unknown variables are also expanded using Taylor expansion and the terms of the same order of FE equation are equated (each order is equated to zero) to obtain the derivatives of the responses [14]. As the Taylor series expansion is used in perturbation approach, with the increase in COV of input the random variables, more and more number of terms are needed to be included in the series. Another method of solution is Neumann expansion method [5] where the inverse of the stiffness matrix (deterministic and stochastic) is given by a series expansion. The advantage of Neumann expansion is that the deterministic part of the stiffness matrix is required to be inverted only once, and the components of responses are evaluated as recurring process. Reduced basis method [33] is another method of evaluation of stochastic responses. The method is known as reduced basis method as the number of unknowns to be evaluated is lesser than DOF. The responses are approximated using a linear combination of stochastic basis vector with unknown coefficients. The method uses Bubnov-Galerkin scheme to evaluate the unknown coefficient.

Based on the work of Wiener [34], Ghanem and Spanos [22, 35] introduced PC-based method to solve structural mechanics problems with Gaussian randomness in the input variables. The method approximates the response using known random orthogonal function and unknown coefficients. Thus, evaluations of statistical responses become equivalent to evaluation of the coefficients of the expansion. PC-based method has gained lot of attention from research community and has been used in different fields of engineering. However, the method has some drawback like the curse of dimensionality. A detailed discussion on PC-based method and curse of dimensionality is discussed in subsequent sections.

## 8.4 POLYNOMIAL CHAOS-BASED STOCHASTIC FINITE ELEMENT METHOD

The KL expansion is considered to discretize the random field of Young's modulus with $Q$ terms. The stochastic finite element equation can be written as [22, 35],

$$\left[ \bar{K}_{N \times N} + \sum_{n=1}^{Q} \xi_n(\theta) K_{n_{N \times N}} \right] u_{N \times 1} = q_{N \times 1} \tag{8.7}$$

$\bar{K}$ and $K_n$ are the deterministic and stochastic parts of the stiffness matrix. $N$ appearing in the subscript is equal to the number of DOF. Ghanem and Spanos [22, 35] used the concept of homogeneous chaos [34] and approximated the responses of structural system using Hermite polynomial of Gaussian process. A second-order random variable can be represented by a mean square convergent series using PC expansion [22]. PC is an infinite series and often truncated at suitable number ($P$). The responses are approximated as

$$u = \sum_{i=0}^{P} c_i \Psi_i \left[ \{ \xi_r(\theta) \} \right] \tag{8.8}$$

where $c_i$ is unknown coefficient. Thus, $\Psi_i \left[ \{ \xi_r(\theta) \} \right]$ is Hermite polynomial of input random variables $\xi_i(\theta)$, $r$ is the number of random variables. The basic property of PC is that polynomial of different orders is orthogonal to each other and also true for same order with different argument, $\left\langle \Psi_i \left[ \{ \xi_r(\theta) \} \right] \Psi_j \left[ \{ \xi_r(\theta) \} \right] \right\rangle = \delta_{ij} \left\langle \Psi_i^2 \left[ \{ \xi_r(\theta) \} \right] \right\rangle$. The mean of PC higher than zeroth order is zero $\left\langle \Psi_i \left[ \{ \xi_r(\theta) \} \right] \right\rangle = 0$, $i > 0$, and mean of zeroth-order polynomial is one $\left\langle \Psi_0 \left[ \{ \xi_r(\theta) \} \right] \right\rangle = 1$. The total number of terms, $(P + 1)$, can be calculated using [36]

$$(P+1) = \frac{(Q+p)!}{Q! \, p!} \tag{8.9}$$

where $p$ is the order of expansion and $Q$ is the number of random variables considered.

There are two approaches for the evaluation of the coefficient $c_i$ of Eq. (8.8), known as *intrusive* and *nonintrusive* methods and discussed in Section 8.4.1. One major drawback of the PC-based method is the curse of dimensionality. For each DOF, $P + 1$ number of unknown coefficients is required to be evaluated. A system with $N$ DOFs will thus have total number of unknowns as $N(P + 1)$. As the order of expansion and/or the number of variables increase, the size of expansion increases exponentially and is discussed in Section 8.4.3.

### 8.4.1 INTRUSIVE AND NONINTRUSIVE FORMULATION

As discussed before, the stochastic problem becomes equivalent of finding coefficients of PC expansion. There are two approaches as intrusive and nonintrusive formulation for the evaluation of these coefficients. In the case of intrusive formulation, the stochastic equations (Eq. 8.7) are converted to a set of equivalent deterministic simultaneous equations using Galerkin projection as,

$$\bar{K}\left\langle \Psi_m^2 \right\rangle c_m + \sum_{i=0}^{P} Y_{im} c_i = \left\langle q\Psi_m \right\rangle, \ m = 0,1,2,\ldots, P \tag{8.10}$$

where $Y_{im} = \sum_{n=1}^{Q} K_n X_{nim}$ and $X_{nim} = \left\langle \xi_n(\theta)\Psi_i \Psi_m \right\rangle$. The system matrix size of the above equation is $N(P+1) \times N(P+1)$, and can be solved using any standard deterministic solver. In Eq. (8.8), $\Psi_i\left[\{\xi_r(\theta)\}\right]$ are known quantities and random in nature. $c_i$ is unknown and can be obtained by solving Eq. (8.10). $\Psi_i\left[\{\xi_r(\theta)\}\right]$ are functions of random variables $\{\xi_r(\theta)\}$ and form an orthogonal polynomial base. In case of Gaussian $\xi_r(\theta)$, these polynomials are Hermite polynomial and are optimal for Gaussian random field.

In case of nonintrusive method, the coefficient of PC expansion is evaluated using regression method. Berveiller et al. [37] evaluated the coefficient using a least square regression, which is based on a least squares minimization between the exact solution and the approximate solution. In this method, first the input random variables are transformed into standard Gaussian random variables. Once the random variables are transformed, PCs are formulated with unknown coefficients. These coefficients can be evaluated by minimizing the error $(\varepsilon)$ in the differences,

$$\varepsilon = \sum_{k=1}^{n}\left[ u^{(k)} - \sum_{j=0}^{P} c_j \Psi_j\left(\theta^k\right) \right]^2 \tag{8.11}$$

where $u$ is response evaluated by solving stochastic equation for $n$ number of support points. The unknown coefficients are calculated by minimizing the error $(\varepsilon)$ with respect to the unknown coefficients. The minimized equation can be written as

$$\sum_{j=0}^{P}\sum_{k=1}^{n} \Psi_j\left(\theta^k\right)\Psi_l\left(\theta^k\right)c_j = \sum_{k=1}^{n} u^k \Psi_l\left(\theta^k\right), \ l = 0,1,2,\ldots, P \tag{8.12}$$

The left-hand side of this equation is a $(P+1)\times(P+1)$ matrix multiply with $(P+1)$ unknown coefficients. The matrix in left side is independent of the DOF considered, which however, depends on the number of support point. Thus, only once the matrix needs to be inverted. Major concerns of this method are the number and quality of support points to be considered. Moreover, responses for these support points need to be calculated using a MCS scheme.

### 8.4.2 GENERALIZED POLYNOMIAL CHAOS

Ghanem and Spanos [22, 35] considered Hermite PC for Gaussian input quantities. The concept of generalized polynomial chaos (gPC) was given by Xiu and Karniadakis [36] by considering different types of orthogonal polynomials from the Askey scheme for different random processes. Different polynomials were suggested for different random processes and showed that a particular polynomial is more appropriate than the others for attaining an exponential convergence and these polynomial are optimal. The basic idea for selection of optimal polynomials is that when the random variables are considered as weight function to the polynomial, these are orthogonal. In case of random variables, which do not belong to any of standard category of polynomials, the authors recommended to project the input random process onto the Wiener-Askey PC directly in order to solve the differential equation. Wan and Karniadakis [38] and Witteveen et al. [39] proposed PC expansion for an arbitrary probability input. Multi-element generalized polynomial chaos (ME-gPC) was proposed by Wan and Karniadakis [38], where the stochastic space was discretized into mutually exclusive space. The orthogonal polynomials are generated in each space with the corresponding PDF. Witteveen et al. [39] studied PC with arbitrary probability based on the corresponding statistical input of arbitrary probability, and constructed 1D PC using Gram-Schmidt orthogonalization. The multidimensional PCs were generated using tensor product of 1D PCs. For generation of multidimensional PC from 1D PC using tensor product, the random variables should be uncorrelated, which is difficult to construct.

In most of the methods, it is assumed that the input random variables are statistically independent in nature. However, there may be some amount of dependency or correlation among the input random variables and generally, some transformation like Cholesky decomposition is used for dependent random Gaussian variables. Navarro et al. [40] constructed PC using Gram-Schmidt orthogonalization for correlated variables by considering tensor product before orthogonalization process, thus avoiding requirement of ensuring uncorrelation among random variables. The same is considered by Nath et al. [28, 41] for structural mechanics problems with both Gaussian and non-Gaussian system properties and named as modified PC (mPC). The authors showed convergence of responses with increase in order of expansion. The basis vectors for the Gram-Schmidt process are formulated by considering the tensor product of each of the random variables. Thus, the basis vectors for Gram-Schmidt process for a two random variables problem are $V(\xi) = G(\xi_i) \otimes G(\xi_j)$, where $G(\xi_i) = \{1, \xi_i, \xi_i^2, \xi_i^3, \ldots, \xi_i^p\}$, and $p$ is the order of expansion. The Gram-Schmidt process is carried out as [42]

$$\Psi_k(\xi) = V_k(\xi) - \sum_{l=0}^{k-1} \frac{\langle V_k(\xi), \Psi_l(\xi) \rangle}{\langle \Psi_l(\xi), \Psi_l(\xi) \rangle} \Psi_l(\xi), \ \Psi_0(\xi) = 1, \ k = 1, 2, 3, \ldots, P \tag{8.13}$$

where the dimension $(P+1)$ is given by the dimension of $V(\xi)$. The advantage of the mPC formulated using Gram-Schmidt process over gPC is that it can form an orthogonal polynomial even for arbitrary random variables.

### 8.4.3 THE CURSE OF DIMENSIONALITY

One of the major drawbacks of PC-based method is the exponential increase in number of terms in the PC expansion with the increase in number of random variable and/or order of expansion. The number of terms in PC expansion is given by Eq. (8.9) and presented for different number of random variables as well as order of expansion in Table 8.1. The exponential growth of number of terms is

**TABLE 8.1**

**Total Number of Terms ($P + 1$) in PC for $p$ Order of Expansion and $Q$ Number of Random Variables**

| | | | $p$ | | | |
|---|---|---|---|---|---|---|
| $Q$ | 0 | 1 | 2 | 3 | 4 | 5 |
| 1 | 1 | 2 | 3 | 4 | 5 | 6 |
| 2 | 1 | 3 | 6 | 10 | 15 | 21 |
| 3 | 1 | 4 | 10 | 20 | 35 | 56 |
| 4 | 1 | 5 | 15 | 35 | 70 | 126 |
| 5 | 1 | 6 | 21 | 56 | 126 | 252 |
| 6 | 1 | 7 | 28 | 84 | 210 | 462 |

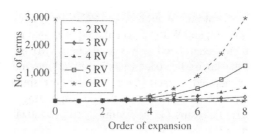

**FIGURE 8.1**  Number of terms in PC expansion corresponds to different number of random variables and order of expansion.

also shown in Figure 8.1. It can thus be observed that as number of random variables and/or order increases, the number of unknown coefficients also increases exponentially leading to increase in the computational cost. Researchers have been working on to address the curse of dimensionality.

Doostan et al. [43] addressed curse of dimensionality of PC-based method. The authors considered analysis using different meshes, starting with a coarse mesh. An optimal order of PC expansion was evaluated with an acceptable accuracy level based on the coarse mesh analysis. The analysis on fine mesh was carried out with PC generated from responses of coarse mesh. Hermite PC of Gaussian random variables was considered to evaluate the responses.

Blatman and Sudret [44] studied structural mechanics problem in non-institutive framework and considered hyperbolic PC expansion, where higher order cross-terms among random variables were not considered in the expansion. Jakeman et al. [45] considered $l_1$ minimization to find the dominant modes of PC expansion. Thus, construction of sparse PC is not a straightforward procedure and requires additional post-processing. Further, the solution accuracy depends on the initially chosen order of expansion, which depends on the randomness of the considered problem.

One of ways of reducing the matrix size and computation cost is by reducing the stiffness matrix size. A reduced PC expansion method was proposed by Pascual and Adhikari [46], where eigenvalue decomposition of stiffness matrix was considered to reduce the stiffness matrix size. Considering the deterministic part, the stiffness matrix of SFE equation (Eq. 8.7), the eigenvalue decomposition of $\bar{K}u = q$ was carried out. The responses of this equation can be expressed as a sum of responses from individual modes and the series can be truncated at a suitable term depending upon the ratio of first eigenvalue to the considered eigenvalue. Considering these reduced eigenvectors as a coordinate transformation, Eq. (8.7) was converted to equation of smaller size. These reduced sized stochastic equations were solved using PC-based method, similar to the strategy adopted for full matrix. The size of the final deterministic equation (Eq. 8.10) for reduced stiffness approach is lesser than that of original stiffness matrix leading to reduction in computational cost for solution of the problem.

The accuracy of representation of simulated random field depends on the ratio of length of the process (domain of the random field) to the correlation length. A higher correlation length compared to domain of the problem requires lesser number of terms. Based on this properties, Chen et al. [47] decomposed the physical domain to nonoverlapping domain so that the higher correlation length can be achieved. The problem is solved for each domain, thus reducing the curse of dimensionality of PC expansion. This was further studied by Pranesh and Ghosh [48], where it was generalized to any covariance function with domain shape independence and used finite element tearing and interconnecting solver.

A transformation-free gPC was introduced by Rahman [49] with multivariate Hermite orthogonal polynomial in dependent Gaussian random variables. The method was further improved by Rahman [50], which addressed non-product-type probability measure, thus a true generalized PC expansion can deal with arbitrary and dependent probability distributions.

Pranesh and Ghosh [51] studied the curse of dimensionality of PC expansion for elliptical equation by adaptively selecting the PC bases. The unknown coefficient of PC expansion is evaluated from the set of algebraic deterministic equations that are obtained after stochastic Galerkin projection. The curse of dimensionality is addressed by considering an adaptive scheme for selection of dominant chaos bases. The adaptive selection is done during the iterative process of preconditioned conjugate gradient (PCG) for solution of the algebraic equations. During initial few PCG iterations, all the terms of PC expansion are considered and later only dominant terms are considered. The authors further increased the computational efficiency by considering a reformulation of stochastic Galerkin method as generalized Sylvester equation.

Cheng and Lu [52] studied the curse of dimensionality of PC expansion by proposing an adaptive sparse PC expansion. A full PC model is established using support vector regression. Based on the contribution to variance of the model output, nonsignificant terms are deleted and significant terms are retained. Further, an iterative algorithm of forward adding and backward deleting of PC bases was considered to obtain the desired level of accuracy. The major strength of the proposed method is that it could detect a group of basis functions simultaneously, thus making it as efficient for high-dimensional problems. Another prominent approach to address the curse of dimensionality is polynomial dimension decomposition (PDD) method [53], which is based on a hierarchical decomposition of a multivariate response function in terms of variables with increasing dimensions.

### 8.4.3.1 Iterative Polynomial Chaos for Problems under Static Loading (ImPC)

Nath et al. [28, 41] addressed curse of dimensionality at multiple levels. First, the size of the stiffness matrices is reduced by considering the random field at Gauss points rather than MPs. This way a more accurate representation of random field is obtained without increasing the size of stiffness matrix. This is achieved by considering the eigenvectors at Gauss point of elements. As the random field is considered at Gauss points, in the case of non-Gaussian random fields, the integration in iterative calculation of random variables $(\xi_n(\theta))$ is spaced unevenly. Algorithm proposed by Gill and Miller [54] was considered to evaluate the integral in the generation of non-Gaussian random fields using KL expansion. These random variables are further considered for product-moment orthogonalization to reduce the dependencies among them. As discussed earlier, these random variables may not be independent and Nath et al. [28] considered ICA to reduce the dependencies among these random variables. The random variables after ICA are further transformed to standard Gaussian random variables. This transformation to standard Gaussian variables is applicable for generation of PC only, while the random variables after ICA considered to represent material properties remain unchanged. Second, reduction of size of stiffness matrices was attempted by considering eigenvector decomposition as proposed by Pascual and Adhikari [46], which is discussed in Section 8.4.3. Third, proposed an iterative method that addresses the curse of dimensionality of PC-based method and termed as ImPC. The method is based on the observation made by Xiu and Karniadakis [36] that orthogonal polynomials are optimal for representation of a random process when the weight functions for some orthogonal polynomials are identical to the probability functions. The method iteratively finds polynomials that are optimal to represent the responses.

The first step of the iterative method (ImPC) is to solve the stochastic problem using first-order mPC with the desired random variables. Depending on the SD of the random field, the responses are likely to be of different probability distribution. However, these evaluated responses are not accurate as only first-order expansion is considered initially. The authors subsequently solved the problem in a iterative process using higher order PC (mPC) generated using random variables from responses of previous iteration. The number of random variables is reduced by considering only dominant component of responses. These were evaluated using KL expansion on responses.

### 8.4.4 TIME-DEPENDENT GENERALIZED POLYNOMIAL CHAOS
### FOR PROBLEMS UNDER DYNAMIC LOADING

As Young's modulus and sectional dimension are considered as random quantities, in the case of stochastic mechanics problems under dynamic loading, the stiffness as well as the mass matrices become random. The random fields for these quantities can be discretized using KL expansion as discussed in Section 8.2.3. Thus, mass matrix can be written in a general form as $M = \bar{M} + \sum_{i=1}^{Q_M} \xi_i^{(M)}(\theta) M_i$ and stiffness matrix as $K = \bar{K} + \sum_{i=1}^{Q_K} \xi_i^{(K)}(\theta) K_i$, respectively, where $\bar{K}$ and $\bar{K}_i$ are the deterministic and stochastic parts of stiffness matrix with $Q_K$ terms in the KL expansion of Young's modulus. Similarly, $\bar{M}$ and $\bar{M}_i$ are the deterministic and stochastic parts of mass matrix with $Q_M$ terms in KL expansion of area (a parameter responsible for stochastic component of mass). Considering damping matrix, $D$ as deterministic for simplicity, the discretized structural equation for linear system under dynamic loading, $M\ddot{u} + D\dot{u} + Ku = q(t)$, can be written as

$$\left( \bar{M} + \sum_{i=1}^{Q_M} \xi_i^{(M)}(\theta) M_i \right) \ddot{u} + D\dot{u} + \left( \bar{K} + \sum_{i=1}^{Q_K} \xi_i^{(K)}(\theta) K_i \right) u = q(t) \tag{8.14}$$

where $u, \dot{u}, \ddot{u}$, and $q(t)$ are displacement, velocity, acceleration, and the applied force vector, respectively.

The stochastic dynamic equation (Eq. 8.14) can be solved using MCS for each of the realizations of $\xi_n(\theta)$. Similar to static problems, the responses can be approximated using PC as $u = \sum_{j=0}^{P} c_j \Psi_j [\xi_r(\theta)], \dot{u} = \sum_{j=0}^{P} d_j \Psi_j [\xi_r(\theta)]$, and $\ddot{u} = \sum_{j=0}^{P} e_j \Psi_j [\xi_r(\theta)]$, where the random variable $\xi_r(\theta)$ comprises of random variables from mass as well as stiffness matrices $\{\xi_i^{(M)}(\theta), \xi_i^{(K)}(\theta)\}$ and number of random variable $r = 0,1,2,\ldots,(Q_M + Q_K)$. It is also important to note that $d$ and $e$ are the first and second derivative of $c$ with respect to time. Similar to the static problem, the stochastic equation (Eq. 8.14) can be converted to simultaneous deterministic equations having $N(P+1)$ equations following Galerkin projection. The statistical responses of the structural system can be obtained by solving these equations using time integration scheme like Newmark-$\beta$ method. Thus, evaluation of $c, d$, and $e$ will provide the stochastic responses. The PCs, $\Psi_j [\xi_r(\theta)]$, can be evaluated using Gram-Schmidt orthogonalization process as discussed in Section 8.4.2. Though the problem can be solved using PC-based method, it fails to approximate the responses for long duration. The statistical properties of responses change over time, while these are approximated using initially considered PC bases, thus the initially considered PC bases loss its optimality [55].

Gerritsma et al. [55] studied time-dependent problem with stochastic parameters and observed that the mean responses of stochastic ordinary differential equation deviate from the deterministic solution as time progresses. Kundu and Adhikari [56] studied structural mechanics problem with random parameter and observed that the mean responses deviate from deterministic responses with an effect that could be considered as equivalent to that of damping on the mean responses. TDgPC was proposed by Garristma et al. [55], which can be considered in the case of time-dependent problem where PC expansion is observed to loss its optimality. An updating in the PC expansion was proposed, when the expansion is no more suitable to approximate the responses with an assumed accuracy criterion. These updated PCs are formulated based on the responses. The authors studied first-order ODE (ordinary differential equation) with uniformly distributed random decay rate and Kraichnan-Orszag three modes problems with random initial conditions.

Nath et al. [57] studied linear structural mechanics problems with random material properties under long duration dynamic loading using TDgPC. The authors modeled Young's modulus and area (parameter related to mass matrix) as random variable for a truss problem and Young's modulus as non-Gaussian random field for beam problem. However, damping matrix is considered as deterministic. It was shown that TDgPC could be effectively implemented in case of linear structural mechanics problem where PC-based method failed. A desired order of PC expansion is considered and PC (mPC) is formulated using Gram-Schmidt orthogonalization processes for the associated random variables. Responses are approximated using PC and the stochastic equations are transformed to set of simultaneous deterministic equation using Galerkin projection. These equations are solved using Newmark-$\beta$ method with appropriate initial conditions. The accuracy of responses was checked by adopting a root-mean-square (RMS) based accuracy criteria, where ratio of the RMS value of the nonlinear terms to the RMS of linear terms is checked, which shall be within a threshold limit. If the results are within the accuracy limit analysis continues otherwise the PC bases are changed, which is known as TDgPC as the PC changes with time. In order to change the PC bases, random variables from responses and transformed loadings are evaluated. The dominant component of responses and loadings is evaluated using KL expansion. Once the random variables are evaluated, these are considered to generate PC using Gram-Schmidt orthogonalization. Next, using these PCs, the responses are approximated and Galerkin projection is performed. In order to solve these equations, the initial conditions are updated and analysis using Newmark-$\beta$ is continued.

## 8.5　CONCLUSIONS

SFEM is a branch of FEM, where solution of stochastic system is evaluated. In this chapter, SFEM for structural mechanics problems is discussed in detail. Discretization of random field for SFEM can be carried out using KL expansion. Methods of SFEM are discussed, particularly PC-based method is discussed in detail. The curse of dimensionality of PC-based method and approaches proposed by various researchers to overcome it are also discussed. Another drawback of PC-based method is the loss of optimality of PC approximation in the case of time-dependent problems, which is addressed using TDgPC for structural mechanics problems.

## REFERENCES

[1] Stefanou, G. (2009). "The stochastic finite element method: Past, present and future". *Computer Methods in Applied Mechanics and Engineering* 198(9), 1031–1051.

[2] Vanmarcke, E., Shinozuka, M., Nakagiri, S., Schuller, G., and Grigoriu, M. (1986). "Random fields and stochastic finite elements". *Structural Safety* 3(3), 143–166.

[3] Graham, L. and Deodatis, G. (2001). "Response and eigenvalue analysis of stochastic finite element systems with multiple correlated material and geometric properties". *Probabilistic Engineering Mechanics* 16(1), 11–29.

[4] Huang, S. P., Quek, S. T., and Phoon, K. K. (2001). "Convergence study of the truncated Karhunen – Loève expansion for simulation of stochastic processes". *International Journal for Numerical Methods in Engineering* 52(9), 1029–1043.

[5] Yamazaki, F., Shinozuka, M., and Dasgupta, G. (1988). "Neumann expansion for stochastic finite element analysis". *Journal of Engineering Mechanics* 114(8), 1335–1354.

[6] Gurley, K. R., Tognarelli, M. A., and Kareem, A. (1997). "Analysis and simulation tools for wind engineering". *Probabilistic Engineering Mechanics* 12(1), 9–31.

[7] Grigoriu, M. (1984). "Crossings of non-Gaussian translation processes". *Journal of Engineering Mechanics* 110(4), 610–620.

[8] Yamazaki, F. and Shinozuka, M. (1988). "Digital generation of non-Gaussian stochastic fields". *Journal of Engineering Mechanics* 114(7), 1183–1197.

[9] Deodatis, G. and Micaletti, R. C. (2001). "Simulation of highly skewed non-Gaussian stochastic processes". *Journal of Engineering Mechanics* 127(12), 1284–1295.

[10] Sakamoto, S. and Ghanem, R. (2002). "Polynomial chaos decomposition for the simulation of non-Gaussian nonstationary stochastic processes". *Journal of Engineering Mechanics* 128(2), 190–201.

[11] Phoon, K. K., Huang, S. P., and Quek, S.T. (2002). "Simulation of second-order processes using Karhunen-Loeve expansion". *Computers & Structures* 80(12), 1049–1060.

[12] Bocchini, P. and Deodatis, G. (2008). "Critical review and latest developments of a class of simulation algorithms for strongly non-Gaussian random fields". *Probabilistic Engineering Mechanics* 23(4). Dedicated to Professor Ove Ditlevsen, 393–407.

[13] Vanmarcke, E. and Grigoriu, M. (1983). "Stochastic finite element analysis of simple beams". *Journal of Engineering Mechanics* 109(5), 1203–1214.

[14] Liu, W. K., Belytschko, T., and Mani, A. (1986). "Random field finite elements". *International Journal for Numerical Methods in Engineering* 23, 1831–1845.

[15] Kiureghian, A.D. and Ke, J.-B. (1988). "The stochastic finite element method in structural reliability". *Probabilistic Engineering Mechanics* 3(2), 83–91.

[16] Matthies, H. G., Brenner, C. E., Bucher, C. G., and Soares, C. G. (1997). "Uncertainties in probabilistic numerical analysis of structures and solids-stochastic finite elements". *Structural Safety* 19(3). Devoted to the work of the Joint Committee on Structural Safety, 283–336.

[17] Loeve, M. (1977). *Probability Theory I*. New York: Springer-Verlag.

[18] Spanos, P. and Ghanem, R. (1989). "Stochastic finite element expansion for random media". *Journal of Engineering Mechanics* 115(5), 1035–1053.

[19] Zhang, J. and Ellingwood, B. (1994). "Orthogonal series expansions of random fields in reliability analysis". *Journal of Engineering Mechanics* 120(12), 2660–2677.

[20] Li, C. and Der Kiureghian, A. (1993). "Optimal discretization of random fields". *Journal of Engineering Mechanics* 119(6), 1136–1154.

[21] Brenner, C. E. and Bucher, C. (1995). "A contribution to the SFE-based reliability assessment of nonlinear structures under dynamic loading". *Probabilistic Engineering Mechanics* 10(4), 265–273.

[22] Ghanem, R. and Spanos, P. (1991). *Stochastic Finite Element: A Spectral Approach*. New York: Springer-Verlag.

[23] Phoon, K., Huang, H., and Quek, S. (2005). "Simulation of strongly non-Gaussian processes using Karhunen-Loève expansion". *Probabilistic Engineering Mechanics* 20(2), 188–198.

[24] Betz, W., Papaioannou, I., and Straub, D. (2014). "Numerical methods for the discretization of random fields by means of the Karhunen-Loève expansion". *Computer Methods in Applied Mechanics and Engineering* 271, 109–129.

[25] Florian, A. (1992). "An efficient sampling scheme: Updated Latin hypercube sampling". *Probabilistic Engineering Mechanics* 7(2), 123–130.

[26] Comon, P. (1994). "Independent component analysis: A new concept?" *Signal Processing* 36(3). Higher Order Statistics, 287–314.

[27] Hyvärinen, A. and Oja, E. (2000). "Independent component analysis: Algorithms and applications". *Neural Networks* 13(4), 411–430.

[28] Nath, K., Dutta, A., and Hazra, B. (2019). "An iterative polynomial chaos approach toward stochastic elastostatic structural analysis with non-Gaussian randomness". *International Journal for Numerical Methods in Engineering* 119(11), 1126–1160.

[29] Melchers, R. (1989). "Importance sampling in structural systems". *Structural Safety* 6(1), 3–10.

[30] Sobol, I. M. (1998). "On quasi-Monte Carlo integrations". *Mathematics and Computers in Simulation* 47(2), 103–112.

[31] Bjerager, P. (1988). "Probability integration by directional simulation". *Journal of Engineering Mechanics* 114(8), 1285–1302.

[32] Rahman, S. and Xu, H. (2004). "A univariate dimension-reduction method for multidimensional integration in stochastic mechanics". *Probabilistic Engineering Mechanics* 19(4), 393–408.

[33] Nair, P. B. and Keane, A.J. (2002). "Stochastic reduced basis methods". *American Institute of Aeronautics and Astronautics* 40(8), 1653–1664.

[34] Wiener, N. (1938). "The homogeneous chaos". *American Journal of Mathematics* 60(4), 897–936.

[35] Ghanem, R. and Spanos, P.D. (1990). "Polynomial chaos in stochastic finite elements". *Journal of Applied Mechanics* 57(1), 197–202.

[36] Xiu, D. and Karniadakis, G.E. (2002). "The Wiener-Askey polynomial chaos for stochastic differential equations". *SIAM Journal on Scientific Computing* 24(2), 619–644.

[37] Berveiller, M., Sudret, B., and Lemaire, M. (2006). "Stochastic finite element: A non intrusive approach by regression". *European Journal of Computational Mechanics* 15(1–3), 81–92.

[38] Wan, X. and Karniadakis, G. E. (2006). "Beyond Wiener-Askey expansions: Handling arbitrary PDFs". *Journal of Scientific Computing* 27(1–3), 455–464.

[39] Witteveen, J.A., Sarkar, S., and Bijl, H. (2007). "Modeling physical uncertainties in dynamic stall induced fluid-structure interaction of turbine blades using arbitrary polynomial chaos". *Computers & Structures* 85(11), 866–878.

[40] Navarro, M., Witteveen, J., and Blom, J. (2014). "Polynomial chaos expansion for general multivariate distributions with correlated variables". *arXiv preprint arXiv*:1406.5483.

[41] Nath, K., Dutta, A., and Hazra, B. (2019). "An iterative polynomial chaos approach for solution of structural mechanics problem with Gaussian material property". *Journal of Computational Physics* 390, 425–451.

[42] Golub, G. H. and Van Loan, C. F. (1983). *Matrix Computations*. Baltimore, MD: The Johns Hopkins University Press.

[43] Doostan, A., Ghanem, R. G., and Red-Horse, J. (2007). "Stochastic model reduction for chaos representations". *Computer Methods in Applied Mechanics and Engineering* 196(37). Special Issue Honoring the 80th Birthday of Professor Ivo Babuka, 3951–3966.

[44] Blatman, G. and Sudret, B. (2011). "Adaptive sparse polynomial chaos expansion based on least angle regression". *Journal of Computational Physics* 230(6), 2345–2367.

[45] Jakeman, J., Eldred, M., and Sargsyan, K. (2015). "Enhancing $l_1$-minimization estimates of polynomial chaos expansions using basis selection". *Journal of Computational Physics* 289, 18–34.

[46] Pascual, B. and Adhikari, S (2012). "A reduced polynomial chaos expansion method for the stochastic finite element analysis". *Sadhana* 37(3), 319–340.

[47] Chen, Y., Jakeman, J., Gittelson, C., and Xiu, D. (2015). "Local polynomial chaos expansion for linear differential equations with high dimensional random inputs". *SIAM Journal on Scientific Computing* 37(1), A79–A102.

[48] Pranesh, S. and Ghosh, D. (2016). "Addressing the curse of dimensionality in SSFEM using the dependence of eigenvalues in KL expansion on domain size". *Computer Methods in Applied Mechanics and Engineering* 311, 457–475.

[49] Rahman, S. (2017). "Wiener-Hermite polynomial expansion for multivariate Gaussian probability measures". *Journal of Mathematical Analysis and Applications* 454(1), 303–334.

[50] Rahman, S. (2018). "A polynomial chaos expansion in dependent random variables". *Journal of Mathematical Analysis and Applications* 464(1), 749–775.

[51] Pranesh, S. and Ghosh, D. (2018). "Cost reduction of stochastic Galerkin method by adaptive identification of significant polynomial chaos bases for elliptic equations". *Computer Methods in Applied Mechanics and Engineering* 340, 54–69.

[52] Cheng, K. and Lu, Z. (2018). "Adaptive sparse polynomial chaos expansions for global sensitivity analysis based on support vector regression". *Computers and Structures* 194, 86–96.

[53] Rahman, S. (2008). "A polynomial dimensional decomposition for stochastic computing". *International Journal for Numerical Methods in Engineering* 76(13), 2091–2116.

[54] Gill, P. E. and Miller, G. F. (1972). "An algorithm for the integration of unequally spaced data". *The Computer Journal* 15(1), 80–83.

[55] Gerritsma, M., Steen, J.-B. van der Vos, P., and Karniadakis, G. (2010). "Time-dependent generalized polynomial chaos". *Journal of Computational Physics* 229(22), 8333–8363.

[56] Kundu, A. and Adhikari, S. (2014). "Transient response of structural dynamic systems with parametric uncertainty". *Journal of Engineering Mechanics* 140(2), 315–331.

[57] Nath, K., Dutta, A., and Hazra, B. (2020). "Long duration response evaluation of linear structural system with random system properties using time dependent polynomial chaos". *Journal of Computational Physics* 418, 109596.

# 9 Structural Health Monitoring-Integrated Reliability Assessment of Engineering Structures

*Çağlayan Hızal and Engin Aktaş*
Izmir Institute of Technology, Urla, Izmir, Turkey

## CONTENTS

## 9.1 INTRODUCTION

Probabilistic performance and/or condition assessment of existing structures are extensively investigated in the civil engineering community since those types of applications are capable of providing a confidence interval for any predefined failure parameter. In this context, classical reliability methods aim at identifying a probabilistic range to assess the limit state conditions. During this assessment, the uncertainty of regular (induced by dead, live loads) and/or extreme (such as wind and/or seismic load, or temperature effects) loading effects as well as the resistance of the structural members are considered as leading order parameters based on the assumed analytical models. Here, a limit state is defined either regarding the strength of a particular structural member or load resisting capacity of the whole system (Frangopol et al. 2008). In addition to these, a failure criterion can also be defined as displacement and/or strain limit for any partition of the considered structure (Döhler and Thöns 2016; Liu et al. 2019).

Although the resistance, material, and loading parameters are generally considered as random variables in most reliability-based structural monitoring, the importance of modeling uncertainty might be overlooked during this process. In this regard, structural health monitoring (SHM) integrated reliability analysis provides an extensive and convenient tool to overcome such kinds of complex problems. Different from the classical reliability applications, SHM-integrated approaches handle the main problem as a prior-to-posterior information transfer for the failure probability to be updated (Beck and Au 2002; Catbas et al. 2008; Ozer et al. 2015; Straub and Papaioannou 2015). Therefore, such methodologies require an updated probability density function (*pdf*) based on the modeling assumptions and uncertainty dependent on the observed (or measured) data. The key problem at this point arises in the calculation of the updated *pdf* using field observations. This requirement also increases the complexity of the problem due to the lack of information for the initial finite element (FE) model of the structure. To overcome this problem, Markov Chain Monte

DOI: 10.1201/9781003194613-9

Carlo (MCMC) based simulation methods are presented to the literature to achieve more reliable solutions for the considered models (Papadimitriou et al. 2001).

Although different applications and modifications are available for MCMC methods in order to increase the effectiveness of the applied methodology (Beck and Au 2002; Straub and Papaioannou 2015), some limitations still persist regarding the computational time and effort during the implementation phase. The reason for this limitation lies in the difficulties in the estimation of likelihood function during the updating procedure of posterior *pdf* for model parameters. To overcome this problem, two-stage approaches can also be adapted to the model updating problem (Yuen and Kuok 2011; Yan and Katafygiotis 2015; Au and Zhang 2016; Zhang and Au 2016; Zhang et al. 2017; Hızal and Turan 2020). By making use of this, the application of time-consuming MCMC becomes unnecessary since the likelihood function can be directly estimated based on the measurement data, for a specific model class.

The two-stage approaches present a thorough methodology including both modal identification and model updating procedures. At the first stage, the modal parameters (e.g., natural frequencies, modal shape vectors) are identified by using the Bayesian operational modal analysis (BAYOMA) technique that makes it possible to obtain uncertainty information for those identified parameters. Here, the estimated uncertainties contain both modeling errors (due to the applied modal identification methodology) and measurement noise (due to the biased errors in the data acquisition system). Then, the likelihood function for the second step, in which the modal identification process is undertaken, can be simply constructed through the identification uncertainties estimated at the previous stage. Although additional modeling errors still persist at the second stage (model updating stage), this two-stage methodology provides rather reasonable results for FE model updating.

Motivated from the previous implementations summarized above, a reliability assessment procedure that is connected by the two-stage Bayesian FE model updating method is presented in this study. For this purpose, the reliability problem is handled by considering the posterior *pdf* of the FE model updated by using the two-stage approach. By making use of this, it is aimed to mitigate the computational effort in the reliability analysis as well as to improve the parameter estimation quality. A numerical example is conducted to see the variations in the failure probabilities based on the FE model of a two-story shear frame structure subjected to stochastic ground motion. The obtained results indicate that the proposed methodology provides reasonable estimations for the SHM-integrated reliability analysis.

## 9.2 TWO-STAGE BAYESIAN FE MODEL UPDATING

Assuming that the actual values of mass components are known (or can be well estimated), the structural model parameters can be updated by using the Bayes' theorem, as follows:

$$p(\theta|\chi,\mathbf{D}) = c_0\, p(\chi|\theta,\mathbf{D}) \times p(\varepsilon_m|\theta) \times p(\theta) \tag{9.1}$$

where $\theta$, $\chi$, $\mathbf{D}$, and $\varepsilon_m$ show the sets of stiffness scaling parameters, identified modal parameters, vibration measurements, and a model error vector respectively. In addition, $c_0 = \left[\int p(\chi|\theta,\mathbf{D}) \times p(\varepsilon_m|\theta) \times p(\theta)\, d\theta\right]^{-1}$ represents a normalizing constant (Yuen 2010). Here, $\chi$ and $\varepsilon_m$ are defined as:

$$\chi = \left[\chi_1^T, \quad \cdots \quad ,\chi_{N_m}^T\right]^T \in \mathbb{R}^{(N_d+1)N_m \times 1}$$

$$\chi_n = \left[\lambda_n, \quad \frac{\Phi_n^{\mathrm{T}} \mathbf{L}_0^{\mathrm{T}}}{\|\mathbf{L}_0 \Phi_n\|}\right]^T \in \mathbb{R}^{(N_d+1) \times 1} \tag{9.2}$$

$$\varepsilon_m = \left[ \varepsilon_{m_1}^T, \quad \cdots \quad, \varepsilon_{m_{N_m}}^T \right]^T \in \mathbb{R}^{N_m N \times 1}$$

(9.3)

$$\varepsilon_{m_n} = \left\| \left( \mathbf{K}(\theta) - \lambda_n \right) \mathbf{\Phi}_n \right\| \in \mathbb{R}^{N \times 1}$$

where $\lambda_n$ and $\mathbf{\Phi}_n \in \mathbb{R}^{N \times 1}$ represent the eigenvalues and eigenvectors of the parametric FE model, $\mathbf{L}_0 \in \mathbb{R}^{N \times N_d}$ is a selection matrix that extracts the measured degrees of freedom (DOFs) from $\mathbf{\Phi}_n$, $N$ = number of DOFs of FE model, and $N_d$ = number of measured DOFs. Here, the parametric stiffness matrix of the FE model can be written as below:

$$\mathbf{K}(\theta) = \mathbf{K}_0 + \sum_{i=1}^{N_\theta} \theta_i \mathbf{K}_i \in \mathbb{R}^{N \times N}$$

(9.4)

where $N_\theta$ = number of stiffness scaling parameters, $\mathbf{K}_0$ = stiffness components that do not depend on $\theta$, and $\mathbf{K}_i$ represents the non-parametric component of the $i^{th}$ sub-structural stiffness matrix.

In Eq. (9.1), $p(\chi|\theta, \mathbf{D})$ and $p(\varepsilon_m|\theta)$ represent the *likelihood functions* for modal parameters and *modeling error vector*, which are given by:

$$p(\chi|\theta, \mathbf{D}) = \prod_{n=1}^{N_m} p(\chi_n|\theta, \mathbf{D})$$

$$p(\chi_n|\theta, \mathbf{D}) = \frac{(2\pi)^{-(N_d+1)/2}}{\left| \mathbf{C}_{\hat{\chi}_n} \right|^{1/2}} exp \left( -\frac{1}{2} \left[ \chi_n - \hat{\chi}_n \right]^T \mathbf{C}_{\hat{\chi}_n}^{-1} \left[ \chi_n - \hat{\chi}_n \right] \right)$$

(9.5)

$$p(\varepsilon_m|\theta) = \prod_{n=1}^{N_m} p(\varepsilon_{m_n}|\theta)$$

$$p(\varepsilon_{m_n}|\theta) = \frac{(2\pi)^{-N/2}}{\mathbf{S}_{\varepsilon_m}^{N/2}} exp \left( -\frac{1}{2} \varepsilon_{m_n}^T \mathbf{S}_{\varepsilon_m}^{-1} \varepsilon_{m_n} \right)$$

(9.6)

where $\hat{\chi}_n = \left[ \hat{\lambda}_n, \quad \dfrac{\hat{\phi}_n^T}{\|\hat{\phi}_n\|} \right]^T \in \mathbb{R}^{(N_d+1) \times 1}$ represents the most probable $n$th modal eigenvalue and mode

shape vector ($\hat{\phi}_n \in \mathbb{R}^{N_d \times 1}$). Moreover, $p(\theta)$ and $p(\theta|\chi, \mathbf{D})$ indicate the *prior* and *posterior* probability distributions for $\theta$, respectively. Here, $p(\theta)$ can be selected as a truncated normal or lognormal distribution using the nominal stiffness scaling values, considering large variances (Das and Debnath 2018; Hızal and Turan 2020). Then, the posterior *pdf* can be updated by solving Eq. (9.1). The major challenge in this procedure is solving the integral of $\int p(\chi|\theta, \mathbf{D}) \times p(\varepsilon_m|\theta) \times p(\theta) d\theta$ since the posterior covariance matrix ($\mathbf{C}_{\hat{\chi}_n}$) is an unknown parameter here. Therefore, the application of Markov Chain Monte Carlo Simulation (MCMS) becomes necessary to estimate the likelihood functions correctly, which may significantly increase the computation time and effort. However, an alternative and fast solution procedure can be implemented by connecting the FE model updating problem with BAYOMA. In this context, various two-stage approaches are available in the literature that calculates $\mathbf{C}_{\hat{\chi}_n}$ by implementing the BAYOMA in the modal identification phase (Yuen 2010; Yan and Katafygiotis 2015; Au and Zhang 2016; Zhang and Au 2016; Hızal and Turan 2020). Hızal and Turan (2020) have shown that $\mathbf{C}_{\hat{\chi}_n}$ can be calculated as a block diagonal matrix by the inclusion of norm constraints for modal shape vectors, as below:

$$\mathbf{C}_{\hat{\chi}_n} = \begin{bmatrix} \mathbf{C}_{\hat{\lambda}_n} & 0 \\ 0 & \mathbf{C}_{\hat{\phi}_n} \end{bmatrix}$$

(9.7)

where $\mathbf{C}_{\hat{\lambda}_n}$ and $\mathbf{C}_{\hat{\phi}_n}$ indicate the posterior covariance matrices for eigenvalues and mode shape vectors identified by BAYOMA, respectively. After updating the most probable values (MPVs) of system modal and model parameters, the posterior marginal *pdfs* can be well approximated by the Gaussian distribution based on the second-order Taylor series expansion of $p(\theta|\chi,\mathbf{D})$, as below (Yan and Katafygiotis 2015; Au and Zhang 2016; Hızal and Turan 2020):

$$p(\Gamma) = \frac{(2\pi)^{-((N_d+1)N_m+N)/2}}{\left|\mathbf{C}_{\hat{\Gamma}}\right|^{1/2}} exp\left(-\frac{1}{2}\left[\Gamma-\hat{\Gamma}\right]^T \mathbf{C}_{\hat{\Gamma}}^{-1}\left[\Gamma-\hat{\Gamma}\right]\right) \tag{9.8}$$

where $\Gamma = \left[\chi^T \theta^T\right]^T$, and $\mathbf{C}_{\hat{\Gamma}}$ is given by:

$$\mathbf{C}_{\hat{\Gamma}} = \begin{bmatrix} \mathbf{C}_{\hat{\chi}} & \mathbf{C}_{\hat{\chi}\hat{\theta}} \\ \mathbf{C}_{\hat{\theta}\hat{\chi}} & \mathbf{C}_{\hat{\theta}} \end{bmatrix} \tag{9.9}$$

## 9.3  ROBUST RELIABILITY ESTIMATION BASED ON UPDATED FE MODEL

The robust methodology that was first introduced by Papadimitriou et al. (2001) proposes an efficient framework for reliability assessment integrating the probabilistic FE model updating approach. According to this strategy, the reliability integral is defined as (Papadimitriou et al. 2001):

$$P(\mathcal{F}|\mathbf{D}) = \int P(\mathcal{F}|\theta,\mathbf{D}) \times p(\theta|\chi,\mathbf{D}) d\theta \tag{9.10}$$

Here, $P(\mathcal{F}|\theta,\mathbf{D})$ can be estimated for a specific $\theta$ by using random vibration theory (Papadimitriou et al. 2001; Yuen 2010). The most important challenge in the solution of Eq. (9.8), however, lies in the fact that $p(\theta|\chi,\mathbf{D})$ is generally unknown in reliability-based model updating problems (Beck and Au 2002; Straub and Papaioannou 2015; Betz et al. 2018). In such a case, MCMC-based algorithms become necessary to achieve reasonable solutions. Here, enhanced methodologies such as adaptive MCMC simulation (Beck and Au 2002), which is developed based on the Metropolis-Hastings algorithm, provide very useful and effective approximations for the solution of reliability integral given in Eq. (9.8). When the posterior distribution of model parameters is known or can be well approximated, however, the solution of Eq. (9.8) might be much more practical. In this case, it is possible to generate $P(\mathcal{F}|\theta)$ based on the parameter space of $\theta$ using the updated $p(\theta|\chi,\mathbf{D})$. Therefore, using the classical Monte Carlo Simulation (MCS) methodology, the reliability integral given in Eq. (9.8) can be simply calculated as follows:

$$P(\mathcal{F}|\mathbf{D}) = \frac{1}{N_s}\sum_{r=1}^{N_s} P(\mathcal{F}|\theta_r,\mathbf{D}) \tag{9.11}$$

Two-stage approaches are paid attention to constitute a consistent link between the modal identification and model updating procedures. From this perspective, integrating the BAYOMA into the first phase of the procedure makes it possible to obtain a closed-form solution for the posterior distribution for FE model parameters. Thus, time-consuming methods such as adaptive MCMC simulations can be eliminated from the solution process of high-dimensional reliability integrals.

## 9.4  NUMERICAL ANALYSIS

FE model of a two-story shear frame is utilized in this section for numerical verification. The considered structure is modeled with a constant interstory stiffness and story mass, which are defined

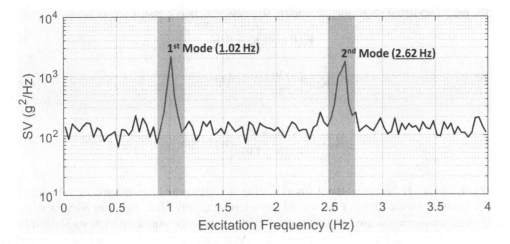

**FIGURE 9.1** Selected prior probability distributions for stiffness scaling parameters.

as $k = 130 \times 10^3$ N/m and $m = 1250$ kg, respectively. In the modal identification phase, it has been assumed that the structure is subjected to independent and identically distributed (*i.i.d.*) Gaussian distribution so that the modal one-sided power spectral density (PSD) will be 1 $\mu g^2$/Hz. Under these ambient loading effects, the dynamic response of the structure is numerically obtained by the Newmark-beta integration scheme (assuming constant acceleration) for $dt = 1/100,000$ s. Then, the acceleration response at the floor levels has been down sampled to 1000 Hz sampling frequency. Subsequently, the acquired response is contaminated with *i.i.d.* Gaussian white noise with a one-sided PSD of 100 $\mu g^2$/Hz.

Figure 9.1 presents the singular value (SV) spectrum of the measured response data. Here, only the maximum SVs calculated at the discrete excitations are presented since the vibration modes are well separated. As a first view, the possible modes can be detected around the frequencies at 1.02 and 2.62 Hz. At the next step, the most probable modal parameters, as well as the posterior uncertainties of the considered structure, are identified by using the Fast Bayesian Fast Fourier Transform approach (BFFTA), which has been developed by Au (2011). The identification results and their actual values are presented in Table 9.1. Here, the identified values have a good match with actual ones. Considering the posterior uncertainties, one can be deduced that the quality of identified modal shape vectors and damping ratios remains relatively less than the modal frequencies. This result also indicates that the weighting of the modal frequencies in the constructed *pdf* will be larger than the modal shape vector.

**TABLE 9.1**
**Identified Modal Parameters with Posterior Coefficient of Variation (CoV)**

| Parameter | Actual | Identified (MPV) | CoV |
|---|---|---|---|
| $f_1$ (Hz) | 1.00 | 1.01 | $2.97 \times 10^{-3}$ |
| $f_2$ (Hz) | 2.63 | 2.63 | $2.21 \times 10^{-3}$ |
| $\xi_1$ (%) | 1.00 | 0.98 | $3.28 \times 10^{-1}$ |
| $\xi_2$ (%) | 1.00 | 1.02 | $2.67 \times 10^{-1}$ |
| $\Phi_1$ | 0.5257 | 0.4623 | $4.17 \times 10^{-2}$ |
|  | 0.8507 | 0.8867 |  |
| $\Phi_2$ | −0.8507 | −0.8453 | $3.17 \times 10^{-2}$ |
|  | 0.5257 | 0.5342 |  |

In the model updating stage, the parametric stiffness matrix is constructed as below:

$$\mathbf{K}(\theta) = \theta_1 \mathbf{K}_1 + \theta_2 \mathbf{K}_2 \tag{9.12}$$

$$\mathbf{K}_1 = \begin{bmatrix} 130 & 0 \\ 0 & 0 \end{bmatrix} \times 10^3 \, \text{N/m} \tag{9.13a}$$

$$\mathbf{K}_2 = \begin{bmatrix} 130 & -130 \\ -130 & 130 \end{bmatrix} \times 10^3 \, \text{N/m} \tag{9.13b}$$

where $\theta_1$, $\theta_2$ and $\mathbf{K}_1$, $\mathbf{K}_2$ represent the stiffness scaling parameters and non-parametric sub-structural stiffness matrices for the first and the second stories, respectively. Here, the prior *pdf*s for stiffness scaling parameters are selected so that their nominal values (or initial estimation) show a large deviation from the actual values. In this context, the selected prior *pdf*s are also presented in Figure 9.2. Here, prior MPVs (nominal values) and variances are selected as 2.00, 4.00 and 0.37, 1.34 for $\theta_1$ and $\theta_2$, respectively.

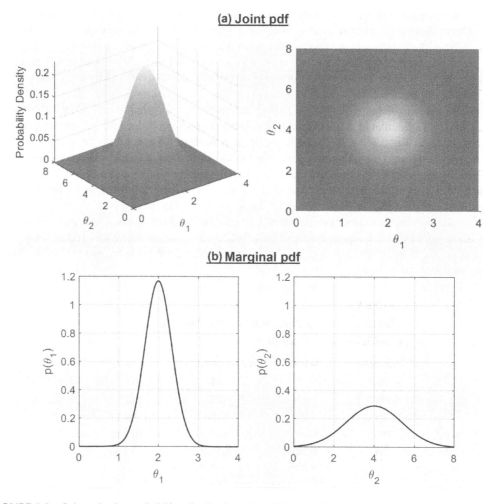

**FIGURE 9.2**   Selected prior probability distributions for stiffness scaling parameters.

**TABLE 9.2**

**Updated Modal and Model Parameters with Posterior Coefficient of Variation (CoV)**

| Parameter | Actual | Updated (MPV) | CoV |
|---|---|---|---|
| $\theta_1$ | 1.00 | 1.00 | $0.0113 \times 10^{-3}$ |
| $\theta_2$ | 1.00 | 1.02 | $0.0056 \times 10^{-3}$ |
| $\lambda_1$ (rad²/s²) | 6.30 | 6.32 | $0.0017 \times 10^{-3}$ |
| $\lambda_2$ (rad²/s²) | 16.50 | 16.61 | $0.0009 \times 10^{-3}$ |
| $\Phi_1$ | 0.5257 | 0.5296 | $0.21 \times 10^{-3}$ |
|  | 0.8507 | 0.8483 |  |
| $\Phi_2$ | −0.8507 | −0.8475 | $0.13 \times 10^{-3}$ |
|  | 0.5257 | 0.5308 |  |

The updated system model and modal parameters as well as their posterior uncertainties in terms of posterior coefficient of variation (CoV) values are presented in Table 9.2. Here, one can observe that the updated parameters show a good match with the actual values. In addition, the calculated joint *pdf* for stiffness scaling parameters, which are generated as zero-mean Gaussian distributions based on the second-order Taylor series expansion, is shown in Figure 9.3.

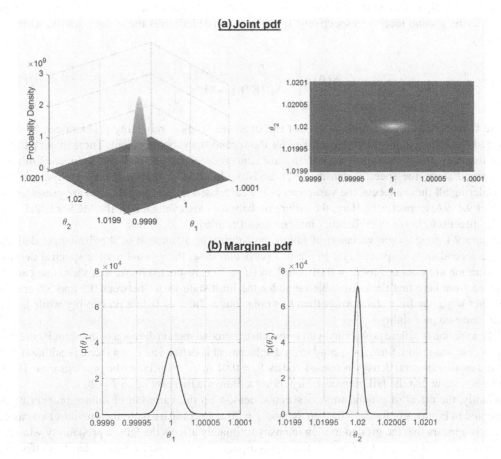

**FIGURE 9.3**   Approximated posterior *pdf*, $p(\theta_1, \theta_2) \propto p(\theta | \chi = \hat{\chi}, D)$.

The reliability problem considered in this section is constructed in a similar manner with an adaptive MCMC simulation-based numerical application undertaken by Yuen (2010). In this context, a limit state is defined as a failure criterion for the second story displacement of the considered structure, based on a stationary stochastic ground motion. Here, the probability of failure is defined by a Poisson process whose *pdf* is given by Yuen (2010):

$$P(\mathcal{F}|\theta,\mathbf{D}) \approx 1 - exp(-2Tv(\theta))  \tag{9.14}$$

where $v(\theta)$ represents the model-dependent mean up-crossing rate, which is defined for the second story displacement of the structure (Yuen 2010):

$$v(\theta) = \sqrt{\frac{S_{\dot{y}2}}{4\pi^2 S_{y2}}} exp\left(-\frac{y_{2,\lim}^2}{2S_{y2}}\right)  \tag{9.15}$$

where $S_{\dot{y}2}$ and $S_{y2}$ represent the covariance values for the second story velocity and displacement values that are obtained by using any random vibration theory approach. In this context, the Lyapunov's equation can be employed to determine the covariance of the state response vector (Yuen 2010):

$$\mathbf{A}(\theta)\Sigma_x^T + \Sigma_x \mathbf{A}(\theta)^T + 2\pi S_g \mathbf{1}_{N\times N} = 0  \tag{9.16}$$

where $\Sigma_x$ and $S_g$ represent the covariance matrix of the state vector, $x = \left[y^T, \dot{y}^T\right]^T$, and the PSD of stochastic ground motion, respectively. In addition, $\mathbf{A}(\theta)$ indicates the system matrix, which is given by:

$$\mathbf{A}(\theta) = \begin{bmatrix} 0 & \mathbf{I}_{N\times N} \\ -\mathbf{M}^{-1}\mathbf{K}(\theta) & -\mathbf{M}^{-1}\mathbf{C} \end{bmatrix}  \tag{9.17}$$

where C represents the damping matrix of the structure. Thus, a reliability problem can be constructed based on a failure criterion defined for the second story displacement. Three major parameters that may affect the failure probability are concerned: the predefined displacement limit for the second story, the spectral density level, and the duration of the stationary ground motion. Considering all these aspects, the variations of the calculated failure probabilities are presented in Figures 9.4–9.6, respectively. Here, the failure probabilities are evaluated by the MCS method, and by the direct solution of the reliability integral (exact results).

Figure 9.4 presents the variation of failure probability as a function of the limit state defined for the second story displacement, $y_2$. In this particular case, the ground motion spectral density and duration are assumed as $S_g = 0.01$ m$^2$/s$^3$ and $T = 20$ s, respectively. At first view, one can be deduced from here that the reasonable range for the limit state varies between 3.5 and 5.5 cm. In the other word, the limit states lower than this range has a 100% of failure probability while larger values have no probability.

The variation of failure probability with respect to the ground motion duration is shown in Figure 9.5. In this case, the failure limit for second story displacement is defined as $y_2 = 4.0$ cm. In addition, the ground motion spectral density is considered as $S_g = 0.01$ m$^2$/s$^3$, similar to the previous case. Here, the results show that the failure probability shows a dramatic increase up to $T = 20$ s.

Finally, the effect of ground motion spectral density on the variation of failure probability is presented in Figure 9.6, for $S_g = 0.01$ m$^2$/s$^3$ and $T = 20$ s. In comparison to the previous two cases, it clearly appears that the ground motion intensity ultimately affects the failure probability when $S_g$ varies between 0.005 and 0.013 m$^2$/s$^3$. From this aspect, one can observe that $S_g$ has less effect on the variation range of the failure probability, compared to the previous two cases.

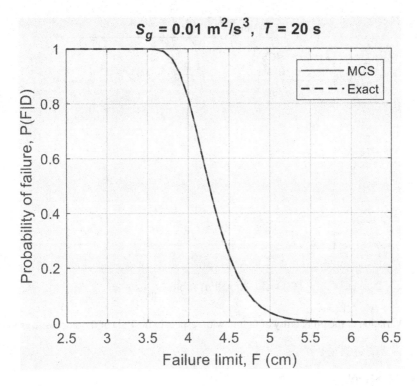

**FIGURE 9.4**    Variation of the probability of failure with respect to failure limit for second story displacement.

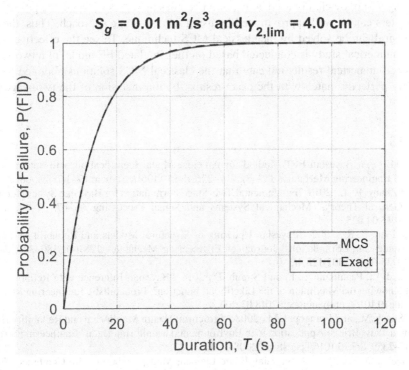

**FIGURE 9.5**    Variation of the probability of failure with respect to the duration of stochastic ground excitation.

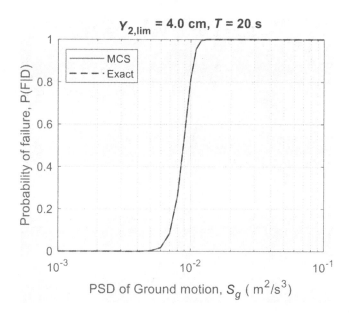

**FIGURE 9.6** Variation of the probability of failure with respect to the PSD of the stochastic ground motion.

## 9.5 CONCLUSION

In this study, an SHM-integrated reliability analysis application is undertaken by using a two-stage Bayesian FE model updating approach. The main advantage of this application lies in the fact that performing a Markov chain-based simulation becomes unnecessary since the posterior *pdf* for model parameters can be well approximated by using the two-stage approach. Thus, the resulting reliability integral can be solved by the classical MCS technique. To see the effectiveness of this application, a numerical study is conducted based on the simulated FE model of a two-story shear frame. Obtained numerical results indicate that the classical MCS solutions obtained for the reliability integral perfectly match with the exact results, by the inclusion of the two-stage Bayesian approach.

## REFERENCES

Au, S. -K., 2011. "Fast Bayesian FFT Method for Ambient Modal Identification with Separated Modes." Journal of Engineering Mechanics 137 (3): 214–226. doi:10.1061/(asce)em.1943-7889.0000213.

Au, S. -K., and Zhang, F. -L., 2016. "Fundamental Two-Stage Formulation for Bayesian System Identification, Part I: General Theory." Mechanical Systems and Signal Processing 66–67: 31–42. doi:10.1016/j.ymssp.2015.04.025.

Beck, J. L., and Au, S. K., 2002. "Bayesian Updating of Structural Models and Reliability Using Markov Chain Monte Carlo Simulation." Journal of Engineering Mechanics 128 (4): 380–391. doi:10.1061/(ASCE)0733-9399(2002)128:4(380).

Betz, W., Beck, J. L., Papaioannou, I., and Straub, D., 2018. "Bayesian Inference with Reliability Methods without Knowing the Maximum of the Likelihood Function." Probabilistic Engineering Mechanics 53: 14–22. doi:10.1016/j.probengmech.2018.03.004.

Catbas, F. N., Susoy, M., and Frangopol, M., 2008. "Structural Health Monitoring and Reliability Estimation: Long Span Truss Bridge Application with Environmental Monitoring Data." Engineering Structures 30 (9): 2347–2359. doi:10.1016/j.engstruct.2008.01.013.

Das, A., and Debnath, N., 2018. "A Bayesian Finite Element Model Updating with Combined Normal and Lognormal Probability Distributions Using Modal Measurements." Applied Mathematical Modelling 61: 457–483. doi:10.1016/j.apm.2018.05.004.

Frangopol, D. M., Strauss, A., and Kim, S., 2008. "Bridge Reliability Assessment Based on Monitoring." Journal of Bridge Engineering 13 (3): 258–270. doi:10.1061/(ASCE)1084-0702(2008)13:3(258).

Hızal, Ç., and Turan, G., 2020. "A Two-Stage Bayesian Algorithm for Finite Element Model Updating by Using Ambient Response Data from Multiple Measurement Setups." Journal of Sound and Vibration 469: 115139. doi:10.1016/j.jsv.2019.115139.

Liu, H., He, X., Jiao, Y., and Wang, X., 2019. "Reliability Assessment of Deflection Limit State of a Simply Supported Bridge Using Vibration Data and Dynamic Bayesian Network Inference." Sensors (Switzerland) 19 (4), 837. doi:10.3390/s19040837.

Döhler, M., and Thöns, S., 2016. "Efficient Structural System Reliability Updating with Subspace-Based Damage Detection Information." 8th European Workshop on Structural Health Monitoring, EWSHM 2016 1: 566–575.

Ozer, E., Feng, M. Q., and Soyoz, S., 2015. "SHM-Integrated Bridge Reliability Estimation Using Multivariate Stochastic Processes." Earthquake Engineering and Structural Dynamics 44: 601–618. doi:10.1002/eqe.2527.

Papadimitriou, C., Beck, J.L., and Katafygiotis, L. S., 2001. "Updating Robust Reliability Using Structural Test Data." Probabilistic Engineering Mechanics 16: 103–113.

Straub, D., and Papaioannou, I., 2015. "Bayesian Updating with Structural Reliability Methods." Journal of Engineering Mechanics 141 (3): 1–13. doi:10.1061/(ASCE)EM.1943-7889.0000839.

Yan, W. J., and Katafygiotis, L.S., 2015. "A Novel Bayesian Approach for Structural Model Updating Utilizing Statistical Modal Information from Multiple Setups." Structural Safety 52 (PB). Elsevier Ltd: 260–271. doi:10.1016/j.strusafe.2014.06.004.

Yuen, K. V., 2010. Bayesian Methods for Structural Dynamics and Civil Engineering. John Wiley & Sons (Asia) Pte. Ltd., Singapore.

Yuen, K. V., and Kuok, S.C., 2011. "Bayesian Methods for Updating Dynamic Models." Applied Mechanics Reviews 64 (1): 010802. doi:10.1115/1.4004479.

Zhang, F. L., and Au, S. K., 2016. "Fundamental Two-Stage Formulation for Bayesian System Identification, Part II: Application to Ambient Vibration Data." Mechanical Systems and Signal Processing 66–67. Elsevier: 43–61. doi:10.1016/j.ymssp.2015.04.024.

Zhang, F. L., Ni, Y.C., and Lam, H. F., 2017. "Bayesian Structural Model Updating Using Ambient Vibration Data Collected by Multiple Setups." Structural Control and Health Monitoring 24 (12): 1–18. doi:10.1002/stc.2023.

# 10 Time-Dependent Reliability of Aging Structures Exposed to Imprecise Deterioration Information

*Cao Wang*

University of Wollongong, Wollongong, NSW, Australia

## CONTENTS

## 10.1 INTRODUCTION

In-service civil structures are typically subjected to the impacts of severe operating or environmental conditions such as chloride ingress resulting in the corrosion of steel bars. Thus, the resistances (e.g., strength and stiffness) of these structures may degrade below an acceptable level as assumed for newly-constructed ones. This deterioration will unavoidably impair the structural serviceability and safety level. Nonetheless, most of these degraded structures are still in use due to some socio-economic considerations. As a result, it is essentially important to evaluate the safety and remaining service life of aging structures in a quantitative manner, which can be further used to guide the decision-makings regarding the repair or maintenance strategies for these structures (Mori and Ellingwood, 1993; Hong, 2000; van Noortwijk et al., 2007; Stewart and Mullard, 2007; Wang 2021). The time-dependent reliability assessment of aging structures is a powerful tool for estimating the structure's ability of fulfilling desired services under a probabilistic framework, where the uncertainties arising from both the resistance deterioration and the external load actions are incorporated. Mathematically, for a reference period of [0, T], let R(t) and S(t) denote the resistance and load effect at time $t \in [0,T]$, respectively. The structural time-dependent reliability, L(T), is determined as follows,

$$L(T) = P\{R(t) \geq S(t), \forall t \in [0,T]\} \tag{10.1}$$

in which P{} denotes the probability of the event in the brackets. Eq. (10.1) implies that the resistance deterioration and external load processes are the key elements in the estimate of structural

DOI: 10.1201/9781003194613-10

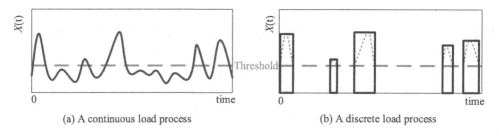

(a) A continuous load process               (b) A discrete load process

**FIGURE 10.1**   A discrete or continuous load process.

time-dependent reliability. For the former, a deterioration function $G(t)$ is often used to describe the deterioration process, which is defined as the ratio of $R(t)$ to $R_0$, where $R_0$ is the initial resistance. For the latter, the load process can be typically classified into two categories: a discrete process and a continuous process, as illustrated in Figure 10.1 (Wang et al., 2019). When a discrete load process is considered, the structural failure is defined as the case where the load effect exceeds the resistance at any time of load occurrence (Mori and Ellingwood, 1993; Li et al., 2015). In the presence of a continuous load process, however, the failure is due to the occurrence of the first upcrossing of the load process with respect to the resistance degradation process. In the following, the time-dependent reliability problem with a discrete load process is presented in Section 10.2.1, while that associated with a continuous load process is discussed in Section 10.2.2.

For either type of load process, the resistance deterioration process is always continuous by nature. Practically, the low-order moments of resistance deterioration parameters (e.g., the mean value and variance) can be determined through experimental data or in-situ inspection, while the probability distribution types of the deterioration parameters are often difficult to identify due to the limit of available data (Ellingwood, 2005; Wang et al., 2016b). In such cases, a reliability-bounding approach can be used to handle the imprecise information of resistance deterioration, that is, to determine the envelope (lower and upper bounds) of structural reliability associated with unknown distribution functions. This topic will be further discussed in Section 10.3.

## 10.2 STRUCTURAL TIME-DEPENDENT RELIABILITY

### 10.2.1 DISCRETE LOAD PROCESS

Consider the reliability of a structure within a reference period of $[0, T]$. Extreme load events that may impair structural safety dramatically occur randomly in time with random intensities, which can be described by a discrete load process. A widely-used model is the Poisson point process. If the load-effect sequence is denoted by $S(t_1), S(t_2), ..., S(t_n)$, occurring at times $t_1, t_2, ..., t_n$, respectively, then Eq. (10.1) becomes,

$$L(T) = P[R(t_1) \geq S(t_1) \cap R(t_2) \geq S(t_2) \cap ... \cap R(t_n) \geq S(t_n)] \tag{10.2}$$

where $R(t_1), R(t_2), ... R(t_n)$ are the resistances at times $t_1, t_2, ... t_n$, respectively. The time-dependent failure probability, $P_f(T)$, is equal to $1 - L(T)$.

If the sequence of load effects, $S(t_1), S(t_2), ..., S(t_n)$, is modeled as a stationary Poisson process, let $\lambda$ denote the occurrence rate, and $F_S(s)$ the cumulative density function (CDF) of each load effect. With this, the time-dependent reliability is as follows (Mori and Ellingwood, 1993),

$$L(T) = \int_0^\infty \exp\left\{ -\int_0^T \lambda \left\{ 1 - F_S[r \cdot g(t)] \right\} dt \right\} f_{R_0}(r) dr \tag{10.3}$$

where $g(t)$ is the deterioration function (deterministic), and $f_{R_0}(r)$ is the probability distribution function (PDF) of the initial resistance $R_0$. Eq. (10.3) has been adopted in the international standard ISO 13822: Bases for design of structures – Assessment of existing structures (ISO, 2010).

The deterioration function $g(t)$ in Eq. (10.3) may take different forms depending on the dominate deterioration mechanism. For example, when the loss of rebar diameter due to active general corrosion in concrete structures is governed by Faraday's Law, the deterioration process of the flexural capacity is approximately linear (Clifton and Knab, 1989; Mori and Ellingwood, 1993). Accordingly, $g(t)$ takes a form of $g(t) = 1 - kt$, where $k$ is a constant that reflects the deterioration rate. With this regard, if further considering the uncertainty arising from the deterioration process, Eq. (10.3) becomes

$$L(T) = \int_0^\infty \int_0^\infty \exp\left\{-\int_0^T \lambda\{1 - F_S[r \cdot (1 - kt)]\}dt\right\} f_K(k) f_{R_0}(r) dk dr \tag{10.4}$$

in which $f_K(k)$ is the PDF of the deterioration rate $K$. An approximate method was proposed in Wang et al. (2016a) to solve Eq. (10.4) in an efficient manner. Suppose that the intensity of load effects can be described by an Extreme Type I (Gumbel) distribution, and the CDF is as follows (Melchers, 1999),

$$F_S(x) = \exp\left(-\exp\left(-\frac{x-u}{\alpha}\right)\right) \tag{10.5}$$

where $u$ and $\alpha$ are the location and scale parameters of the load effect. With this, Eq. (10.4) is simplified as follows,

$$L(T) = \int_0^\infty \int_0^\infty \exp(\lambda\xi) f_K(k) f_{R_0}(r) dk dr \tag{10.6}$$

where

$$\xi = \begin{cases} -\exp\left(-\dfrac{r-u}{\alpha}\right) \cdot T, & k = 0 \\ -\exp\left(-\dfrac{r-u}{\alpha}\right) \cdot \dfrac{\alpha}{kr} \cdot \left[\exp\left(\dfrac{krT}{\alpha}\right) - 1\right], & k \neq 0 \end{cases} \tag{10.7}$$

Compared with Eq. (10.4), the computational cost of using Eq. (10.6) is greatly reduced.

As a generalized case of Eq. (10.3), if the load process is a non-stationary Poisson process, let $\lambda(t)$ denote the time-variant occurrence rate, and $F_S(s,t)$ the time-dependent CDF of load effect at time $t$ (c.f. Figure 10.2). With this, Eq. (10.3) becomes (Li et al., 2015),

$$L(T) = \int_0^\infty \exp\left\{-\int_0^T \lambda(t)\{1 - F_S[r \cdot g(t), t]\}dt\right\} f_{R_0}(r) dr \tag{10.8}$$

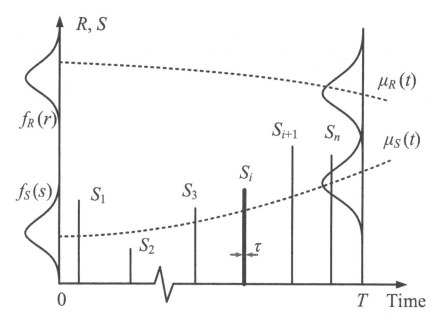

**FIGURE 10.2**  Time-dependent reliability with a discrete load process.

Modeling the deterioration process as linear, similar to Eq. (10.4), Eq. (10.8) becomes the following if incorporating the uncertainty associated with the deterioration process,

$$L(T) = \int_0^\infty \exp\left\{-\int_0^T \lambda(t)\{1 - F_S[r \cdot (1 - kt), t]\} dt\right\} f_K(k) f_{R_0}(r) dk dr \qquad (10.9)$$

In terms of an efficient approach to solve Eq. (10.9), suppose that the mean value (or equivalently, the location parameter) of the load effect increases with time with a rate of $\kappa_m > 0$ (that is, $u(t) = u(0) + \kappa_m t$, where $u(t)$ is the location parameter of the load effect at time $t$), and the load effect follows a Gumbel distribution with a constant scale parameter [c.f. Eq. (10.5)]. If the occurrence rate is time-invariant, denoted by $\lambda$, Eq. (10.9) is approximated by Eq. (10.6) but with a different expression of $\xi$ as follows (Wang et al., 2016a),

$$\xi = -\exp\left(-\frac{r - u(0)}{\alpha}\right) \cdot \frac{\alpha}{kr + \kappa_m} \cdot \left[\exp\left(\frac{T(kr + \kappa_m)}{\alpha}\right) - 1\right] \qquad (10.10)$$

On the other hand, if the load intensity is time-invariant in Eq. (10.9) while the occurrence rate increases linearly with time and with a rate of $\kappa_\lambda > 0$ (i.e., $\lambda(t) = \lambda(0) + \kappa_\lambda t$, where $\lambda(t)$ is the occurrence rate of loads at time $t$), then Eq. (10.9) is rewritten as follows,

$$L(T) = \int_0^\infty \int_0^\infty \exp(\lambda(0)\xi + \kappa_\lambda \psi) f_K(k) f_{R_0}(r) dk dr \qquad (10.11)$$

where $\xi$ is same as in Eq. (10.7), and

$$\psi = \begin{cases} -\exp\left(-\dfrac{r-u}{\alpha}\right) \cdot \dfrac{1}{2}T^2, & k=0 \\[3mm] -\exp\left(-\dfrac{r-u}{\alpha}\right) \cdot \left\{ \exp\left(\dfrac{krT}{\alpha}\right) \cdot \left[ \dfrac{\alpha T}{kr} - \left(\dfrac{\alpha}{kr}\right)^2 \right] + \left(\dfrac{\alpha}{kr}\right)^2 \right\}, & k \neq 0 \end{cases} \qquad (10.12)$$

Eqs. (10.6) and (10.11) imply that the probability distribution of the deterioration model (reflected by the PDFs of $K$ and $R_0$) is the key element in the estimation of structural time-dependent reliability. In practice, the performance of in-service structures is seldom evaluated partially due to the relatively high costs. Thus, the resistance deterioration models interpreted from limited data are often associated with large epistemic uncertainty. Empirically selecting a PDF for the deterioration function cannot fully reflect the realistic structural safety level. In such cases, one needs to consider a set of potential distribution functions for the deterioration function rather than an individual, and correspondingly, the structural reliability (or failure probability) would vary within an interval. The reliability-bounding method will be later discussed in Section 10.3.

## 10.2.2 Continuous Load Process

As illustrated in Figure 10.1, the structural load process can be categorized as either discrete or continuous. In the presence of a continuous load process, the structural reliability problem can be converted into solving the first-passage probability, as shown in Figure 10.3 (Rice, 1944; Zheng and Ellingwood, 1998; Hu and Du, 2013; Wang et al., 2019; Wang, 2020). Consider a case where the load process is Gaussian with a deterministic resistance deterioration process. Notice that, using equivalent representations, this case can be further extended to a non-Gaussian load process (Wang et al., 2019). Recall Eq. (10.1), which is rewritten as follows,

$$L(T) = P\left\{ \Omega(t) \geq X(t), \forall t \in [0,T] \right\} \qquad (10.13)$$

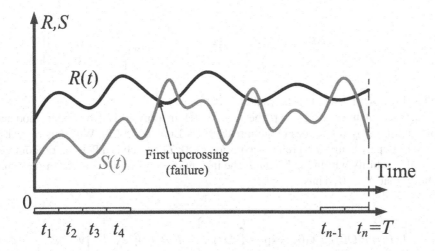

**FIGURE 10.3** Structural reliability converted into a first-passage probability problem.

where $\Omega(t) = R(t) - \mu_{S(t)}$, $X(t) = S(t) - \mu_{S(t)}$, and $\mu_{S(t)}$ is the mean value of $S(t)$. In such a way, the load process $S(t)$ is normalized to a zero-mean stationary Gaussian process, denoted by $X(t)$, with a standard deviation of $\sigma_X = \sigma_S$, where $\sigma_X, \sigma_S$ are the standard deviations of $X(t)$, $S(t)$, respectively.

The upcrossing rate of $X(t)$ over $\Omega(t)$, $\nu^+(t)$, is determined by

$$
\lim_{dt \to 0} \nu^+(t)dt = P\{\Omega(t) > X(t) \cap \Omega(t+dt) < X(t+dt)\}
$$
$$
= \int_{\dot{\Omega}(t)}^{\infty} [\dot{X}(t) - \dot{\Omega}(t)] f_{X\dot{X}}[\Omega(t), \dot{X}(t)] d\dot{X}(t) dt \tag{10.14}
$$

where the dot over a variable means the differential of the variable, and $f_{X\dot{X}}(\dot{x}, x)$ is the joint PDF of $X(t)$ and $\dot{X}(t)$. With Eq. (10.14), one has,

$$
\nu^+(t) = \int_{\dot{\Omega}(t)}^{\infty} [\dot{X} - \dot{\Omega}] f_{X\dot{X}}[\Omega, \dot{X}] d\dot{X} \tag{10.15}
$$

Note that $X(t)$ is a Gaussian process, with which $\dot{X}(t)$ is also a Gaussian process. Assume that $\dot{X}(t)$ is independent of $X(t)$. With this,

$$
f_{X\dot{X}}(\dot{x}, x) = \frac{1}{2\pi\sigma_X\sigma_{\dot{X}}} \exp\left[-\frac{1}{2}\left(\frac{x^2}{\sigma_X^2} + \frac{\dot{x}^2}{\sigma_{\dot{X}}^2}\right)\right] \tag{10.16}
$$

where $\sigma_{\dot{X}}$ is the standard deviation of $\dot{X}(t)$, which is equal to the standard deviation of $\dot{S}(t)$, $\sigma_{\dot{X}}$. Substituting Eq. (10.16) into Eq. (10.15) yields

$$
\nu^+(t) = \frac{\exp\left(-\dfrac{\Omega^2(t)}{2\sigma_X^2}\right)}{2\pi\sigma_X} \cdot \left\{\sigma_{\dot{X}} \exp\left(-\frac{\dot{\Omega}^2(t)}{2\sigma_{\dot{X}}^2}\right) - \sqrt{2\pi}\dot{\Omega}(t)\left[1 - \Phi\left(\frac{\dot{\Omega}(t)}{\sigma_{\dot{X}}}\right)\right]\right\} \tag{10.17}
$$

where $\Phi()$ is the CDF of a standard normal distribution.

The occurrence of upcrossings [c.f. the term $\nu^+(t)$ in Eq. (10.17)] has been modeled as a Poisson process in many researches (Vanmarcke, 1975; Li et al., 2005). With this regard, let $N_T$ be the number of upcrossings for a reference period of $[0, T]$, which is a Poisson random variable. The structural reliability within $[0, T]$ equals the instantaneous reliability at the initial time times the probability that $N_T = 0$. Thus,

$$
L(T) = P[\Omega(0) \geq X(0)] \cdot \exp\left\{-\int_0^T \nu^+(t)dt\right\} = F_X[\Omega(0)] \cdot \exp\left\{-\int_0^T \nu^+(t)dt\right\} \tag{10.18}
$$

where $F_X(x)$ is the CDF of $X(t)$. In Eq. (10.18), if modeling the resistance deterioration process as linear (that is, $g(t) = 1 - kt$ as before), one has $\Omega(t) = r(1 - kt) - \mu_{S(t)}$ and $\dot{\Omega}(t) = -kr$. With this, the upcrossing rate $v^+(t)$ in Eq. (10.17) becomes,

$$v^+(t) = A(r,k)\exp\left[-\frac{[r(1-kt)-\mu_{S(t)}]^2}{2\sigma_S^2}\right] \tag{10.19}$$

where

$$A(r,k) = \frac{1}{2\pi\sigma_S} \cdot \left\{\sigma_{\dot{S}}\exp\left(-\frac{k^2 r^2}{2\sigma_{\dot{S}}^2}\right) + \sqrt{2\pi}kr\Phi\left(\frac{kr}{\sigma_{\dot{S}}}\right)\right\} \tag{10.20}$$

Considering the non-stationarity in the load process, assume that $\mu_{S(t)}$ varies linearly with time according to $\mu_{S(t)} = \mu_{S(0)} + \kappa_S t$, where $\kappa_S$ is the changing rate of $\mu_{S(t)}$. Based on Eq. (10.19), the time-dependent reliability in Eq. (10.18) becomes

$$L(T) = \Phi\left(\frac{\Omega(0)}{\sigma_S}\right) \cdot \exp\left\{-\int_0^T v^+(t)\,dt\right\}$$

$$= \Phi\left(\frac{r - \mu_{S(0)}}{\sigma_S}\right) \cdot \exp\left\{-A(r,k)\int_0^T \exp\left[-\frac{[r - \mu_{S(0)} - (kr + \kappa_S)t]^2}{2\sigma_S^2}\right]dt\right\}$$

$$= \Phi\left(\frac{r - \mu_{S(0)}}{\sigma_S}\right) \cdot \exp\left\{-A(r,k)\frac{\sqrt{2\pi}\sigma_S}{kr + \kappa_S}\left[\Phi\left(\frac{r - \mu_{S(0)}}{\sigma_S}\right) - \Phi\left(\frac{r - \mu_{S(0)} - (kr + \kappa_S)T}{\sigma_S}\right)\right]\right\} \tag{10.21}$$

Furthermore, in the presence of the uncertainties associated with both the initial resistance $R_0$ and the deterioration rate, Eq. (10.21) becomes

$$L(T) = \int_0^\infty\int_0^\infty \Phi\left(\frac{r - \mu_{S(0)}}{\sigma_S}\right) \cdot \exp\left\{-A(r,k)\frac{\sqrt{2\pi}\sigma_S}{kr + \kappa_S}\left[\Phi\left(\frac{r - \mu_{S(0)}}{\sigma_S}\right) - \Phi\left(\frac{r - \mu_{S(0)} - (kr + \kappa_S)T}{\sigma_S}\right)\right]\right\}$$

$$\cdot f_K(k)f_{R_0}(r)\,dk\,dr \tag{10.22}$$

Recall that in Eqs. (10.18) and (10.22), the occurrence of upcrossings has been modeled as an independent process. For cases where the occurrence of each upcrossing is rare and weakly dependent, this assumption is reasonable (Zhang and Du, 2011). However, when the upcrossings are associated with large occurrence rate and correlation, the Poisson approximation could result in significant errors (Yang and Shinozuka, 1971; Madsen and Krenk, 1984). In an attempt to improve the accuracy of structural reliability, an analytical solution was proposed in Wang (2020), where the correlation between the upcrossings is taken into account.

For a service period of $[0, T]$, the period is first subdivided into $n$ identical sections, namely $[t_0 = 0, t_1), [t_1, t_2), .. [t_{n-1}, t_n = T]$, where $n$ is sufficiently large. For each $i = 0, 1, 2, ..., n$, a Bernoulli variable $Y_i$ is introduced, which returns one if $\Omega(t_i) \geq X(t_i)$ and zero otherwise. Using a Markov process to describe the occurrence of upcrossings, it follows that

$$L(T) = P(Y_0 = 1, Y_1 = 1, ... Y_n = 1) = P(Y_0 = 1) \prod_{i=0}^{n-1} P(Y_{i+1} = 1 | Y_i = 1)$$

$$= P(Y_0 = 1) \prod_{i=0}^{n-1} \frac{P(Y_{i+1} = 1 \cap Y_i = 1)}{P(Y_i = 1)}$$

(10.23)

where

$$P(Y_0 = 1) = P[\Omega(0) \geq X(0)] = F_X[\Omega(0)]$$

(10.24)

and

$$P(Y_{i+1} = 1 \cap Y_i = 1) = P[X(t_i) \leq \Omega(t_i) \cap X(t_{i+1}) \leq \Omega(t_{i+1})]$$

$$= P[X(t_i) \leq \Omega(t_i) \cap X(t_i) + [\dot{X}(t_i) - \dot{\Omega}(t_i)]\Delta \leq \Omega(t_i)]$$

(10.25)

in which $\Delta = T/n$. Note that $X(t)$ is a stationary Gaussian process, with which Eq. (10.23) is further derived as follows,

$$L(T) = F_X[\Omega(0)] \cdot \exp\left\{ -\int_0^T \frac{v^+(t)}{F_X[\Omega(t)]} dt \right\}$$

(10.26)

where $v^+(t)$ is as in Eq. (10.17).

Using the same configuration as in Eq. (10.21) (that is, linear deterioration process and linear mean load effect), Eq. (10.26) becomes,

$$L(T) = \Phi\left( \frac{\Omega(0)}{\sigma_s} \right) \cdot \exp\left\{ -\int_0^T \frac{V^+(t)}{F_X[\Omega(t)]} dt \right\}$$

$$= \Phi\left( \frac{r - \mu_{S(0)}}{\sigma_s} \right) \cdot \exp\left\{ -A(r,k) \int_0^T \frac{\exp\left[ -\dfrac{\left[ r - \mu_{s(0)} - (kr + \kappa_s)t \right]^2}{2\sigma_s^2} \right]}{\Phi\left( \dfrac{r - \mu_{s(0)} - (kr + \kappa_s)t}{\sigma_s} \right)} dt \right\}$$

$$= \Phi\left( \frac{r - \mu_{S(0)}}{\sigma_s} \right) \cdot \exp\left\{ -A(r,k) \frac{\sqrt{2\pi}\sigma_s}{kr + \kappa_s} \left[ \ln \Phi\left( \frac{r - \mu_{s(0)}}{\sigma_s} \right) - \ln \Phi\left( \frac{r - \mu_{s(0)} - (kr + \kappa_s)T}{\sigma_s} \right) \right] \right\}$$

(10.27)

Subsequently, considering the uncertainties arising from both the initial resistance and the deterioration rate ($R_0$ and $K$), Eq. (10.27) becomes

$$L(T) = \int_0^\infty \int_0^\infty \Phi\left(\frac{r - \mu_{S(0)}}{\sigma_S}\right) \cdot \exp\left\{-A(r,k)\frac{\sqrt{2\pi}\sigma_S}{kr + \kappa_S}\left[\ln\Phi\left(\frac{r - \mu_{S(0)}}{\sigma_S}\right) - \ln\Phi\left(\frac{r - \mu_{S(0)} - (kr + \kappa_S)T}{\sigma_S}\right)\right]\right\}$$

$$\cdot f_K(k)f_{R_0}(r)dkdr \tag{10.28}$$

Similar to Eqs. (10.6) and (10.11), the probability model of the deterioration process plays a key role in the estimate of structural reliability in Eqs. (10.22) and (10.28). The reliability-bounding approach in the presence of imprecise deterioration information will be discussed in the next section.

## 10.3 RELIABILITY BOUNDING WITH IMPRECISE DETERIORATION INFORMATION

The result of structural reliability assessment could be significantly sensitive to the probability models of the random variables involved. When the probabilistic information is imprecise, the incompletely-informed random variable(s) should be described by a set of possible probability distributions instead of an individual. One illustrative case of imprecise probability information is that the standard deviation of a random variable, say, $X$, can be roughly estimated from limited observations by 0.25 times the difference between the maxima and minima of the samples, if the observed values are believed to vary with a range of mean $\pm 2 \times$ standard deviation of $X$; however, the determination of the specific distribution type of $X$ depends on further probabilistic information. With this regard, a practical way of dealing with imprecise variables is to employ a probability-bounding technique that considers the envelope of the imprecise probability functions (Ferson et al., 2003; Tang and Ang, 2007; Zhang, 2012; Wang et al., 2018).

Recall that in Eqs. (10.6), (10.11), (10.22), and (10.28), it is an essential step to determine the probability models of the resistance deterioration process, reflected by the PDFs of the initial resistance $R_0$ and the deterioration rate $K$. When $R_0$ and $K$ are incompletely informed, the structural time-dependent reliability would vary within an interval, featured by the lower and upper bounds. The reliability-bounding approach will be discussed in this section.

To start with, consider a simple case with one imprecise variable $X$, whose mean value and standard deviation (denoted by $\mu_X$ and $\sigma_X$) are known whilst the probability distribution type is unavailable. The variable $X$ is scaled to vary within [0, 1]. The structural reliability is expressed as follows,

$$L(T) = \int_0^1 \zeta(x) \cdot f_X(x)dx \tag{10.29}$$

where $f_X(x)$ is the PDF of $X$, and $\zeta(x)$ is the conditional reliability on $X = x$. The computation of tight bounds of Eq. (10.29) can be performed employing the algorithm of linear programming (Wang et al., 2018). A linear programming problem is expressed as follows,

$$\begin{aligned} &\min & &\mathbf{c}^T\mathbf{x}, \\ &\text{subjected to} & &\mathbf{A}_{eq}\mathbf{x} = \mathbf{b}_{eq}, \mathbf{A}_{in}\mathbf{x} \le \mathbf{b}_{in} \quad \text{and} \quad \mathbf{x} \ge \mathbf{0} \end{aligned} \tag{10.30}$$

where $\mathbf{x}$ is a variable vector (to be determined), $\mathbf{c}$ is the coefficient vector, $\mathbf{b}_{eq}$ and $\mathbf{b}_{in}$ are two known vectors, $\mathbf{A}_{eq}$ and $\mathbf{A}_{in}$ are two coefficient matrices, and the superscript "T" denotes the transpose of a matrix. Some well-established algorithms for linear programming-based optimization are available in the literature (Inuiguchi and Ramık, 2000; Lofberg, 2004; Jansen et al., 2017).

In Eq. (10.29), since the distribution type of $X$ is unknown, one cannot uniquely determine the value of $f_X(x)$ for each $x$. The domain of $X$, $[0, 1]$, is discretized into $n$ identical sections, namely $[x_0 = 0, x_1], [x_1, x_2], \ldots [x_{n-1}, x_n = 1]$, where $n$ is sufficiently large such that $\left| f_X(x) - f_X\left(\dfrac{x_{i-1} + x_i}{2}\right) \right|$ is negligible for $\forall\ i = 1, 2, \ldots n$ and $\forall x \in [x_{i-1}, x_i]$. The sequence $f_X\left(\dfrac{x_{i-1} + x_i}{2}\right)$, $\forall\ i = 1, 2, \ldots n$ is denoted by $\{f_1, f_2, \ldots f_n\}$ for simplicity. With this, Eq. (10.29) is rewritten as follows,

$$L(T) = \lim_{n \to \infty} \sum_{i=1}^{n} \zeta\left(\frac{i - 0.5}{n}\right) \cdot f_i \cdot \frac{1}{n} \tag{10.31}$$

The definition of the mean value and standard deviation of $X$, along with the basic characteristics of a distribution function, gives the following physical constraints for $\{f_1, f_2, \ldots f_n\}$,

$$\begin{cases} \displaystyle\sum_{i=1}^{n} f_i \cdot \frac{1}{n} = 1 \\[3mm] \displaystyle\sum_{i=1}^{n} f_i \cdot \frac{1}{n} \cdot \frac{i}{n} = \mu_X \\[3mm] \displaystyle\sum_{i=1}^{n} f_i \cdot \frac{1}{n} \cdot \left(\frac{i}{n}\right)^2 = \mu_X^2 + \sigma_X^2 \\[3mm] 0 \le f_i \le n, \forall i = 1, 2, \ldots n \end{cases} \tag{10.32}$$

It is observed from Eqs. (10.31) and (10.32) that the lower bound estimate of L(T) can be converted into a linear programming problem [c.f. Eq. (10.30)], that is, Eq. (10.31) is the objective function to be optimized with $\mathbf{f} = [f_1, f_2, \ldots f_n]^T$ being the vector of variables to be determined, and Eq. (10.32) gives the linear constraints. Similarly, the upper bound of L(T) can also be obtained by solving a linear programming problem, which is equivalent to find $\mathbf{f} = [f_1, f_2, \ldots f_n]^T$ that minimizes −L(T).

Next, the reliability-bounding problem with two imprecise random variables (denoted by $X_1$ and $X_2$) will be discussed (Wang et al., 2018). This corresponds to the case that in Eqs. (10.6), (10.11), (10.22), and (10.28), both $R_0$ and $K$ are imprecise variables. Assume that $X_1$ and $X_2$ are scaled to vary within $[0, 1]$ and are statistically independent. With this regard, the structural reliability is expressed as follows,

$$L(T) = \int_0^1 \int_0^1 \zeta(x_1, x_2) \cdot f_{X_1}(x_1) f_{X_2}(x_2) dx_1 dx_2 \tag{10.33}$$

in which $f_{X_1}(x_1)$ and $f_{X_2}(x_2)$ are the PDFs of $X_1$ and $X_2$ respectively, and $\zeta(x_1, x_2)$ is the conditional reliability on $(X_1, X_2) = (x_1, x_2)$. In the presence of the imprecise information of $X_1$ and $X_2$, the structural reliability in Eq. (10.33) is a function of $f_{X_1}(x_1)$ and $f_{X_2}(x_2)$, denoted by

$$L(T) = h(f_{X_1}, f_{X_2}) \tag{10.34}$$

Consider the lower bound of L($T$) in Eq. (10.34). There exists a family of possible distribution types for both $f_{X_1}(x_1)$ and $f_{X_2}(x_2)$, denoted by $\Omega_{X_1}$ and $\Omega_{X_2}$, respectively. To begin with, one can first assign an arbitrary distribution type for $X_1$ and $X_2$, whose PDFs are $_0 f_{X_1} \in \Omega_{X_1}$ and $_0 f_{X_2} \in \Omega_{X_2}$. Next, $_1 f_{X_2} \in \Omega_{X_2}$ is determined, which minimizes $h(_0 f_{X_1}, f_{X_2})$ for $\forall f_{X_2} \in \Omega_{X_2}$, followed by finding $_1 f_{X_1} \in \Omega_{X_1}$ which minimizes $h(f_{X_1}, _1 f_{X_2})$ for $\forall f_{X_1} \in \Omega_{X_1}$. The approach to determine $_1 f_{X_2}$ and $_1 f_{X_1}$ has been discussed earlier, where only one imprecise variable is involved. With this, it follows that

$$h(_1 f_{X_1}, _1 f_{X_2}) \leq h(_0 f_{X_1}, _1 f_{X_2}) \leq h(_0 f_{X_1}, _0 f_{X_2}) \tag{10.35}$$

which implies that the pair $(_1 f_{X_1}, _1 f_{X_2})$ leads to a reduced L($T$) compared with $(_0 f_{X_1}, _0 f_{X_2})$. In a similar manner, one can further determine the subsequent sequences $(_2 f_{X_1}, _2 f_{X_2})$ through to $(_l f_{X_1}, _l f_{X_2})$, in which $l$ is the number of iteration. By noting that $h(f_{X_1}, f_{X_2})$ is bounded, $h(_l f_{X_1}, _l f_{X_2})$ converges to the lower bound of L($T$) with an increasing $l$. Furthermore, the upper bound of L($T$) can be found using a similar procedure.

## 10.4  NUMERICAL EXAMPLES

### 10.4.1  EXAMPLE 1: DISCRETE LOAD PROCESS

Consider the time-dependent reliability of an aging structure with imprecise deterioration information. The structure is designed according to the following criterion: $0.9R_n = 1.2D_n + 1.6L_n$, where $R_n$ is the nominal resistance, $D_n$ and $L_n$ are the nominal dead load and live load respectively. It is assumed that $D_n = L_n$. The dead load, $D$, is deterministic and equals $D_n$. The live load is modeled as a Poisson process with an occurrence rate of 1/year. Conditional on occurrence, the live load effect follows a Gumbel distribution with a standard deviation of $0.12L_n$ and a time-variant mean value of $(0.4 + 0.005t)L_n$ at time $t$ (in years). The initial resistance of the structure, $R_0$, has a mean value of $1.05R_n$ and a standard deviation of $0.1R_n$. The deterioration function of the structure takes a form of $G(t) = 1 - Kt$, where $t$ is in years, and $K$ has a mean value of 0.005 so that the resistance linearly degrades by 20% on average over a reference period of 40 years. The coefficient of variation (COV) of $K$ is 0.2. The structural reliability (failure probability) can be calculated based on Eq. (10.11).

Without introducing additional constraints in regard to the distribution function of $R_0$ and $K$, the lower and upper bounds of the time-dependent failure probability for reference periods up to 40 years are presented in Figure 10.4. For comparison purpose, the probabilities of failure with lognormally distributed $R_0$ and additional assumptions on the distribution type of $K$, including normal, Gamma, lognormal, Beta, and uniform distributions, are also shown in Figure 10.4. It can be seen from Figure 10.4 that lower and upper bounds construct an envelope for the time-dependent failure probabilities. These bounds have reflected all the possible distribution types for $R_0$ and $K$ and thus enclose the failure probabilities with additional assumptions for the imprecise variables. The results in Figure 10.4 demonstrate that if simply assigning some specific distribution types for imprecise variables without reasonable justification, the failure probability could be significantly underestimated.

### 10.4.2  EXAMPLE 2: CONTINUOUS LOAD PROCESS

Consider the time dependent reliability of an aging structure subjected to continuous load process and imprecise deterioration information. The structural resistance degrades with time according to $R(t) = R_0(1 - Kt)$, where $R_0$ is the initial resistance, $K$ is the deterioration rate, and $t$ is in years. The initial resistance has a mean value of 3.5 and a COV of 0.1, while $K$ has a mean value of 0.005 and a COV of 0.2. The dead load is deterministically 0.5, while the live load process $S(t)$ is Gaussian with mean $(0.5 + 0.01t)$ in year $t$ and standard deviation 0.3. Suppose that the autocorrelation function of $S(t)$ takes a form of $\exp(-\gamma x^2)$ so that the standard deviation of $\dot{S}(t)$ equals $\sqrt{2\gamma}$ for a positive $\gamma$. Let $\gamma = 1$ in the following. The structural reliability is computed by Eq. (10.28).

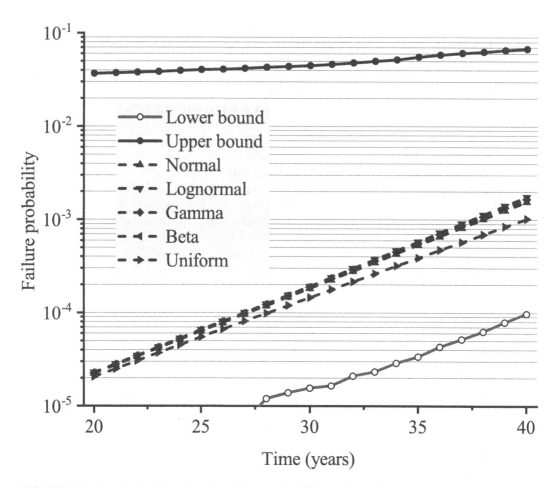

**FIGURE 10.4** Envelope of time-dependent failure probabilities in Example 1.

Without information on the distribution type of $R_0$ and $K$, Figure 10.5 presents the lower and upper bounds of the time-dependent failure probability for reference periods up to 40 years. The probabilities of failure with lognormally distributed $R_0$ and specifically-selected distribution types for $K$ (normal, lognormal, Gamma, Beta, and uniform) are also shown in Figure 10.5. Similar to Figure 10.4, the lower and upper bounds enclose the failure probabilities with specific distribution types of random variables. It hence suggests the importance of properly interpreting the imprecise probabilistic information in structural reliability assessment.

## 10.5 CONCLUDING REMARKS

This chapter has discussed the approaches for structural time-dependent reliability assessment, where the resistance deterioration process and the load process are two essential elements. While the former is continuous by nature, the latter can be categorized as either a discrete process or a continuous one, depending on how the reliability problem is modeled. For both types of load process, closed-form solutions are available for structural time-dependent reliability assessment considering resistance deterioration and non-stationary load process. These solutions can be further simplified, yielding an improved calculation efficiency, for some specific but typical distribution types of load effect.

When the probabilistic information on the resistance deterioration process is imprecise, which is often the case in practice due to the limit of available data, a reliability-bounding approach can be used to find the envelope of structural time-dependent reliability, featured by the lower and upper bounds.

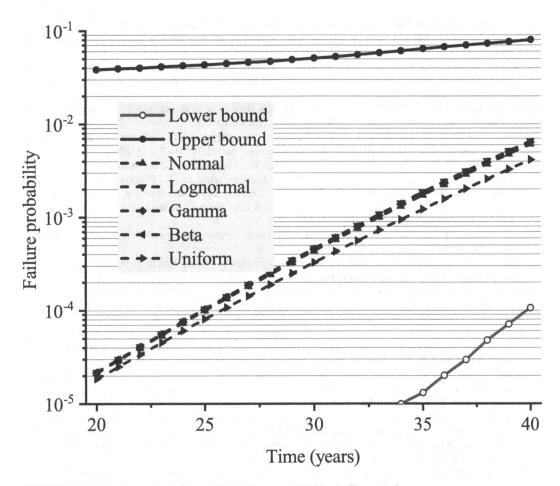

**FIGURE 10.5** Envelope of time-dependent failure probabilities in Example 2.

Specifically, when the mean value and standard deviation of the imprecise random variable(s) are known while the distribution type is unavailable, a linear programming-based approach can be used to bound the time-dependent reliability. The applicability of the reliability bounding technique is illustrated through two numerical examples, where a discrete and a continuous load process are considered, respectively. It is shown that the estimate of structural reliability may be non-conservative if simply assuming some common distribution types for the imprecise variables.

## ACKNOWLEDGMENTS

The research described in this chapter was supported by the Vice-Chancellor's Postdoctoral Research Fellowship from the University of Wollongong. This support is gratefully acknowledged.

## REFERENCES

Bruce R Ellingwood. Risk-informed condition assessment of civil infrastructure: State of practice and research issues. *Structure and Infrastructure Engineering*, 1:7–18, 2005.

C-Q Li, W Lawanwisut, and JJ Zheng. Time-dependent reliability method to assess the serviceability of corrosion-affected concrete structures. *Journal of Structural Engineering*, 131 (11):1674–1680, 2005.

Cao Wang. Stochastic process-based structural reliability considering correlation between up-crossings. *ASCE-ASME Journal of Risk and Uncertainty in Engineering Systems, Part A: Civil Engineering*, 6(4):06020002, 2020.

Cao Wang. *Structural Reliability and Time-Dependent Reliability.* Switzerland AG: Springer Nature, 2021.

Cao Wang, Quanwang Li, and Bruce R Ellingwood. Time-dependent reliability of ageing structures: An approximate approach. *Structure and Infrastructure Engineering*, 12(12):1566–1572, 2016a.

Cao Wang, Quanwang Li, Long Pang, and Aming Zou. Estimating the time-dependent reliability of aging structures in the presence of incomplete deterioration information. *Journal of Zhejiang University Science A*, 17:677–688, 2016b.

Cao Wang, Hao Zhang, and Michael Beer. Computing tight bounds of structural reliability under imprecise probabilistic information. *Computers & Structures*, 208:92–104, 2018.

Cao Wang, Hao Zhang, and Michael Beer. Structural time-dependent reliability assessment with new power spectral density function. *Journal of Structural Engineering*, 145(12): 04019163, 2019.

Christoph Jansen, Georg Schollmeyer, and Thomas Augustin. Concepts for decision making under severe uncertainty with partial ordinal and partial cardinal preferences. In *Proceedings of the Tenth International Symposium on Imprecise Probability: Theories and Applications*, pp. 181–192, 2017.

Erik H Vanmarcke. On the distribution of the first-passage time for normal stationary random processes. *Journal of Applied Mechanics*, 42(1):215–220, 1975.

H Hong. Assessment of reliability of aging reinforced concrete structures. *Journal of Structure Engineering*, 126:1458–1465, 2000.

Hao Zhang. Interval importance sampling method for finite element-based structural reliability assessment under parameter uncertainties. *Structural Safety*, 38:1–10, 2012.

ISO. *Bases for design of structures – Assessment of existing structures.* ISO 13822: 2010(E), 2010.

J-N Yang and Masanobu Shinozuka. On the first excursion probability in stationary narrow- band random vibration. *Journal of Applied Mechanics*, 38:1017–1022, 1971.

J.R Clifton and L.I. Knab. Service life of concrete. Technical report. Division of Engineering, National Institute of Standards and Technology, Gaithersburg, USA, Technical Report No. NUREG/CR-5466 and NISTIR-89-4086. Washington, USA, Nuclear Regulatory Commission, 1989.

Jan M van Noortwijk, Johannes AM van der Weide, Maarten-Jan Kallen, and Mahesh D Pandey. Gamma processes and peaks-over-threshold distributions for time-dependent re-liability. *Reliability Engineering & System Safety*, 92(12):1651–1658, 2007.

Johan Lofberg. *YALMIP: A toolbox for modeling and optimization in Matlab.* In *IEEE International Symposium on Robotics and Automation*, pp. 284–289, IEEE, 2004. doi: 10.1109/CACSD.2004.1393890.

Junfu Zhang and Xiaoping Du. Time-dependent reliability analysis for function generator mechanisms. *Journal of Mechanical Design*, 133(3):031005, 2011.

M G Stewart and J A Mullard. Spatial time-dependent reliability analysis of corrosion damage and the timing of first repair for RC structures. *Engineering Structures*, 29:1457–1464, 2007.

Masahiro Inuiguchi and Jaroslav Ramık. Possibilistic linear programming: A brief review of fuzzy mathematical programming and a comparison with stochastic programming in portfolio selection problem. *Fuzzy Sets and Systems*, 111(1):3–28, 2000.

Peter Hauge Madsen and Steen Krenk. An integral equation method for the first-passage problem in random vibration. *Journal of Applied Mechanics*, 51(3):674–679, 1984.

Quanwang Li, Cao Wang, and Bruce R Ellingwood. Time-dependent reliability of aging structures in the presence of non-stationary loads and degradation. *Structural Safety*, 52:132–141, 2015.

R. E. Melchers. *Structural Reliability Analysis and Prediction.* Wiley, New York, 1999.

Ruohua Zheng and Bruce R Ellingwood. Stochastic fatigue crack growth in steel structures subject to random loading. *Structural Safety*, 20(4):303–323, 1998.

Scott Ferson, Vladik Kreinovich, Lev Ginzburg, Davis S Myers, and Kari Sentz. Constructing probability boxes and Dempster-Shafer structures. Technical Report SAND2002–4015. Sandia National Laboratories, 2003.

Stephen O Rice. Mathematical analysis of random noise. *The Bell System Technical Journal*, 23(3):282–332, 1944.

Wilson H Tang and A Ang. *Probability Concepts in Engineering: Emphasis on Applications to Civil and Environmental Engineering.* 2nd Ed. Wiley, New York, 2007.

Yasuhiro Mori and Bruce R Ellingwood. Reliability-based service-life assessment of aging concrete structures. *Journal of Structural Engineering*, 119(5):1600–1621, 1993.

Zhen Hu and Xiaoping Du. Time-dependent reliability analysis with joint upcrossing rates. *Structural and Multidisciplinary Optimization*, 48(5):893–907, 2013.

# 11 Nonlinear Numerical Analyses of Reinforced Concrete Structures

## Safety Formats, Aleatory and Epistemic Uncertainties

*Paolo Castaldo, Diego Gino, and Giuseppe Mancini*
Politecnico di Torino, Turin, Italy

## CONTENTS

## 11.1 INTRODUCTION

Reliability analysis of reinforced concrete (RC) members and structures requires methodologies able to fulfill safety requirements expected by the society. These requirements, as defined by international codes (EN 1990, 2013; *fib* Model Code 2010, 2013; ISO 2394, 2015), are represented by limits on the likelihood that structural failure may occur in a given reference period. These limits are dependent on the typology, the use, and the lifetime along which a structure should carry out its serviceability. In this context, approaches and methodologies, aimed to design and assess RC systems in compliance with target reliability levels, are provided. With the progress of the last 20 years, nonlinear numerical analyses (NLNAs) have become the preferred instrument devoted to simulate the actual physical response of civil RC structures and infrastructures under the relevant loading configurations. A huge number of modeling hypotheses to perform NLNAs of RC structures and members are available for engineers and practitioners, who more and more should become confident with such complex calculation methods (*fib* Bulletin 45, 2008). As discussed within this chapter, a modeling hypothesis collects all the assumptions in terms of equilibrium, kinematic compatibility, and constitutive laws related to a specific nonlinear numerical model. In the near future, design codes of practice will allow to perform design and assessment using NLN simulations and, for instance, appropriate methodologies are necessary to include safety issues.

This chapter introduces and describes how the main sources of uncertainty affecting structural engineering problems (i.e., aleatory and epistemic), with particular care to assessment and design of RC structures, can be accounted for when NLNAs are used. First of all, the global resistance format (GRF) for evaluation of structural reliability is described and, then, the corresponding safety formats (SFs) are introduced. The comparison among the different SFs is proposed in order to highlight their limits of applicability. To overcome these limits, specific proposals are introduced with particular reference to the aleatory uncertainty influence. Next, a methodology to characterize the epistemic uncertainties associated with the establishment of nonlinear numerical models of RC structures is described and appropriate partial safety factors are proposed to fulfill specific target levels of reliability.

## 11.2   ALEATORY AND EPISTEMIC UNCERTAINTIES WITHIN THE GLOBAL RESISTANCE FORMAT

The GRF (*fib* Model Code 2010, 2013; Allaix, Carbone & Mancini, 2013) deals with the uncertainties associated with structural behavior in line with the limit states design method (EN 1990, 2013; *fib* Model Code 2010, 2013) at the level of global structural behavior (i.e., structural resistance). The main sources of uncertainties are accounted for to determine the design structural resistance $R_d$ and are included within the calculation through appropriate partial safety factors. In detail, the aleatory uncertainties relate to the intrinsic randomness of both materials (e.g., concrete compressive strength and reinforcement yielding strength) and geometric (e.g., members size, concrete cover, and reinforcement location) properties, while the epistemic ones are mainly referred to missing knowledge, simplification, and hypotheses adopted establishing the resistance numerical model itself (see also Der Kiureghian & Ditlevsen, 2009; Castaldo, Gino & Mancini, 2019). The GRF has been proposed by design codes and literature (*fib* Model Code 2010, 2013; Allaix, Carbone & Mancini, 2013) with the scope to perform structural assessment and design of RC structures by means of NLNAs. In general, the evaluation of the safety of existing and new structures is performed through the local verification of cross sections design capacity ($R_d$) compared to the design value of the effect of the actions ($E_d$) deriving from the linear elastic analysis: $E_d \leq R_d$ (EN 1990, 2013). This procedure for the structural reliability assessment is denoted as "local," as it requires just verification of different cross sections of the structural elements composing the structure disregarding from the overall structural response.

For instance, if the structural reliability is evaluated using refined NLNAs, the possibility of RC structural systems to develop internal redistribution of stresses within the selected loading configuration should not be disregarded. In detail, the adoption of NLN requires to consider a "global" evaluation of structural safety, performing the comparison between the actions simultaneously present on the structure to the associated overall capacity, denoted as structural resistance. The main features of the mentioned above approaches are shown in Figure 11.1. In accordance with the GRF, the safety condition can be formulated as follows:

$$R_d \geq F_d \tag{11.1}$$

where the design value of actions $F_d$ can be evaluated in accordance with the EN 1990 (2013) according to the proper combination, while the design structural resistance $R_d$ can be assessed

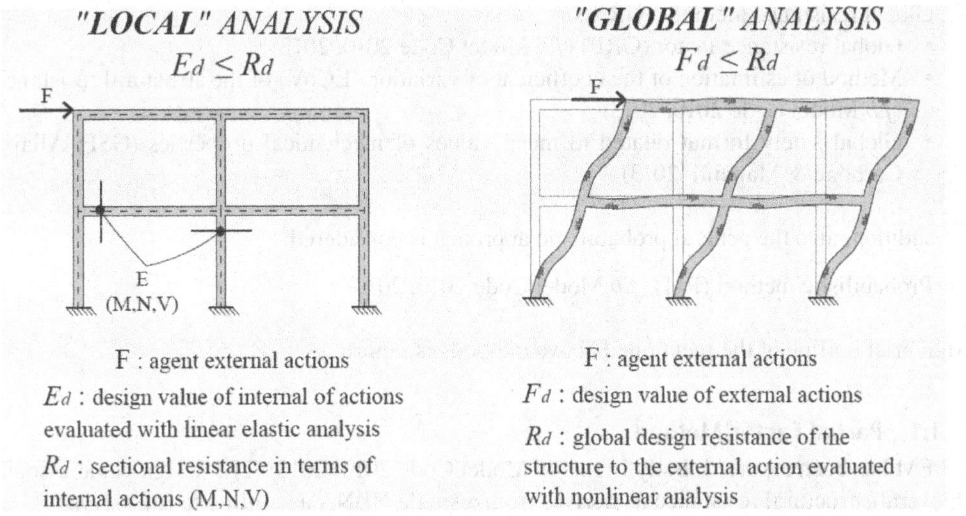

"LOCAL" ANALYSIS
$E_d \leq R_d$

"GLOBAL" ANALYSIS
$F_d \leq R_d$

F : agent external actions

$E_d$ : design value of internal of actions evaluated with linear elastic analysis

$R_d$ : sectional resistance in terms of internal actions (M,N,V)

F : agent external actions

$F_d$ : design value of external actions

$R_d$ : global design resistance of the structure to the external action evaluated with nonlinear analysis

**FIGURE 11.1** Local and global approaches for structural analysis. (Modified from Castaldo, Gino & Mancini, 2019, with permission.)

through NLNAs adopting the SFs (*fib* Model Code 2010, 2013; Allaix, Carbone & Mancini, 2013) based on the GRF as follows:

$$R_d = \frac{R_{NLNA}\left(X_{rep}; a_{rep}\right)}{\gamma_R \gamma_{Rd}} \tag{11.2}$$

In Eq. (11.2), $R_{NLNA}(X_{rep}; a_{rep})$ denotes the structural resistance computed by means of NLNAs using the representative values $X_{rep}$ and $a_{rep}$ for material and geometrical properties, respectively, in accordance with the adopted SF, as commented in the next. The level of structural reliability is involved by means of two different partial safety factors:

- The *global resistance partial safety factor* $\gamma_R$, which takes into account, at the level of global structural behavior, the aleatory uncertainties influence associated with the material properties and even geometry. This partial safety factor may be evaluated in line with the specific target level of reliability according to the methodology described by the selected SF (*fib* Model Code 2010, 2013; Allaix, Carbone & Mancini, 2013).
- The *resistance model uncertainty partial safety factor* $\gamma_{Rd}$, which takes into account the level of epistemic uncertainty associated with simplifications, assumptions, and choices performed to define the NLN model. This partial safety factor is horizontal and independent on the SF (*fib* Model Code 2010, 2013; Allaix, Carbone & Mancini, 2013) adopted to carry out the global structural verification.

In the next sections, the principal features of the most common SFs based on the GRF are described.

### 11.2.1 SAFETY FORMATS FOR NLNAS OF RC STRUCTURES

As previously discussed, the safety verification by means of NLNAs can be performed according to the GRF. In scientific literature and codes of practice, different SFs based on the GRF have been proposed. In the present chapter, the following SFs are considered:

1. Partial factor method (PFM) (*fib* Model Code 2010, 2013)

2. Global resistance methods (GRMs):
   - Global resistance factor (GRF) (*fib* Model Code 2010, 2013)
   - Method of estimation of the coefficient of variation (ECoV) of the structural resistance (*fib* Model Code 2010, 2013)
   - Global safety format related to mean values of mechanical properties (GSF; Allaix, Carbone & Mancini, 2013)

In addition, also the general probabilistic approach is considered:

3. Probabilistic method (PM) (*fib* Model Code 2010, 2013)

Next, a brief outline of the mentioned above methods is reported.

### 11.2.1.1   Partial Factor Method

The PFM has been reported firstly in the *fib* Model Code 2010 (2013). The design value associated with overall structural resistance $R_d$ derives from a single NLNA according to Eq. (11.3):

$$R_d = \frac{R_{NLNA}(X_d; a_d)}{\gamma_{Rd}} \tag{11.3}$$

where $R_{NLNA}(X_d; a_d)$ represents the structural resistance estimated with a NLNA defined adopting the design values of both geometric ($a_d$) material ($X_d$) characteristics; $\gamma_{Rd}$ is the safety factor associated with uncertainty in the resistance model definition. Within PFM, the influence of the aleatory uncertainties is accounted for applying partial safety factors $\gamma_M$ to divide, and then, reduce the characteristic values of materials properties ($X_k$) (e.g., characteristic value of cylinder concrete strength $f_{ck}$ and reinforcement characteristic tensile strength $f_{yk}$), and the adoption of design values of geometric properties ($a_d$). The partial safety factors for material properties $\gamma_M$ may be defined accounting for the appropriate target level of reliability for new structures, see EN 1990 (2013), *fib* Model Code 2010 (2013) and existing structures, see *fib* Bulletin 80 (2016). The design value of geometrical properties ($a_d$) can be defined starting from their nominal values $a_{nom}$ modified with appropriate deviation ($\Delta a$) to account for construction imperfections (EN 1990, 2013).

### 11.2.1.2   Global Resistance Methods

The GRMs are the SFs based on the evaluation of the design structural resistance $R_d$ by means of partial safety factors applied directly to the global structural resistance estimated throughout NLNAs. In particular, three main SFs can be recognized:

I. *GRF method*
   As for the GRF method (EN1992-2, 2005; *fib* Model Code 2010, 2013), the design structural resistance $R_d$ is defined as follows:

$$R_d = \frac{R_{NLNA}(f_{cmd}, f_{ym}; a_{nom})}{\gamma_{GL}} \tag{11.4}$$

In Eq. (11.4), the global safety factor $\gamma_{GL}$ is set to 1.27 with reference to target reliability level identified by reliability index $\beta_t = 3.8$ referred to 50 years of working life. With the scope to link the GRF method with the other GRMs and SFs, the fixed value of $\gamma_{GL}$ can be represented by the product between global resistance partial safety factor $\gamma_R$ and the partial safety factor for model uncertainty $\gamma_{Rd}$. To evaluate the value representing the structural resistance $R_{NLNA}(f_{cmd}, f_{ym}; a_{nom})$, the mean value of the reinforcement tensile yield stress

$f_{ym}$ and the reduced value of concrete cylinder strength $f_{cmd}$ in compression have to be employed as reported below:

$$f_{ym} = 1.1 f_{yk} \; ; \; f_{cmd} = 0.85 f_{ck} \tag{11.5}$$

where $f_{yk}$ and $f_{ck}$ represent, respectively, the 5% characteristic values of the reinforcement tensile yield stress and cylinder concrete compressive strength.

II. *Method of ECoV of the structural resistance*

The method of ECoV, as suggested by *fib* Model Code 2010 (2013), allows the estimation of the design value of structural resistance $R_d$ as follows:

$$R_d = \frac{R_{NLNA}(X_m; a_{nom})}{\gamma_R \cdot \gamma_{Rd}} \tag{11.6}$$

where $R_{NLNA}(X_m; a_{nom})$ is the structural resistance determined through a NLN simulation established adopting the mean material properties $(X_m)$ and nominal geometric ones $(a_{nom})$ to define the numerical structural model; $\gamma_R$ is the partial safety factor taking into account uncertainties associated with the properties of materials; $\gamma_{Rd}$ denotes the partial safety factor associated with uncertainty in definition of the numerical model. The ECoV method is founded on the assumptions that the aleatory variability of the structural resistance can be represented by a lognormal probabilistic distribution and that the related mean value is almost equal to the value of $R_{NLNA}(X_m; a_{nom})$. According to these hypotheses, the global resistance partial safety factor $\gamma_R$ may be evaluated as follows:

$$\gamma_R = \exp(\alpha_R \cdot \beta_t \cdot V_R) \tag{11.7}$$

where $V_R$ denotes the CoV (i.e., coefficient of variation) of the probabilistic distribution associated with structural resistance; $\alpha_R$ represents the FORM (first-order reliability method) sensitivity coefficient, adopted as 0.8 according to *fib* Model Code 2010 (2013) for dominant resistance variables; $\beta_t$ is the target value of the reliability index that allows to consider the required level of reliability for the investigated structure. In accordance with the hypothesis of lognormally distributed variable for structural resistance, the estimate of the CoV $V_R$ can be easily determined according to the expression:

$$V_R = \frac{1}{1.65} \ln\left(\frac{R_{NLNA}(X_m; a_{nom})}{R_{NLNA}(X_k; a_{nom})}\right) \tag{11.8}$$

where $R_{NLNA}(X_k; a_{nom})$ denotes the value of structural resistance estimated using NLNA established adopting characteristic material properties $(X_k)$ and nominal $(a_{nom})$ for geometric ones.

III. *Global safety format (GSF) related to mean values of material properties*

The method of Allaix, Carbone and Mancini (2013), denoted as global safety format (GSF), allows to compute the design value of overall structural resistance $R_d$ basing NLN simulation on mean material properties $(X_m)$ and nominal geometric ones $(a_{nom})$ grounding on the same assumptions of the ECoV method. In fact, the estimation of the value $R_d$ can be performed according to Eq. (11.6), as well as, the global resistance partial safety factor $\gamma_R$ can be determined according to Eq. (11.7). The GSF is different from the method of ECoV. In fact, the level of refinement in the estimation of $V_R = \sigma_R / \mu_R$ in the hypothesis of lognormal distribution (with $\mu_R$ and $\sigma_R$ the mean value and the standard deviation of the probabilistic distribution representing structural resistance, respectively). Specifically, within the GSF, the statistical parameters are evaluated through reduced Monte Carlo simulation technique employing the Latin hypercube sampling (LHS) technique (from Mckey, Conover & Beckman, 1979) with a proper number of samples (in general, 30 samples are sufficient if $V_R \leq 0.20$). The probabilistic model for the random variables representing material properties and even geometrical ones can be assumed with reference to JCSS Probabilistic Model Code (2001) or *fib* Model Code 2010 (2013).

### 11.2.1.3 Probabilistic Method

Finally, the PM reported in the *fib* Model Code 2010 (2013) is founded on several NLNAs delineated adopting Latin hypercube or Monte Carlo's sampling techniques to select input data from the relevant random variables characterizing the aleatory uncertainties (i.e., material and geometric). The results from numerical analyses represented by the overall structural resistance should be statistically processed to characterize the most likely probabilistic model assessing the statistics of the distribution of structural resistance. The statistical parameters can be represented by the mean value $\mu_R$ and CoV $V_R$. Successively, it can be suitable to directly derive the quantile associated with the design value $R_d$, which should correspond to a specific target value of the reliability index $\beta_t$ as follows:

$$R_d = \frac{F_R^{-1}\left[\Phi\left(-\alpha_R\beta_t\right)\right]}{\gamma_{Rd}} \tag{11.9}$$

where $F_R$ is the cumulative probabilistic distribution representing structural resistance; $\Phi$ denotes the cumulative standard normal distribution; $\alpha_R$ represents the FORM sensitivity coefficient, adopted the value of 0.8 in accordance with *fib* Model Code 2010 (2013) for dominant variables associated with structural resistance; $\gamma_{Rd}$ denotes the partial safety factor associated with the uncertainty when the numerical model is defined. The PM differs from the method of GSF for the hypothesis of lognormal distribution on the latter one. Moreover, the GSF grounds on mean material properties and nominal geometric ones adopting, as a simplification, the first order a Taylor expansion of the function describing the structural response (Allaix, Carbone & Mancini, 2013), while, the PM is able to ensure the estimation of the quantile of the most likely probabilistic distribution as reported by Eq. (11.9).

## 11.3   INFLUENCE OF ALEATORY UNCERTAINTIES IN THE USE OF THE SAFETY FORMATS

The present section discusses and highlights relevant aspects associated with the use of the SFs in relation to the aleatory uncertainties influence (i.e., material and geometric) on the computation of the design value associated with structural resistance $R_d$.

### 11.3.1 Comparison between the Safety Formats Based on the GRF

In the next, the comparison of the outcomes in terms of the design value of structural resistance $R_d$ deriving from the application of the different SFs based on the GRF is discussed highlighting limits and possible enhancements. In particular, the investigation of Castaldo, Gino and Mancini (2019) reports the outcomes from NLN models established to simulate the ultimate behavior of different structural members and assess the design structural resistance in accordance with previously described methods. In addition, particular attention is devoted to the accuracy of each SF to predict the relevant failure modes associated with the computation of the design structural resistance $R_d$.

### 11.3.1.1 Structural Members Considered to Compare the Safety Formats

Castaldo, Gino and Mancini (2019) report the application of the several SFs for estimation of the design value of structural resistance $R_d$ of four simply supported beams with transverse openings in

**TABLE 11.1**

**Modeling Hypotheses Adopted by Castaldo, Gino and Mancini (2019) to Define the NLN Models**

| | **Software: ATENA 2D** |
|---|---|
| *Equilibrium* | • Solution method based on standard Newton-Raphson |
| | • Convergence criteria referred to strain energy |
| | • Definition of the loading process in line with the experimental tests set |
| *Kinematic compatibility* | • Iso-parametric finite elements for plane stress analysis with four nodes (the integration grid of $2 \times 2$ Gauss points using linear interpolation is used) |
| | • Discrete reinforcements |
| | • Mesh refinement evaluated by validation process to get numerical accuracy |
| *Constitutive laws* | *Model for concrete properties* |
| | • Fixed crack model, smeared cracking, constant shear retention factor assumed as 0.2 |
| | • Uni-axial model extended to the biaxial stress state |
| | • Compressive response: non-linear presenting post-peak linear softening |
| | • Tensile response: elastic presenting post-peak linear tension softening (LTS) |
| | *Model for reinforcement properties* |
| | • Bi-linear relationship for tensile and compressive response |

the web tested in the laboratory by Aykac et al. (2013) and one isostatic "T" beam designed to fulfill the requirements of EN1992-1-1 (2014). Aykac et al. (2013) report the results from experimental tests performed on four beams having a rectangular cross section with dimensions $15 \times 40$ cm and span between supports of 390 cm. The first three beams present n°12 $20 \times 20$ cm transverse web holes having squared shape. These beams have been realized with increasing values of the reinforcement ratio and are denoted with SL, SM, and SH. The last beam of Aykac et al. (2013) presents n°12 circular transverse holes having a radius of 10 cm and casted with crossed reinforcements between openings in the web. The concrete compressive strength ranges between 20 and 22 MPa. The details of geometrical properties and reinforcements arrangements are acknowledged in Aykac et al. (2013) and Castaldo, Gino and Mancini (2019). All the specimens present a symmetrical "six points bending" statically determined test scheme. The last beam considered from Castaldo, Gino and Mancini (2019) has been designed according to EN1992-1-1 (2014). The beam is defined with a "T" cross section (i.e., T-Beam) with a height of 50 cm, top flange width of 50 cm, and width of the web equal to 15 cm. The thickness of the top flange is 10 cm. The concrete compressive strength, representing the mean value, is 28 MPa. More details may be found in Castaldo, Gino and Mancini (2019). The NLN models have been realized by Castaldo, Gino and Mancini (2019) using the software ATENA 2D (2014). The main modeling hypotheses adopted by Castaldo, Gino and Mancini (2019) to define the numerical models are listed in Table 11.1. The NLN models have been calibrated with the purpose to reproduce the response of the mentioned above RC members adopting experimental/mean values for materials properties. The comparison of the experimental investigation of Aykac et al. (2013) with the numerical simulations of Castaldo, Gino and Mancini (2019) is reported in Figure 11.2 as well as the identification of the main features of the failure modes of the RC members in relation to the systems denoted as SL and T-Beam.

### 11.3.1.2 Results from Probabilistic Analysis and Application of the Safety Formats

The NLN models so far introduced have been adopted to realize the probabilistic analysis of the structural resistance. The probabilistic model has been defined as suggested by JCSS Probabilistic Model Code (2001) in relation to the description of the main random variables (i.e., concrete cylinder compressive strength, reinforcements yielding strength, and Young's modulus). Probabilistic analysis has been carried out using the LHS technique (Mckey, Conover & Beckman, 1979) to

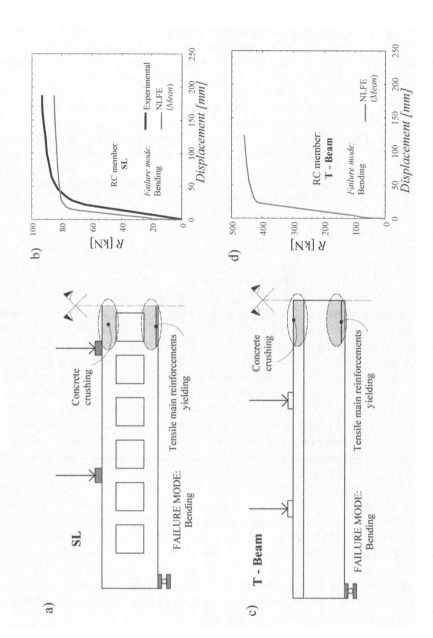

**FIGURE 11.2**  Results from numerical simulations and experimental load-displacement curves with identification of the failure modes. (Modified from Castaldo, Gino & Mancini, 2019, with permission.)

achieve 30 sampled NLN models for each one of the investigated RC members. More information about the probabilistic analysis may be found in Castaldo, Gino and Mancini (2019). The results of the 30 NLN simulations are briefly commented in the following and reported in Figures 11.3 and 11.4 for RC members SL and T-Beam, respectively. The failure mechanism of beam SL is denoted by a bending failure presenting crushing of concrete in compression and yielding of the tensed reinforcements.

The T-Beam shows two different possible mechanisms of failure denoted by a bending failure with crushing of concrete and yielding of the tensed reinforcement and a shear failure caused by crushing of concrete close to support devices. According to Castaldo, Gino and Mancini (2019), the lognormal probabilistic distributions can be used to represent the structural resistance of the considered RC members. The characterization of the design structural resistance for the SL and T-Beam RC members has been evaluated in accordance with the methods and SFs described in this chapter. The design value of structural resistance has been estimated adopting the resistance model uncertainty safety factor $\gamma_{Rd}$ set to 1.00, whereas the target value of reliability index $\beta_t$ is set to 3.8 for structures with normal consequences of failure for a working life of 50 years (EN 1990, 2013; *fib* Model Code 2010, 2013; ISO 2394, 2015). The sensitivity coefficient $\alpha_R$ is set to 0.8 according to the assumption of the dominant variable associated with structural resistance (*fib* Model Code 2010, 2013). Table 11.2 lists the assumptions adopted to apply the SFs. Figure 11.5 shows, for members SL and T-Beam, the cumulative distribution functions (CDFs) estimated according to probabilistic analysis. These CDFs derived from the PM are adopted as reference curves and the design values of structural resistance obtained by the other SFs based on the GRF are marked and compared. The values of the achieved reliability index $\beta$ are also reported on the ordinates axis and compared with

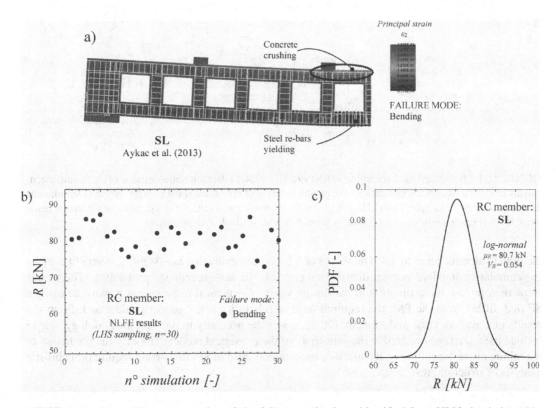

**FIGURE 11.3** Beam SL: representation of the failure mechanisms identified from NLN simulations (a); results in terms of structural resistance for the NLN simulations deriving from the 30 samples from LHS (b); probabilistic distribution suitable to represent the structural resistance R (c). (Figure modified from Castaldo, Gino & Mancini, 2019, with permission.)

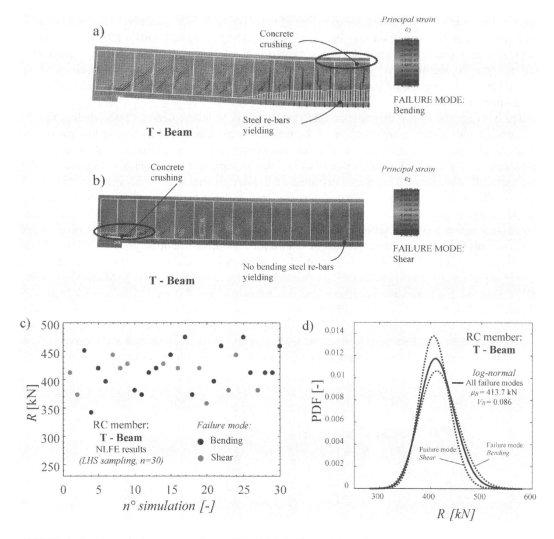

**FIGURE 11.4** Beam designed according to EN1992-1-1 (2014) (T-Beam): representation of the failure mechanisms identified from NLN simulations (a); results in terms of structural resistance for the NLN simulations deriving from the 30 samples from LHS (b); probabilistic distribution suitable to represent the structural resistance R (c). (Figure modified from Castaldo, Gino & Mancini, 2019, with permission.)

the target value set equal to 3.8. The values of β have been evaluated as $-\Phi(p)/\alpha_R$, where $\Phi$ denotes the cumulative standard normal distribution and $p$ is the not-exceedance probability. The results show that, in case the estimation of the design value of structural resistance is performed using a SF that differs from the PM, the required level of reliability is not properly addressed. From the results of Castaldo, Gino and Mancini (2019), arises the necessity to define a methodology able to include the uncertainty related to the sensitivity of the numerical model to modify the prediction of the failure mode depending on the aleatory uncertainties also in SFs that do not require probabilistic analysis of structural resistance.

## 11.3.2 Sensitivity of the Numerical Model to Aleatory Uncertainties within the GRF

With the scope to use for practical cases, the SFs defined according to the GRF, Castaldo, Gino and Mancini (2019) propose a methodology able to identify the mentioned above sensitivity of the

## TABLE 11.2

## Assumptions for Application of the Safety Formats in Castaldo, Gino and Mancini (2019)

| Safety Format | $R_{NLNA}(X_{rep};a_{rep})$ | $\gamma_R$ [-] | $\gamma_{Rd}$ [-] |
|---|---|---|---|
| PFM | $R_{NLNA}(f_{cd}f_{yd};a_{nom})^a$ | 1.00 | 1.00 |
| ECOV | $R_{NLNA}(f_{cm}f_{ym};a_{nom})$ | $\exp(\alpha_R\beta V_R)^b$ | |
| GRF | $R_{NLNA}(f_{cmd}{}^c f_{ym};a_{nom})$ | 1.27 | |
| GSF | $R_{NLNA}(f_{cm}f_{ym};a_{nom})$ | $\exp(\alpha_R\beta V_R)^d$ | |
| PM | $R_{NLNA,m}{}^d$ | $\exp(\alpha_R\beta V_R)^d$ | |

a   The partial safety factors used to calculate $f_{cd}$ and $f_{yd}$ are respectively set equal to 1.5 and 1.15 according to EN1992-1-1 (2014), while the design value of geometrical properties, $a_d$, is set equal to the nominal one $a_{nom}$ that corresponds to the size of the experimental setup.

b   Coefficient of variation of global resistance $V_R$ is estimated with a simplified approach according to the hypothesis of lognormal distribution performing two NLNAs using mean ($R_{NLNA}(f_{cm}f_{ym};a_{nom})$) and characteristic values ($R_{NLNA}(f_{ck}f_{yk};a_{nom})$) of material properties. The geometrical properties are set equal to the nominal one $a_{nom}$ that corresponds to the size of the experimental setup.

c   The representative value for concrete compressive strength is evaluated as: $f_{cmd} = 0.85 f_{ck}$.

d   The mean value $R_{NLNA,m}$ and coefficient of variation $V_R$ of structural resistance can be estimated by performing a reduced Monte Carlo simulation such as the Latin hypercube sampling with at least 30 samples. Lognormal distribution has been assumed to represent the structural resistance.

numerical model. In particular, two preliminary NLNAs are suggested with the scope to understand if the PFM and GRMs are able to provide results consistent with the more accurate PM. The two mentioned above preliminary numerical simulations can be defined as follows:

1. The first NLN analysis may be performed adopting the mean concrete properties and the design reinforcement properties.
2. The second NLN analysis may be performed adopting the design concrete properties and the mean reinforcement properties.

The design values associated with material characteristics should be evaluated according to EN 1990 (2013), EN1992-1-1 (2014), *fib* Model Code 2010 (2013), and *fib* Bulletin 80 (2016) with reference to

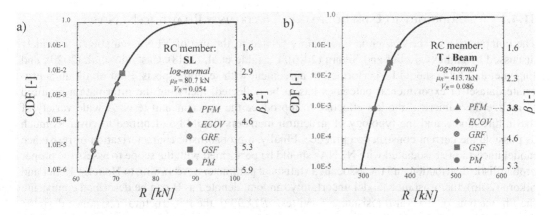

**FIGURE 11.5**   Representation of probability plot of the design value of structural resistance associated with the several safety formats: Beam SL (a); T-Beam (b). ($\gamma_{Rd}$ = 1.00; $\beta_t$ = 3.8; $\alpha_R$ = 0.8). (Figure modified from Castaldo, Gino & Mancini, 2019, with permission.)

the preferred target safety level. In case the results in terms of failure mode (i.e., characterized by the region of the structure where the local failure occurs and by the material that governs the failure mechanism) of the two preliminary NLN simulations are different, the numerical model is sensitive to modify the prediction of the failure mode depending on the aleatory uncertainties. In this circumstance, this additional source of uncertainty, which is not included in the partial safety factors $\gamma_R$ and $\gamma_{Rd}$, should be accounted for when methods different from the PM (i.e., PFM and GRMs) are adopted to compute the design structural resistance. In particular, a further failure mode-based safety factor denoted as $\gamma_{FM}$ is suggested and applied according to the following expression:

$$R_d = \frac{R_{rep}}{\gamma_R \cdot \gamma_{FM} \cdot \gamma_{Rd}} \qquad (11.10)$$

This safety factor $\gamma_{FM}$ has been defined in order to meet the outcomes of $R_d$ computed with the PFM and GRMs in line with the values of $R_d$ estimated with the PM. The results of Castaldo, Gino and Mancini (2019) lead to values of $\gamma_{FM}$ ranging between 1.00 and 1.18. The value of 1.15 is proposed for $\gamma_{FM}$ adopting the hypotheses of $\beta_t = 3.8$ with $\alpha_R = 0.8$, normal consequences of failure, working life of 50 years.

## 11.4    EPISTEMIC UNCERTAINTIES ACCORDING TO THE SAFETY FORMATS FOR NLNAs

The numerical models used for NLNAs with the prediction of structural behavior are merely estimations of the actual structural response. In general, the structural numerical model definition grounds on physical principles associated with constitutive relationships, equilibrium of forces, and kinematic compatibility that, all together, denote a specific modeling hypothesis. Specifically, focusing on NLNAs, the mentioned above basic principles are met by means of iterative solution methods (e.g., Newton-Raphson algorithm, arch-length algorithm) that lead, inevitably, to an approximation of the exact solution for the structural response. Then, the multiplicity of choices that may be performed defining a NLN model induces to an increasing level of uncertainty having an epistemic nature. It implies that, in accordance with the methodologies proposed by the methods and SFs for NLNAs described in this chapter, the reliability-based evaluation of the modeling uncertainty safety factor associated with structural resistance $\gamma_{Rd}$ is also an important topic for future codes implementation.

### 11.4.1   METHODOLOGY TO ESTIMATE EPISTEMIC UNCERTAINTY RELATED TO NLNAs

The computation of the epistemic uncertainty related to the use of NLN simulations, as widely discussed by Holický, Retief and Sikora (2016), Castaldo et al. (2018), Castaldo et al. (2020), and Engen et al. (2017) should be performed in agreement with several aspects. First of all, an appropriate dataset of experimental outcomes has to be collected, including the information required to realize the numerical models effective to reproduce the experimental tests. A wide variety of the failure modes and the typology of structural members should be identified to cover as much as possible the current construction practice. Finally, a probabilistic characterization of resistance model uncertainty associated with NLNAs should be performed with the scope to assess the proper probabilistic distribution and the associated statistical parameters. According to Holický, Retief and Sikora (2016), the resistance model uncertainty random, denoted as $\theta$, can be described comparing the $i$th outcome of structural resistance from tests $R_i(X,Y)$ to the $i$th structural resistance deriving from NLNA $R_{NLNA,i}(X)$ as follows:

$$R_i(X,Y) \approx \vartheta_i R_{NLNA,i}(X) \qquad (11.11)$$

where $X$ collects the variables explicitly included in the resistance model definition, $Y$ collects variables that relate to the resistance mechanism but are disregarded by the numerical model. In order to estimate the partial safety factor $\gamma_{Rd}$, the following steps can be followed, as suggested by Castaldo et al. (2018) and Castaldo et al. (2020):

1. *Selection of the benchmark experimental tests*: the collection of the benchmark set of experimental results should be performed collecting members having different nature, geometry, and different failure modes.
2. *Differentiation between modeling hypotheses*: the plausible modeling hypotheses available to engineers and practitioners able to simulate specific RC structures by means of NLNAs should be involved in the calibration procedure.
3. *Probabilistic calibration (Bayesian approach)*: accounting for the differentiation between modeling hypotheses, the probabilistic distribution devoted to represent the random variable $\vartheta$ should be characterized and the mean value $\mu_\vartheta$ and the variance $\sigma^2_\vartheta$ evaluated. The processing of the model uncertainties can be done in accordance with the Bayes' theorem grounding on the assumption of equiprobable modeling hypotheses (Castaldo et al., 2018, 2020).
4. *Characterization of the partial safety factor $\gamma_{Rd}$*: grounding on lognormality hypothesis for the random variable representing the model uncertainty $\vartheta$, the corresponding partial safety factor $\gamma_{Rd}$ is determined in accordance with the following expression:

$$\gamma_{Rd} = \frac{1}{\mu_\vartheta \exp(-\alpha_R \beta_t V_\vartheta)} \tag{11.12}$$

where $\mu_\vartheta$ denotes the mean value of the resistance model uncertainty $\vartheta$; $V_\vartheta$ denotes the CoV of the variable $\vartheta$ computed as $\sigma_\vartheta/\mu_\vartheta$; $\alpha_R$ is set to 0.32 as reported by *fib* Model Code 2010 (2013) for non-leading random variables; $\beta_t$ denotes the target value of the reliability index.

### 11.4.1.1 Partial Safety Factor $\gamma_{Rd}$ for Uncertainties in Definition of Numerical Model

The methodology so far described to characterize the partial safety factor $\gamma_{Rd}$ associated with uncertainty in the definition of the numerical model has been established by Castaldo et al. (2018) and Castaldo et al. (2020) in order to quantify the epistemic uncertainties corresponding to NLNAs in case of incremental monotonic and cyclic loading processes. The main outcomes related to the mentioned investigations are present in Table 11.3.

In the literature, also other studies have been devoted to the characterization of the resistance model uncertainty random variable $\vartheta$ with reference to NLNAs of RC structures. In particular, the results of Cervenka, Cervenka and Kadlek (2018) and Engen et al. (2017) leadto the characterization of the resistance model uncertainty random variable $\vartheta$ distinguishing between different failure mechanisms. Table 11.3 also reports the main results of these other investigations. The partial safety factors $\gamma_{Rd}$ that are listed in Table 11.3 are calculated from the statistical parameters according to Eq. (11.12) and are related to the value of 3.8 in terms of target reliability index $\beta_t$ for RC structure having 50 years of working life with the assumption of non-dominant resistance variable (i.e., $\alpha_R = 0.32$).

## 11.5  CONCLUSIONS

The present chapter reports the description and the discussion of the methodologies proposed by literature to perform safety verifications using NLN simulations. These methodologies, denoted as SFs, are able to include, more or less explicitly, the influence of both aleatory (i.e., material and geometric) and epistemic (i.e., missing knowledge, simplifications, and approximations related to the definition of the numerical model) uncertainties. A comparison between different SFs has been reported highlighting the necessity of a further partial safety factor, denoted as failure mode-based partial safety factor $\gamma_{FM}$, which is able to consider the uncertainty associated with the sensitivity

**TABLE 11.3**

**Characterization of Model Uncertainty Random Variable $\vartheta$ from Different Literature References; the Resistance Model Uncertainty Partial Safety Factor $\gamma_{Rd}$ Is Calculated Adopting Target Reliability Index $\beta_t$ Set Equal to 3.8 for RC Structures Having 50 Years of Reference Period with the Assumption of Non-Dominant Resistance Variable (i.e., $\alpha_R = 0.32$)**

| Ref. | Type of NLNA | Failure Mode | Resistance Model Uncertainty Random Variable $\vartheta$ | | | Resistance Model Uncertainty Partial Safety Factor |
|---|---|---|---|---|---|---|
| | | | Probabilistic Distribution | Mean Value $\mu_\vartheta$ | Coefficient of Variation $V_\vartheta$ | $\gamma_{Rd}$ |
| Castaldo et al. (2018) | Plane stress non-linear finite element analysis, incremental monotonic loading | Several failure modes inclusive of concrete crushing and reinforcement yielding | Lognormal | 1.01 | 0.12 | 1.15 |
| Castaldo et al. (2020) | Plane stress non-linear finite element analysis, cyclic loading | Several failure modes inclusive of concrete crushing and reinforcement yielding | Lognormal | 0.88 | 0.13 | 1.35 |
| Cervenka, Cervenka and Kadlek (2018) | 3D non-linear finite element analysis, incremental monotonic loading | Punching | Lognormal | 0.97 | 0.08 | 1.16 |
| | | Shear | | 0.98 | 0.07 | 1.13 |
| | | Bending | | 1.07 | 0.05 | 1.01 |
| | | All | | 0.98 | 0.08 | 1.16 |
| Engen et al. (2017) | 3D non-linear finite element analysis, incremental monotonic loading | Ductile (significant reinforcements yielding) | Lognormal | 1.04 | 0.05 | 1.02 |
| | | Brittle (concrete crushing without reinforcements yielding) | | 1.14 | 0.12 | 1.02 |
| | | All | | 1.10 | 0.11 | 1.04 |

of the numerical structural model in the prediction of the failure mechanism depending on aleatory uncertainty. The sensitivity of the numerical model is suggested to be checked through two preliminary simulations. If the numerical model turns out to be sensitive, a failure mode-based partial safety factor $\gamma_{FM}$ set to 1.15 is proposed to be included for evaluation of design value of global structural resistance.

Moreover, the general methodology to estimate epistemic uncertainties and partial safety factor $\gamma_{Rd}$ related to uncertainty in the numerical model definition is also described. Finally, appropriate values of $\gamma_{Rd}$ have been proposed grounding on several literature results.

## REFERENCES

Allaix DL, Carbone VI, Mancini G. Global safety format for non-linear analysis of reinforced concrete structures. Structural Concrete 2013; 14(1): 29–42.

ATENA 2D v5. Cervenka Consulting s.r.o.. Prague. Czech Republic. 2014.

Aykac B, Kalkan I, Aykac S, Egriboz EM. Flexural behaviour of RC beams with regular square or circular web openings. Engineering Structures 2013; 56: 2165–2174.

Castaldo P, Gino D, Bertagnoli G, Mancini G. Partial safety factor for resistance model uncertainties in 2D non-linear finite element analysis of reinforced concrete structures. Engineering Structures 2018; 176: 746–762.

Castaldo P, Gino D, Bertagnoli G, Mancini G. Resistance model uncertainty in non-linear finite element analyses of cyclically loaded reinforced concrete systems. Engineering Structures 2020; 211(2020): 110496, https://doi.org/10.1016/j.engstruct.2020.110496

Castaldo P, Gino D, Mancini G. Safety formats for non-linear analysis of reinforced concrete structures: discussion, comparison and proposals. Engineering Structures 2019; 193: 136–153, https://doi.org/10.1016/j.engstruct.2018.09.041.

Cervenka V, Cervenka J, Kadlek L. Model uncertainties in numerical simulations of reinforced concrete structures. Structural Concrete 2018; 19: 2004–2016.

Der Kiureghian A, Ditlevsen O. Aleatory or epistemic? Does it matter? Structural Safety 2009; 31: 105–112.

EN 1990. CEN. EN 1990: Eurocode – Basis of structural design. CEN 2013. Brussels.

EN1992-1-1. CEN EN 1992-1-1: Eurocode 2 – Design of concrete structures. Part 1-1: general rules and rules for buildings. CEN 2014. Brussels.

EN1992-2. CEN EN 1992-2 Eurocode 2 – Design of concrete structures, Part 2: concrete bridges. CEN 2005. Brussels.

Engen M, Hendriks MAN, Köhler J, Øverli JA, Åldtstedt E. A quantification of modelling uncertainty for non-linear finite element analysis of large concrete structures. Structural Safety 2017; 64: 1–8.

fib Bulletin 80. Partial factor methods for existing concrete structures, Lausanne, Switzerland; 2016.

fib Bulletin N°45. Practitioner's guide to finite element modelling of reinforced concrete structures – State of the art report. Lausanne; 2008.

fib Model Code 2010. fib Model Code for Concrete Structures 2010. fib 2013. Lausanne.

Holický M, Retief JV, Sikora M. Assessment of model uncertainties for structural resistance. Probabilistic Engineering Mechanics 2016; 45: 188–197.

ISO 2394. General principles on reliability for structures. Genéve. 2015.

JCSS. JCSS Probabilistic Model Code. 2001.

Mckey MD, Conover WJ, Beckman RJ. A comparison of three methods for selecting values of input variables in the analysis from a computer code. Technometrics 1979; 21: 239–245.

# 12 Risk-Based Seismic Assessment of Curved Damper Truss Moment Frame

*S. F. Fathizadeh and S. Dehghani*
Shiraz University, Shiraz, Iran

*T. Y. Yang*
The University of British Columbia, Vancouver, Canada

*E. Noroozinejad Farsangi*
Graduate University of Advanced Technology, Kerman, Iran

*Mohammad Noori*
California Polytechnic State University, San Luis Obispo, California, USA

*P. Gardoni*
University of Illinois at Urbana-Champaign, Urbana, Illinois, USA

*C. Malaga-Chuquitaype*
Imperial College London, London, UK

*A. R. Vosoughi*
Shiraz University, Shiraz, Iran

## CONTENTS

DOI: 10.1201/9781003194613-12

## 12.1    INTRODUCTION

The 1989 Loma Prieta and 1994 Northridge earthquakes showed that the structures designed to just satisfy the life safety (LS) performance objective could undergo a large amount of economic losses [1]. The Pacific Earthquake Engineering Research Center (PEER) introduced the performance-based earthquake engineering (PBEE) framework to design structures to meet various performance goals by considering the monetary losses, downtime, and casualty rate [2, 3]. PBEE methodology is performed using nonlinear dynamic analysis under a wide range of earthquake ground motion records considering various uncertainty sources such as seismic hazard, structural modeling, structural responses, and component performance [4–6].

Cornell et al. [7] introduced a probabilistic basis to seismically design and assess the steel moment-resisting structures. In this approach, demand and capacity concepts were used to quantify performance levels [7–9]. Yang et al. [10] proposed a procedure to implement the PBEE methodology by considering the seismic hazard, structural response, damage, and repair costs. Bai et al. [11] related fragility estimates with probabilities of being in different damage states. Eads et al. [12] investigated the impacts of various factors in the calculation of the annual frequency of collapse. Pragalath et al. (2014) investigated the reliability evaluation of four-story concrete structure through two methods of a series of nonlinear dynamic analyses and incremental dynamic analysis (IDA). Comparison of reliability indices demonstrated that both methodologies lead to almost the same results [13]. Bai et al. [14] defined story-specific demand models and estimated the fragility of the multi-story building by system reliability. More recently, Xu and Gardoni [15] developed probabilistic capacity and seismic demand models and fragility estimates considering three-dimensional analyses. The work showed how traditional two-dimensional analyses are generally inaccurate [15].

In recent years, knee-braced frames (KBFs) have been extensively developed and investigated experimentally and numerically by many researchers. Leelataviwat et al. [16] seismically designed KBFs and assessed their performance. A new buckling-restrained knee bracing (BRKB) system was introduced by Shin et al. [17]. In this system, steel channel sections were utilized to restrict the central steel plate of BRKB [17]. Yang et al. [18] and Wongpakdee et al. [19] introduced the buckling restraint knee-braced truss moment frame (BRKBTMF) that employs the buckling-restrained braces (BRBs) to absorb the seismic energy, whereas the remaining members are designed to behave elastically. BRKBTMF has the advantage of spanning a long-distance interior spaces [18, 19]. Doung [20] designed KBFs using BRBs and single-plate shear connections (SPSCs). The effectiveness of the BRKBs as a recent structural system was experimentally and analytically demonstrated by Junda et al. [21]. Dehghani et al. [22] developed a novel seismic structural system called curved damper truss moment frame (CDTMF) system using a two-phased energy dissipation mechanism including the curved dampers (CDs) and semi-rigid connections as primary and secondary structural fuses. Fathizadeh et al. [23] optimally designed the recently developed CDTMF system using the NSGA-II (non-dominated sorting genetic algorithm-II) technique. Interstory drift and floor acceleration were considered as objective functions to minimize the structural and non-structural damage, respectively [23].

In this research, the seismic risk assessment of a new force-resisting CDTMF system is investigated. The CDTMF system consists of the CDs as the primary fuses the semi-rigid connections as secondary fuses. More specifically, in lower seismic intensities, the CDs are designed to become inelastic and absorb seismic forces, whereas the structural members and semi-rigid connections are designed to be elastic. In high earthquake intensity, the semi-rigid connections (secondary fuses) along with CDs are set to yield and absorb energy, whereas the remaining members are designed to remain elastic [22, 23]. In order to achieve this goal, the novel design approach named equivalent energy design procedure (EEDP) proposed by Yang et al. [24] is applied to design a nine-story CDTMF structure. The contribution of a secondary phase of energy dissipation has been recognized to mitigate against extreme demands [25]. A seismic-based risk assessment is conducted based on the methodology proposed by PEER [26]. According to this framework, the seismic fragility curve is multiplied with the seismic hazard curve to calculate the annual frequency of exceeding of structure in a certain damage state. The seismic fragility curve is derived using the IDA and the seismic hazard curves of Log Angeles and Seattle sites are extracted from the USGS (United States Geological Survey) website [27]. The results show that the probabilities of collapse for the nine-story CDTMF system at both locations satisfy the collapse prevention (CP) limit states.

## 12.2   ASSESSMENT OF SEISMIC RISK

According to the PEER-PBEE assessment framework, two important components make a contribution to the collapse risk. The first component is the seismic hazard curve that represents the mean annual frequency of exceedance of an intensity level at a specific site. The hazard curve is derived from the probabilistic seismic hazard analysis (PSHA). Another essential component is the fragility curve that is described as the conditional probability of attaining or exceeding a specified damage level for a given ground motion intensity [28, 29]. When an accurate reliability analysis cannot be performed as suggested in Gardoni et al. [28, 30], a fragility curve can be roughly approximated with the shape of a lognormal cumulative distribution function [8, 9, 31]. Mathematically, the collapse risk is assessed using the following Eq. (12.1) [12]:

$$\lambda_c = \int_0^\infty P(C|e).\left|\frac{d\lambda(e)}{d(e)}\right| d(e) \qquad (12.1)$$

where $P(C|e)$ is the collapse probability of structure conditioned on earthquake intensity $e$, $\dfrac{d\lambda(e)}{d(e)}$ is the slope of the seismic hazard curve, and $\lambda_c$ is the annual frequency of collapse.

It was recommended that the numerical integration methods are preferable to closed-form solution from computational cost and accuracy perspectives [32, 33]. Hence, the numerical approach introduced by Williams et al. [34] and Eads and Miranda (2012) [35] is used to solve Eq. (12.1) as follows (Eq. 12.2):

$$\lambda_c = \sum_{i=1}^\infty P(C|e_i).\left|\frac{d\lambda(e_i)}{d(e)}\right| \Delta e \qquad (12.2)$$

From Eq. (12.2), the fragility curve and hazard curve should be split into a number of equal intervals of small intensities. As it can be seen from Figure 12.1, each quantity on the $\lambda_c$ deaggregation curve is computed as the product of the collapse probability at a given intensity, the slope of hazard curve at the same intensity, and the width of that intensity range ($\Delta e$). The $\lambda_c$ is the summation of products for all intensity levels or the area under the deaggregation curve. In fact, the $\lambda_c$ deaggregation curve is used to show the contribution of different intensities to collapse. Finally, by describing the earthquake occurrence in time with Poisson distribution, the collapse probability in $t$ years is calculated using Eq. (12.3):

$$P_c \left(\text{in } t \text{ years}\right) = 1 - \exp\left(-\lambda_c t\right) \qquad (12.3)$$

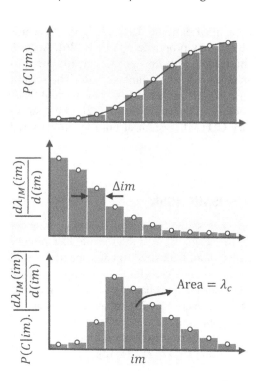

**FIGURE 12.1** The schematic of the mean annual frequency of collapse ($\lambda_c$) calculation.

## 12.3 CURVED DAMPER TRUSS MOMENT FRAME SYSTEM AND DESIGN PROCEDURE

The CDTMF system is an innovative seismic structural system introduced by Dehghani et al. [22] and Fathizadeh et al. [23]. The details of the CDTMF system are depicted in Figure 12.2. This system is a fused structural system that benefits from a two-phased energy dissipation mechanism. The CDs are the primary structural fuses and the semi-rigid connections play the role of the secondary fuses. In the CDTMF, the beam elements are constructed using the beam-truss assembly that allows the CDTMF to span a large interior space. The use of the beam-truss assembly also provides a high stiff floor that forces the deformation within CDs. Hence, making the yielding mechanism simple.

The seismic response of the CDTMF system is schematically drawn by an idealized trilinear force-deformation relationship in Figure 12.3 [18, 19]. The EEDP [24] is used to ensure the CDTMF system can reach different performance objectives at various hazard levels. According to EEDP, the CDTMF system is designed to behave elastically at the service-level earthquake (SLE), so the immediate occupancy (IO) performance level is met. At the design-based earthquake (DBE) level, the CDs are designed to act as the primary dissipating elements to dissipate the input energy, whereas the secondary fuses are set to be elastic. Being separated from the gravity load allows the CDs to be effectively examined and changed in the case of failure, hence the functionality of the structure is not disrupted and the CDTMF reaches the rapid return (RR) performance level at the DBE. This makes the CDTMF very resilient toward future earthquakes. At the maximum credible earthquakes (MCEs), the secondary fuses along with CDs yield and absorb the imposed energy by the earthquake, hence the CDTMF can achieve the CP performance level. Moreover, in the CDTMF system, the beams, columns, and truss members are capacity designed to behave elastically. Hence, these elements do not need to be repaired. The readers are recommended to refer to Yang et al. [24] and Dehghani et al. [22] for more information about the EEDP approach and the design process of the CDTMF system, respectively.

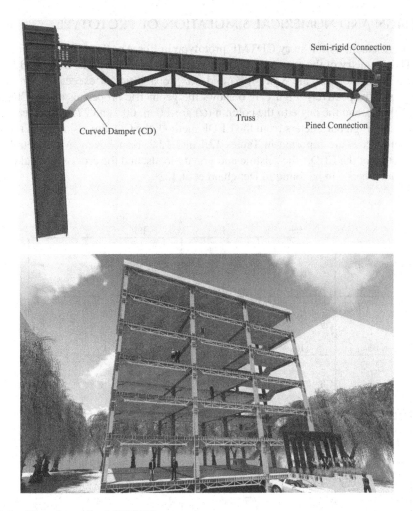

**FIGURE 12.2**   The schematic of CDTMF system.

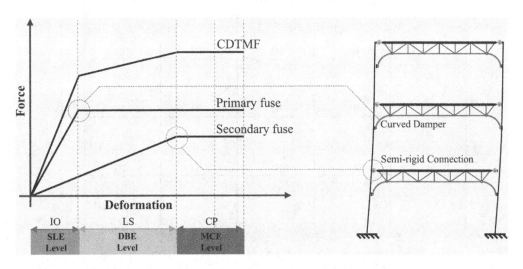

**FIGURE 12.3**   The idealized trilinear force-deformation relationship of the CDTMF system. (Adapted from Dehghani et al. [22].)

## 12.4 DESIGN AND NUMERICAL SIMULATION OF PROTOTYPE

In the present research, a nine-story CDTMF prototype in Los Angeles is adopted from Dehghani et al. [22]. The geometry of the structure is shown in Figure 12.4. The SLE, DBE, and MCE hazard levels, shown in Figure 12.5, are 87%, 10%, and 2% probabilities of exceedance in 50 years for the prototype site, respectively. Figure 12.6 shows the geometric properties of the CDs where the length, its angle ($\theta$), and the angle to the column ($\alpha$) are 1.0 m, 60°, and 27°, respectively.

The obtained design parameters from the EEDP method, the CD dimensions, and the semi-rigid connections properties are reported in Tables 12.1 and 12.2, respectively. More details about the design parameters of the EEDP, the seismic and gravity loads, and the cross-sectional properties of the structural members can be found in Dehghani et al. [22].

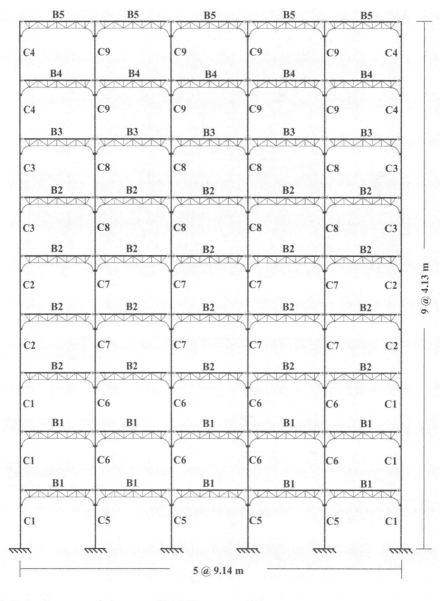

**FIGURE 12.4** Geometry of nine-story CDTMF structure [22].

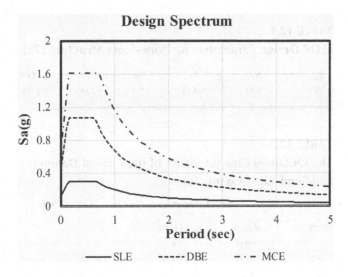

**Design Spectrum**

**FIGURE 12.5**  The selected hazard levels [41].

The OpenSees [36] is used to numerically simulate the prototype. The beams and columns are constructed with the *forceBeamColumn*. The truss members are modeled using the truss element. P-Δ effect is considered and the behavior of the steel is described with uniaxial *Giuffre-Menegotto-Pinto (Steel02)* material and strain hardening ratio of 2%. The *zero-length* element and *Steel02* material are applied to model the semi-rigid connections. The CDs were simulated with *forceBeam-Column* elements, *Giuffre-Menegotto-Pinto* material, and corotational coordinate transformation. Furthermore, the stress-strain behavior of the primary and secondary fuses is limited using the *Minmax* material in OpenSees. Detailed information about the verification of modeling accuracy is found in Dehghani et al. [22]. Figure 12.7 shows the obtained pushover curve and the trilinear curve for the nine-story prototype. As it can be seen, the drifts at which the CDs and semi-rigid connections yield ($\Delta_y$ and $\Delta_p$) matched the design assumptions of the EEDP tabulated in Table 12.1.

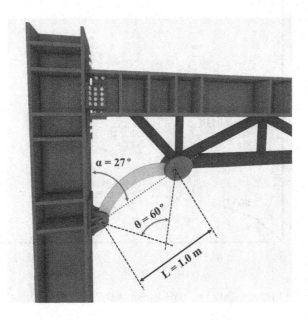

**FIGURE 12.6**  The geometric characteristics of the CDTMF system.

**TABLE 12.1**

**EEDP Design Parameters for Nine-Story Structure [22]**

| $\Delta_y$ | $\Delta_p$ | $\Delta_u$ | $C_0$ | $\gamma_a$ | $\gamma_b$ |
|------------|------------|------------|-------|------------|------------|
| 0.36%      | 1.10%      | 2.00%      | 1.4   | 2.00       | 2.05       |

**TABLE 12.2**

**The Obtained Characteristics of the Curved Dampers and Semi-Rigid Connections**

| | Curved Damper | | Connection | |
|-------|--------|--------|-------------|--------|
| Story | $d$ (cm) | $t$ (cm) | $\alpha^a$ | $\beta^b$ |
| 1 | 20.0 | 5.0 | 2.0 | 0.40 |
| 2 | 20.0 | 5.0 | 2.8 | 0.40 |
| 3 | 24.0 | 6.0 | 2.8 | 0.35 |
| 4 | 24.0 | 6.0 | 2.8 | 0.35 |
| 5 | 20.0 | 5.0 | 2.6 | 0.40 |
| 6 | 20.0 | 5.0 | 2.6 | 0.40 |
| 7 | 19.0 | 5.0 | 2.5 | 0.40 |
| 8 | 19.0 | 5.0 | 2.5 | 0.40 |
| 9 | 14.0 | 4.0 | 2.0 | 0.30 |

[a] The ratio of initial stiffness of connection to the elastic bending stiffness of the connected beam.

[b] The ratio of plastic-moment capacity of connection to plastic-moment capacity of connected beam.

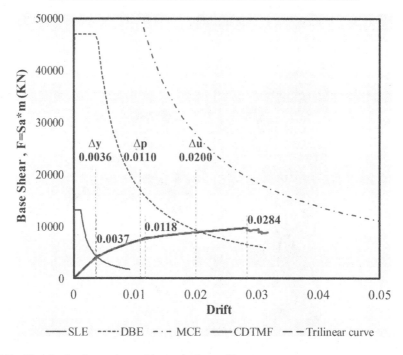

**FIGURE 12.7**  The obtained capacity and its equivalent trilinear curves.

## 12.5  RISK ASSESSMENT RESULTS

### 12.5.1  SEISMIC FRAGILITY CURVE

The IDA is performed to attain the seismic fragility curves of the nine-story prototype at different limit states in accordance with the recommendations presented in FEMA P695 [37]. The IDA is implemented considering 22 far-field earthquake records in FEMA P695 [37]. The intensity measure (IM) is 5%-damped spectral acceleration at the first natural period, Sa (T1, 5%). The record-to-record uncertainty is assumed to be 0.35 [37]. The maximum interstory drift ratios (ISDRs) of 0.7%, 2.5%, and 5% are considered as the limit states corresponding to IO, LS, and CP performance objectives based on FEMA 356 [38] and the 10% ISDR is set to be the criteria for collapse [37]. The obtained IDA curves and the seismic fragility curves of different limit states are drawn in Figure 12.8. Figure 12.8 shows that the median collapse intensity (S(CT)) is 1.68 g for the nine-story CDTMF prototype.

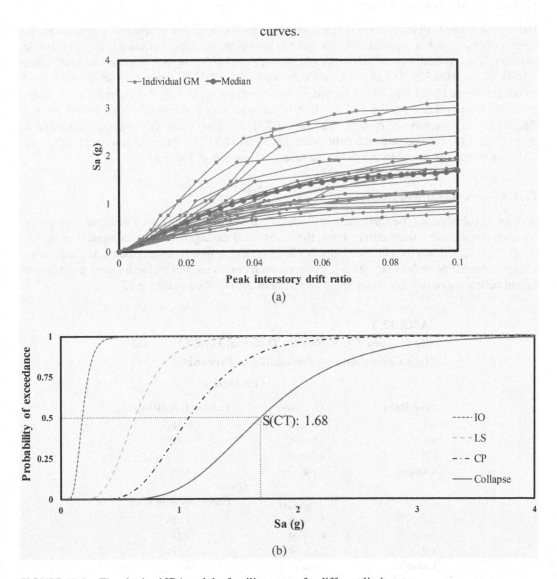

**FIGURE 12.8**  The obtained IDA and the fragility curves for different limit states.

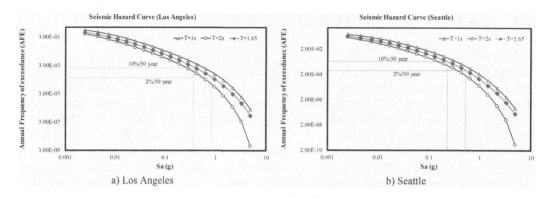

a) Los Angeles                                                   b) Seattle

**FIGURE 12.9**    The seismic hazard curves of considered sites. (a) Los Angeles. (b) Seattle.

## 12.5.2  Seismic Hazard Curve

The seismic hazard curves provide information about the mean annual frequency of exceedance of various intensity levels at structure sites. In order to investigate the impact of seismic hazard curve on seismic risk assessment, two different sites of Los Angeles (34.053 °N and −118.245 °W) and Seattle (47.604 °N and −122.329 °W) are selected for the nine-story CDTMF structure. The seismic hazard curves are adopted from the USGS Earthquake Hazards Program [27] for $T = 1$ s and $T = 2$ s. Then, it should be interpolated for $T = 1.65$ s, which is the structure's first natural period as it is seen in Figure 12.9. The structure site is categorized into the C/D boundary class. The intensities corresponding to 2% and 10% exceedance probabilities are 0.845 g and 0.377 g for Los Angeles and 0.527 g and 0.23 g for Seattle. These points are shown on hazard curves of both locations.

## 12.5.3  Risk-Based Assessment

In order to calculate $\lambda$, the *CurveExport Professional* software [39] is used to fit an appropriate curve on the seismic hazard curve. Then, the $\lambda$ different damage states are computed using Eq. (12.2). Table 12.3 shows the obtained values of $\lambda$ for different limit states and their equivalent probability of exceeding in 50 years. The $\lambda$ deaggregation curves as well as the fragility and seismic hazard curves for all damage states and both locations are depicted in Figure 12.10.

**TABLE 12.3**

**The Obtained $\lambda$ Values for Different Limit States and Their Corresponding Probability of Exceeding**

| | Los Angeles | |
| --- | --- | --- |
| Limit States | $\lambda$ (1/g/year) | Probability in 50 Years |
| IO | 1.40e-02 | 50.24% |
| LS | 1.95e-03 | 9.28% |
| CP | 6.45e-04 | 3.18% |
| Collapse | 2.10e-04 | 1.04% |
| | Seattle | |
| | $\lambda$ (1/g/year) | Probability in 50 Years |
| IO | 5.66e-03 | 24.64% |
| LS | 6.97e-04 | 3.43% |
| CP | 2.03e-04 | 1.01% |
| Collapse | 6.33e-05 | 0.32% |

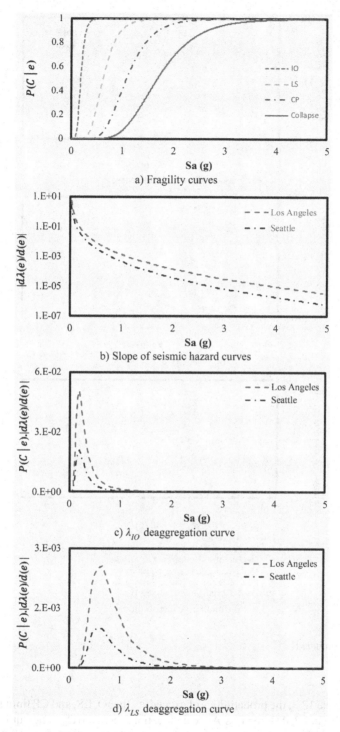

**FIGURE 12.10** The details of $\lambda$ deaggregation curves for different limit states. (a) Fragility curves. (b) Slope of seismic hazard curves. (c) $\lambda_{IO}$ deaggregation curve. (d) $\lambda_{LS}$ deaggregation curve. (e) $\lambda_{CP}$ deaggregation curve. (f) $\lambda_{CP}$ deaggregation curve. (g) Cumulative distribution to $\lambda_C$.

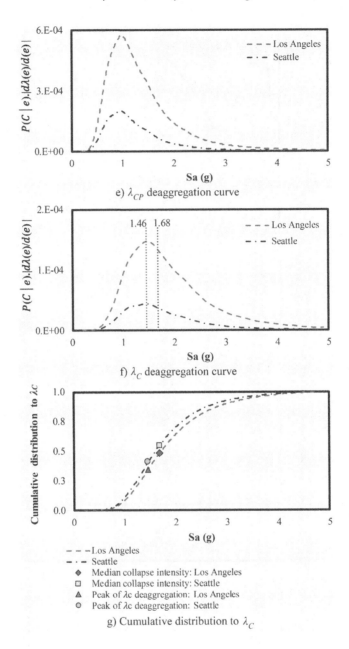

e) $\lambda_{CP}$ deaggregation curve

f) $\lambda_C$ deaggregation curve

g) Cumulative distribution to $\lambda_C$

**FIGURE 12.10**   (Continued)

According to Table 12.3, the probabilities of exceeding the IO, LS, and CP limit states in 50 years are 50.24%, 9.28%, and 3.18% for Los Angeles location, respectively. The equivalent figures for Seattle are 24.64%, 3.43%, and 1.01% for damage states related to IO, LS, and CP objectives in 50 years, respectively. Based on ASCE 7-16 [6], the probability of exceeding the MCE hazard level should be less than 10% and the nine-story CDTMF structure meets this criterion. The computed mean annual frequencies of collapse for Los Angeles and Seattle are $2.10 \times 10^{-4}$ and $6.33 \times 10^{-5}$ that correspond to 1.04% and 0.32% collapse probability in 50 years, respectively. The quantified values almost satisfy the 1% collapse probability in 50 years mentioned in ASCE 7-10 [40].

Considering that the fragility curve is assumed to be the same for both locations, the differences in probabilities are rooted in seismic hazard curves, meaning that Los Angeles is an area with higher seismicity than Seattle. It is obvious that Sa (T1) = 1.46 g makes the greatest contribution to $\lambda_C$ deaggregation curve for both locations, which corresponds to 34% probability of collapse in the fragility curve. The cumulative distribution of $\lambda_C$ is displayed in Figure 12.10. As it can be seen 48% and 55% of $\lambda_C$ for Los Angeles and Seattle come from earthquakes with intensities less than median collapse intensity (1.68 g), respectively. This result highlights the importance of weak ground motion intensities (located in the lower parts of the fragility curve) when it comes to collapse risk prediction, meaning that their higher frequencies of exceedance outweigh their small collapse probabilities.

## 12.6 SUMMARY AND CONCLUSION

The present research investigates the seismic performance of a recently developed resilient structural system named CDTMF in terms of seismic risk assessment. The CDTMF dissipates the earthquake energy via a two-phased energy dissipation mechanism containing the CDs (primary structural fuses) and semi-rigid connections (secondary fuses). A nine-story CDTMF prototype is considered and numerically modeled in OpenSees. The two-phased energy dissipation mechanism is achieved by employing the novel EEDP. At the SLE hazard level, the whole structure behaves elastically. At the DBE hazard level, the CDs are considered to yield and absorb the input energy, while the remaining structure is still elastic. The CDs are easily replaceable since they are decoupled from the gravity loads. At the MCE level, the semi-rigid connections begin to yield and absorb the energy, so the CDTMF does not collapse. The PEER framework to assess the seismic risk is used to seismically demonstrate the performance of the CDTMF. The IDA is implemented in order to derive the fragility curve and the seismic hazard curves are extracted for two locations of Los Angeles and Seattle. The obtained mean annual frequencies of exceeding various limit states and their corresponding probabilities show that Los Angeles is a site with higher seismicity than Seattle. The obtained probabilities of exceeding CP and collapse at both locations meet the structural codes criteria that demonstrate the desirable performance of the CDTMF system and capability of the EEDP approach to design fused structures. According to the cumulative distribution of the mean annual frequency of collapse, the intensity that makes the most significant contribution to the $\lambda_c$ deaggregation curve is less than the median collapse intensity, meaning that the weak earthquake intensities are as important as high-intensity ground motions due to their higher frequencies of exceedance.

## NOTES

1. Maximum yield displacement ratio.
2. Plastic roof drift ratio.
3. Ultimate roof drift ratio.
4. Displacement modification factor.
5. Energy modification factor.
6. Energy modification factor.

## REFERENCES

[1] L. Eads, "Seismic collapse risk assessment of buildings: effects of intensity measure selection and computational approach [Ph. D. dissertation]," ed, Stanford, CA: Department of Civil and Environmental Engineering, Stanford University, USA, 2013.
[2] H. Krawinkler and E. Miranda, "Performance-based earthquake engineering," *Earthquake Engineering: From Engineering Seismology to Performance-Based Engineering*, vol. 9, pp. 1–9, 2004.

[3] G. Deierlein, "Overview of a Comprehensive Framework for Earthquake Performance Assessment, PEER Rep. 2004/05," ed, Berkeley, CA, 2004.

[4] C. Molina Hutt, "Risk-based seismic performance assessment of existing tall steel framed buildings," UCL (University College London), 2017.

[5] H. Aslani, "Probabilistic earthquake loss estimation and loss disaggregation in buildings," Stanford University, USA, 2004.

[6] B. U. Gokkaya, "Seismic reliability assessment of structures incorporating modeling uncertainty and implications for seismic collapse safety," Stanford University, 2015.

[7] C. A. Cornell, F. Jalayer, R. O. Hamburger, and D. A. Foutch, "Probabilistic basis for 2000 SAC federal emergency management agency steel moment frame guidelines," *Journal of Structural Engineering*, vol. 128, no. 4, pp. 526–533, 2002.

[8] S. K. Ramamoorthy, P. Gardoni, and J. M. Bracci, "Probabilistic demand models and fragility curves for reinforced concrete frames," *Journal of Structural Engineering*, vol. 132, no. 10, pp. 1563–1572, 2006.

[9] S. K. Ramamoorthy, P. Gardoni, and J. M. Bracci, "Seismic fragility and confidence bounds for gravity load designed reinforced concrete frames of varying height," *Journal of Structural Engineering*, vol. 134, no. 4, pp. 639–650, 2008.

[10] T. Yang, J. Moehle, B. Stojadinovic, and A. Der Kiureghian, "Seismic performance evaluation of facilities: methodology and implementation," *Journal of Structural Engineering*, vol. 135, no. 10, pp. 1146–1154, 2009.

[11] J.-W. Bai, M. B. D. Hueste, and P. Gardoni, "Probabilistic assessment of structural damage due to earthquakes for buildings in Mid-America," *Journal of Structural Engineering*, vol. 135, no. 10, pp. 1155–1163, 2009.

[12] L. Eads, E. Miranda, H. Krawinkler, and D. G. Lignos, "An efficient method for estimating the collapse risk of structures in seismic regions," *Earthquake Engineering & Structural Dynamics*, vol. 42, no. 1, pp. 25–41, 2013.

[13] P. D. Haran, R. Davis, and P. Sarkar, "Reliability evaluation of RC frame by two major fragility analysis methods," 2015.

[14] J. -W. Bai, P. Gardoni, and M. B. D. Hueste, "Story-specific demand models and seismic fragility estimates for multi-story buildings," *Structural Safety*, vol. 33, no. 1, pp. 96–107, 2011.

[15] H. Xu and P. Gardoni, "Probabilistic capacity and seismic demand models and fragility estimates for reinforced concrete buildings based on three-dimensional analyses," *Engineering Structures*, vol. 112, pp. 200–214, 2016.

[16] S. Leelataviwat, B. Suksan, J. Srechai, and P. Warnitchai, "Seismic design and behavior of ductile knee-braced moment frames," *Journal of Structural Engineering*, vol. 137, no. 5, pp. 579–588, 2011.

[17] J. Shin, K. Lee, S. -H. Jeong, and H. -S. Lee, "Experimental Study on Buckling-Restrained Knee Brace with Steel Channel Sections," in *Proceedings of the 15th World Conference on Earthquake Engineering*, Lisbon, Portugal, 2012.

[18] T. Yang, Y. Li, and S. Leelataviwat, "Performance-based design and optimization of buckling restrained knee braced truss moment frame," *Journal of Performance of Constructed Facilities*, vol. 28, no. 6, p. A4014007, 2014.

[19] N. Wongpakdee, S. Leelataviwat, S. C. Goel, and W. -C. Liao, "Performance-based design and collapse evaluation of buckling restrained knee braced truss moment frames," *Engineering Structures*, vol. 60, pp. 23–31, 2014.

[20] M. P. Doung, "Seismic Collapse Evaluation of Buckling-Restrained Knee-Braced Frames with Single Plate Shear Connections," 2015.

[21] E. Junda, S. Leelataviwat, and P. Doung, "Cyclic testing and performance evaluation of buckling-restrained knee-braced frames," *Journal of Constructional Steel Research*, vol. 148, pp. 154–164, 2018.

[22] S. Dehghani *et al.*, "Performance evaluation of curved damper truss moment frames designed using equivalent energy design procedure," *Engineering Structures*, vol. 226, p. 111363, 2021.

[23] S. Fathizadeh *et al.*, "Trade-off Pareto optimum design of an innovative curved damper truss moment frame considering structural and non-structural objectives," *Structures*, vol. 28, pp. 2338–1353.

[24] T. Yang, D. P. Tung, and Y. Li, "Equivalent energy design procedure for earthquake resilient fused structures," *Earthquake Spectra*, vol. 34, no. 2, pp. 795–815, 2018.

[25] C. Málaga-Chuquitaype, A. Y. Elghazouli, and R. Enache, "Contribution of secondary frames to the mitigation of collapse in steel buildings subjected to extreme loads," *Structure and Infrastructure Engineering*, vol. 12, no. 1, pp. 45–60, 2016.

[26] C. Cornell and H. Krawinkler, "Progress and challenges in seismic performance assessment, PEER Center News 3," University of California, Berkeley, 2000.

[27] J. G. M. P. Calculator, "Version 5.1. 0, United States Geological Survey (USGS)," ed, 2011.

[28] P. Gardoni, A. Der Kiureghian, and K.M. Mosalam, "Probabilistic capacity models and fragility estimates for reinforced concrete columns based on experimental observations," *Journal of Engineering Mechanics*, vol. 128, no. 10, pp. 1024–1038, 2002.

[29] P. Gardoni, "Risk and reliability analysis," in *Risk and Reliability Analysis: Theory and Applications*: Springer, Germany, 2017, pp. 3–24.

[30] P. Gardoni, K.M. Mosalam, and A. Der Kiureghian, "Probabilistic seismic demand models and fragility estimates for RC bridges," *Journal of Earthquake Engineering*, vol. 7, no. spec01, pp. 79–106, 2003.

[31] D. Vamvatsikos and C. A. Cornell, "Incremental dynamic analysis," *Earthquake Engineering & Structural Dynamics*, vol. 31, no. 3, pp. 491–514, 2002.

[32] H. Aslani and E. Miranda, "Probability-based seismic response analysis," *Engineering Structures*, vol. 27, no. 8, pp. 1151–1163, 2005.

[33] B. A. Bradley and R.P. Dhakal, "Error estimation of closed-form solution for annual rate of structural collapse," *Earthquake Engineering & Structural Dynamics*, vol. 37, no. 15, pp. 1721–1737, 2008.

[34] R. J. Williams, P. Gardoni, and J. M. Bracci, "Decision analysis for seismic retrofit of structures," *Structural Safety*, vol. 31, no. 2, pp. 188–196, 2009.

[35] L. Eads and E. Miranda, "Seismic collapse risk assessment of buildings: effects of intensity measure selection and computational approach," Stanford University, California, 2013.

[36] F. McKenna and G. Fenves, "The OpenSees Command Language Manual, 1.2," ed, Pacific Earthquake Engineering Center, University of California, Berkeley ..., USA, 2001.

[37] FEMA-P695, "Quantification of building seismic performance factors," ed, Washington, DC, 2009.

[38] FEMA-356, "FEMA 356: Prestandard and Commentary for the Seismic Rehabilitation of Buildings, Report No," ed, FEMA, 2000.

[39] D. Hyams, "CurveExpert Professional. CurveExpert Software," ed, 2010.

[40] ASCE/SEI-7-10, *Minimum Design Loads for Buildings and Other Structures, Standard ASCE/SEI 7-10*. American Society of Civil Engineers, USA, 2013.

[41] FEMA-355C, "FEMA 355 State of the Art Report on Systems Performance of Steel Moment Frames Subject to Earthquake Ground Shaking," ed, FEMA, Washington, DC, 2000.

# 13 Risk-Target-Based Design of Base-Isolated Buildings

*Donatello Cardone, Amedeo Flora, Giuseppe Perrone, and Luciano R.S. Viggiani*
University of Basilicata, Italy

## CONTENTS

## 13.1 INTRODUCTION

Conventional earthquake-resistant structures are designed relying on their strength and ductility capacity. A completely different approach is pursued with seismic isolation, i.e., limiting the seismic effects that a structure undergoes during an earthquake. This result is achieved by introducing a horizontal disconnection between the superstructure and the foundation, where a suitable isolation system, characterized by an adequate deformability, is implemented, thus shifting the fundamental period of vibration of the structure beyond the dominant period of the earthquake. The isolation system is composed by a certain number of devices combined together to get the required behavior. Different solutions have been proposed and engineered in the last 30 years. Some of them resulted in several applications worldwide.

The isolation devices currently used can be divided into two main categories: (a) sliding isolators, based on the low frictional resistance that develops between different materials (e.g., stainless steel and PTFE [Polytetrafluoroethylene]) and (b) elastomeric devices, based on the high elastic deformability of rubber. High damping rubber bearings (HDRBs) are certainly the most widely used rubber-based isolation device. HDRBs exploit the mechanical properties of special rubber compounds with excellent damping features (higher than 10%–15%), vulcanized with steel shims to increase the vertical stiffness. It is worth noting that often, hybrid isolation systems obtained by combining HDRBs with flat sliding bearings (FSBs) are proposed, in order to further elongate the fundamental period of vibration of the building without penalizing the gravity load capacity of HDRBs.

DOI: 10.1201/9781003194613-13

Generally speaking, linear models are adopted in the technical practice for the design of the isolation system. However, nonlinear models are needed to capture the nonlinear behavior of the isolation devices and to perform accurate performance assessment of base-isolated buildings. In this optic, the main challenge of the numerical modeling is represented by the realistic simulation of the cyclic behavior of the isolation devices, beyond the design conditions, up to the collapse of the devices. Recently, a new model for HDRB, referred to as *Kikuchi bearing element* (Ishii and Kikuchi 2019), has been proposed and implemented in OpenSees. The main advantage of the *Kikuchi bearing element* is associated with its ability to capture the axial-shear load interaction for small and large displacements and, as a consequence, the associated pre- and post-buckling behavior.

Most of the current seismic codes allow engineers to design seismic-resistant structures (included base-isolated structures) with a certain amount of safety with respect to the onset of given limit states (such as serviceability, life safety, or collapse), for certain earthquake intensity levels, with a given return period. However, the effective risk toward given performance levels is not explicitly stated. Unknown values of failure probability, for instance, are then implicitly accepted (Spillatura 2018). Moreover, the resulting risk is not constant among different structural types and/or locations (Iervolino et al. 2018). In this context, the results of the RINTC (Rischio Implicito di strutture progettate secondo le Norme Tecniche per le Costruzioni) (implicit risk of code-conforming Italian buildings) research project, funded by the Italian Department of Civil Protection within the ReLuis/DPC 2015-2018 research program, have outlined that, although the compliance with the design requirements ensures higher performance in terms of protection of structural and non-structural elements from damage onset, a little margin toward global collapse (beyond the design earthquake intensity level) is found for code-conforming base-isolated buildings (Ragni et al. 2018). As a consequence, the failure rate of collapse for buildings with seismic isolation results very similar to (or even greater than) those for fixed-base structures (Iervolino et al. 2018).

Recently, a number of risk-target design procedures have been proposed for fixed-base buildings, in order to achieve a given uniform risk level for the structure (Luco et al. 2007; Gkimprixis et al. 2019) and an acceptable resilience objective. In this context, Luco et al. (2007) proposed a systematic design approach based on the use of theoretical (generic) collapse fragility functions and on the definition of location-specific risk factors. The risk factors are then defined as the ratio between the design ground motion level that guarantees an "acceptable" risk and the design ground motion level prescribed by the code (e.g., that exceeded with 10% probability in 50 years). Such procedures have been also implemented in advanced seismic codes (e.g., ASCE [American Society of Civil Engineers] 7-16). In the study by Luco et al (2007), a collapse rate of $2 \times 10^{-4}$ (i.e., 1% probability in 50 years) is assumed as an "acceptable" risk level.

So far, specific risk-target-based design (RTBD) approaches for base-isolated buildings are still missing. In this chapter, a RTBD approach for buildings with seismic isolation is presented, considering different sources of uncertainty (including record-to-record variability, modeling assumptions, and limit state definition). The chapter is divided into four parts. In the first part, the proposed RTBD approach for base-isolated buildings is described. In the second part, the failure modes and collapse conditions of reinforced concrete (RC) frame buildings with seismic isolators are examined. Then, a refined numerical model for the description of the nonlinear cyclic behavior of HDRBs is presented and calibrated against experimental results of type tests on commercial devices. The calibrated model is used to derive collapse fragility curves and risk index for a six-story base-isolated RC building designed by the Italian seismic code, based on results of comprehensive multi-stripe nonlinear time-history analyses (MSAs). In the last part of the paper, the RTBD approach is applied to the same case study building, assuming a target collapse rate of $2 \times 10^{-4}$. MSAs are then performed to validate the envisaged RTBD approach for base-isolated buildings. Finally, some preliminary practice-oriented design recommendations are drawn.

## 13.2 RISK-TARGET-BASED DESIGN OF BASE-ISOLATED BUILDINGS

The RTBD framework proposed for base-isolated buildings is inspired to previous works by Žižmond and Dolšek (2019) for fixed-base buildings. Obviously, some adjustments have been implemented to make the procedure by Žižmond and Dolšek (2019) suitable for base-isolated buildings.

The probability of collapse $\lambda_C$ can be expressed as (Gkimprixis et al. 2019):

$$\lambda_c = \int_0^{+\infty} P_f[\text{failure} \mid IM = x] |d\lambda_{IM}(x)| \tag{13.1}$$

where $\lambda_C$ is the mean annual frequency of collapse, briefly named as collapse risk, $\lambda_{IM}$ is the seismic hazard function expressed in terms of mean annual frequency of exceedance (MAFE), corresponding to the selected seismic intensity measure (IM), $P_f[\text{failure} \mid IM = x]$ represents the collapse fragility function. The latter is typically expressed by a lognormal cumulative distribution function governed by two parameters: the median intensity causing collapse ($S_{a,C}$) and the associated dispersion (standard deviation) ($\beta_C$):

$$P_f[\text{failure} \mid IM = x] = \Phi\left[(1/\beta_c)\ln\left(S_a(T)/S_{a,C}\right)\right] \tag{13.2}$$

Assuming that ground motions corresponding to $IM(TR) > IM(TR = 100{,}000 \text{ years})$ will certainly cause failure, the mean annual failure rate ($\lambda_c$) can be conservatively approximated as (Iervolino et al. 2018)

$$\lambda_c = \int_0^{100000} P_f[\text{failure} \mid IM = x] |d\lambda_{IM}(x)| + 10^{-5} \tag{13.3}$$

A step-by-step risk-targeted procedure is then needed. The procedure presented in the following includes five main steps. In the first step, the target (acceptable) collapse risk $\lambda_{C,T}$ is defined. Herein, a target collapse rate of $2.0 \times 10^{-4}$ has been assumed according to Luco et al. (2007). In the second step, the target median (50% probability of exceedance) collapse spectral acceleration corresponding to the fundamental period of the structure, $S_{a,c,T}$, is derived. For this purpose, some specific data or assumptions are required. First of all, the fundamental period of the structure ($T = T_{iso}$) should be set. Moreover, a proper value of the total dispersion ($\beta_{c,T}$) of the fragility curve should be assumed. The hazard curve of the building site is then needed. Assuming a linear representation of the seismic hazard function $H(S_a)$ in the log-space:

$$H(S_a) = k_0 S_a(T)^k \tag{13.4}$$

where $k$ is the slope of the hazard curve in the log-log domain, and $k_0$ is the annual rate of exceedance of $S_a(T)$ equal to 1 g. The probability of collapse, $\lambda_C$, can be computed as follows (Gkimprixis et al. 2019):

$$\lambda_C = k_0 S_{a,C}^{-k} e^{\left(k^2 \beta_{c,T}^2/2\right)} \tag{13.5}$$

As a consequence:

$$S_{a,C,T} = \left(k_0/\lambda_{C,T}\right)^{1/k} e^{\left(k\beta_{c,T}^2/2\right)} \tag{13.6}$$

As shown in Gkimprixis et al. (2019), the best approximation is obtained using a linearization approach.

In order to account for different sources of uncertainty (due to record-to-record variability, $\beta_{rtr}$, modeling variability, $\beta_m$, and uncertainty in collapse definition, $\beta_{ls}$), a total standard deviation $\beta_{c,T}$ can be calculated using the square root of the sum of the squares (SRSS) rule:

$$\beta = \sqrt{\beta_{rtr}^2 + \beta_m^2 + \beta_{ls}^2} \qquad (13.7)$$

Some authors (Cardone et al. 2019a) showed that, for base-isolated buildings, a dispersion due to record-to-record variability between 0.2 and 0.3 can be reasonably assumed. Moreover, according to ATC-58 (ATC 2012), the modeling dispersion can be taken in the range 0.1–0.4, depending on the quality/completeness of the numerical model. Finally, a value of $\beta_{ls}$ equal to 0.2 has been proposed by Spillatura (2018) for the uncertainty related to the collapse definition. All that considered, a total dispersion ranging between 0.3 and 0.8 is obtained.

In modern seismic codes, the near-collapse (NC) limit state is considered for the safety verification toward collapse conditions. The transition between collapse and NC limit state can be expressed through a reduction factor $\gamma_{ls}$, which relates the median spectral acceleration corresponding to collapse, $S_{a,C,T}$, to the median spectral acceleration associated with the attainment of the NC limit state, $S_{a,NC,T}$, i.e.,

$$\gamma_{ls} = S_{a,C,T}/S_{a,NC,T} \qquad (13.8)$$

The value of $\gamma_{ls}$ significantly depends on the definition of the NC limit state. It is also influenced by the structural typology (Žižmond and Dolšek 2019). In the present study, the coefficient $\gamma_{ls}$ has been calculated in terms of displacement assuming a constant value of the equivalent viscous damping and a direct proportionality between spectral acceleration intensities and response displacements, passing from NC to collapse limit state. Considering different collapse failure modes for a hybrid isolation system (detailed are provided in the next paragraph), the following general formulation is proposed:

$$\gamma_{ls} = \min\left(t_e \gamma_u; D_{ult}; D_{buck}\right)/D_{cap} \qquad (13.9)$$

where $t_e \gamma_{max}$ represent the collapse displacement associated with the attainment of the ultimate shear strain ($\gamma_u$) for an HDRB device featuring a rubber height equal to $t_e$; $D_{buck}$ represents the collapse displacement corresponding to the buckling failure of the device; $D_{ult}$ is the ultimate displacement capacity of FSBs; $D_{cap}$ is the displacement capacity of the isolation system at NC limit state. For the sake of simplicity, herein the tensile failure of the HRDBs is not considered and the collapse of the superstructure is neglected.

Considering that at this step of the RTBD procedure, the isolation system is still unknown, a trial value of $\gamma_{ls}$ should be chosen as the first attempt. At the end of the procedure, once the isolation system has been effectively designed, a check on $\gamma_{ls}$ must be done and if needed a new analysis, with an adjusted value of $\gamma_{ls}$, be performed. First attempt values of $\gamma_{ls}$ have been derived, as a function of the prevalent failure mode of the isolation system, examining the results reported in Ragni et al. (2018), which refer to four case studies buildings equipped with hybrid isolation systems. In particular, $\gamma_{ls}$ equal to 1.75, 1.25, and 1.4 can be assumed in case of shear failure of HDRBs, buckling collapse, and attainment of the ultimate displacement capacity of FSBs, respectively. Once the transient coefficient $\gamma_{ls}$ has been defined, $S_{a,NC,T}$ can be derived as follows:

$$S_{a,NC,T} = S_{a,C,T}/\gamma_{ls} \qquad (13.10)$$

Generally speaking, it could be useful to define the risk-targeted safety factor coefficient (Žižmond and Dolšek 2019) relating the value $S_{a,NC,T}$ to the corresponding seismic demand defined by the traditional (code-based) uniform hazard map, for the associated limit state return period, $T_R$ ($S_{a,TR}$):

$$\gamma_{im} = \left( S_{a,NC,T} / S_{a,TR} \right) \geq 1 \tag{13.11}$$

In the European seismic code, a return period of 975 years is typically associated with the collapse prevention limit state, for ordinary buildings. In order to respect the code limitation at the selected limit state, a lower bound of $\gamma_{im}$ equal to 1 is proposed. The following relationship can be then obtained by combining Eq. (13.10) with Eq. (13.11):

$$S_{a,NC,T} = S_{a,C,T} / \gamma_{ls} = \gamma_{im} S_{a,TR} \tag{13.12}$$

A proper reduction factor $r_{NC}$ is used to derive the design risk-targeted spectral acceleration $S_{a,D,T}$ from the 5%-damping elastic counterpart ($S_{a,NC,T}$). For fixed-base buildings, the reduction factor depends on the available ductility and overstrength ratio of the structure (Žižmond and Dolšek 2019). For base-isolated buildings, $r_{NC}$ can be assumed equal to $1/\eta$, where $\eta$ is the damping reduction factor of the base-isolated building:

$$S_{a,D,T} = S_{a,NC,T} / r_{NC} = S_{a,NC,T} \ \eta \tag{13.13}$$

According to the seismic Italian/Eurocode, the damping reduction factor is expressed by the following relationship:

$$\eta = \sqrt{10 / (5 + \xi)} \tag{13.14}$$

where $\xi$ is the effective damping ratio of the isolation system.

The design spectral displacement can be then evaluated as:

$$S_{d,D,T} = S_{a,D,T} / \omega^2 = S_{a,D,T} \left( T_{iso} / 2\pi \right)^2 \tag{13.15}$$

Once the fundamental period ($T_{iso}$) has been set (Step 1) and the maximum displacement at the NC limit state ($S_{d,D,T}$) has been obtained, the design of the isolation system can be done. Assuming a reasonable value for the superstructure mass ($M_{tot}$), the total stiffness of the isolation system is obtained. Next, a suitable isolation device can be identified by entering the manufacturer's catalog with the relevant displacement demand ($S_{d,D,T}$), the axial load capacity (from gravity load analysis), and the effective stiffness of the single device ($K_{iso}$). The latter can be evaluated as follows:

$$K_{iso} = \left( M_{tot} / N_{HDRB} \right) \left( 2\pi / T_{iso} \right)^2 \tag{13.16}$$

where $N_{HDRB}$ is the number of elastomeric isolators of the isolation system.

In the last step, the performances of the designed building are checked in accordance with the reference seismic code, by means of traditional structural analysis.

## 13.3 FAILURE MODES AND COLLAPSE CONDITIONS OF RC FRAME BUILDINGS WITH SEISMIC ISOLATORS

Generally speaking, three different failure modes can be identified for HDRBs, namely: cavitation, shear failure, and buckling. Cavitation occurs when tensile strain (in the post-elastic regime) exceeds a given limit due to the formation of small air bubbles in the rubber layers. Herein, the cavitation failure mode of a single device is deemed to occur in the nonlinear post-cavitation stage, for an axial tensile strain equal to 50%, in accordance with Keerthana et al. (2014). The buckling failure mode is attained when the device is no longer able to withstand gravitational loads. Herein, the collapse criterion associated with buckling has been expressed in terms of compressive strain; for any single device, buckling is supposed to occur when an axial compressive strain of 50% is reached (Monzon et al. 2016). Finally, shear failure occurs when the device reaches a limit shear strain corresponding to the occurrence of significant cracks in the rubber layers or the detachment between steel shims and rubber layers. Based on Nishi et al. (2019), a conventional limit shear strain equal to 350% has been assumed. It is worth remarking that the latter value is somehow a safety-side choice considering that most of HDRBs experience shear failure for shear strains higher than 400%.

As mentioned before, hybrid isolation systems obtained by combining HDRBs and FSBs can be adopted to further elongate the fundamental period of vibration of the building without penalizing the gravity load capacity of HDRBs. This choice has been made also in this study. For FSBs, two failure criteria have been considered: (i) the attainment of a critical uplift value corresponding to $H/2$, where $H$ is the total height of the device, and (ii) the attainment of a critical value of compression rate, equal to 60 MPa (CEN-EN 2000). As a matter of fact, indeed, as the horizontal displacement increases, the effective resisting area of the device reduces, and as a consequence the contact pressure increases. Numerical simulations (Cardone et al. 2019b) showed that the aforesaid compression rate limit threshold is usually reached for values of the horizontal displacement approximately equal to the design displacement capacity of the device ($d_{max,FSB}$) increased by an extra-displacement $\alpha\Phi_p$, with $\alpha \approx 1/2$. As a consequence, the ultimate displacement capacity of the FSBs has been conventionally assumed equal to $d_{max,FSB} + 1/2\Phi_p$.

The global collapse of the entire isolation system has been conventionally associated with the attainment of any failure mode in (at least) 50% of the devices. Since the lateral strength of the connections must be (at least) two times greater than the maximum shear force transmitted by the isolation device (NTC 2018), the connections have been considered as rugged (not vulnerable) elements.

Considering that the isolation system and the superstructure represent two elements of the same in-series system, in first approximation, it can be assumed that the global collapse of a base-isolated building occurs when either the superstructure or the isolation system fails. For what concerns the collapse of the superstructure, a simplified (practice-oriented) assumption has been made. The collapse of the superstructure is supposed to occur when a limit roof displacement, corresponding to a 50% drop of the maximum lateral strength (on the pushover curve), is attained. In other words, the superstructure collapse occurs when, in one of the two main directions, the maximum value of the roof displacement derived from dynamic analyses exceeds the aforementioned threshold value.

## 13.4 MODELING APPROACH

The structural model of the base-isolated building has been implemented in the OpenSees finite element program. The finite elements used for rubber isolators and RC members of the superstructure are described below.

### 13.4.1 RUBBER-BASED ISOLATION SYSTEM

The nonlinear behavior of HDRBs is characterized by many important features such as coupled motion in the horizontal directions, coupling between vertical and horizontal response, cavitation under tension, change of buckling load capacity due to lateral displacements, post-buckling effects, shear strength degradation due to cyclic loading (Mullin's effect), shear failure, etc.

The latest version of the *Kikuchi bearing element* (Ishii and Kikuchi 2019), recently implemented in OpenSees, has been chosen to capture the cyclic nonlinear behavior of HDRBs. The *Kikuchi bearing element* is a 3D model with a large displacement formulation that, differently to others, takes into

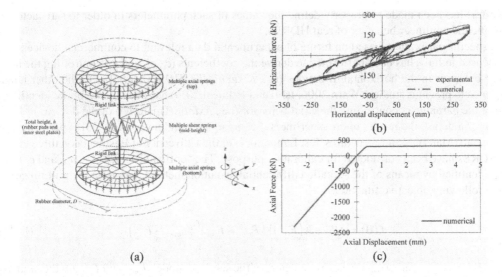

**FIGURE 13.1** (a) Multi-springs numerical model for circular elastomeric isolators (Ishii and Kikuchi 2019); (b) shear behavior; (c) axial behavior.

account the coupled behavior in the vertical and horizontal direction. In particular, it is able to capture the "*P-Δ*" effects (i.e., geometric nonlinearity) that imply a reduction of the horizontal stiffness due to an increase of vertical load and/or horizontal displacements. Such peculiarity appears very important especially for medium-high level of vertical pressure (Brandonisio et al. 2017).

The *Kikuchi bearing element* includes two sub-models: the *multi-shear spring model* (*MSS*) governing the shear bidirectional behavior and the *multi-normal spring model* (*MNS*) governing the Euler column buckling behavior.

The *KikuchiAikenHDR* and the *AxialSP* uniaxial materials, already implemented in OpenSees, have been used for the MSS and MNS models, respectively. The main parameters controlling the constitutive laws of the mentioned uniaxial materials are summarized in Table 13.1. In this study, a

## TABLE 13.1
## Kikuchi Model's Parameters

| Parameter | Controlled Properties | Calibration | Value |
|---|---|---|---|
| $a_r$ | Area of rubber | Geometry | $\pi D^2/4$ |
| $h_r$ | Total thickness of rubber | Geometry | $t_e$ |
| $c_g$ | Equivalent shear modulus | Experimental | 1.15 |
| $c_h$ | Equivalent viscous damping ratio | Experimental | 0.75 |
| $c_u$ | Ratio of shear force at zero displacement | Experimental | 0.75 |
| $s_{ce}$ | Compressive modulus | Ishii and Kikuchi 2019 | From eq. (9) of Ishii and Kikuchi (2019) |
| $f_{ty}$ | Cavitation stress | Warn (2006) | 2G |
| $f_{cy}$ | Compression yielding stress | Ishii and Kikuchi (2019) | 100 MPa |
| $b_{te}$ | Reduction rate for tensile elastic range | Warn (2006) | 1.00 |
| $b_{ty}$ | Post-cavitation stiffness | Warn (2006) | 0.43% |
| $b_{cy}$ | Reduction rate for compressive yielding | Ishii and Kikuchi (2019) | 0.5 |
| $f_{cr}$ | Target point stress | Warn (2006) | 0.00 MPa |

big effort has been made to properly define the values of such parameters in order to satisfactorily describe the nonlinear behavior of real HDRBs.

A specific calibration, based on fitting of experimental data relevant to commercial devices currently used in Italy, has been performed to derive the coefficients ($c_g$, $c_h$, and $c_u$) controlling the nonlinear behavior in the horizontal directions (i.e., *KikuchiAikenHDR* material). On the other hand, reference to previous studies (Warn 2006; Ishii and Kikuchi 2019) has been made for the identification of the parameters controlling the axial behavior (i.e., *Axial SP* material) of the device.

More in detail, the results of an experimental campaign (Brandonisio et al. 2017) performed at the SisLab (Materials and Structures Test Laboratory) of the University of Basilicata on three circular HDRBs have been considered for calibration purpose. The "optimal" values of $c_g$, $c_h$, and $c_u$ have been identified by means of the so-called differential evolution method, based on the minimization of the following objective function:

$$f(\theta) = \left(1/S \cdot \mathrm{var}\left(F_{exp}\right)\right)\left(F_{num} - F_{exp}\right)^T \left(F_{num} - F_{exp}\right) \tag{13.17}$$

where $\theta$ is a vector gathering the key parameters of the selected model, $F_{num}$ and $F_{exp}$ are the numerical and experimental force values, while $S$ indicates the total number of samples and var($F_{exp}$) is the variance of the experimental values of the force.

The cyclic behavior of FSBs has been described by the *FlatSliderBearing* element, which features a velocity-dependent and axial load-dependent friction material model (Constantinou et al. 1990). A friction coefficient (at the maximum load capacity) equal to 1%, for fast velocities, and 0.5%, for low velocities, has been assumed.

### 13.4.2 SUPERSTRUCTURE

A lumped plasticity model has been chosen for RC beams and columns. The Ibarra-Medina-Krawinkler (Ibarra et al. 2005) model has been used to define the flexural nonlinear cyclic behavior of the plastic hinges. For the short columns of the stairs (which are liable to premature shear failure), a fictitious ultimate rotation capacity has been defined. Masonry infill panels have been modeled with an equivalent compression-only strut element. The skeleton curves of the infills have been described by a modified version of the Decanini model. More details about the modeling strategy of the RC frame can be found in Ragni et al. (2018).

## 13.5  CASE STUDY BUILDING

A six-story RC frame building for residential use, located in L'Aquila (central Italy) on soil type C (medium-soft soil), has been selected as a case study building (see Figure 13.2). It was designed by the Italian seismic code (NTC 2018) through modal response spectrum analysis (MRSA). The building features a regular plan of 240 m² and a constant interstory height of 3.05 m (except for the ground level, which is 3.4 m). The building is characterized by four frames in the long (X-) direction and six frames in the short (Y-) direction (see Figure 13.2). The staircase features a knee beams structure. Cross-section dimensions and reinforcement ratios of beams/columns can be found in Ragni et al. (2018).

Masonry infills are regularly distributed (in plan and elevation). They are realized with hollow clay bricks with 300 mm thickness. Masonry infills are characterized by different percentages of openings in the two facades. An average cylindrical compressive strength of 28 MPa is assumed for concrete (concrete class C28/35). Similarly, a yield strength of 430 MPa is assumed for steel reinforcement (steel class B450C). The examined building is equipped with a hybrid isolation system composed by 16 HDRBs (equivalent viscous damping ratio equal to 15%), arranged below the perimeter columns of the building, and 8 FSBs (friction coefficient equal to 1%), arranged below

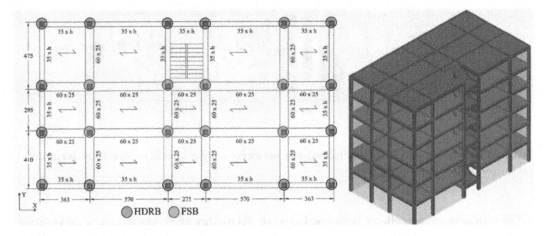

**FIGURE 13.2** Floor plan for new buildings.

**TABLE 13.2**

**Main Characteristics of the Isolation Systems Designed within the RINTC Project (First Line) and Following the Proposed RTBD Approach (Second Line) for the Same Case Study Building**

| Case Study | Isolation System | HDRB* | | | | | | FSB* | | | | |
| | | $G$ | $\xi$ | $V_2$ | $D$ | $t_e$ (mm) | $d_c$ | $V_2$ | $d_c$ | $\mu$ | $T_{iso}$ | $T_{iso}/T_{fb}$ |
|---|---|---|---|---|---|---|---|---|---|---|---|---|
| RINT | 16HDRB+8FS | 0.4 | 15% | 880 | 600 | 176 | 350 | 3500 | 350 | 1% | 3.04 | 3.27 |
| RTDB | 16HDRB+8FS | 0.4 | 15% | 820 | 650 | 207 | 400 | 3500 | 450 | 1% | 3.04 | 3.27 |

*Note:* * $G$: Dynamic shear modulus of rubber; $\xi$: Equivalent damping ratio; $V_2$: Maximum gravity load capacity at the displacement $d_c$; $d_c$: Maximum displacement capacity of HDRB/FSB; $D$: Diameter of HDRB; $t_e$: Total thickness of rubber layers; $\mu$: Friction coefficient of FSB at the maximum vertical load capacity.

the inner columns. The first line of Table 13.2 summarizes the main characteristics of the isolation system designed by the Italian seismic code (NTC 2018). The isolation ratio of the building (i.e., the ratio between the fundamental period of the base-isolated building, $T_{iso}$, and the period of the same building in the fixed-base configuration, $T_{fb}$) is reported.

## 13.6 SEISMIC RISK ASSESSMENT

### 13.6.1 Fragility Curves and Seismic Risk Index of the Code-Conforming Building

The seismic performance at the collapse of the case study building has been assessed by MSAs carried out considering 10 earthquake intensity levels and 20 ground motion pairs per stripe (Ragni et al. 2018). Probabilistic seismic hazard analysis (PSHA) has been carried out with OpenQuake (Iervolino et al. 2018) to derive the hazard curve at the building site as a function of the selected IM, corresponding to the spectral acceleration, $Sa(T^*)$, at the fundamental period of vibration of the building ($T \approx 3$ sec). Ten earthquake intensity levels, with return period equal to 10, 50, 100, 250, 500, 1000, 2500, 5000, 10,000, 100,000 years, respectively, have been considered. For each of them, 20 seismic ground motion pairs have been selected, considering suitable conditional mean spectra (CMS) for the building site. More details can be found in Iervolino et al. (2018).

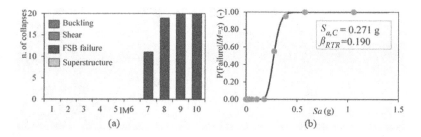

**FIGURE 13.3**  Code-conforming building model: (a) number of analysis cases in which collapse is registered and (b) associated collapse fragility curve.

For each analysis, the collapse is assumed to occur when either the superstructure or the isolation system fails. Figure 13.3 shows the number of analysis cases, for each IM, in which the collapse of the code-conforming base-isolated building occurs and the associated collapse fragility curve. As can be seen, the prevalent (unique) collapse mode is associated with the attainment of the ultimate displacement of FSBs. Integrating the product between seismic hazard and fragility curve, an annual failure rate, $\lambda_c$, equal to $5.91 \times 10^{-4}$ is found. A slight reduction of the annual failure rate, of the order of 20% (i.e., $\lambda_c = 4.72 \times 10^{-4}$), can be obtained by increasing a little (e.g., by 50 mm) the design displacement of FSBs compared to that of HDRBs. Therefore, despite the minimum requirements of the Italian seismic code to ensure an acceptable level of safety for the design seismic intensity (return period $\approx$ 1000 years), the resilience objective against global collapse (assumed equal to $2.0 \times 10^{-4}$) is far from being achieved.

### 13.6.2 Application of the Risk-Target-Based Design Approach

The RTBD approach appears to be the best solution to guarantee the desired resilience objective. Such approach has been applied to the selected case study building assuming a risk-targeted value ($\lambda_{C,T}$) equal to $2.0 \times 10^{-4}$ and a target isolation period ($T_{iso}$) equal to 3 sec.

The main steps of the design procedure are summarized in Table 13.3. For the sake of clarity, the following preliminary assumptions have been made:

- Step 1: $S_{a,C,T}$ is evaluated with Eq. (13.6) interpolating the hazard curve between 1000 and 10,000 years and assuming a total dispersion $\beta_{C,T}$ equal to 0.4.
- Step 2: $S_{a,NC,T}$ is evaluated with Eq. (13.10) assuming in first approximation $\gamma_{ls} = 1.75$; as a matter of fact, the collapse of the isolation system is dominated by the shear failure of HDRBs.
- Step 3: $S_{a,D,T}$ is evaluated with Eq. (13.13) assuming a damping ratio of 15%.

---

**TABLE 13.3**

**Application of the RTBD Approach to the Selected Case Study Building**

| Step 1 | Step 2 | Step 3 | Step 4 |
|---|---|---|---|
| $\lambda_{C,T} = 2 \times 10^{-4}$ | $\beta_{C,T} = 0.4$ | $\gamma_{ls} = 1.75$ | $\eta = 0.71$ |
| $T_{iso} = 3.0$ s | $k_0 = 3.63 \times 10^{-5}$ $k = 1.84$ | $\gamma_{im} = 1.48$ | $S_{a,D,T(1)} = 0.185$ g |
| - | $S_{a,C,T} = 0.46$ g | $S_{a,NC,T} = 0.26$ g | $S_{d,D,T(1)} = 0.412$ m |
| - | - | $\gamma_{ls} = 1.81$ | $S_{a,D,T(2)} = 0.179$ g |
| - | - | $S_{a,NC,T} = 0.26$ g | $S_{d,D,T(2)} = 0.399$ m |

---

The design displacement ($S_{d,D,T}$) at the end of the first iteration is equal to 412 mm (see Table 13.3). Entering the commercial catalog of one of the main Italian manufacturers with a nominal displacement capacity of 400 mm and an axial load capacity from gravity load analysis, the solution reported in the second line of Table 13.2 is found. Basically, the new isolation system features the same number of HDRBs but with larger dimensions (650 mm vs. 600 mm diameter, 207 mm vs. 176 mm total rubber thickness). The corresponding coefficient $\gamma_{ls}$ turns out to be equal to 1.81. Repeating Step 2 and Step 4 assuming $\gamma_{ls} = 1.81$ (second iteration), a new design displacement, $S_{d,D,T}$, equal to 399 mm is found. Therefore, the selected device appears adequate (399mm<400mm). In line with the current practice and according to the Italian seismic code requirements (NTC 2018), sliding bearings (FPBs) should feature a displacement capacity at least 20% greater than the design displacement derived from the analysis, in order to account for possible residual displacements that may jeopardize the ultimate displacement capacity of not-recentering isolation systems (Cardone et al. 2015).

As a consequence, the displacement capacity of FSBs has been set equal to 450 mm. The greater displacement capacity of HDRBs (i.e., 400 mm vs. 350 mm), with approximately the same horizontal stiffness, implies greater values of the shear force transmitted to the superstructure. In that case, a premature collapse of the superstructure may be observed. As a consequence, two alternative values of the behavior factor ($q$) have been considered for the superstructure: (i) $q = 1.5$ and (ii) $q = 1$. Both comply with the requirements of the code.

### 13.6.3 Fragility Curves and Seismic Risk Index of the RTBD Building

MSAs have been carried out to validate the proposed RTBD approach. Obviously, the failure modes and modeling assumptions adopted the same as those presented in Sections 13.3 and 13.4, respectively. Figures 13.4a and 13.5a summarize the number of collapses and the failure modes observed assuming $q = 1.5$ and $q = 1$, respectively. Figures 13.4b and 13.5b show the associated collapse fragility functions. Comparing Figure 13.4a with Figure 13.3a, one can note that the number of collapse cases reduces (particularly at IM7, passing from 10 to 2) and the number of cases involving the collapse of the superstructure increases.

The median value of the $S_{a,C}$ thus derived (see Figure 13.4b) is equal to 0.356 g, which significantly differs from the expected value $S_{a,C,T}$, equal to 0.46 g shown in Table 13.3. The reason is that, in first approximation, during the application of the RTBD procedure, the collapse of the superstructure is neglected, while Figure 13.4a clearly shows that the collapse of the superstructure plays a not negligible role, especially when it is designed assuming a behavior factor equal to 1.5. The annual failure rate, $\lambda_c$, thus obtained is equal to $2.68 \times 10^{-4}$, which is relatively close to the target values selected at Step 1 of the RTBD process. Assuming $q = 1$ (see Figure 13.5), the number of cases involving the collapse of the superstructure reduces, however, new collapse cases involving the

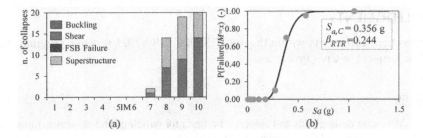

**FIGURE 13.4** RTBD ($q = 1.5$) building model: (a) number of analysis cases in which collapse is registered and (b) associated collapse fragility curve.

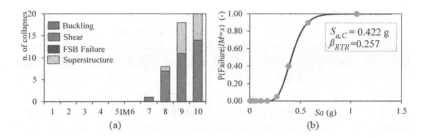

**FIGURE 13.5** RTBD ($q$ = 1) building model: (a) number of analysis cases in which collapse is registered and (b) associated collapse fragility curve.

isolation system arise. Ultimately, the value of the annual failure rate ($\lambda_c$) turns out to be $1.97 \times 10^{-4}$, which perfectly matches the target value assumed at Step 1 of the RTBD procedure.

From a practical point of view, the risk-target objective has been attained increasing, by approximately 20%, the collapse displacement capacity of HDRBs, compared to that obtained following a "code-conforming design" and assuming a behavior factor equal to 1 for the superstructure.

It's worth remarking that the results of this study are site-specific and also depend on the risk-targeted safety factor assumed for the NC limit state ($\gamma_{im,NC}$). In principle, if $\gamma_{im,NC}$ results lower than 1, the target objective can be attained without increasing the collapse displacement of the isolation system (of the order of 20%) or the lateral strength of the superstructure ($q$ = 1). This may occur when the engineer assumes (intentionally or not) significant extra margins in the design of the isolation system and/or superstructure.

## 13.7 CONCLUSIONS

The results of the RINTC research project (Ragni et al. 2018) dealing with Italian code-conforming base-isolated buildings outlined that the current design approach cannot guarantee an acceptable resilience objective in terms of probability of collapse, especially for high seismicity areas. In this chapter, a RTBD approach has been developed for buildings with seismic isolation. The proposed RTBD approach has been applied to a RC frame building equipped with rubber-based seismic isolation, assuming a target failure rate of $2 \times 10^{-4}$. The proposed approach has been validated through multiple-stripe analysis. An annual failure rate of $1.97 \times 10^{-4}$ has been then obtained, which fully complies with the target objective of the design. As a first approximation, and in a simplified way, the same resilience objective could be attained, following current design approaches, increasing by approximately 20% the maximum displacement of the isolation system at the NC limit state, and assuming, at the same time, a behavior factor ($q$) equal to 1, for the verification of the superstructure at the life-safety limit state.

## ACKNOWLEDGMENTS

This study has been carried out within the ReLUIS-DPC 2019-2021 research program, funded by the Italian Department of Civil Protection.

## REFERENCES

ASCE (2017). Minimum design loads and associated criteria for buildings and other structures, ASCE/SEI 7–16. American Society of Civil Engineers, Reston, USA.

ATC – Applied Technology Council FEMA P-58. 2012. Next-generation Seismic Performance Assessment for Buildings, Volume 1-2, 2012. Federal Emergency Management Agency, Washington, DC, USA.

Brandonisio, G., Ponzo, F., Mele, E., Luca, A. D. 2017. Experimental tests of elastomeric isolators: influence of vertical load V and of secondary shape factor, Proc. of 17th ANIDIS Conference, Pistoia, Italy.

Cardone, D., Gesualdi, G., Brancato, P. 2015. Restoring capability of friction pendulum seismic isolation systems. Bulletin of Earthquake Engineering. 13(8), 2449–2480.

Cardone, D., Perrone, G., Piesco, V. 2019a. Developing collapse fragility curves for base-isolated buildings. Earthquake Engineering and Structural Dynamics. 48, 78–102. https://doi.org/10.1002/eqe.3126

Cardone, D., Gesualdi, G., Perrone, G. 2019b. Cost-benefit analysis of alternative retrofit strategies for RC frame buildings. Journal of Earthquake Engineering. 23(2), 208–241, doi: 10.1080/13632469.2017.1323041.

CEN-EN (2000). Structural Bearings – Part 1: General Design Rules, 1337–1, Brussels.

Constantinou, M. C., Mokha, A., Reinhorn, A. 1990. Teflon bearings in base isolation. II: Modeling. Journal of Structural Engineering (ASCE). 1990. 116(2), 455–474.

DM. LL. PP. 17/01/2018: Norme tecniche per le costruzioni (NTC 2018). Rome, Italy (in Italian).

Gkimprixis, A., Tubaldi, E., Douglas, J. 2019. Comparison of methods to develop risk-targeted seismic design maps. Bulletin of Earthquake Engineering. 17(7), 3727-3752. doi: 10.1007/s10518-019-00629-w.

Ibarra, L. F., Medina, R. A., Krawinkler, H. 2005. Hysteretic models that incorporate strength and stiffness deterioration. Earthquake Engineering and Structural Dynamics. 34, 1489–1511. https://doi.org/10.1002/eqe.495

Iervolino, I., Spillatura, A., Bazzurro, P. 2018. Seismic reliability of code-conforming Italian buildings. Journal of Earthquake Engineering. 22, 5–27. https://doi.org/10.1080/13632469.2018.1540372

Ishii, K., Kikuchi, M. 2019. Improved numerical analysis for ultimate behavior of elastomeric seismic isolation bearings. Earthquake Engineering and Structural Dynamics. 48, 65–77. https://doi.org/10.1002/eqe.3123

Keerthana, S., Kumar, K. S., Balamonica, K., & Jagannathan, D. S. 2014. Seismic response control using base isolation strategy. International Journal of Emerging Technology and Advanced.

Luco, N., Ellingwood, B. R., Hamburger, R. O., Hooper, J. D., Kimballm, J. K., Kirchner, C. A. 2007. Risk-targeted versus current seismic design maps for the conterminous United States SEAOC 2007 Convention Proceedings, California.

Monzon, E. V., Buckle, I. G., Itani, A. M. 2016. Seismic performance and response of seismically isolated curved steel I-girder bridge. Journal of Structural Engineering. 142, 04016121. https://doi.org/10.1061/(ASCE)ST.1943-541X.0001594

Nishi, T., Suzuki, S., Aoki, M., Sawada, T., Fukuda, S. 2019. International investigation of shear displacement capacity of various elastomeric seismic-protection isolators for buildings. Journal of Rubber Research. 22, 33–41. https://doi.org/10.1007/s42464-019-00006-x

Ragni, L., Cardone, D., Conte, N., Dall'Asta, A., Cesare, A. D., Flora, A., Leccese, G., Micozzi, F., Ponzo, C. 2018. Modelling and seismic response analysis of Italian code-conforming base-isolated buildings. Journal of Earthquake Engineering. 1–33. 22(sup2), 198–230 https://doi.org/10.1080/13632469.2018.1527263

Spillatura, A. 2018. From Record Selection to Risk Targeted Spectra for Risk based Assessment and Design. Ph.D. Thesis, Dipartimento di Costruzioni e Infrastrutture, Istituto Universitario degli Studi Superiori (IUSS), Pavia, Italy.

Warn, G. P. 2006. The coupled horizontal-vertical response of elastomeric and lead rubber seismic isolation bearings. Ph.D. Dissertation, The State University of New York at Buffalo, Buffalo, NY.

Žižmond, J., Dolšek, M. 2019. Formulation of risk-targeted seismic action for the force-based seismic design of structures, Earthquake Engineering and Structural Dynamics. 48(12), 1406–1428.

# 14 Structural Performance Evaluation System for Improved Building Resilience and Reliability

*Maria João Falcão Silva*
Laboratório Nacional de Engenharia Civil (LNEC), Lisboa, Portugal

*Nuno Marques de Almeida*
University of Lisbon (IST), Lisboa, Portugal

*Filipa Salvado*
Laboratório Nacional de Engenharia Civil (LNEC), Lisboa, Portugal

*Hugo Rodrigues*
University of Aveiro, Aveiro, Portugal

## CONTENTS

DOI: 10.1201/9781003194613-14

## 14.1  INTRODUCTION

Since the beginning of the industrial revolution, and particularly after the end of the Second World War, three major conceptual approaches have been gradually crystallized: quality, performance, and risk. From the 1960s, the construction sector, and the building subsector, began a process of interpretation and adherence to these three conceptual approaches.

During the first decade of the 21st century, these three approaches started to be reasonably consolidated. This new reality allows those three interrelated approaches to be reconciled and integrated, in the context of the building's subsector, aiming at the renovation and revaluation of conventional management philosophies of this type of construction undertakings (Hollnagel, 2014).

This chapter builds upon and expands a building management framework based on the hypothesis that the approaches of quality, performance, and risk are complementary and reconcilable, according to the technical perspective of engineering, in the context of the construction sector and the buildings subsector (Almeida, 2015a). This generic framework includes three interrelated components (Almeida et al., 2010): (i) principles of technical building management based on performance and risk: satisfaction of end users (society and individuals), focus on the built product (building), effective accountability of the main players, guarantees against non-compliant buildings, and availability of technical information; (ii) strategic component of the technical building management model that provides the bases and organizational arrangements necessary for the full implementation of the principles of the technical building management model, as well as its monitoring, review, and continuous improvement; and (iii) the operational component of the technical building management model, consisting on a set of elements that promote the systematic application of policies, procedures, and practices to the building management activities (activities for the development of technical regulations, promotion and commercialization of buildings, project delivery, construction works, tests and inspections, etc.) (Mehmood, 2016; Falcão Silva et al., 2020).

The overall success of the technical building management model depends on the environment that is provided either by the strategic management decisions that provide the foundations and the organizational arrangements necessary to implement the principles of the technical building management philosophy or by the methodologies for its practical operationalization (Foliente, 2000; Foliente and Becker, 2001; Szigeti and Davis, 2005a).

This research work aims to contribute to the systematic implementation of the performance-based building's concept, which implies the transition from prescriptive to performance-based building environments, as can be seen in Figure 14.1, and thus higher freedom for determining technological solutions, construction processes, dimensions, and/or materials used in buildings (Preiser et al., 2018).

More specifically, it proposes a structural performance evaluation system for improved building resilience and reliability (Holický and Vrouwenvelder, 2005; Achour et al., 2015; Wholey, 2015; Cerè et al., 2017; Hassler and Kohler, 2014; Re Cecconi et al., 2018; Wilkinson and Osmond, 2018; Roostaie et al., 2019). This system is based on the guidelines of the first three parts of the ISO 15928 standard (aimed at describing the structural performance of buildings), the standards EN 1990 series (Eurocodes), ISO 22111 (which establishes the general requirements for the design of

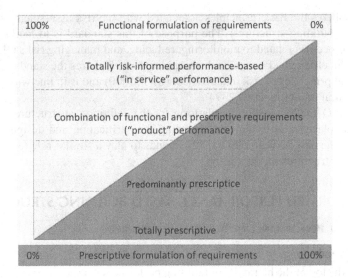

**FIGURE 14.1**   Formulation strategies for building requirements. (Adapted from Foliente, 2000; Szigeti and Davis, 2005a.)

structures buildings and civil engineering works), ISO 13824 (which establishes the basic principles associated with the assessment of the risk of building structures), ISO 2394 (which has played a unifying role with regard to the verification and design of structures for safety and use, over the past 30 years), and ISO 13823 (which aims to play a similar role to ISO 2394 with regard to the assessment and design of structures for durability).

The several parts of the ISO 15928 standard, which aim to harmonize the methods used to describe the performance of housing, are relevant to direct, among others, the technical performance evaluation of buildings. Those parts dedicated to describing structural performance in terms of structural safety (ISO 15928-1), structural serviceability (ISO 15928-2), and structural durability (ISO 15928-3) are particularly relevant. These standards follow the structure recommended by ISO 22539 and, therefore, are compatible with performance-based regulatory or normative environments, namely with the Construction Products Regulation and with the Structural Eurocodes. Structural Eurocodes[1] are an important set of building design codes and civil engineering works that articulate with ISO 15928 and the Construction Products Regulation. It is relevant to note that, in the case of construction works designed using the calculation methods described in Eurocodes, there is a presumption of conformity with essential requirement no. 1, "mechanical strength and stability" – including aspects of essential requirement no. 4, "safety in use," related to mechanical resistance and stability – and part of essential requirement no. 2, "fire safety," as defined in the Construction Products Regulation.

ISO 22111 establishes the general requirements for the design of building structures and civil and industrial engineering works. The objectives that guided its development and publication were to: (i) facilitate the activity of structural design at an international level; (ii) normalize the process of establishing rules for structural design at the international level (although each economy may specify the levels of performance or risk it deems most appropriate); (iii) ensure greater international acceptance of the activities carried out by regulatory and standardizing entities, by designers, and by the technical and scientific environment of each country; (iv) encourage the regulatory authorities of each country to describe their mandatory requirements in an internationally accepted format; (v) facilitate coordination between the various specialists who collaborate in the development of international standards related to structures; (vi) promote transparency in the process of comparing national standards related to structures.

ISO 13824 stabilizes the basic principles associated with the risk assessment of structural subsystems (including building structures). The purpose of this standard is to facilitate and improve decision-making processes related to monitoring, reducing, and managing risks related to this type of subsystems, in an efficient, effective, and transparent way. It shares the standardized vocabulary (ISO 73 Guide) and principles of risk management (ISO 31000) and is in line with the international standard on structural reliability (ISO 2394).

ISO 2394 and ISO 13823 constitute a platform for the definition of structure design rules. ISO 2394 standard has played a unifying role regarding the verification and design of structures for safety and use, for the past 30 years. ISO 13823 intends to play a similar role regarding the assessment and design of structures for durability.

## 14.2  RISK-INFORMED PERFORMANCE-BASED BUILDING STRUCTURES

### 14.2.1  TECHNICAL PERFORMANCE OF BUILDING STRUCTURES

#### 14.2.1.1  User Needs

User needs are at the top of the hierarchy of building performance requirements (Vale, 2014). These needs include, in a general and definitive way, the essential interests and expectations of society in general (which are reflected in technical regulations) and others related to individual expectations. User needs can be formulated in the following ways:

1. In the form of a generic statement that can include the fundamental aspects valued by end users (society in general and the stakeholders, including individuals).[2]
2. In such a way that characterizes the performance of the building required to be considered by the user as satisfactory, namely through the naming of qualitative attributes.[3]

The needs of users, in terms of structural performance, can be formulated by using generic statements including relevant aspects valued by end users, or by using qualitative attributes to be met by structures.

Generic declarations are usually included in the legislation, standards, or other documents. These statements can, for example, have the following formulations:

1. Safety of structures, protection of people and goods, trustworthiness of commercial transactions (*Decreto-Lei* No. 301/2007, Portuguese legislation).
2. Protection of human lives, limitation of economic losses, and maintenance of important civil protection facilities (Eurocode 8).
3. Safety of the occupants of the house (ISO 15928-1).
4. Acceptance, by part of the occupants, of the functioning and appearance of the dwelling and its components, of the activities of the other occupants, of the functioning of the equipment in the dwelling, the comfort provided, and the real state value of the dwelling (ISO 15928-2).
5. Acceptance, by part of the occupants, of the level of safety and structural serviceability of the dwelling throughout the agreed life span (ISO 15928-3).
6. Protecting the lives of the occupants, preventing injuries to occupants, safeguarding property and property (Meacham, 2004a).
7. Protection of human life, safeguarding property, maintaining functionalities, and other objectives that are expected of the building (Fujitani et al., 2005).

ISO 15928 standard reflects the needs of users in terms of structural performance, namely through three qualitative attributes: structural safety (ISO 15928-1), structural serviceability (ISO 15928-2), and structural durability (ISO 15928-3).

### 14.2.1.2 Description of Structural Performance

User needs should be detailed in terms of the technical ("in service") performance intended for the building. To do this, one must use the technical and engineering perspective to describe the behavior of the building in use.[4] This description can follow two possible qualitative formulations:

1. Definition of a set of qualitative attributes, of a predominantly technical nature, that reflect the needs of users.
2. Declaration that identifies the agents that act on and/or that qualitatively affect the technical ("in service") performance of the building (ISO 22539; ISO 6241).

The preparation of the technical performance description must take into account all formal and informal references, explicit or implicit, that are related to the building process and that detail the user needs, namely (Foliente, 2000; Huovila, 2005; Spekkink, 2005): (i) legislation (mainly performance-based regulations) of mandatory adoption; (ii) contractual documents; (iii) the client's preliminary program; (iv) standards (mainly performance-based standards) of voluntary adoption; (v) technical publications; (vi) other ways of transmitting the needs, explicit or implicit, of the various stakeholders.

Attributes (other equivalent terms for translating this concept include functional requirements or essential requirements, among others) are fundamental topics that constitute the requirements for the building's "in-service" technical performance. To safeguard the objectivity of the description of the technical performance of the building, the attributes can be replaced and/or complemented by a statement that qualitatively describes that performance, in order to eliminate any ambiguities that may exist when the transposition of the user needs is based exclusively on attributes (Nibel et al., 2004). Qualitative does not necessarily mean subjective, as a qualitative attribute can and should be objective. For example, the attribute "fire safety" can be complemented by a qualitative statement such as "the escape route from the building should lead directly to the outside." This formulation is objective but has a qualitative character as opposed to a quantitative formulation of the type "the distance from any point of the building to a protected staircase must not exceed 15 meters," which has a subjective character (depends on the existence of a staircase protected and does not establish conditions to access the outside).[5]

To describe the structural performance, it is appropriate to use the qualitative attributes (essential requirements) established in the Construction Products Regulation and the first attribute (first essential requirement) established in this mandatory document – mechanical strength and stability. This attribute is at the top of the hierarchy of structural performance requirements. It is from this attribute that derives the three sub-attributes of structural safety, structural serviceability, and structural durability referred to in the first three parts of ISO 15928.

Parallel to the naming of these three qualitative attributes, the description of structural performance can also be formulated by reference to a specific set of limit states. These limit states distinguish between desirable and undesirable states of a structure (ISO 2394). The effect of exceeding a limit state can be irreversible, in which case the resulting damage or malfunction remains until the structure is repaired, or reversible, in the case where the damage or malfunction remains only as long as the cause of exceeding the limit state is present (and because the cessation of the cause allows the transition from the unwanted state back to the desired state) (ISO 2394). The categories of limit states that have direct correspondence with those three qualitative attributes are, respectively (Deierlein and Hamilton, 2003; Fujitani et al., 2005)[6]:

1. Ultimate limit states, which correspond to the compromising of the structure safety (ISO 19338; ISO 22111; ISO 13823; ISO 2394; EN 1990).
2. Serviceability limit states, which correspond to the compromising of the normal use of the structure (ISO 19338; ISO 22111; ISO 13823; ISO 2394; EN 1990).

3. Durability or initiation limit states, which correspond to not reaching the ultimate and/or serviceability limit states due to deterioration during the life period agreed for the structure (ISO 19338; ISO 22111; ISO 13823).

The complementary description of the structural safety attribute is the expression of the capacity of the building as a whole or of its parts to maintain, with an adequate degree of reliability, the resistance and stability under all actions that may occur during the agreed lifetime (ISO 15928-1). The complementary description of the structural serviceability attribute is the expression of the capacity of the building as a whole or of its parts to behave, with an adequate degree of reliability, within parameters established for normal use and under all actions (ISO 15928-2). The complementary description of the structural durability attribute is the expression of the capacity of the building as a whole or of its parts to achieve, with an adequate degree of reliability, the desired structural safety and serviceability within the environment in which it is found throughout its life span agreed for the building and when subjected to the intended use (ISO 15928-3; ISO 13823; ISO 22111; ISO 2394).

### 14.2.1.3 Principles for Describing Structural Performance

The principles for describing technical performance determine the method by which the required performance will be described (ISO 22539). There are, fundamentally, two methods for describing the technical performance of the building[7]:

1. Break down of technical performance requirements into progressively more concrete and measurable performance requirements, possibly expressed in terms of an expanded and detailed requirement statement and/or physical parameters with a significant degree of autonomy,[8] which jointly satisfy the performance requirements established higher in the hierarchy.
2. Definition of the concrete parameters that describe the technical performance and that express in quantitative and objective terms the agents that act on and/or that qualitatively affect the building technical behavior.

In the first case, the performance description method is based mainly on building a hierarchy of performance requirements that must ensure the following overall characteristics (ibel et al., 2004; Szigeti and Davis, 2005a, 2005b; Becker, 2008): (i) Retain the user's needs (or the general purpose of the building); (ii) be positive and not negative; (iii) can be used as indicators; (iv) be precise and unambiguous; (v) be operational (maintain the possibility of being reached); (vi) be comparable; (vii) be quantifiable (measurable/valuable).

The application of this first method is only possible when the characteristics of the building are well defined (prescribed), so that it is possible to use typified descriptions or physical parameters capable of quantitatively evaluating the building, depending on the results of tests or other types of findings related to the technical solutions adopted – therefore, it is mainly valid for describing the performance of "the product." This first method may not be compatible with the description of "in service" performance (description of buildings without well-defined or prescribed physical characteristics) since both the standard descriptions and the physical parameters underlying this method are subjective (e.g., for a structure, it varies depending on whether the material selected is wood, concrete, or steel).

The second method uses parameters of a technical nature that objectively characterize the agents that influence the technical "in service" performance of the building. These parameters must be measurable according to current engineering discipline procedures. The agents (and the parameters for the description of the technical performance that characterize them) can be favorable or unfavorable in relation to the performance requirements established higher in the hierarchy (which must be positive and non-negative, as mentioned above). These agents that affect the technical "in service" performance (of the building or its constituent parts) can be typified, namely, in terms of the nature of its mechanisms of action or in terms of their origin or specific cause. One of the possible types is the following (Vanier et al., 1996; Sousa et al., 2008; ISO 6241): (i) mechanical agents (e.g., structural actions, structural resistance, deformations, etc.); (ii) physical agents (e.g., temperature, humidity, etc.); (iii) chemical agents (e.g., water and solvents, oxidizing and reducing agents, acids and bases, etc.); (iv) biological agents (e.g., vegetables and microbiotics, insects, etc.)

The description of structural performance should allow the determination of the in-service performance of building structures, that is, the determination of the performance of structural solutions without prescribed characteristics. Principles for describing structural performance can include the following:

1. Identification of qualitative attributes to be satisfied by the structures of buildings (structural performance requirements), namely structural safety, structural serviceability, and structural durability.
2. Determination of the agents that are associated with these qualitative attributes and that can somehow alter the structural performance of the building.
3. Characterization of these agents through parameters of engineering disciplines that allow expressing structural performance in technical, objective, and measurable terms.

The principle for describing the performance of structural safety is the definition of structural actions in the building and the strength of the structure under the effect of these actions (ISO 15928-1: 2003). The principle for describing the performance of structural serviceability is the definition of structural actions in the building that are relevant under conditions of normal use and the response of the structure subjected to the effect of these actions (ISO 15928-2: 2005). The principle for describing the performance of structural durability is the definition of the environment, the agreed service life for the components, and the maintenance program (ISO 15928-3; ISO 13823; ISO 22111).

## 14.2.2 Technical Risk of Building Structures

### 14.2.2.1 Establishing the Context for the Technical Risk of Building Structures

Pretending to consider all the risks that affect a building is a counterproductive ambition (May, 2007; El-Sayegh, 2008) and, from the start, doomed to failure (Flanagen and Norman, 1993; ANDI, 2006), due to insufficient resources and time to do it. It is, therefore, essential to contextualize and define the critical situations (Baccarini and Archer, 2001; Barkley, 2004; Öztaş and Ökmen, 2005).

Some authors have for long considered that the technical risk of buildings offers a limited perspective of undesirable effects for the society and criticize this perspective for reducing those effects to numerical values (Kolluru et al., 1996). Some other authors in the same line of thinking admitted that the disregard of strictly technical considerations is acceptable, in view of certain sets of values or concerns (Stahl et al., 2000).

However, technical risks should not be disproportionately disregarded, and one must be aware and informed about the most relevant technical risks, particularly when there is a lack of specific technical skills at the decision-making level. In line with other authors that have in the past made this point explicit (Klinke and Renn, 2001), the importance of a technical approach to risk is advocated, particularly in cases where the probability of occurrence and/or the extent of the consequences are well known and where there is sufficient information (situation of reduced uncertainty).

The technical risk of a building structure can be seen as the effect of technical uncertainties on the objectives of that structure, estimated from the engineering perspective.[9] The objectives of the building structure can be expressed in terms of values such as the protection of property, the safeguarding of safety and health levels, and environmental protection, among others (ISO 13824) or, more objectively, in terms of the following set of attributes that reflect the needs of end users: (i) structural safety, (ii) structural serviceability, and (iii) structural durability.

The technical risk management of the building's structural subsystem implies the use of the structural engineering discipline, namely for[10]:

1. Establish design bases: when it is necessary to define or use parameters that reflect the levels of structural reliability associated with the risk of relevant limit states being exceeded.
2. Prepare information to support decisions: when it is necessary to inform decisions related to the allocation of limited resources or the optimization of investments to prevent risks.

### 14.2.2.2   Categories of Technical Risk for Building Structures

There is no standardized way for organizing and categorizing risk-related information in construction projects, but this effort is needed (Tah and Carr, 2000) and should be preceded by an understanding of the context within which risk is to be managed (e.g., decennial insurance against buildings defects). Multiple approaches have been proposed, over the years, to carry out organization and categorization of risk in construction projects (Perry and Hayes, 1985; Cooper and Chapman, 1987; Flanagen and Norman, 1993; Kangari, 1995; Wirba, 1996; Chapman, 1997; Edwards and Bowen, 1998; Ahmed et al., 1999; Klinke and Renn, 2001; Tah and Carr, 2000; Wang and Chou, 2003; Fang et al., 2004; Barber, 2005; ANDI, 2006; Ling and Hoi, 2006; El-Sayegh, 2008).

These proposals tend, more or less explicitly, to distinguish two contexts (external and internal) and to adopt one of two partially overlapping points of view as can be seen in Figure 14.2: (i) centered on the systems organized by humans and (ii) centered on organizations.

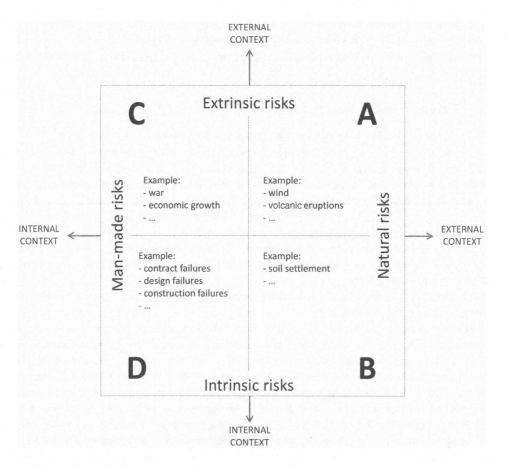

**FIGURE 14.2**   Forms of risk organization in construction projects.

In cases where the risk categories take a point of view centered on systems organized by humans, two major groups of risks are usually distinguished (Edwards and Bowen, 1998; Klinke and Renn, 2001; ISO 6241):

1. Natural hazards (quadrants A and B in Figure 14.2): risks that occur outside systems organized by humans, such as climate-related risks (winds, floods, etc.) and geological risks (soil settlements, earthquakes, volcanic eruptions, etc.), among others.
2. Human or man-made risks (quadrants C and D in Figure 14.2): risks that occur within systems organized by humans, such as social risks (criminal acts, riots, etc.), political risks (wars, civil disorder, etc.), economic risks (inflation, unemployment, fluctuations in currency values, etc.), financial risks (changes in interest rates, cash flows, etc.), legal risks (changes in regulation or contractual aspects, aspects related to licensing and/or patents, etc.), health risks (epidemics, etc.), management risks (related to quality assurance, cost control, human resources, planning, etc.), and technological risks (design or operation failures, etc.), among others (e.g., related to cultural and religious behaviors).

In cases where the risk categories take a point of view centered on organizations (namely interested parties of a building project), and where it is usually formulated based on the respective causes and potential consequences (cause-effect determination), there are usually two main groups (Vanier et al., 1996; Tah and Carr, 2001; Barber, 2005; Palomo et al., 2007; El-Sayegh, 2008; ISO 6241):

1. Intrinsic risks (quadrants D and B in Figure 1.2): risks related to the internal resources of the organizations, namely those that are under the responsibility of the various interested parties in carrying out a building project, including owners (delays in payments, stipulation of unrealistic deadlines, etc.), designers (design errors, delays in execution, etc.), builders and subcontractors (accidents, defects in the service provided, etc.), and suppliers (non-compliance with deadlines, etc.), among others (including those that can be associated with the end user himself).
2. Extrinsic risks (quadrants C and A in Figure 14.2): risks related to resources that are external to the organization and which are typically not under the responsibility of the main interested parties in a building project, such as political risks (strikes, changes in the law, corruption, delays in approvals, etc.), social and cultural risks (criminal acts, local protectionism, racial conflicts, etc.), economic risks (inflation, scarcity of basic resources, etc.), and natural risks (weather and other unforeseen events), among others.

Note that, considering the forms of the organization presented in Figure 14.2, technical risks such as those related to the wind, floods, or earthquakes belong to quadrant A. Technical risks related to the condition of the terrain can be allocated to quadrant B. Risks arising from a situation of alteration of technical regulations belong to quadrant C. Technical risks associated with the design and execution of the building belong to quadrant D.

This organizational ambiguity can hamper the technical risk management activities of buildings. For management purposes, it is thus convenient to categorize the technical risks according to the level of effort needed to control and manage those risks. It is the possibility of risk control that dictates the nature and the format of the risk management process to be implemented in a building project. The following categories of technical risk can thus be considered, as can be seen in Figure 14.3:

1. Inherent technical risk (more difficult to manage): includes technical risks that are difficult or even impossible to control (e.g., risks related to earthquakes).
2. Aggravation factors (easier to manage): includes technical risks that can be controlled (e.g., risks related to the design or the physical execution of the construction works).

**FIGURE 14.3**    Risk categories based on the level of effort needed to control the risk. (Adapted from Barber, 2005.)

The inherent technical risk is generally assumed to be somewhat involuntary (Flanagen and Norman, 1993) and is typically (but not exclusively) extrinsic (Chapman, 1997, 2001) to the organization (Barber, 2005; El-Sayegh, 2008). It can encompass both natural and human risks.

The aggravation factors that accumulate over the inherent technical risk, in turn, are generally assumed voluntarily (Flanagen and Norman, 1993) and are typically intrinsic (Chapman, 2001) to the organization (Barber, 2005; El-Sayegh, 2008). They mainly encompass the technical risks that arise from human activity and that arise within the rules, policies, processes, organizational structures, actions, decisions, behaviors, or cultures of the organization and/or the various stakeholders in that organization (Barber, 2005; El-Sayegh, 2008). These aggravating factors are usually called "gross human errors."

For the case of building structures, the following two categories of technical risk apply, as shown in Figure 1.3 (Barber, 2005; Martinez, 2008; Montull, 2008):

1. Technical risk inherent in the structure, corresponding to the technical risk of the building structure, the source of which does not lie in the internal resources of the interested parties in the building structure, even though the consequences may depend on these resources.
2. Aggravation factors augmenting the technical risk inherent in the structure, the source of which resides within the internal resources of the interested parties in the building project and whose consequences depend exclusively on these resources.

### 14.2.2.3   Principles for Describing Building Structural Risk

The need to describe the technical risk of the building can be deduced from the step of risk identification in the ISO 10006 and ISO 31000 risk management process. The output from the identification of the building technical risk should be the description of technical risks that can generate, optimize, prevent, degrade, accelerate, or delay the fulfillment of building objectives.[11] This description should cover both the inherent technical risks and the aggravation factors.

The description of risks can consist of a structured statement that includes the following elements (ISO 73 Guide): (i) source, (ii) events, (iii) causes, and (iv) consequences. The description of the technical risk of building structures must cover the effects of uncertainties in meeting the structural performance requirements (structural safety, structural serviceability, and structural durability). This overall description should cover both the inherent technical risks and the aggravation factors.

The principle for describing the inherent technical risk of the structure consists of defining the concepts and technical parameters of the discipline of structural engineering that best express uncertainty in quantitative and concrete terms. For example, structural reliability can be defined following the indications of ISO 2394 (ISO 15928-1).

The principle for the description of the aggravating factors of the structure is to characterize in a technically relevant way the factors that aggravate the identified inherent technical risk.

The modeling of the technical risk of building structures includes the considerations of the standard of generic principles for the risk assessment of systems involving structures (ISO 13824), as well as other relevant references produced by the technical and scientific community. The proposed modeling solution is restricted to technical risks that affect new and current structures. It does not cover situations associated with existing structures, exceptional structures, and/or extraordinary events, although it can be adapted to these contexts.

## 14.3  PERFORMANCE EVALUATION SYSTEM

### 14.3.1  Purpose and Systematization of Building Performance Evaluations

A performance evaluation system for buildings aims at establishing reproducible procedures to measure the technical characteristics of the engineering solutions of a particular building. It should measure both the levels of technical performance and the inherent technical risk. The output of performance evaluations is an understanding of whether (or how well) the technical characteristics of a building correspond to the programmed expectations.

A building performance evaluation system includes the following aspects:

1. Technical parameterization: the groups of variables or options, usually used by the applicable engineering disciplines, capable of quantitatively describing the technical performance and the inherent technical risk of the building.
2. Evaluation criteria: terms of reference (which can take the form of data sets, acceptance intervals, or pass/fail limits) to measure, calculate, predict, evaluate, and/or compare technical performance and the inherent technical risk of the building.
3. Evaluation and design methods: guidelines for linking the method for evaluating the technical performance of the building with the procedures and rules for designing the engineering solutions that pertain to that building.
4. Additional information: set of observations that detail and facilitate the understanding and practical application of the technical evaluation method.

These four components of the proposed technical evaluation can be incorporated directly into the engineering design processes, unlike conventional evaluation systems that are usually applied by specialists external to the design process, but that often have a limited influence on the optimization of solutions projected. They enable the practical application of the performance-based building design (Becker, 2008), which constitutes a concrete response to performance-based building procurement.

It is worth noting that the proposed performance evaluation system seeks to inform the interested parties in a building project of the inherent technical risk of buildings (see Figure 1.3). This system must be complemented by appropriate risk management procedures to bring the aggravation factors under control (Almeida et al., 2015b).

### 14.3.2  Technical Parametrization

Performance-based building procurement depends on indicators that objectively measure expectations on the "demand" side (May, 2007; ENISA, 2011). Similarly, performance-based building design depends on parameters that are directly and explicitly quantifiable or measurable (Meacham, 2004a). Procurement indicators are necessarily derived from these parameters, although the general meaning of the indicators may transcend that of the parameters from which they are derived

(ISO 21929-1). Specifically, the indicators that define the profile of the building's technical performance are derived from two categories of technical parameters:

1. Parameters for the description of technical performance.
2. Parameters for the description of the inherent technical risk.

The first of these two categories is naturally associated with the performance approach and conforms to the principles for describing the building technical performance. Similarly, the second category is associated with the risk approach and conforms to the principles for describing the technical risk of the building. In both cases, the parameterization of the indicators must include the variables or options that best reflect, in engineering projects, the expectations established in the building program. These variables or options must be measurable and observable, and as such they must be able to express quantitatively, both the agents that act on and/or qualitatively affect the technical behavior of the building (technical parameters for the description of technical performance), as well as the uncertainty associated with the achieving of this performance and/or behavior (technical parameters for describing the inherent technical risk).

### 14.3.2.1 Technical Performance Description Parameters

The indicators that define the profile of the building's technical performance must be based on parameters for the description of the technical performance "in service" that reflect the expected conditions of use to which the building will be subjected. It should not be based on the physical characteristics of the alternative engineering solutions that may constitute the building or on its conditions of use. It should not be based on technical parameters that describe the technical performance of the "product." In general terms, the number of parameters must be sufficiently reduced so that it is practical to incorporate them into the design process of the building. However, the final parameterization must fully reflect the user needs (ISO 22539) and be linked with the high-level technical performance requirements of the building. The parameters for describing technical performance may coincide with some of the "parameters for describing performance" established in ISO 15928.

### 14.3.2.2 Inherent Technical Risk Description Parameters

As mentioned, the description of technical performance "in service" should be based on information about technical risks that are difficult or impossible to control, that is, information on the technical risk inherent in the building. Therefore, the parameters for the description of technical performance "in service" must be complemented by specific parameters that describe this category of technical risk. The most appropriate strategy to parameterize information about inherent technical risks depends on the options taken to express the level of this type of risk (risk matrix technique, concept of maximum tolerable impacts, concept of reliability, or others), on the one hand, and the design procedures associated with the engineering discipline in question, on the other hand. As an indication, the following strategies for parameterization of the inherent technical risk can be mentioned: (i) parameterization of the casualty of events, which is appropriate when the level of the inherent technical risk is expressed through risk matrices (example of parameters: probability or frequency occurrences, return period, magnitude of the event, etc.); (ii) parameterization of reliability, which is appropriate when expressing the level of inherent technical risk through the concept of reliability (example of parameters: probability of failure, reliability index, safety factors, multiplicative factors for strategic differentiation of reliability, etc.). The parameterization of the inherent technical risk must cover the information to be provided to the designer, and that he must bear in mind during the design activities, but not the information generated because of his activity. That is, it must cover the basic data or hypotheses of the project (where the level of technical risk inherent in the building materializes). Human errors that may occur during the design of the project should be factors that aggravate the inherent technical risk.[12] Considerations about the appropriate number of parameters used to describe technical performance are also applicable to the case of parameters used to describe the inherent technical risk. Informative annexes and bibliographic references of the various parts of ISO 15928 are useful sources of information to parameterize the inherent technical risk (Meacham and Van Straalen, 2018).

### 14.3.3 EVALUATION CRITERIA

The technical evaluation of the building fundamentally involves the comparison between the levels of technical performance intended for the building and the levels achieved by the engineering solutions designed to respond to these claims. This comparison is only possible if, in addition to an adequate technical parameterization, there are criteria that make it concrete. The criteria used in the technical evaluation of the building are terms of reference (which can take the form of data sets, acceptance intervals, or pass/fail limits) to measure, calculate, predict, evaluate, and/or compare the technical performance and the inherent technical risk of the building (Tubbs, 2004; Meacham, 2004b, 2007, 2010, 2016).[13] These criteria are thresholds that guide the range of acceptability of engineering solutions conceived by the designer, according to the performance profile programmed for the building by the building procurement entity. The evaluation criteria must be properly calibrated.[14] Decisions related to this calibration must consider technical regulations and other applicable legal or regulatory requirements, theoretical knowledge, and/or experience gained.[15] These decisions may depend on studies on costs and benefits, on socio-economic and/or environmental aspects, on the motivations of the various stakeholders seeking the success of the building project, and on the priorities of limited resource allocation (ISO 31000). Ideally, the evaluation criteria should be objective. An example of objective criteria is those that make the attribution of the performance class dependent on the imposition of ranges of values on the relevant engineering variables, that is, on the parameters that describe the technical performance and the inherent technical risk. When the ideal situation is not possible, guidelines should be established that allow the designer to develop criteria embedded in the building design activities (criteria based on examples of acceptable solutions and/or solutions based on performance). The formulation of the evaluation criteria must not compromise the principles established for the description of the technical performance of the building or the principles established for the description of the inherent technical risk of the building.

### 14.3.4 INTEGRATION OF DESIGN AND EVALUATION METHODS

The incorporation of performance evaluation methods in engineering design activities is often ineffective (Meacham et al., 2002), especially in cases where these evaluation methods are applied at the end of design activities. Ideally, in the context of performance-based building design, the performance evaluation method should allow the systematic estimation and evaluation of the levels of technical performance (and the inherent technical risk) of various alternative engineering solutions (Figure 14.4). The evaluation method must be linked to the design procedures and rules common to the several alternative engineering solutions.

Some engineering disciplines still have knowledge gaps that prevent the full application of the performance-based design concept (Foliente, 2000; Tubbs, 2004; Meacham, 2007),[16] namely due to the lack of technical parameterization strategies and/or adequate assessment. In these cases, the assessment can be made using "valid methods" (methods based on well-established engineering principles and experience) that help to recognize the technical characteristics of the engineering solutions that most influence the expectations programmed for the building (this fills the technical parameterization gaps), which guide the designer in the practical interpretation of those expectations (this fills the evaluation criteria gaps), and which ensure the conformity of the engineering solution with those expectations.[17] These "valid methods" can be based, for example, on regulations, design guides, or standards of a prescriptive nature. Among the disciplines adhering to the concept

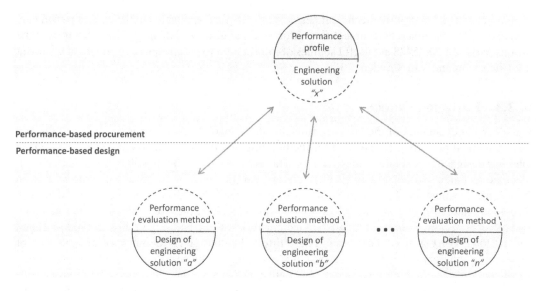

**FIGURE 14.4** Design of performance-based engineering solutions.[29]

of performance-based design include, there are structural engineering (ISO 15928-1; ISO 15928-2; ISO 15928-3), fire safety (ISO 15928-4), and building physics, namely in the field of energy efficiency (ISO 15928-5).

### 14.3.5 COMPLEMENTARY INFORMATION

The technical evaluation system can be accompanied by an explanation that guides and facilitates its direct implementation in design activities (ISO 22539; ISO 21929-1) and/or for complementary information that is useful to carry out the evaluation (Hattis and Becker, 2001). These explanations and/or complementary information must be available to all interested parties, namely those associated with the design, and may include, for example, the following contents[18]: (i) justifications for the options adopted in the parameterization of technical performance and risk; (ii) clarifications on how to assign technical performance classes to the building; (iii) details of the interaction between the evaluation procedures and the design of alternative engineering solutions; (iv) practical recommendations and examples of particularly advantageous engineering solutions; (v) information on relevant databases; (vi) references to technical support bibliography.

## 14.4 APPLICATION OF THE EVALUATION SYSTEM TO BUILDING STRUCTURES

### 14.4.1 PARAMETERIZATION OF STRUCTURAL PERFORMANCE

The parameterization of structural performance describes, in quantitative terms, the three attributes of structural performance (Francis and Bekera, 2014): (i) structural safety (related to the ultimate limit states); (ii) structural serviceability (related to the serviceability limit states); and (iii) structural durability (related to the initiation limit states). For each attribute, the description is based on two interdependent categories of technical parameters: (i) parameters for the description of technical performance; and (ii) parameters for the description of the inherent technical risk. The parameters presented are compatible with the Structural Eurocodes and derive from: (i) the international standards ISO 15928-1, ISO 15928-2, and ISO 15928-3; and (ii) the international standards with principles for the design of structures ISO 22111, ISO 2394, and ISO 13823.

**TABLE 14.1**

**Description and Parameterization for the Attribute Structural Safety**

| Agent | | Code | Parametrization |
|---|---|---|---|
| Actions | Permanent actions (G) | Sa.A1 | Magnitude of permanent actions other than self-weight<br>Location of imposed loads<br>Reliability parameters (partial safety factor $\gamma G$; multiplication factor $K$) |
| | Variable Imposed actions (Q) | Sa.A2 | Magnitude of uniformly distributed floor or roof loads<br>Magnitude of concentrated floor or roof loads over a specified area<br>Magnitude of a concentrated wall impact load applied at a specified height above the floor<br>Magnitude of uniformly distributed horizontal line load applied at a specified height above the floor<br>Applicable information on the area or height of application of the loads<br>Parameters for differentiating reliability (partial safety factor $\gamma Q$, multiplication factor $K$) |
| | Wind | Sa.A3 | Representative value of wind speed (or dynamic wind pressure)<br>Annual probability of occurrence (or average return period)<br>Parameters for differentiating reliability (partial safety factor $\gamma Q$, multiplication factor $K$) |
| | Snow | Sa.A4 | Representative values of snow accumulation (meteorological data or characteristic snow load at ground level)<br>Annual probability of occurrence (or average return period)<br>Parameters for differentiating reliability (partial safety factor $\gamma Q$, multiplication factor $K$) |
| | Seismic ($A_E$) | Sa.A5 | Representative value of seismic activity (effective peak ground acceleration, ground acceleration response spectrum, or others)<br>Probabilities of occurrence (or average return periods)<br>Parameters for differentiating reliability (importance coefficient $\gamma I$, multiplication factor $K$) |
| | Accidental (A) | Sa.A6 | Basic parameters representing accidental actions (to be defined for each individual project)<br>Parameters for differentiating reliability (aggravation coefficient for the characteristic value, affectation coefficient of the nominal value, multiplication factor $K$ and/or others) |
| | Other actions (Q, A) | Sa.A7 | Representative parameters of other actions |
| | Combination of actions | Sa.A8 | Set of action combination situations (persistent or transitory and accident)<br>Factors for combination of actions in a combination situation<br>Set of loading situations |
| | Soil condition and movements | Sa.A9 | Same as Table 1.2 |
| Structural resistance (R) | | Sa.B | Characteristic value of the strength of the materials<br>Test methods for determining the characteristic strength values of the materials<br>Parameters for differentiating reliability (partial safety factor $\gamma R$, multiplication factor $K$) |

The principle for describing structural safety is to determine the agents that can somehow modify the structural safety of buildings. The parameterization of performance proposed in Table 14.1 is based on ISO 15928-1 and that of the inherent technical risk derives from multiple sources with high degree of consensus (ISO 2103; ISO 4354; ISO 3010; ISO 4355; ISO 22111; ISO 2394; EN 1990; EN 1991; EN 1997; EN 1998).

Similarly, to structural safety, the principle for describing structural serviceability is to determine the agents that can modify the structural serviceability of the building. In practice, this description consists of determining and parameterizing the structural actions that affect the building under normal use and the structural response under the effect of these actions. Only relevant actions and responses should be considered when describing structural serviceability (ISO 15928-2). The performance parameters presented in Table 14.2 are based on ISO 15928-2 and those for describing the inherent technical risk take into consideration the guidelines established in ISO 2103, ISO 4354, ISO 3010, ISO 4355, ISO 10137, ISO 22111, ISO 4356, ISO 2394, EN 1990, EN 1991, EN 1997, and EN 1998.

**TABLE 14.2**

**Description and Parameterization for the Attribute Structural Serviceability**

| | Agent | Code | Parametrization |
|---|---|---|---|
| Actions | Permanent actions ($G$) | Se.A1 | Same as Table 1.1 |
| | Variable Imposed | Se.A2 | Same as Table 1.1 |
| | actions Wind | Se.A3 | Same as Table 1.1 |
| | ($Q$) Snow | Se.A4 | Same as Table 1.1 |
| | Seismic ($A_E$) | Se.A5 | Same as Table 1.1 |
| | | | Floor acceleration response (for equipment and fittings) |
| | Vibration sources ($G, Q$) | Se.A6 | Parameters representative of vibration sources (forces, displacements, speeds, accelerations, energy, or others) |
| | | | Parameters for differentiating reliability (partial safety factor, multiplication factors $K$, and/or others) |
| | Impacts sources ($Q$) | Se.A7 | Parameters representative of impact sources (mass, energy, or others) |
| | | | Parameters for differentiating reliability (partial safety factor, multiplication factors $K$, and/or others) |
| | Derived from fittings ($G, Q$) | Se.A8 | Parameters representative of imposed actions, actions originating from vibration sources, and/or actions originating from impacts sources |
| | Other actions ($Q, A$) | Se.A9 | Same as Table 1.1 |
| | Combination of actions | Se.A10 | Set of situations of combination of actions (characteristic, frequent, and quasi-permanent) |
| | | | Factors for combination of actions in a combination situation |
| | | | Set of loading situations |
| | Soil condition and movements | Se.A11 | Representative value of bearing capacity |
| | | | Expected nature and magnitude of ground movements |
| | | | Representative stiffness of the soil |
| | | | Diameter and distance/location of any soft spot that may cause loss of support to the building |
| | | | Test methods for determining the values of representative parameters. |
| | | | Parameters for differentiating reliability (partial safety factors $\gamma\varphi'$, $\gamma c'$, $\gamma cu$, $\gamma qu$, and $\gamma\gamma$ and multiplication factors $K$) |
| Structural response | Deformations | Se.B1 | Representative parameters of deflections (deflection ratios, absolute values) and tilt (angular distortions) |
| | Vibrations | Se.B2 | Parameters representative of the level of vibration in the receiver (frequency together with acceleration, speed, displacement, and/or others) |
| | | | Other parameters (adjustment factors for human vibration response, vibration mitigation strategies, others) |
| | Local damage | Se.B3 | Representative cracking parameters (crack opening, location, and frequency) |
| | | | Representative parameters of spalling (depth, area, location, and frequency of spalling) |
| | Response to impact | Se.B4 | Representative parameters of the impact response (impact energy, deformation, displacement, and local damage) |
| | | | Restrictions for different impact energies |
| | Fittings | Se.B5 | Parameters representative of the ability to respond to actions from fittings |

**TABLE 14.3**

**Description and Parameterization for the Attribute Structural Durability**

| Agent | | Code | Parameterization |
|---|---|---|---|
| Structure environment | Geographic location | Du.A1 | Parameters for describing the geographical location |
| | Influences and environmental agents | Du.A2 | Representative parameters of influences and environmental agents (depends on the structural materials used) |
| Standard service life for components and maintenance program | | Du.B | Parameters for the description of the agreed component service life |
| | | | Parameters for the description of the maintenance program |

The durability of structures and their constituent parts is relevant to structural performance and is related to both structural safety and structural serviceability. The durability of the structures is strongly correlated with the environment in which the structure and its constituent parts fit, with the duration of the period to which are subjected to the intended use and with the maintenance program associated with that period (ISO 22111). These aspects are agents, in some way, capable of altering structural performance. As can be seen in Table 14.3, the description of structural durability consists precisely in the determination and parameterization of these agents (ISO 15928-3). The suggested parameterization is based on the guidelines of ISO 15928-3, ISO 13823, and ISO 22111.

### 14.4.2 STRUCTURAL PERFORMANCE EVALUATION CRITERIA

The decisions taken during the design of the engineering solutions for the building structure must be based on an evaluation of their conformity against the programmed structural performance profile (Falcão Silva et al., 2020). In this evaluation, the designer must be governed by the predetermined terms of reference (ISO 13824) for structural safety (related to the ultimate limit states), structural serviceability (related to the serviceability limit states), and structural durability (related to the initiation limit states). The terms of reference presented in Tables 14.4–14.6 follow two formats (Almeida, 2011): (i) objective evaluation criteria (see Sa.A1 to Sa.A5 in Table 14.4 and Se.A6 to Se.A7 in Table 14.5), based on numerical values to be assigned to engineering or technical parameters that describe structural performance; and (ii) qualitative guidelines (see Se.B1, Se.B2, and Se.B4 in Table 14.5 and Du.B in Table 14.6) that instruct the subsequent formulation of objective evaluation criteria.

The evaluation criteria presented in Tables 14.4–14.6 are proposed for class II (CC2) structures in new buildings (Almeida et al., 2015a, 2015b). These proposals cover performance classes A+, A, and B. Lower performance classes C and D may be admitted for existing buildings but are considered unacceptable structural performance for new building structures (Falcão Silva et al., 2020). These evaluation criteria must be reviewed and calibrated by the technical and scientific community before being incorporated into technical regulations and standards.

The practical meaning of the criteria proposed in Table 14.4 is that, e.g., if permanent and imposed actions are factored as suggested, class A+ buildings shall withstand design loads 10% higher than legal minimum values and class A or B buildings shall withstand design loads equal to the legal minimum values. Or, when considering wind actions, class A+ buildings shall withstand wind loads 10% higher than those with a return period higher than the legal minimum (e.g., 2500 years), while class A buildings building shall withstand wind loads equal to those generated with a return period higher than the legal minimum (e.g., 2500 years) and class B buildings shall withstand wind loads equal to those generated with the legal minimum return period (e.g., 1000 years). Likewise, class A+ buildings shall withstand seismic loads at least 50% higher than legal minimum values, class A

**TABLE 14.4**

**Terms of Reference for Evaluating the Structural Safety of New Buildings**

| Code | Agent | Class | Evaluation Criteria | |
|------|-------|-------|----|----|
| Sa.A1 | Permanent actions ($G$) | A+ | Magnitude and location of all permanent loads other than self-weight, duly considered | Use of the calculation value of the permanent load $G$ obtained by multiplying the representative value of that load by the legal partial safety factor $\gamma G$ compounded by the multiplicative factor $K_{A+} = 1.1$ |
| | | A | | Use of the calculation value of the permanent load $G$ |
| | | B | | obtained by multiplying the representative value of that load by the legal partial safety factor $\gamma G$ ($K_A = K_B = 1.0$) |
| Sa.A2 | Variable actions ($Q$) – Imposed | A+ | | Use of the calculation value of the variable action $Q$ ($q$) obtained by multiplying the representative value of that action by the legal partial safety factor $\gamma Q$ compounded by the multiplicative factor $K_{A+} = 1.1$ |
| | | A | | Use of the calculation value of the variable action $Q$ ($q$) obtained by multiplying the representative value of that action by the legal partial safety factor $\gamma Q$ ($K_A = K_B = 1.0$) |
| | | B | | |
| Sa.A3 | Variable actions ($Q$) – wind | A+ | | Use of the representative value of the wind speed ($w$) corresponding to a very rare event (for example, equivalent to an average return period between 1000 and 2500 years) and a design value $Qd,w$ obtained by increasing the legal partial safety coefficient $\gamma Q$ compounded with a multiplicative factor $K_{A+} = 1.1$ |
| | | A | | Use of the representative value of the wind speed ($w$) corresponding to a very rare event (for example, equivalent to an average return period between 1000 and 2500 years) and a design value $Qd,w$ obtained by using the legal partial safety factor $\gamma Q$ ($K_A = 1.0$) |
| | | B | | Use of the representative value of the legal wind speed ($w$) (for example, equivalent to an average return period between 500 and 1000 years) and a design value $Qd,w$ obtained by using the legal partial safety factor $\gamma Q$ ($K_B = 1.0$) |
| Sa.A4 | Variable actions ($Q$) – snow | A+ | | Use of the characteristic value of the snow load at ground level ($sk$) corresponding to a very rare event (for example, equivalent to an average return period of 750 years) and a design value $Qd,s$ obtained by increasing the legal partial safety factor $\gamma Q$ compounded with a multiplicative factor $K_{A+} = 1.1$ |
| | | A | | Use of the characteristic value of the snow load at ground level ($sk$) corresponding to a very rare event (e.g., equivalent to an average return period of 750 years) and a design value $Qd,s$ obtained by using the legal partial safety factor $\gamma Q$ ($K_A = 1.0$) |
| | | B | | Use of the legal characteristic value of the snow load at ground level ($sk$) (for example, equivalent to an average return period of 500 years) and a design value $Qd,s$ obtained by using the legal partial safety factor $\gamma Q$ ($K_B = 1.0$) |
| Sa.A5 | Seismic action ($A_E$) | A+ | | Use of seismic activity calculation values ($A_{Ed}$), obtained by using a coefficient of importance $\gamma I$ aggravated by a multiplicative factor $K_{A+} = 1.5$, corresponding to an average return period of approximately 1500 years |
| | | A | | Use of seismic activity calculation values ($A_{Ed}$), obtained by using a coefficient of importance $\gamma I$ aggravated by a multiplicative factor $K_{A+} = 1.25$, corresponding to an average return period of approximately 1000 years |
| | | B | | Use of seismic activity calculation values ($A_{Ed}$), obtained by using the legal coefficient of importance $\gamma I$ (multiplication factor $K_B = 1.0$), corresponding to an average reference period (e.g., 475 years) |

## TABLE 14.5
### Reference Terms for Evaluating the Structural Serviceability of New Buildings

| Code | Agent | Class | Evaluation Criteria |
|------|-------|-------|---------------------|
| Se.A1 | Permanent actions ($G$) | | Same as Se.A1 |
| Se.A2 | Variable actions ($Q$) – Imposed | | Same as Se.A2 |
| Se.A3 | Variable actions ($Q$) – Wind | | Same as Se.A3 |
| Se.A4 | Variable actions ($Q$) – Snow | | Same as Se.A4 |
| Se.A5 | Seismic action ($A_E$) | | Same as Se.A5 |
| Se.A6 | Actions from vibration sources ($G$, $Q$) | A+ | Use of the calculation value of the action originating from vibration sources obtained by multiplying the representative value of that action by the legal/recommended partial safety factor compounded with a multiplicative factor $K_{A+} = 1.1$ |
| | | A | Use of the calculation value of the action originating from vibration sources obtained by multiplying the representative value of that action by the legal/recommended partial safety factor ($K_A = K_B = 1.0$) |
| | | B | |
| Se.A7 | Actions from impact sources | A+ | Use of the calculation value of the actions originating from sources of impacts obtained by multiplying the representative value of that action by the legal/recommended partial safety factor compounded with a multiplicative factor $K_{A+} = 1.1$ |
| | | A | Use of the calculation value of the actions from sources of impact obtained by multiplying the representative value of that action by the legal/recommended partial safety factor ($K_A = K_B = 1.0$) |
| | | B | |
| Se.B1 | Deformations | A+ | Evaluation criteria that reflect a clearly higher likelihood of safeguarding the appearance and/or conditions of use of the structural subsystem |
| | | A | Evaluation criteria that reflect the threshold of acceptability of appearance and/or conditions of use of the structural subsystem |
| | | B | |
| Se.B2 | Vibrations | A+ | Evaluation criteria that translate vibration levels clearly lower than the threshold of acceptance and human perception and that do not compromise the restrictions of the contents and structures of buildings |
| | | A | Evaluation criteria that translate acceptable vibration levels for occupants and that do not compromise the restrictions of the contents and structures of buildings |
| | | B | |
| Se.B4 | Response to impact | A+ | Evaluation criteria that reflect exceptionally demanding restrictions for different predefined impact energy levels |
| | | A | Evaluation criteria that translate restrictions that are clearly more demanding than normal, but not exceptionally demanding, for different predefined impact energy levels |
| | | B | Evaluation criteria that translate normal restrictions for different predefined impact energy levels |

## TABLE 14.6
### Reference Terms for Evaluating the Structural Durability of New Buildings

| Code | Factor | Class | Evaluation Criteria |
|------|--------|-------|---------------------|
| Du.B | Standard service life for components and maintenance program | A+ | Use of engineering solutions and/or strategies that comply with the programmed level of structural safety and structural serviceability, for a period of time greater than twice the life span normally agreed for the building as a whole (for example, 100 years or more) |
| | | A | Use of engineering solutions and/or strategies that comply with the programmed level of structural safety and structural serviceability, for a period of time between the life span normally agreed for the building as a whole and twice that value (for example, between 50 and 100 years old) |
| | | B | |

buildings shall withstand seismic loads at least 25% higher than legal minimum values, and class B buildings shall withstand legal minimum seismic loads.

These proposals allow designers to systematically make use of metrics (e.g., parameters describing structural loads and structural resistance, among others) against which structural performance is to be ranked in accordance with harmonized evaluation criteria (e.g., using an equation that compares these metrics with limiting or acceptable values for applicable engineering parameters; Almeida et al., 2015b). This provides the basis for performance-based structural design because the proposed evaluation system is applicable to an abstract functional unit that represents the structure of the building without a concrete link to the unique context-specific engineering solution into which this abstract functional unit may be materialized (e.g., concrete structure, steel structures, wood structure, etc.). Designers thus benefit with a mechanism that transforms the performance required for the building structure as an abstract functional unit (i.e., structural performance +, A or B) into a tangible engineering output. This enables performance-based thinking to be deeply embedded into the design process.

### 14.4.3 Integrating Performance Evaluations in Structural Design

The structural performance evaluation is based on the comparison of the basic options assumed in the design of the engineering solutions, namely regarding the agents that can somehow modify the levels of structural safety, structural serviceability, and the structural durability, against the terms of reference that determine the acceptability of these options. In performance-based environments, this evaluation method must be incorporated into the engineering procedures used to design structures (Akiyama et al., 2000; Almeida et al., 2010). The evaluation method must link to the procedures and design rules that are common to various types of structures (reinforced concrete, mixed, wood, etc.), namely: (i) design procedures and rules for structural safety (method of ultimate limit state) and structural serviceability (method of serviceability limit states), according to ISO 2394 and ISO 22111, and; (ii) design procedures and rules for structural durability (method of limit states of initiation), according to ISO 13823.[19]

Calculation models can basically take on two different formats: the probabilistic format and the partial safety factor format (ISO 2394). The probabilistic format aims to ensure that the probability of failure of a structure $P_f$ does not exceed a certain probable failure probability $P_{target}$, during a given period (ISO 2394). For most of the ultimate limit states and for some serviceability limit states, this failure probability can be described in the generic form $P_f = P[g(\underline{V}) < 0]$, where $g(\underline{V})$ is the function of the limit state, $g(\underline{V}) < 0$ is the condition that characterizes the undesired state of the structure, and $\underline{V}$ represents the basic variables relevant to the problem (ISO 2394). The probabilistic format is convenient in special situations and is also used for the calibration of these safety factors, but for reasons of practical applicability, the format commonly used is that of partial safety factors (ISO 2394). According to this latter format, the condition that defines the desired state of a structure can take the form $g(F_d, f_d, a_d, \theta_d, C, \gamma_n) \geq 0$, where $F_d$ are the design values for the actions, $f_d$ are the calculation values for the material properties, $a_d$ are the calculation values for the geometric variables, $\theta_d$ are the calculation values for the variables $\theta$ that translate the uncertainties in the modeling, $C$ are restrictions that define the limit states, and $\gamma_n$ is a coefficient that reflects the importance of the structure and the consequences of failures (ISO 2394).

In many cases, it is possible to separate the basic variables and the $\theta$ coefficients into groups that represent the effects of the actions and the responses to those actions (ISO 2394). In these cases, the condition that defines the desired state of a structure can take the generic form $g(E_d, R_d) \geq 0$, where $E_d = E(F_d, a_d, \theta_{Ed})$ and $R_d = R(f_d, a_d, \theta_{Rd})$.[20] In practice, the verification of structural safety (related to the ultimate limit states) involves demonstrating the satisfaction of the response requirement $R_d - E_d \geq 0$,[21] the static balance requirement $E_{d,stab} + R_d - E_{d,estab} \geq 0$,[22] and the structural robustness requirement[23] (ISO 22111). The verification of structural serviceability in normal use (related to the serviceability limit states) involves demonstrating the requirement $C_d - E_d \geq 0^{24}$ (ISO 22111). This means that, for example, the resistance in the ultimate limit states can be verified through the condition $P_f = P(R_d - E_d < 0) \leq P_{target}$, and the same reasoning can be applied for the other requirements. The failure probability can be expressed through a reliability index $\beta = \Phi^{-1}(P_f)$, where $\Phi^{-1}$ is the inverse function of the normalized normal distribution (ISO 2394). The ISO 2394 standard contains provisions for calculating these failure probabilities and the respective degrees of reliability. It has for long been noted that the general public seems to be more prepared to accept the concept of an adequate degree of reliability (positive expression of risk) than probabilities of failure (negative expression of risk).

The recommended method for the design and verification of structural durability is also that of the limit states (ISO 13823). The design for structural durability implies the knowledge and treatment of the specific reality of a given structural component or structure, during the expected time of use. It is important to understand the environment of the structure, the transfer mechanisms, the environmental actions, and the respective effects that can cause component or structural failure (ISO 13823). The verification of structural durability, that is, that the structure and its components do not compromise the safety and comfort of the occupants during the agreed design life $(t_D)$, can be done according to two formats (ISO 13823): (i) service-life format (ISO 13823); (ii) limit-states format (ISO 13823). Verification according to the service-life format consists of guaranteeing the basic requirement that the expected service life $(t_{SP})$ must equal or exceed the agreed design life $(t_D)$, that is, $t_{SP} \geq t_D$. For a probabilistic forecast of service life $(t_S)$,[25] that basic requirement takes the form $P(t_S \leq t_D) \leq P_{target}$. Alternatively, a partial safety factor $\gamma_S \geq 1,0$ that affects a characteristic value of the service life $(t_{Sk}/\gamma_S \geq t_D)$ can also be used. The verification according to the format of the limit states involves the verification of the basic requirement of the ultimate limit states $R(t) \geq E(t)$,[26] the basic requirement of the serviceability limit states $C_{lim} > E(t)$,[27] and the basic requirement of the initiation or durability limit states.[28] The verification of these basic requirements is ensured as long as the probability of failure respects the condition $P_f(t) \leq P_{target}$, both in the case of the ultimate limit states $P_f(t) = P(R(t) - E(t) < 0$ as in the case of the serviceability limit states $P_f(t) = P(C_{lim} - E(t) < 0)$. In practice, the verification of these basic requirements involves the use of partial safety factors $(\gamma_F \text{ e } \gamma_M)$ calibrated in the function of these probabilistic conditions.

### 14.4.4 ADDITIONAL INFORMATION ON STRUCTURAL PERFORMANCE EVALUATION

Information that supports the understanding of the structural performance evaluation may include a set of comments on the structural safety (Sa) and the structural serviceability (Se) evaluations and another set of comments on the evaluation of structural durability (Du). Within these two sets of comments, aspects related to the parameterization of structural performance can be found, with the structural performance evaluation criteria and with the structural performance evaluation and design methods.

## 14.5 CONCLUSIONS

Despite being a concept being discussed for several decades, the practical implementation of the philosophy of performance-based buildings still depends on the resolution of inconsistencies of different nature. For example, difficulties in objectively measuring expectations on the "demand" side and establishing a clear link to the "supply" side. This chapter aims at tackling this inconsistency by exploring and reconciling the approaches to quality, performance, and risk, from the technical viewpoint of engineering in the context of the construction sector and the buildings subsector. The output of this contribution is the foundations of a performance evaluation system for building structures that are compatible with the Structural Eurocodes, the first three parts of the ISO standard

15928 (ISO 15928-1; ISO 15928-2; ISO 15928-3) and the standards developed by the ISO TC98/SC2 technical committee (ISO 13824; ISO 13823; ISO 2394; ISO 22111).

The proposed performance evaluation system requires the prior programming of a structural performance profile on the side of demand (i.e., on the side of the building owner), in which the desired levels of structural safety, structural serviceability, and structural durability are set, together with the corresponding levels of inherent technical risk. This evaluation system comprises the following aspects:

1. Parameterization of structural performance: definition of engineering variables or options, used in structural design procedures, which allow the quantification of the building performance in terms of structural safety, structural serviceability, and structural durability.
2. Criteria for evaluating structural performance: terms of reference that distinguish between the desirable and undesirable states for a structure, namely with regard to structural safety (ultimate limit states), structural serviceability (serviceability limit states), and structural durability (durability or initiation limit states).
3. Structural performance evaluation and design methods: instructions for incorporating structural performance evaluation into structural design procedures for structural safety, structural serviceability, and structural durability.
4. Complementary information on structural performance evaluation: complementary observations on the parameterization strategies followed, on the evaluation criteria adopted, and on the applicable evaluation and design methods.

The principles of the proposed evaluation system can be applied to evaluate the technical performance of any attribute for new buildings (e.g., any of the six essential requirements set out in the Construction Products Directive), regardless of the type of occupation they are intended for (residential and non-residential) and the type promotion method adopted (public or private). It can also be adapted and expanded to evaluate the performance of existing buildings.

## NOTES

1. As part of the elimination of technical barriers to trade and the harmonization of technical specifications in the European construction sector, the development of a set of harmonized technical rules for the structural design of construction works began in 1975, with a view to replace the different rules existing in the different Member States (Gulvanessian et al., 2002). The first set of documents was published between 1984 and 1988 (ENV pre-standards), and its elaboration was transferred to the European Committee for Standardization (CEN) with a view to adapt to the format of European standards currently known as Structural Eurocodes (Gulvanessian et al., 2002). It is the technical documentation that explains the pre-normative bases that supported the development of these documents (JCSS, 2009) and the guidelines for their implementation (JRC, 2009).
2. For example, as in the Hammurabi code.
3. For example, as in the Vitruvian triad.
4. Adapted ISO 6240 and ISO 6241.
5. Example adapted from Nibel et al. (2004). In this example, the agent that qualitatively affects technical performance is the "escape route."
6. In addition to these categories of limit states, the following could also be mentioned: reparability limitable states, which correspond to the compromise of the facility to repair damage caused by external agents (ISO 22111; Fujitani et al., 2005); fire resistance limit states, which correspond to the impairment of structural sufficiency during and after a fire (ISO 19338; Deierlein and Hamilton, 2003); and fatigue limit states, which correspond to the compromise of the other limit states due to fatigue reasons (ISO 19338). The attributes corresponding to these limit states are outside the scope of the present work, but need to be considered, together with the qualitative attribute of robustness constant in the norm ISO 22111.
7. In either method, performance should always be viewed from the perspective of the end user. This allows to avoid repetition and overlapping of descriptions (ISO 21929-1).

8. The choice for one or another type of expression depends on the flexibility that is intended to be given to the method or on limitations related to the situation to be described (Meacham, 2004a).

9. Adapted ISO 31000; ISO 13824.

10. Adapted ISO 13824.

11. Adapted ISO 31000.

12. The factors of aggravation of the inherent technical risk are addressed within the scope of the fifth element of the proposed model (technical control).

13. Based on the definition of risk criteria established in the ISO 73 Guide, the definition of performance or risk criteria established by the Immigration, Refugees and Citizenship Canada (IRCC) (Meacham, 2010; Meacham, 2007; Meacham, 2004b; IRCC, 2010) and the definition of criteria established by the group of work CIB TG37 (Meacham, 2007; Meacham, 2004b; Tubbs, 2004; CIB, 2004).

14. Adapted from the ISO 21929-1 generic considerations on sustainability indicators. See, for example, calibration adopted in Japanese building regulations (BCJ, 2005).

15. Adapted from the guidelines established in the ISO 31000 and ISO 10006 risk assessment processes.

16. The full adherence of a particular engineering discipline to the philosophy of performance-based buildings implies the existence of clear and unambiguous guidelines on the estimation, calculation, and mediation of the levels of technical performance and the technical risk inherent in the engineering solutions addressed by these disciplines (not only at the project level but also at the level of the production of construction materials and laboratory tests).

17. Adapted from the level of verification in the model by (Horvat, 2005; Hattis et al., 2001; Foliente et al., 1998; Oleszkiewicz, 1994), the level of acceptable solutions in the model proposed by the IRCC (Meacham, 2010; Meacham, 2007; Meacham, 2004b), and the assessment level established for ISO 15928 (ISO 22539).

18. Adapted ISO 21929-1.

19. The ISO 2394 standard, which has been used in the preparation and harmonization of national and regional structural design standards and regulations worldwide for over 30 years (the first version of this standard was published in 1973), was responsible for the consolidation and dissemination generalized use of the limit state method currently used by the discipline of structural engineering. The ISO 13823 standard, first published in 2008, aims mainly to harmonize and improve the design of structures for durability through the promotion and incorporation of building physics principles in the structural engineering procedures (namely, in the limit state method, such as defined in ISO 2394).

20. For $E$, the properties of materials $f$ can also be considered in special cases (for example, second-order calculations). The same is true for $R$, in relation to $F$ actions (ISO 2394).

21. Related to collapse, cracking, or excessive deformation of the structure, structure element, or connection (ISO 22111). $R_d$ and $E_d$ are, respectively, the design values of the resistance and the effect of the action. Both can be modeled mathematically (ISO 2394).

22. Related to destabilization situations (for example, overturn) (ISO 22111 - 11.2.2). $E_d$, corresponds to the calculation value of the effect of shares that have a stabilizing effect, $R_d$ is the calculation resistance of the elements that prevent destabilization, if any, and $E_d$, corresponds to the calculation value of shares with destabilizing effect (ISO 22111 - 11.2.2).

23. Compliance with this requirement is associated with the adoption of design and detailing rules and methods (ISO 22111). For example, Structural Eurocodes deal with this aspect in the context of unidentified accidental situations that can be related to local failures arising from unspecified causes (EN 1991-1-7).

24. $C_d$ corresponds to the calculation value of the parameters that define the serviceability limit and $E_d$ is the design-action effect (ISO 22111).

25. The service life $(t_S)$ can be modeled mathematically according to a function of agent transfer for $t_{start}$ and a function of damage or resistance for $t_{exposed}$ (ISO 13823).

26. $R(t)$ and $E(t)$ are, respectively, the resistant capacity of the structural component at time $t$ and the representation of the effects of the action at time $t$. Both can be modeled mathematically (ISO 2394).

27. $C_{lim}$ is the limit constant value of the parameters that define the serviceability limit state and $E(t)$ represents the action effect at time $t$ (ISO 22111; ISO 2394).

28. Verification is here identical to the ultimate and serviceability limit states. The ISO 13823 standard presents an example of verifying the durability of a concrete structure subject to induced corrosion through the initiation limit state method.

29. The engineering solution "x," selected by the designer, must respect the programmed performance profile.

## REFERENCES

Achour, N., Pantzartzis, E., Pascale, F., Price, A. (2015). Integration of resilience and sustainability: from theory to application. International Journal of Disaster Resilience in the Built Environment, Vol. 347–362, ISSN: 1759–5908.

Ahmed, S., Ahmad, R., Saram, D. (1999). Risk management trends in Hong Kong construction industry: a comparison of contractors and owners perspectives. Engineering Construction and Architectural Management, Vol. 6, 3, pp. 225–234.

Akiyama, H., Teshigawara, M., Fukuyama, H. (2000). A Framework of Structural Performance Evaluation System. CD Proceedings of the 12WCEE, 1–8. Retrieved from http://www.iitk.ac.in/nicee/wcee/article/2171.pdf.

Almeida, N. (2011). Technical model of building management performance and risk based: Design, development, and example of application to structures. PhD in Civil Engineering, IST-ULisboa (in Portuguese).

Almeida, N., Sousa, V., Alves Dias, L., Branco, F. (2010). A framework for combining risk-management and performance-based building approaches. Building Research & Information, Vol. 38, 2, pp. 157–174. https://doi.org/10.1080/09613210903516719.

Almeida, N., Sousa, V., Alves Dias, L., Branco, F. (2015a). Managing the technical risk of performance-based building structures. Journal of Civil Engineering and Management, Vol. 21, 3, pp. 384–394. https://doi.org/10.3846/13923730.2014.893921.

Almeida, N., Sousa, V., Alves Dias, L., Branco, F. (2015b). Engineering risk management in performance-based building environments. Journal of Civil Engineering and Management, Vol. 21, 2, 218–230. https://doi.org/10.3846/13923730.2013.802740.

ANDI. (2006). The importance and allocation of risks in Indonesian construction projects. Construction Management and Economics, Vol. 24, 1, pp. 69–80.

Baccarini, D., Archer, R. (2001). The risk ranking of projects: a methodology. International Journal of Project Management, Vol. 19, pp. 139–145.

Barber, R. (2005). Understanding internally generated risks in projects. International Journal of Project Management, Vol. 23, pp. 584–590.

Barkley, B. (2004). Project Risk Management. Nova Iorque: McGraw-Hill.

BCJ. (2005). The Housing Quality Assurance Act and Japan Housing Performance Indication Standards. Tokyo: The Building Center of Japan.

Becker, R. (2008). Fundamentals of performance-based building design. Building Simulation, Vol. 1, pp. 356–371.

Cerè, G., Rezgui, Y., Zhao, W. (2017). Critical review of existing built environment resilience frameworks: Directions for future research. International Journal of Disaster Risk Reduction, Vol. 25, pp. 173–189, https://doi.org/10.1016/j.ijdrr.2017.09.018.

Chapman, C. (1997). Project risk analysis and management – PRAM the generic process. International Journal of Project Management, Vol. 15, 5, pp. 273–281.

Chapman, R. (2001). The controlling influences on effective risk identification and assessment for construction design management. International Journal of Project Management, Vol. 19, pp. 147–160.

Cooper, D., Chapman, C. (1987). Risk Analysis for Large Projects. Chichester: Wiley.

Deierlein, G. G. (2004, April). White Paper 3 Framework for Structural Fire Engineering and Design Methods Gregory G. Deierlein & Scott Hamilton. In NIST-SFPE Workshop for Development of a National R&D Roadmap for Structural Fire Safety Design and Retrofit of Structures: Proceedings, p. 75.

Edwards, P., Bowen, P. (1998). Risk and risk management in construction: a review and future directions for research. Engineering, Construction and Architectural Management, Vol. 5, p. 4.

El-Sayegh, S. (2008). Risk assessment and allocation in the UAE construction industry. International Journal of Project Management, Vol. 26, pp. 431–438.

ENISA. (2011). Measurement Frameworks and Metrics for Resilient Networks and Services: Technical report. In European Network and Information Security Agency. https://doi.org/10.2495/SI080261.

Falcão Silva, M. J., Almeida, N., Salvado, F., Rodrigues, H. (2020). Modelling structural performance and risk for enhanced building resilience and reliability. Innovative Infrastructure Solutions, vol. 5, 1, pp. 1–20. https://10.1007/s41062-020-0277-1.

Fang, D., Li, M., Fong, P., Shenet, L. (2004). Risks in Chinese construction market – Contractors' perspective. Journal of Construction Engineering and Management, Vol. 130, 6, pp. 853–861.

Flanagen, R., Norman, G. (1993). Risk Management and Construction. Oxford: Blackwell Scientific Publications.

Foliente, G. (2000). Developments in performance-based building codes. Forest Products Journal, Vol. 5, 7/8, pp. 12–21.

Foliente, G., Becker, R. (2001). CIB PBBCS Proactive Programme – Task 1. In Compendium of Building Performance Models. Victoria: CSIRO.

Foliente, G., Leicester, R., November, L. (1998). Development of the CIB Proactive Program on Performance Based Building Codes and Standards. Highett: CSIRO Building Construction and Engineering. BCE Doc 98/232.

Francis, R., Bekera, B. (2014). A metric and frameworks for resilience analysis of engineered and infrastructure systems. Reliability Engineering & System Safety, Vol. 121, pp. 90–103. https://doi.org/10.1016/J.RESS.2013.07.004.

Fujitani, H., Teshigawara, M., Gojo, W., Hirano, Y., Saito, T., Fukuyama, H. (2005). Framework for Performance-Based Design of Building Structures. In Computer-Aided Civil and Infrastructure Engineering (Vol. 20). Retrieved from https://onlinelibrary.wiley.com/doi/pdf/10.1111/j.1467-8667.2005.00377.x.

Gulvanessian, H., Calgaro, J., Holický, M. (2002). Eurocode: Basis of Structural Design. In Designer's Guide to EN 1990. London: Thomas Telford.

Hassler, U., Kohler, N. (2014). Resilience in the built environment. Building Research & Information, Vol. 42, 2, pp. 119–129. https://doi.org/10.1080/09613218.2014.873593.

Hattis, D., Becker, R. (2001). Comparison of the systems approach and the nordic model and their melded application in the development of performance-based building codes and standards. Journal of Testing and Evaluation, JTEVA, Vol. 29, 4, pp. 413–422.

Holický, M., Vrouwenvelder, T. (2005). Chapter I – Basic Concepts of Structural Reliability. Handbook 2 – Reliability Backgrounds: Guide to the basis of structural reliability and risk engineering related to Eurocodes, supplemented by practical examples. Leonardo Da Vinco Pilot Project CZ/02/B/f/PP-134007.

Hollnagel, E. (2014). Resilience engineering and the built environment. Building Research & Information, Vol. 42, 2, 221–228. https://doi.org/10.1080/09613218.2014.862607.

Horvat, M. (2005). Protocol and assessment tool for performance evaluation of light-frame building envelopes used in residential buildings. PhD Thesis. Montreal: Concordia University.

Huovila, P. (2005). Performance Based Performance Based Building Combining Forces-Advancing Facilities Management & Construction through Innovation Series. Retrieved from www.ril.fi.

IRCC. (2010). Performance-based building regulatory systems: Principles and experiences, Report, Inter-jurisdictional Regulatory Collaboration Committee (IRCC).

JCSS. (2009). Joint Committee on Structural Safety. Retrieved from http://www.jcss.ethz.ch/publications/publications.html.

JRC. (2009). Background documents on the Eurocodes. The EN Eurocodes. [Online] European Commission Joint Research Centre. Retrieved from http://eurocodes.jrc.ec.europa.eu/showpage.php?id=BD.

Kangari, R. (1995). Risk management perceptions and trends of US construction. Journal of Construction Engineering and Management, Vol. 121, 4, pp. 422–429.

Klinke, A., Renn, O. (2001). Precautionary principle and discursive strategies: classifying and managing risks. Journal of Risk Research, Vol. 4, 2, pp. 159–173.

Kohler, N. (2018). From the design of green buildings to resilience management of building stocks. Building Research & Information, Vol. 46, 5, 578–593. https://doi.org/10.1080/09613218.2017.1356122.

Kolluru, R., Bartell, S., Pitblado, R., Stricoff, R. (1996). Risk Assessment and Management Handbook. For Environmental, Health, and Safety Professional. Nova Iorque: McGraw-Hill.

Ling, F., Hoi, L. (2006). Risks faced by Singapore firms when undertaking construction projects in India. International Journal of Project Management, Vol. 24, 3, pp. 261–270.

Martinez, A. (2008). Revisión de proyeto: Informe D0. Elementos de apoio do Curso de Técnico Superior en Evaluación de Riesgos Técnicos en Edificación (O.C.T.). Madrid: Universidad Politécnica de Madrid.

May, P. (2007). Thinking About Risk Acceptability. The Use of Risk Concepts in Regulation. San Francisco: ARUP & IRCC.

Meacham, B. (2004a). Performance based-building regulatory system: structure, hierarchy and linkages. Journal of the Structural Engineer Society of New Zealand, Vol. 17, 1, pp. 37–51.

Meacham, B. (2004b). Understanding risk: quantification, perceptions, and characterization. Journal of Fire Protection Engineering, Vol. 14, pp. 199–227.

Meacham, B. (2007). Using Risk as a Basis for Establishing Tolerable Performance: An Approach for Building Regulation. Special Workshop on Risk Acceptance and Risk Communication. Palo Alto, California: Stanford University.

Meacham, B. (2010) Performance-Based Building Regulatory Systems - Principles and Experiences. A Report of the Inter-jurisdictional Regulatory Collaboration Committee. IRCC, 2010.

Meacham, B. (2016). Sustainability and resiliency objectives in performance building regulations. Building Research & Information, Vol. 44, 5–6, pp. 474–489. https://doi.org/10.1080/09613218.2016.1142330.

Meacham, B., Tubbs, B., Bergeron, D., Szigeti, F. (2002). Performance system model – A framework for describing the totality of building performance. 4th International Conference on Performance-Based Codes and Fire Safety Design Methods. Melbourne, VIC: SPFE.

Meacham, B., Van Straalen, J. (2018). A socio-technical system framework for risk-informed performance-based building regulation. Building Research & Information, Vol. 46, 4, 444–462. https://doi.org/10.10 80/09613218.2017.1299525.

Mehmood, A. (2016). Of resilient places: planning for urban resilience. European Planning Studies, Vol. 24, 2, 407–419. https://doi.org/10.1080/09654313.2015.1082980.

Montull, J. (2008). Risk and Insurance Policies. Risk Management in Civil Engineering Advanced Course. Lisboa: LNEC.

Nibel, S., et al. (2004). Evaluation Methods: Theoretical Concepts used by tool designers. IEA BCS Annex 31. Ottawa: Canada Mortgage and Housing Corporation.

Oleszkiewicz, I. (1994). The Concept and Practice of Performance-Based Building Regulations. National Research Council Canada.

Öztaş, A., Ökmen, Ö. (2005). Judgemental risk analysis process development in construction projects. Building and Environment, Vol. 40, pp. 1244–1254.

Palomo, J., Insua, D., Ruggeri, F. (2007). Modeling external risks in project management. Risk Analysis, Vol. 27, 4, pp. 961–978.

Perry, J., Hayes, R. (1985). Risk and it's management in construction projects. Proceedings of the Institution of Civil Engineers. ICE. Vol. 1, pp. 499–521.

Preiser, W., Hardy, A., Schramm, U. (2018). From Linear Delivery Process to Life Cycle Phases: The Validity of the Concept of Building Performance Evaluation. In Building Performance Evaluation (pp. 3–18). https://doi.org/10.1007/978-3-319-56862-1_1.

Re Cecconi, F., Moretti, N., Maltese, S., Dejaco, M. C., Kamara, J., Heidrich, O. (2018). Un rating system per la resilienza degli edifici. TECHNE: Journal of Technology for Architecture & Environment, Vol. 15, pp. 358–365. https://doi.org/10.13128/Techne-22119.

Roostaie, S., Nawari, N., Kibert, C. (2019). Sustainability and resilience: A review of definitions, relationships, and their integration into a combined building assessment framework. Built Environment, Vol. 154, pp. 132–144. https://doi.org/10.1016/j.buildenv.2019.02.042.

Sousa, V., Almeida, N., Dias, L., Matos, J., Branco, F. (2008). Gestão do risco na construção – Aplicação a sistemas de drenagem urbana, 13° Encontro Nacional de Saneamento Básico, Covilhã, 14–17 de outubro.

Spekkink, D. (2005). Performance Based Design of Buildings. PeBBu Domain 3 Final Report. s.l.: CIBdf.

Stahl Jr, R. Bachmann, R., Barton, A., Clark, j. R deFur, P., Ells, S., Pittinger, C., Slimak, M., na Wentsel, R. eds. 2000. EcologicalRisk Management: AFtametoorkforandApproachesto EcologicalRisk-BasedDecisionMaking. Pensacola,FL,SETAePress, inpress. p. 222.

Szigeti, F., Davis, G. (2005a). Performance Based Building: Conceptual Framework. Final Report. s.l: PeBBu Thematic Network.

Szigeti, F., Davis, G. (2005b). What Is Performance Based Building (PBB): In a Nutshell. News Article. s.l.: PeBBu Thematic Network.

Tah, J., Carr, V. (2000). Information modelling for a construction project risk management system. Engineering. Construction and Architectural Management. s.l.: Blackwell Science. Vol. 7, p. 2.

Tubbs, B. (2004). Final report. CIB TG37. Toronto: CIB World Building Congress – Building for the future.

Vale, L. (2014). The politics of resilient cities: whose resilience and whose city? Building Research & Information, Vol. 42, 2, pp. 191–201. https://doi.org/10.1080/09613218.2014.850602.

Vanier, D., Lacasse, M., Parsons, A. (1996). Using product models to represent user requirements. Construction on the Information Highway, Proceedings of CIB Workshop W78. Bled, Slovenia: National Research Council Canada, June 10–12, pp. 511–524.

Wang, M., Chou, H. (2003). Risk allocation and risk handling of highway projects in Taiwan. Journal of Management in Engineering, Vol. 19, 2, pp. 60–68.

Wholey, F. (2015). Building resilience: a framework for assessing and communicating the costs and benefits of resilient design strategies. Perkins + Will Research Journal, Vol. 7, 7-181. Retrieved from https:// perkinswill.com/sites/default/files/ID2_PWRJ_Vol0701_01_BuildingResilience.pdf.

Wilkinson, S., Osmond, P. (2018). Building resilience in urban settlements. International Journal of Building Pathology and Adaptation, Vol. 36, 4, pp. 334–336. https://doi.org/10.1108/IJBPA-08-2018-066.

Wirba, E. (1996). An object-oriented knowledge-based approach to project control. PhD Thesis. London: South Bank University.

# 15 Line Sampling Simulation
## *Recent Advancements and Applications*

*Marcos A. Valdebenito*
Universidad Adolfo Ibáñez, Viña del Mar, Chile


*Marco de Angelis*
University of Liverpool, United Kingdom


*Edoardo Patelli*
University of Strathclyde, United Kingdom

## CONTENTS

## 15.1 INTRODUCTION

The application of numerical models for simulating the behavior of complex systems has become a standard in all branches of engineering. In this way, it is possible to study the performance of systems before being built and take decisions regarding its design. These numerical models are usually highly refined and capable of describing intricate phenomena at the expense of considerable numerical efforts. Their practical implementation demands identifying several parameters such as material properties, loads, degradation processes, etc. However, in practical situations, it can be challenging or even impossible to define such parameters in a crisp way due to inherent uncertainties (Beer et al., 2013). Probability theory (and its recent generalizations) offers the means for characterizing uncertainty in terms of, e.g., random variables, random fields, and stochastic processes. As a consequence, the output response of the numerical model becomes uncertain as well. A possible way to quantify the effect of uncertainty consists of calculating the probability of failure, which measures the chances that one or more responses of the system exceed a prescribed threshold (Melchers & Beck, 2018). Nonetheless, this failure probability can be seldom calculated in closed form, as it depends on the system's performance, which is available from the numerical models for specific values of the input parameters. Simulation techniques such as Monte Carlo methods become most useful, as they are capable of providing estimates of the failure probability by means of sample realizations of the output response, which are calculated based on the numerical solution of the system's model for different realizations of input parameters. However, the downside of sampling-based methods is that they may demand a considerable number of model evaluations, which becomes computationally prohibitive when large, complex models are involved.

The numerical challenges associated with the implementation of Monte Carlo methods have prompted the development of the so-called advanced simulation methods, such as *importance sampling* (Melchers & Beck, 2018), *directional sampling* (Ditlevsen et al., 1988), *line sampling* (Koutsourelakis et al., 2004), and *subset simulation* (Au & Beck, 2001; Au & Patelli, 2016), etc. Among these methods, *line sampling* has played a key role, as it is a simulation approach capable of estimating small failure probabilities, as usually encountered in practical engineering problems. The idea of *line sampling* as an efficient simulation method was proposed by Helmut J. Pradlwarter and first presented in Koutsourelakis et al. (2004). According to Papaioannou and Straub (2021), *line sampling* can be regarded as a generalization of *axis orthogonal sampling*, which was introduced in Hohenbichler and Rackwitz (1988) as a post-process step of first-order reliability method (FORM). While the scope of application of *line sampling* is quite general, it is particularly well suited for treating problems involving a large number of uncertain input parameters (in the order of hundreds or even thousands) where the system's output response is weakly or moderately nonlinear with respect to the input parameters. The key concept associated with *line sampling* consists of estimating failure probabilities by combining random simulation with unidimensional numerical integration along a so-called important direction. In this way, it is possible to generate probability estimates with reduced variability when compared to other simulation approaches (Koutsourelakis et al., 2004). *Line sampling* has received considerable attention from the engineering community, as it has allowed solving challenging engineering problems in aerospace (Pellissetti et al., 2006), automotive (Hinke et al., 2011), hydraulic (Valdebenito et al., 2018), nuclear (Zio & Pedroni, 2012), and structural engineering (Katafygiotis & Wang, 2009; de Angelis et al., 2015), to name a few. Several improvements and extensions of the original *line sampling* technique have been developed, as discussed in detail in Section 15.3. Furthermore, *line sampling* has been implemented in general-purpose software and is available to the research and engineering community, as discussed in detail in Section 15.4.

The purpose of this chapter is to provide a review of the basic theoretical aspects of *line sampling* as well as presenting some selected recent advances. The material presented conveys a summarized yet comprehensive overview of *line sampling*, which sheds light on its current capabilities.

## 15.2 BASIC CONCEPTS

### 15.2.1 Failure Probability Definition

Consider an engineering system, whose behavior is simulated by means of a numerical model. This numerical model comprises a set of $d$ input parameters collected in a vector $x$, whose uncertainty is characterized by means of a vector-valued random variable $X$ with probability density function $f_X(x; \theta)$, where $\theta$ represents a vector of distribution parameters, such as mean, standard deviation, etc. The system's behavior as a function of these input parameters is monitored in terms of a performance function $g(x)$, which assumes a value equal or smaller than zero whenever the output responses exceed prescribed threshold levels. Note that this performance function is usually not known analytically, as it must be calculated through the numerical model for specific realizations of the input parameters. In view of the uncertainty of the input parameters, the performance function itself becomes uncertain as well. One possibility for quantifying this uncertainty is calculating the failure probability $p_F$, which measures the chances of an undesirable system's behavior. Mathematically, the failure probability is calculated in terms of the following integral.

$$p_F = \int_{g(x) \leq 0} f_X(x; \theta) dx \tag{15.1}$$

For ease of implementation, the vector $X$ of *physical* random variables is projected to a vector $Z$ following a standard Gaussian distribution by means of a (possibly nonlinear) iso-probabilistic transformation $T$, that is, $z = T(x; \theta)$, where $z$ denotes a realization of $Z$ (Der Kiureghian, 2004). Thus, the expression for the failure probability in the standard normal space becomes:

$$p_F = \int_{g(T^{-1}(z; \theta)) \leq 0} f_Z(z) dz \tag{15.2}$$

where $T^{-1}(\cdot)$ denotes the inverse iso-probabilistic transformation.

The calculation of the failure probability is far from trivial: on top of the challenge of a performance function, which is not known in closed form, the associated integral may span a very large number of dimensions $d$, in the order of hundreds or even thousands. These two issues favor the application of simulation methods such as line sampling for estimating failure probabilities.

### 15.2.2 *Line Sampling for Probability Estimation*

The basis of *line sampling* consists of identifying a so-called *important direction* $\alpha$, which is a unit vector pointing toward the *failure domain*, which is the set of realizations of the input parameters that cause an undesirable system's behavior (i.e., $g(z) \leq 0$). Figure 15.1 provides a schematic representation of the failure domain as well as a possible important direction. Approaches for selecting an important direction are discussed in detail in Section 15.2.3.

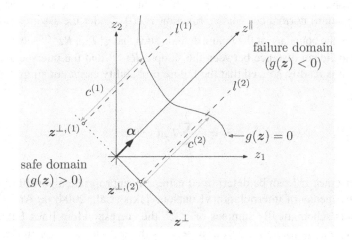

**FIGURE 15.1** Schematic representation of Line Sampling.

*Line sampling* takes advantage of the rotational invariance of the standard normal distribution by introducing a rotated coordinate system for estimating the failure probability:

$$z = Rz^{\perp} + \alpha z^{\|} \tag{15.3}$$

where $z^{\perp}$ is a vector of dimension $(d-1)$ representing coordinates belonging to the hyperplane orthogonal to $\alpha$, $z^{\|}$ is a scalar denoting the coordinate parallel to $\alpha$, and $R$ is a matrix of dimension $d \times (d-1)$. Matrix $[R, \alpha]$ forms an orthonormal basis and thus, $z^{\|} = \alpha^{T} z$ and $z^{\perp} = R^{T} z$. In addition, the probability distributions associated with $z^{\perp}$ and $z^{\|}$ are standard normal distributions in $(d-1)$ dimensions and one dimension, respectively. In view of these properties, the probability integral in Eq. (15.2) can be recast in terms of the rotated coordinate system:

$$p_{F} = \int_{g\left(T^{-1}\left(Rz^{\perp} + \alpha z^{\|}; \theta\right)\right) \le 0} f_{z^{\|}}\left(z^{\|}\right) f_{z^{\perp}}\left(z^{\perp}\right) dz^{\|} dz^{\perp}. \tag{15.4}$$

This integral can be evaluated by performing Monte Carlo simulation on the hyperplane orthogonal to $\alpha$ and then, conducting unidimensional integration along the direction parallel to $\alpha$. For this purpose, $L$ samples are generated according to:

$$Rz^{\perp,(i)} = z^{(i)} - (\alpha^{T} z^{(i)})\alpha, \; i = 1, \ldots, L \tag{15.5}$$

where $z^{(i)}, i = 1, \ldots, L$ denote samples of a $d$-dimensional standard normal distribution on the hyperplane orthogonal to $\alpha$. Note that with this formulation, one does not obtain $z^{\perp,(i)}$ but actually, $Rz^{\perp,(i)}$. This is quite advantageous, as there is no need to calculate the rotation matrix $R$ in explicit form. Then, the Monte Carlo estimate of the probability integral in the rotated coordinate system as shown in Eq. (15.4) is equal to:

$$p_{F} \approx \tilde{p}_{F} = \frac{1}{N} \sum_{i=1}^{N} \int_{g(T^{-1}(Rz^{\perp,(i)} + \alpha z^{\|}; \theta)) \le 0} f_{z^{\|}}(z^{\|}) dz^{\|}. \tag{15.6}$$

The argument of the summation involves a unidimensional integral along the line $i$ that passes through the sample $Rz^{\perp,(i)}$ and which is parallel to the important direction $\alpha$. As this unidimensional integral involves the standard normal probability function, it possesses a closed-form solution in terms of the standard normal cumulative function $F_{z^{\|}}(\cdot)$. Under the assumption that the line $i$ intersects once with the boundary of the failure domain such that $g\left(T^{-1}\left(Rz^{\perp,(i)} + \alpha c^{(i)}; \theta\right)\right) = 0$, where $c^{(i)}$ denotes the Euclidean distance between the sample $Rz^{\perp,(i)}$ and the intersection point with the limit state $g = 0$, it is readily noticed that the failure probability estimator simplifies to the following expression:

$$\tilde{p}_{F} = \frac{1}{L} \sum_{i=1}^{L} F_{z^{\|}}(-c^{(i)}). \tag{15.7}$$

The Euclidean distance $c^{(i)}$ can be determined using any appropriate algorithm for finding roots of equations or by means of interpolation (Koutsourelakis et al., 2004; de Angelis et al., 2015). Figure 15.1 depicts schematically samples of $z^{\perp,(i)}$, the corresponding lines $i$ that pass through those samples and which are parallel to $\alpha$ as well as the distances $c^{(i)}$ for the case where $d = 2$ and $L = 2$.

The accuracy of the estimator generated by *line sampling* can be quantified in terms of its coefficient of variation $\delta$, i.e., the ratio between the standard deviation of the estimate and its expected value. It is calculated according to the formula shown below (Koutsourelakis et al., 2004):

$$\delta = \frac{1}{\tilde{p}_F} \sqrt{\frac{1}{L(L-1)} \sum_{i=1}^{L} \left( F_{z\parallel}\left(-c^{(i)}\right) - \tilde{p}_F \right)^2} \tag{15.8}$$

The coefficient of variation is most useful for judging the quality of the probability estimate and deciding on the number of lines $L$ to be sampled. As a rule of thumb, whenever $\delta \leq 10\%$, then the probability estimate is deemed as sufficiently accurate.

Note that the probability estimator in Eq. (15.7) always leads to the correct result on the long run as $L \rightarrow \infty$, irrespective of the selection of the important direction $\alpha$. In this sense, it should be recalled that *line sampling* combines Monte Carlo simulation and univariate quadrature and hence, no approximations are introduced (other than sampling). However, the level of accuracy of the estimator $\delta$ may be affected (in some cases, to a large extent) by the selection of the important direction $\alpha$. This issue is further discussed below.

### 15.2.3 SELECTION OF THE IMPORTANT DIRECTION

The important direction $\alpha$ should be selected such that the set of sampled Euclidean distances $c^{(i)}$, $i = 1,...,L$ associated with each line exhibit the smallest possible variation. In this way, the coefficient of variation of the probability estimator is minimized. This issue is illustrated schematically in Figure 15.2 for the case where $d = 2$ and $L = 2$. Specifically, Figure 15.2a illustrates a poor selection of the important direction $\alpha_1$, which leads values of $c^{(i)}$, $i = 1$, 2 which are quite different between them. On the contrary, in Figure 15.2b, the selected important direction $\alpha_2$ produces values of $c^{(i)}$, $i = 1$, 2 which are quite similar between them and hence, leads to a probability estimator with a reduced coefficient of variation.

Several different strategies have been proposed to select the important direction $\alpha$ (Koutsourelakis et al., 2004). For example, it can be selected as the unit vector pointing in the opposite direction of the gradient of the performance function at the origin of the standard normal space. This selection strategy is quite simple, and it is expected to work well for problems where the performance function $g$ exhibits a weakly to mildly nonlinear behavior with respect to $z$, see, e.g., Pellissetti et al. (2006). For the practical implementation of this strategy, schemes for estimating the gradient based on sampling are a suitable choice (Pradlwarter, 2007), as they can cope with performance functions available as a black box (without access to analytical gradients) and a large number of input parameters. Another strategy for choosing $\alpha$ is selecting it as the unit vector pointing toward the *design*

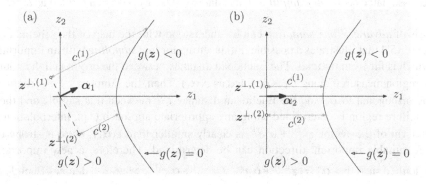

**FIGURE 15.2** Selection of important direction $\alpha$. (a) Poor choice. (b) Good choice.

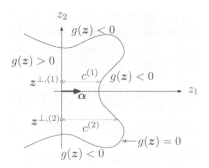

**FIGURE 15.3**   Application of Line Sampling for highly nonlinear performance function.

*point* $z^*$ associated with the performance function. In this context, recall that the design point $z^*$ is the realization of $Z$ with the smallest Euclidean norm such that $g\left(T^{-1}\left(z^*;\theta\right)\right) \leq 0$ (Der Kiureghian, 2004). Yet another strategy for selecting the important direction consists of applying engineering judgment, as failure usually occurs in engineering systems whenever load increases and resistance decreases.

It should be noted that for problems comprising highly nonlinear functions, the variability of the *line sampling* estimator may be considerable, even for properly selected important directions. Such issue is depicted schematically in Figure 15.3. The performance function illustrated in this figure is highly nonlinear. The selected important direction is the one that matches with the direction of the design point. Nonetheless, the variability of the distances $c^{(i)}$, $i = 1$, 2 becomes quite large, as the nonlinearity of the performance function is severe. While line sampling would still lead to a correct estimate of the failure probability on the long run, the variability of the estimate becomes comparable to that of Monte Carlo simulation. Hence, other simulation approaches such as, e.g., subset simulation (Au & Patelli, 2016) may be more appropriate in case a highly nonlinear performance function is encountered.

## 15.3   RECENT DEVELOPMENTS OF *LINE SAMPLING*

### 15.3.1   ADVANCED LINE SAMPLING

Choosing an important direction may be a challenging task for those cases where the value of the performance function is the product of a numerical model, which is available in the form of a black box. An *adaptive* strategy, as developed in de Angelis et al. (2015), can be used to update the important direction on the fly as one collects information about the reliability problem at hand by means of the lines associated with *line sampling*. This adaptive strategy has been termed as *advanced line sampling*.

The basis of *advanced line sampling* can be understood with the help of the schematic illustration in Figure 15.4 that illustrates a possible initial setting for *line sampling* with an important direction $\alpha_1$, which is far from optimal. The Euclidean distance between the origin and the boundary of the failure region measured along $\alpha_1$ is denoted as $c(\alpha_1)$. Then, the sample $z^{\perp,(1)}$ is generated in the hyperplane orthogonal to $\alpha_1$ and the Euclidean distance $c^{(1)}$ between that sample and the boundary of the failure region is determined using any appropriate approach (e.g., interpolation). As the Euclidean norm of the vector $z^{\perp,(1)} + \alpha_1 c^{(1)}$ is clearly smaller than $c(\alpha_1)$, there is strong evidence that a more suitable important direction can be determined. Therefore, a new important direction $\alpha_2$ is defined such that $\alpha_2 = \left(z^{\perp,(1)} + \alpha_1 c^{(1)}\right) / \left\|z^{\perp,(1)} + \alpha_1 c^{(1)}\right\|$, where $\|\cdot\|$ denotes Euclidean norm. Thus, *line sampling* continues with additional lines that are parallel to $\alpha_2$ as shown in Figure 15.4b,

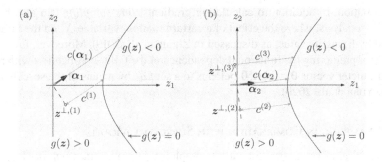

**FIGURE 15.4**  Advanced Line Sampling.

until eventually finding a new, improved important direction. As shown in de Angelis et al. (2015), an unbiased estimate of the failure probability is obtained by means of Eq. (15.7), irrespective of the fact that different distances $c^{(i)}$, $i = 1,\ldots,L$ may have been associated with different important directions.

The advantage of the *advanced line sampling* procedure is that it is possible to start the simulation with an important direction, which is known to be suboptimal. However, as the simulation proceeds and more lines are sampled, the selection of the important direction is automatically updated. This is most convenient from a numerical point of view, as it avoids wasting unnecessary effort on determining an optimal important direction and instead, paying more attention to probability assessment. The advantages of *advanced line sampling* for adaptively selecting the important direction have been well documented for both academic and engineering problems (de Angelis et al., 2015).

### 15.3.2  COMBINATION LINE SAMPLING

*Combination line sampling* (Papaioannou & Straub, 2021) has been developed as a further improvement of *advanced line sampling*. In *combination line sampling*, weights are introduced into Eq. (15.7) in order to account for the fact that different distances $c^{(i)}$, $i = 1,\ldots,L$ may have been calculated with different important directions. These weights are proportional to the product between the probability content (assuming a first-order reliability approximation) and the number of sampled lines associated with each important direction. The advantage of *combination line sampling* is that it is applied following the same steps of *advanced line sampling*, except for the final probability estimator, which is modified by means of the information collected throughout the simulation process, without requiring any additional evaluations of the performance function. Both *advanced line sampling* and *combination line sampling* may offer significant advantages in reliability problems in particular for those involving a small to moderate nonlinearity (Papaioannou & Straub, 2021).

### 15.3.3  PROBABILITY SENSITIVITY ESTIMATION

The failure probability as defined in Eq. (15.1) depends implicitly on the vector of distribution parameters $\theta$. This is an obvious consequence of the structure of that equation, as the vector of distribution parameters affects the probabilistic characterization of the input variables of the system. In certain applications such as, e.g., risk-based optimization (Melchers & Beck, 2018), it is of interest assessing the sensitivity of the failure probability $p_F$ with respect to changes in the value of the distribution parameters $\theta$. Such sensitivity can be evaluated by means of the gradient $\partial p_F / \partial \theta_j, j = 1,\ldots,n_\theta$, where $n_\theta$ denotes the number of distribution parameters involved in the problem. It has been shown that this probability sensitivity can be evaluated in a post-processing step of *line sampling* (Lu et al., 2008; Papaioannou et al., 2013; Valdebenito et al., 2018). This means that once that the probability of failure associated with a certain system has been evaluated, it is possible to obtain the probability sensitivity by re-using the information gathered for the probability estimation. In addition to calculating gradients, *line sampling* can also be applied for *global* sensitivity analysis, where the effect of a certain random variable $X_i$ on the failure probability is measured by fixing its value, as discussed in Zhang et al. (2019). Moreover, *line sampling* can be also applied for predicting the functional dependence of the failure probability with respect to the distribution parameter vector $\theta$, where $\theta$ belongs to a set $\Omega_\theta$, by means of the so-called *augmented line sampling* (Yuan et al., 2020).

### 15.3.4 *LINE SAMPLING* IN COMBINATION WITH SURROGATE MODELS

Most of the numerical effort associated with the implementation of *line sampling* is involved in the exploration of each sampled line in order to determine the point of intersection with the boundary of the failure domain. Such task is usually carried out by means of interpolation or root-finding algorithms with some special modifications, as documented in de Angelis et al. (2015) and Zhang et al. (2019). A possible means for decreasing the numerical cost associated with this step of *line sampling* consists of training a surrogate model for the performance function. As a surrogate model can be evaluated with negligible numerical effort, it is possible to analyze a large number of lines with great efficiency, avoiding full evaluations of the performance function. In this context, the application of *Gaussian process regression* (also known as Kriging) has been investigated in conjunction with *line sampling* in Depina et al. (2016) and Song et al. (2021), allowing to decrease the total number of evaluations of the full numerical model of the system without sacrificing the accuracy of the probability estimates.

### 15.3.5 *LINE SAMPLING* IN THE PRESENCE OF ALEATORY AND EPISTEMIC UNCERTAINTY

The preceding sections have focused on the application of *line sampling* for problems where distribution parameters $\theta$ are known precisely. Nonetheless, in several situations of practical interest, imprecision on distribution parameters $\theta$ may be modeled by means of intervals $\Theta$, leading to a parametric probability box (p-box) characterized by a set of possible probability density functions $f_X(x; \theta)$, $\theta \in \Theta$ (Beer et al., 2013). In such setting, the failure probability as defined in Eq. (15.1) is no longer exact but interval valued, such that $p_F \in \left[ p_F^L, p_F^U \right]$, where $p_F^L$ and $p_F^U$ denote the lower and upper bounds of the failure probability.

The adoption of a parametric probability box requires to compute the failure probability for a collection of different probability models (indexed by the interval variable vector $\theta \in \Theta$) in order to determine its bounds. A straightforward approach is proposed in Li and Lu (2014), where an optimization algorithm determines the extrema of the failure probability for different realizations of $\theta$ by means of *line sampling* but with considerable numerical efforts, as reliability analysis becomes embedded within the optimization. As an alternative, single-loop approaches aim at solving the imprecise reliability problem by means of a single run of *line sampling* and determine the sought probability bounds using sample reweighting (de Angelis et al., 2015; Patelli & De Angelis, 2015). In this context, sample reweighting means that the reliability problem is solved considering certain values $\theta$ for the distribution parameters and then, the probability can be predicted at new values of the distribution parameters $\theta'$ by reweighting the samples taking into account the ratio between probability density functions (i.e., $f_X(x; \theta')/f_X(x; \theta)$). Such strategy has been further extended in order to account for possible weight degeneracy by means of high-dimensional model representation and approximation concepts, leading to a so-called non-intrusive stochastic simulation approach (Song et al., 2020).

## 15.4 SOFTWARE IMPLEMENTATION

*Line sampling* and *advanced line sampling* have been successfully implemented into the Cossan software (Patelli, 2016). Cossan software provides a general-purpose platform for uncertainty quantification available as a Matlab toolbox named OpenCossan (Patelli et al., 2014, 2018) and as a standalone software with user-friendly interface named COSSAN-X (Patelli et al., 2011, 2012). Cossan software has been developed for interacting with existing, third-party deterministic numerical models of complex systems and performing tasks such as reliability analysis, optimization, and robust design, to name a few. A detailed explanation of the software and its capabilities goes beyond the scope of this contribution. The interested reader is referred to the website of software: https://cossan.co.uk

## 15.5 EXAMPLES

### 15.5.1 RANDOM SEEPAGE RELIABILITY ANALYSIS

Consider the seepage problem illustrated schematically in Figure 15.5a, where an impermeable dam lies over a permeable soil layer. The soil permeability is modeled by means of a homogeneous lognormal random field with expected value and standard deviation equal to $10^{-6}$ [m/s] and correlation $e^{-d^2/L_C}$, where $L_C = 5$ [m] represents the correlation length. The objective is to determine the probability that the seepage flow below the dam exceeds the value $q = 40$ [L/h/m] due to stability concerns.

The seepage phenomenon is modeled by means of Poisson's equation, which is discretized by means of a finite element model comprising about 1500 quadratic triangular elements. The random field is discretized by means of the mid-point method. Thus, the discrete representation of the random field by means of the Karhunen-Loève expansion comprises a total of $d = 73$ standard normal random variables. In order to estimate the probability that the seepage flow exceeds its prescribed threshold, line sampling is employed considering $L = 300$ lines. The important direction $\alpha$ is selected as the vector pointing toward the opposite direction of the gradient of the performance function at the origin of the standard normal space.

The evolution probability estimate as a function of the number of lines is shown in Figure 15.5b. It is readily seen that the estimate of the failure probability possesses excellent properties, as it stabilizes with a reduced number of lines. In fact, the coefficient of variation of the probability estimate is equal to $\delta = 1\%$, highlighting the efficiency of *line sampling* for estimating small failure probabilities.

### 15.5.2 TNO BLACK BOX RELIABILITY EXAMPLES

The 2019 TNO black box reliability challenge is a collection of reliability problems already published in the literature and gathered by the Dutch research organization (www.tno.nl/en) to test the performance and accuracy of reliability methods. Only a limited number (1e4) of evaluations of the example functions are allowed to produce an estimate of the failure probability, moreover, the

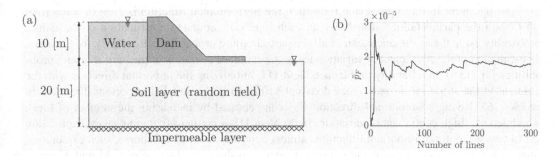

**FIGURE 15.5** Random seepage analysis problem. (a) Dam cross section. (b) Seepage analysis result.

**TABLE 15.1**

**Summary of Reliability Results Obtained with *Line Sampling* for Single-Output Problems**

| Prob. id | 14 | 24 | 28 | 31 | 38 | 53 | 54 | 63[a] | 75 | 107 | 111[a] |
|---|---|---|---|---|---|---|---|---|---|---|---|
| Dist. | Mixed | $\sim N$ | $\sim N$ | $\sim Z$ | $\sim N$ | $\sim N$ | $\sim Exp$ | $\sim Z$ | $\sim Z$ | $\sim Z$ | $\sim Z$ |
| $d$ | 5 | 2 | 2 | 2 | 7 | 2 | 20 | 100 | 2 | 10 | 2 |
| $10^m$ | 1e-4 | 1e-4 | 1e-7 | 1e-4 | 1e-5 | 1e-3 | 1e-4 | 1e-4 | 1e-3 | 1e-7 | 1e-4 |
| $\tilde{p}_{F,\,MC}$ | 5.52 | 1.03 | 1.46 | 1.80 | 1.38 | 2.57 | 9.98 | 3.79 | 3.21 | 2.92 | 1.02 |
| $\tilde{p}_{F,\,LS}$ | **6.94** | **3.49** | **1.86** | **29.4** | **1.63** | **8.14** | **8.47** | **45.7** | **4.60** | **2.87** | **7.29** |
| $\delta$ (1e-2) | **7.49** | **9.75** | **12.37** | **9.96** | **9.19** | **6.64** | **8.19** | **18.9** | **9.30** | **3.01** | **5.63** |
| $N$ | 3970 | 1037 | 2814 | 6519 | 1946 | 672 | 4832 | 8601 | 1874 | 355 | 738 |
| $L$ | 360 | 80 | 400 | 499 | 200 | 45 | 400 | 500 | 120 | 30 | 50 |

[a]   The performance function had to be changed to meet the Monte Carlo reference solution, which was incorrectly reported.

example functions could only be queried remotely by connecting to the challenge server to mimic black box interaction. The example set has 16 reliability problems split into two categories: (1) a set of single-output problems and (2) a set of examples with multi-output *performance function* problems. *Line sampling* has shown competitive results against the other popular reliability methods taking part in the challenge, such as *subset simulation, importance sampling, asymptotic sampling,* and variations thereof. The challenge problems have now been released and live at the following address https://rprepo.readthedocs.io/en/latest/reliability_problems.html, whereby each problem can now be accessed without the need to query the challenge server. In this section, we present the results against the first set of challenge problems using state-of-the-art *line sampling*. And we compare these results against the reference Monte Carlo solution, or *target* solution. We show the stepwise process to achieve these results on one of the example problems.

In Table 15.1, we document the results of using *line sampling* $\tilde{p}_{F,LS}$ to solve the reliability problems against the Monte Carlo reference solution $\tilde{p}_{F,\,MC}$. The number of function evaluations $N$ and number of lines $L$ are also reported for each problem. The failure probability estimation $\tilde{p}_F$ is provided in the scientific form $s*10^m$, where $s$ is the significand and $m$ is the order of magnitude, e.g., $3*10^{-4} := 3*1e-4$. The stopping criterion for ending the simulation is checking that the coefficient of variation is below 1e-2, consistently with the assurance that no better important direction is discovered during the processing of each line. Most of the reliability problems were characterized by normal probability distributions ($\sim N$) or standard normal probability distributions $\sim Z := N(0,1)$. All the random variables were given as mutually independent.

The line sampling method consists of the following six main steps:

(1) Initialize the important direction; (2) Evaluate the performance function on the line samples; (3) Find the point in the input space for which the performance function is zero on each line; (4) Obtain the partial failure probability on each line; (5) Compute an estimation of the failure probability using the arithmetic mean of all the partial failure probabilities Eq. (15.7); (6) Compute an estimation of the coefficient of variation using the standard deviation of the partial failure probabilities Eq. (15.8). In reliability problems **63** and **111**, initializing the important direction with the gradient in the *standard normal space* does not suffice to find a suitable important direction. In problem 63, having a suboptimal direction was compensated by increasing the number of lines, which led to a high coefficient of variation. In problem 111, a gradient-free input space exploration led to two sub-optimal important directions almost opposite oriented. Of course, such exploration implied a further investment in the number of model evaluations. In problems **14** and **38**, one-directional update was necessary to achieve the desired accuracy ($< 10*1e-2$). More details about

each problem solution can be found at https://github.com/marcodeangelis/Line-sampling, where the results presented in Table 15.1 can be replicated.

## 15.6 CONCLUSIONS

*Line sampling* is a robust and efficient simulation method for reliability analysis. From its original formulation, several substantial improvements, such as the selection of an important direction or the exploration of the lines, have been proposed to improve the robustness and applicability of the method. Furthermore, the application of *line sampling* to problems beyond basic reliability assessment has been well documented in the literature, comprising problems of reliability sensitivity analysis, imprecise probabilities, and risk-based optimization. The niche of *line sampling* lies on reliability problems that involve weakly to moderately nonlinear performance functions, where its numerical efficiency excels over other available reliability techniques. The computational implementation of *line sampling* within Cossan software is well tested and is available for the scientific and engineering community. Hence, as a summary, it can be stated that *line sampling* is an already mature approach for reliability analysis and uncertainty quantification. In fact, given its performance on realistic industrial-case examples and its competitiveness against the other popular reliability methods, *line sampling* can be considered a viable choice to tackle challenging large-scale applications, with good scalability in terms of the number of dimensions and failure probability magnitude.

## REFERENCES

Au, S. K., & Beck, J. L. (2001). Estimation of small failure probabilities in high dimensions by subset simulation. *Probabilistic Engineering Mechanics, 16*(4), 263–277.

Au, S. -K., & Patelli, E. (2016). Rare event simulation in finite-infinite dimensional space. *Reliability Engineering & System Safety, 148,* 67–77. https://doi.org/10.1016/j.ress.2015.11.012

Beer, M., Ferson, S., & Kreinovich, V. (2013). Imprecise probabilities in engineering analyses. *Mechanical Systems and Signal Processing, 37*(1–2), 4–29. https://doi.org/10.1016/j.ymssp.2013.01.024

de Angelis, M., Patelli, E., & Beer, M. (2015). Advanced line sampling for efficient robust reliability analysis. *Structural Safety, 52, Part B,* 170–182. https://doi.org/10.1016/j.strusafe.2014.10.002

Depina, I., Le, T. M. H., Fenton, G., & Eiksund, G. (2016). Reliability analysis with metamodel line sampling. *Structural Safety, 60,* 1–15. https://doi.org/10.1016/j.strusafe.2015.12.005

Der Kiureghian, A. (2004). *Engineering Design Reliability Handbook. In* S. Singhal, D.M. Ghiocel, & E. Nikolaidis (Eds.). CRC Press.

Ditlevsen, O., Bjerager, P., Olesen, R., & Hasofer, A.M. (1988). Directional simulation in Gaussian processes. *Probabilistic Engineering Mechanics, 3*(4), 207–217. https://doi.org/10.1016/0266-8920(88)90013-6

Hinke, L., Pichler, L., Pradlwarter, H. J., Mace, B. R., & Waters, T. P. (2011). Modelling of spatial variations in vibration analysis with application to an automotive windshield. *Finite Elements in Analysis and Design, 47*(1), 55–62. https://doi.org/10.1016/j.finel.2010.07.013

Hohenbichler, M., & Rackwitz, R. (1988). Improvement of second-order reliability estimates by importance sampling. *Journal of the Engineering Mechanics Division, ASCE, 114*(12), 2195–2199.

Katafygiotis, L. S., & Wang, J. (2009). Reliability analysis of wind-excited structures using domain decomposition method and line sampling. *Journal of Structural Engineering and Mechanics, 32*(1), 37–51.

Koutsourelakis, P. S., Pradlwarter, H. J., & Schuëller, G. I. (2004). Reliability of structures in high dimensions, part I: algorithms and applications. *Probabilistic Engineering Mechanics, 19*(4), 409–417.

Li, L., & Lu, Z. (2014). Interval optimization based line sampling method for fuzzy and random reliability analysis. *Applied Mathematical Modelling, 38*(13), 3124–3135. https://doi.org/10.1016/j.apm.2013.11.027

Lu, Z., Song, S., Yue, Z., & Wang, J. (2008). Reliability sensitivity method by line sampling. *Structural Safety, 30*(6), 517–532. https://doi.org/10.1016/j.strusafe.2007.10.001

Melchers, R. E., & Beck, A. T. (2018). *Structural Reliability Analysis and Prediction* (3rd ed.). John Wiley & Sons. https://doi.org/10.1002/9781119266105

Papaioannou, I., Breitung, K., & Straub, D. (2013). Reliability sensitivity analysis with Monte Carlo methods. In B. E. G. Deodatis & D. M. Frangopol (Eds.), *11th International Conference on Structural Safety and Reliability (ICOSSAR),* pp. 5335–5342.

Papaioannou, I., & Straub, D. (2021). Combination line sampling for structural reliability analysis. *Structural Safety, 88*, 102025. https://doi.org/10.1016/j.strusafe.2020.102025

Patelli, E. (2016). COSSAN: a multidisciplinary software suite for uncertainty quantification and risk management. In Roger Ghanem, David Higdon, & Houman Owhadi (Eds.), *Handbook of Uncertainty Quantification* (pp. 1–69). Cham: Springer International Publishing. https://link.springer.com/referenceworkentry/10.1007%2F978-3-319-12385-1_59.

Patelli, E., & De Angelis, M. (2015). Line sampling approach for extreme case analysis in presence of aleatory and epistemic uncertainties. *25th European Safety and Reliability Conference (Esrel 2015)*, pp. 2585–2593.

Patelli, E., Broggi, M., De Angelis, M., & Beer, M. (2014). OpenCossan: an efficient open tool for dealing with epistemic and aleatory uncertainties. *Second International Conference on Vulnerability and Risk Analysis and Management (ICVRAM) and the Sixth International Symposium on Uncertainty, Modeling, and Analysis (ISUMA)*, pp. 2564–2573.

Patelli, E., George-Williams, H., Sadeghi, J., Rocchetta, R., Broggi, M., & de Angelis, M. (2018). OpenCossan 2.0: an efficient computational toolbox for risk, reliability and resilience analysis. In André T. Beck, Gilberto F. M. de Souza, & Marcelo A. Trindade (Eds.), *Proceedings of the Joint ICVRAM ISUMA UNCERTAINTIES Conference*. http://icvramisuma2018.org/cd/web/PDF/ICVRAMISUMA2018-0022.PDF

Patelli, E., Panayirci, H. M., Broggi, M., Goller, B., Beaurepaire, P., Pradlwarter, H.J., & Schuëller, G.I. (2012). General purpose software for efficient uncertainty management of large finite element models. *Finite Elements in Analysis and Design, 51*, 31–48. https://doi.org/10.1016/j.finel.2011.11.003

Patelli, E., Valdebenito, M. A., & Schuëller, G. I. (2011). General purpose stochastic analysis software for optimal maintenance scheduling: application to a fatigue-prone structural component. *International Journal of Reliability and Safety, 5*(3–4), 211–228. https://doi.org/10.1504/IJRS.2011.041177

Pellissetti, M. F., Schuëller, G. I., Pradlwarter, H. J., Calvi, A., Fransen, S., & Klein, M. (2006). Reliability analysis of spacecraft structures under static and dynamic loading. *Computers & Structures, 84*(21), 1313–1325.

Pradlwarter, H.J. (2007). Relative importance of uncertain structural parameters. Part I: algorithm. *Computational Mechanics, 40*(4), 627–635.

Song, J., Valdebenito, M., Wei, P., Beer, M., & Lu, Z. (2020). Non-intrusive imprecise stochastic simulation by line sampling. *Structural Safety, 84*, 101936. https://doi.org/10.1016/j.strusafe.2020.101936

Song, J., Wei, P., Valdebenito, M. A., & Beer, M. (2021). Active learning line sampling for rare event analysis. *Mechanical Systems and Signal Processing, 147*, 107113. https://doi.org/10.1016/j.ymssp.2020.107113

Valdebenito, M. A., Jensen, H. A., Hernández, H. B., & Mehrez, L. (2018). Sensitivity estimation of failure probability applying line sampling. *Reliability Engineering & System Safety, 171*, 99–111. https://doi.org/10.1016/j.ress.2017.11.010

Yuan, X., Zheng, Z., & Zhang, B. (2020). Augmented line sampling for approximation of failure probability function in reliability-based analysis. *Applied Mathematical Modelling, 80*, 895–910. https://doi.org/10.1016/j.apm.2019.11.009

Zhang, X., Lu, Z., Yun, W., Feng, K., & Wang, Y. (2019). Line sampling-based local and global reliability sensitivity analysis. *Structural and Multidisciplinary Optimization*. https://doi.org/10.1007/s00158-019-02358-9

Zio, E., & Pedroni, N. (2012). Monte Carlo simulation-based sensitivity analysis of the model of a thermal-hydraulic passive system. *Reliability Engineering & System Safety, 107*, 90–106. https://doi.org/10.1016/j.ress.2011.08.006

# 16 Probabilistic Performance Analysis of Modern RC Frame Structures Subjected to Earthquake Loading

*Andrei Pricopie and Florin Pavel*
Technical University of Civil Engineering Bucharest, Bucharest, Romania

*Ehsan Noroozinejad Farsangi*
Graduate University of Advanced Technology, Kerman, Iran

## CONTENTS

## 16.1 INTRODUCTION: BACKGROUND AND DRIVING FORCES

The assessment of the seismic performance of reinforced concrete structures has been dealt with in a significant number of papers in the literature. The main issues are related to the fact that a large number of nonlinear time-history analyses are required for performing the analysis. Consequently, simplified models have been proposed by various authors in order to reduce the time necessary for performing the analyses. The use of equivalent single-degree-of-freedom (SDOF) systems for the evaluation of the seismic performance and the seismic fragility of reinforced concrete and masonry structures has been proposed by Kuramoto et al. (2000), Decanini, Mollaioli and Mura (2001), Vamvatsikos and Cornell (2005), Oviedo, Midorikawa and Asari (2010), Graziotti, Penna and Magenes (2016), and Suzuki and Iervolino (2019).

Thus, in this study, the primary focus is to perform a fully probabilistic performance analysis in order to evaluate the vulnerability of modern reinforced concrete (RC) frames structures. The secondary focus is to analyse the sensitivity of the results to the number of performed analyses, to the random variables used for the analysis, and whether results obtained on a simplified SDOF system can be adequate from a practitioners' perspective. Four reinforced concrete frame structures situated in both low and high seismicity areas and affected by either crustal or intermediate-depth

DOI: 10.1201/9781003194613-16

earthquake are analysed in this study. After evaluating the seismic performance of the four structures using site-specific ground motion recordings, the seismic fragility is evaluated considering three limit states (serviceability, ultimate, and collapse). Finally, the results in terms of annual probabilities of exceedance of each damage state are evaluated.

## 16.2   STRUCTURAL MODELS

The study addresses four, seven-story reinforced concrete structures designed for different seismic conditions. The design sites are chosen considering different levels of seismic intensity, different soil conditions, and a seismic source that is shallow crustal for two of the structures as well as intermediate depth also on two structures. All of the structures are considered to have three 6-m spans on each direction, and a constant story height of 3 m. A vertical cross-section of the models is illustrated in Figure 16.1.

The design of the structures is carried out according to P100-1/2013 (MDRAP 2013), the current seismic design code in Romania. The code is mainly based on the provisions of the European seismic design code EN 1998-1/2004 (CEN 2004). The structures are designed for two limit states: damage limitation (serviceability limit state [SLS]) and life safety (ultimate limit state [ULS]). The mean return period (MRP) of the peak ground acceleration used for anchoring the design response spectrum specified by the code is 40 years for SLS and 225 years for ULS. All frames were designed using the equivalent lateral force method for the high ductility class (DCH) and a behavior factor $q = 6.75$. The dead and live loads considered for the design are the same for all four structures with a dead load, which accounts for the slab thickness of 15 cm, finishes and partitions of $q_d = 6.5$ kN/m², and a live load of $q_u = 2$ kN/m². The cross-section of the beams and columns is constant along the height of each structure. The base shear coefficients used for the design are in the range 2.3–9.4% of the building weight. The main characteristics of the four frames are listed in Table 16.1.

The design of the structures has been carried out using the equivalent lateral force method, and checked with a static pushover analysis that included determining the displacement demand

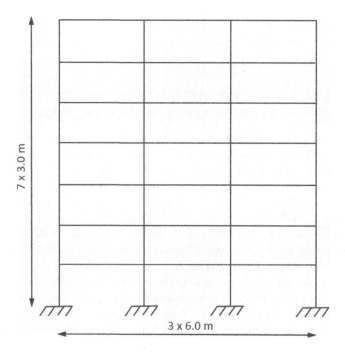

**FIGURE 16.1**   Vertical cross-section of the analysed RC frames.

**TABLE 16.1**
**Characteristics of the Analyzed Structures**

| Structure | Base Shear Coefficient (%) | Fundamental Eigenperiod $T_1$ (s) | Beam Cross-Section (cm) | Column Cross-Section (cm) |
|---|---|---|---|---|
| Bucharest | 9.4 | 0.66 | 30 × 60 | 70 × 70 |
| Iasi | 5.9 | 0.92 | 30 × 45 | 60 × 60 |
| Timisoara | 4.9 | 0.91 | 30 × 45 | 60 × 60 |
| Cluj | 2.3 | 0.95 | 30 × 45 | 55 × 55 |

**TABLE 16.2**
**Beam Top and Bottom Reinforcement for Each Story**

<table>
<tr><th></th><th></th><th colspan="7">Beam Rebar/Story</th></tr>
<tr><th>Structure</th><th></th><th>S1</th><th>S2</th><th>S3</th><th>S4</th><th>S5</th><th>S6</th><th>S7</th></tr>
<tr><td>Bucharest</td><td>Top</td><td colspan="4">2Φ22+1Φ25</td><td colspan="2">3Φ20</td><td>3Φ16</td></tr>
<tr><td></td><td>Bottom</td><td colspan="4">3Φ20</td><td colspan="2">3Φ18</td><td>3Φ16</td></tr>
<tr><td>Iasi</td><td>Top</td><td>3Φ20</td><td colspan="2">3Φ22</td><td colspan="2">3Φ20</td><td>3Φ16</td><td>3Φ14</td></tr>
<tr><td></td><td>Bottom</td><td colspan="7">3Φ16</td></tr>
<tr><td>Timisoara</td><td>Top</td><td colspan="3">2Φ20+1Φ25</td><td colspan="2">3Φ20</td><td colspan="2">3Φ16</td></tr>
<tr><td></td><td>Bottom</td><td colspan="7">3Φ16</td></tr>
<tr><td>Cluj</td><td>Top</td><td colspan="3">2Φ20+1Φ16</td><td colspan="2">2Φ16+1Φ20</td><td colspan="2">3Φ16</td></tr>
<tr><td></td><td>Bottom</td><td colspan="7">3Φ16</td></tr>
</table>

according to the N2 method proposed by Fajfar (2000), which is adopted in the current version of EN 1998-1/2004 (CEN 2004), as well.

For the nonlinear models, the structure has been reduced to a central frame, and the software Seismostruct (Seismosoft 2020) has been employed for the analysis. All of the considered structures have been designed and checked to develop a plastic mechanism which reflects the code requirements of strong columns and weak beams. The column rebar of each of the four structures consists of 12 rebars of 20 mm diameter. The beams situated on the same story have identical top and bottom rebars. All of the rebars are considered to be steel grade BST 500, while the concrete class is C20/25 for the Bucharest and Iasi frames and C25/30 for the Cluj and Timisoara frames. The beam top and bottom reinforcement details for all the analyzed models are given in Table 16.2.

## 16.3 PUSHOVER ANALYSES

The pushover analyses are performed using the software SeismoStruct (Seismosoft 2020). The cross-sections of the structural elements are modeled using fibers, which follow the Menegotto and Pinto (1973) model for reinforcement and the Mander, Priestley and Park (1988) model for the concrete. To model the structure, force-based inelastic plastic hinge frame elements are employed, detailed in Scott and Fenves (2006). The plastic hinges are defined to consider strength deterioration after reaching the rotation capacity as defined per Eq. (A.1) of CEN (2005) without any safety coefficients. In terms of shear strength, the designs are code compliant and the strength degradation of the plastic hinges is specified if the capacity in shear of the cross-sections is surpassed. However, throughout the analyses it has been observed that the shear capacities are not exceeded. The nodes of the frames are assumed to remain in the elastic domain.

## 16.4 STRUCTURAL UNCERTAINTY

The current study aims to address uncertainties given by the material properties, cross-section dimensions, and loads, in the context of seismic design. The study does not address the geometric uncertainty. The uncertainties of the nonlinear model of the structure have been represented initially by considering 11 variables that account for the variability of the materials, the dimensions of the elements, and the loads. The 11 variables are as follows:

The concrete compressive stress $f_c$, the elastic modulus for concrete $E_c$, the elastic modulus for steel $E_s$, the yield strength of steel $f_y$, the cross-section dimensions of the column $b_c$, the concrete cover for the column $a_c$, the height of the beam $h_b$, the width of the beam $b_b$, the effective width of reinforced concrete slab $b_{sl}$, the concrete cover for the beam $a_b$, and the total load on the structure $p$. The load is considered uniformly distributed on the slab and the value accounts for both the permanent and live loads in the seismic design combination. These parameters influence both the elastic response of the structure as well as the nonlinear response of the fiber model hinges through their influence on the properties of the cross-section and on the rotation capacity. An initial analysis aimed to quantify the importance of the variables on the static nonlinear response of the structure. To that extent, for the Iasi frame, a set of 10,000 pushover analyses was carried out. The distribution of the variables and their corresponding values are based on the values given in JCSS (2000) and in the work of Franchin, Mollaioli and Noto (2017). Based on engineering judgment, lower bounds and upper bounds have been assigned to each of the distributions so as not to introduce into the models unreasonable values for the variables. The statistics of all the considered random variables are shown in Table 16.3.

## 16.5 SENSITIVITY ANALYSIS

Considering the relatively high number of variables, the question of whether part of the variables can be omitted from the analysis needs to be investigated. An answer involves treating two problems, firstly, what is a measure of the structures behavior and secondly what algorithm could be used in order to find whether the variables are important or not. In terms of the response of the structure, three measures were considered: the total area of the pushover curve that represents the energy which a structure could potentially absorb during an earthquake, $A_{push}$; the displacement corresponding to

**TABLE 16.3**

**Statistics of the 11 Considered Random Variables for the Iasi and Bucharest Frame. Values in Parentheses are for the Bucharest Frame**

| Random Variable | Mean Value | Coefficient of Variation | Probability Distribution | Lower Bound | Upper Bound |
|---|---|---|---|---|---|
| $f_c$ (MPa) | 28 | 0.15 | Normal | 16 | 40 |
| $E_c$ (MPa) | 25,000 | 0.05 | Normal | 20,000 | 30,000 |
| $f_y$ (MPa) | 550 | 0.05 | Lognormal | 500 | 600 |
| $E_s$ (MPa) | 200,000 | 0.02 | Normal | 195,000 | 205,000 |
| $p$ (kN/m$^2$) | 7.1 | 0.4 | Normal | 2 | 13 |
| $b_{sl}$ (m) | 0.9 | 0.4 | Normal | 0.3 | 1.5 |
| $h_b$ (mm) | 451(601) | 0.015(0.13) | Lognormal | 420(570) | 480(630) |
| $b_b$ (mm) | 301 | 0.019 | Lognormal | 270 | 330 |
| $b_c$ (mm) | 602(702) | 0.013(0.012) | Lognormal | 572(672) | 632(732) |
| $a_b$ (mm) | 25 | 0.4 | Normal | 10 | 40 |
| $a_c$ (mm) | 25 | 0.4 | Normal | 10 | 40 |

50% of the maximum base shear force, $d_{50\%}$, adopted as a conservative and simple proxy for collapse by Franchin, Mollaioli and Noto (2017); and the displacement demand, $d_{dem}$, considering the seismic action with a mean return period of 225 years (as specified in the current Romanian seismic design code P100-1/2013).

In order to assess the importance of the variables on the output, a response sensitivity analysis is carried out. Response sensitivity analysis is a form of Monte Carlo filtering and is applied as described by Wagener et al. (2001) and implemented into the sensitivity analysis software SAFE (Pianosi, Sarrazin and Wagener 2015). The algorithm splits the output $Y_1,...,Y_n$ into ten equal intervals of values. For each interval and each variable $X_1,...,X_n$, the algorithm identifies the variables that are most responsible for driving the output into the target behavior. For each interval, the algorithm splits both the output into terms that fall within (behavioral $B$) and outside (nonbehavioral $\bar{B}$) of the considered interval. The input variables are split as well into two groups based on the output that they produce $X_i|B$, $X_i|\bar{B}$, with the sum of elements in the two groups being the total number of analyses. Then the algorithm computes the probability density functions, $f_n(X_i|B)$ and $f_{\bar{n}}(X_i|\bar{B})$, for each variable. If for a given variable $X_i$ the two distributions are significantly different then, the variable is significant to driving the output in the respective interval. In order to compare the two distributions, the Smirnov two-sample test computes the statistic $d_{n,\bar{n}} = \sup\left\|F_n\left(X_i|B\right) - F_{\bar{n}}\left(X_i|\bar{B}\right)\right\|$. The larger the statistic the more important the parameter. In Table 16.4, the results of the statistic are presented, when the output is split into ten intervals and the maximum statistic is obtained for each one of the variables and the intervals.

From Table 16.4, it is evident that different variables influence different results. The objective of this analysis is not to rank the importance of the variables, as the algorithm can be misleading for this purpose. The statistic represents the maximum difference between the cumulative distribution functions over ten intervals on each response. It might be the case that one interval, although not necessarily important from a design point of view, produces such difference. However, the results can justifiably be used to disregard certain variables from the analysis, as simply not having an influence on the results. In the following four plots, the cumulative distribution functions (CDFs) of $f_y$; yield strength and $h_b$, the beam height, are compared with the two least important variables $E_S$, the steel modulus of elasticity, and $b_b$, the beam width. The CDFs shown in Figure 16.2 correspond to behavioral variables on each of the ten intervals into which the output is divided. For the plots, the area underneath the pushover curve is the output. It is evident from the plots that the CDF for the modulus of elasticity of steel as well as for the beam width are very similar for all the ten considered intervals, as opposed to the other two variables for which the CDFs show some differences. Similar results are obtained for the displacement demand and the displacement corresponding to a 50% decrease in the maximum base shear force.

Analyzing the results, it has been concluded that the modulus of elasticity of steel and the width of the beam, offer little in terms of adding variability to the response, and for this reason only the remaining nine variables have been considered for the subsequent analysis.

**TABLE 16.4**

**Maximum $d_{n,\bar{n}}$ Results Obtained for Each Variable**

|            | $fc$ | $E_e$ | $E_s$ | $f_y$ | $b_c$ | $a_c$ | $h_b$ | $b_b$ | $b_{sl}$ | $a_b$ | $p$ |
|------------|------|-------|-------|-------|-------|-------|-------|-------|----------|-------|-----|
| $A_{push}$ | 0.67 | 0.13 | 0.05 | 0.58 | 0.18 | 0.19 | 0.28 | 0.09 | 0.48 | 0.91 | 0.29 |
| $d_{50\%}$ | 0.48 | 0.08 | 0.08 | 0.18 | 0.11 | 0.12 | 0.06 | 0.06 | 0.57 | 0.82 | 0.77 |
| $d_{dem}$  | 0.06 | 0.08 | 0.06 | 0.03 | 0.06 | 0.04 | 0.12 | 0.04 | 0.08 | 0.14 | 1.00 |
| Max        | **0.67** | **0.13** | 0.09 | **0.58** | **0.18** | **0.19** | **0.28** | 0.09 | **0.57** | **0.91** | **1.00** |

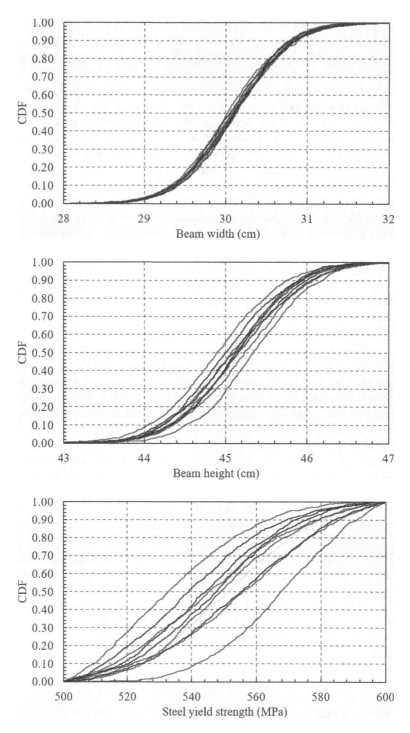

**FIGURE 16.2**   CDF of behavioral variables on the ten considered intervals for the area underneath the push-over curve for: (a) beam width; (b) beam height; (c) steel yield strength; (d) Young's modulus for steel.

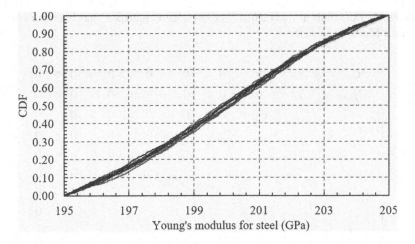

**FIGURE 16.2** (Continued)

If for the first part of the study the variables were considered to be perfectly correlated on the whole model, a more thorough approach, similar to the one of Franchin, Mollaioli and Noto (2017), has been adopted for the second part of the study. To that extent it is assumed that each element of the model is determined by its own set of random variables. Accounting for common practices in the erection of reinforced concrete buildings is likely that there will be a correlation between certain variables of the elements within the same floor as well as between floors. To that extent the following correlation matrix has been used for all of the four structures, with values gathered from sources such as Haselton and Deierlein (2008) and Franchin, Mollaioli and Noto (2017) or assessed based on engineering judgment. In Table 16.5, $\rho_{SF}$ represents the correlation coefficient for members on the same floor and $\rho_{DF}$ represents the correlation coefficient for members on different floors.

It follows that each column will be characterized by a set of five input variables $X_{col} = \{fc, Ec, fy, bc, ac\}$, while for the beam a set of seven variables $X_{beam} = \{fc, Ec, fy, h_b, b_{sl}, a_b, p\}$ is required. The force $p$ is considered to be uniformly distributed on each plate span and is represented in the model through uniformly distributed loads on the beams and concentrated loads on the nodes of the frame. The total number of input variables is $5N_{col} + 7N_{beam} = 287$, where $N_{col}$, $N_{beam}$

**TABLE 16.5**

**Intra-Story and Inter-Story Coefficients of Correlation for the Considered Random Variables**

| | $\rho_{x,fc}$ | $\rho_{x,Ec}$ | $\rho_{x,fy}$ | $\rho_{x,bc}$ | $\rho_{x,ac}$ | $\rho_{x,hb}$ | $\rho_{x,bsl}$ | $\rho_{x,ab}$ | $\rho_{x,p}$ | $\rho_{SF}$ | $\rho_{DF}$ |
|---|---|---|---|---|---|---|---|---|---|---|---|
| $f_c$ | 1 | 0.9 | 0 | 0 | 0 | 0 | 0 | 0 | 0 | 0.6 | 0.6 |
| $E_c$ | 0.9 | 1 | 0 | 0 | 0 | 0 | 0 | 0 | 0 | 0.9 | 0.9 |
| $f_y$ | 0 | 0 | 1 | 0 | 0 | 0 | 0 | 0 | 0 | 0.9 | 0.9 |
| $b_c$ | 0 | 0 | 0 | 1 | 0 | 0 | 0 | 0 | 0 | 0.9 | 0.6 |
| $a_c$ | 0 | 0 | 0 | 0 | 1 | 0 | 0 | 0 | 0 | 0.9 | 0.6 |
| $h_b$ | 0 | 0 | 0 | 0 | 0 | 1 | 0 | 0 | 0 | 0.9 | 0.6 |
| $b_{sl}$ | 0 | 0 | 0 | 0 | 0 | 0 | 1 | 0 | 0 | 0.9 | 0.6 |
| $a_b$ | 0 | 0 | 0 | 0 | 0 | 0 | 0 | 1 | 0 | 0.9 | 0.6 |
| $p$ | 0 | 0 | 0 | 0 | 0 | 0 | 0 | 0 | 1 | 0.8 | 0.6 |

represent the number of columns and beams. The correlation matrix $C$ has been built in the following form using a Matlab[] routine.

$$
C = \begin{bmatrix}
C_{fc} & C_{fc-Ec} & 0 & 0 & 0 & 0 & 0 & 0 & 0 \\
C_{fc-Ec} & C_{Ec} & 0 & 0 & 0 & 0 & 0 & 0 & 0 \\
 & & C_{fy} & 0 & 0 & 0 & 0 & 0 & 0 \\
 & & & C_{bc} & 0 & 0 & 0 & 0 & 0 \\
 & & & & C_{ac} & 0 & 0 & 0 & 0 \\
 & & & & & C_{hb} & 0 & 0 & 0 \\
 & & & & & & C_{bsl} & 0 & 0 \\
 & & & & & & & C_{ab} & 0 \\
sym. & & & & & & & & C_{p}
\end{bmatrix}
\tag{16.1}
$$

Each term in the submatrix $C_{Xk}$, situated on the diagonal of matrix $C$, can be obtained as:

$$
C_{Xk}(i,j) = \begin{cases}
1 \; if \; i = j \\
\rho_{SF} \; if \; i,j \; on \; same \; floor \\
\rho_{DF} \; if \; i,j \; on \; different \; floor
\end{cases}
\tag{16.2}
$$

while correlation in between variables such as $C_{fc-Ec}$ can be obtained as:

$$
C_{Xk-Xl}(i,j) = \begin{cases}
\rho_{Xk-Xl} \; if \; i,j \; on \; the \; same \; element \\
\sqrt{\rho_{SF}\,\rho_{Xk-Xl}} \; if \; i,j \; on \; different \; element \; same \; floor \\
\sqrt{\rho_{DF}\,\rho_{Xk-Xl}} \; if \; i,j \; on \; different \; floor
\end{cases}
\tag{16.3}
$$

## 16.6   NUMBER OF ANALYSES

A brief discussion on the required number of analyses in order to capture the effects of the uncertainty is warranted. Sampling a large number of variables comes with the pitfall of choosing a suitable number of samples for each variable while at the same time taking into account the desired accuracy of the analysis as well as the analysis time. As Soong and Grigoriu (1993) have shown, if one aims to estimate $P_{true}$, a theoretically correct probability, by calculating $\bar{P}$, the number of simulations can be computed using $N$, the number of simulations, and the coefficient of variation of $V_{\bar{P}}$ using the formula $N = \dfrac{1 - P_{true}}{V_{\bar{P}}^2 \, P_{true}}$. Considering a reasonable probability of failure on the order of $10^{-4}$ for a new code conforming structure, and a coefficient of variation of at or below 10%, almost $10^6$ simulations would be required in order to obtain theoretically correct results.

In order to reduce the number of simulations, a Latin hypercube sampling (LHS) method as described by Iman, Helton and Campbell (1981) is employed. The LHS method implies partitioning the range of each $X_i$ into $N$ intervals of equal probability, and randomly selecting just one value from each interval. The method needs to be modified to account for the fact that the variables are

**TABLE 16.6**

**Mean ($\mu$) and Standard Deviation ($\sigma$) as a Function of Sample Size**

| Result | Sample Size | | | |
|---|---|---|---|---|
| | 250 | 500 | 1000 | 10,000 |
| $\mu_{A_{push}}$ | 517.34 | 526.65 | 523.55 | 524.73 |
| $\sigma_{A_{push}}$ | 62.46 | 68.74 | 65.7 | 66.78 |
| $\mu_{d_{80\%}}$ | 0.75 | 0.78 | 0.77 | 0.77 |
| $\sigma_{d_{80\%}}$ | 0.124 | 0.133 | 0.127 | 0.127 |
| $\mu_{d_{dem}}$ | 0.255 | 0.254 | 0.254 | 0.254 |
| $\sigma_{d_{dem}}$ | 0.0046 | 0.0254 | 0.0249 | 0.0247 |

correlated as stated previously. The algorithm used is developed by Iman (2020) and consists of the following steps:

1. Generate Latin hypercube independent samples.
2. Apply for each variable $X_i$ a standard normal inverse cumulative distribution function.
3. Use the Cholensky transformation to introduce the correlation between the variables.
4. Apply the standard normal cumulative distribution function to the correlated distributions.
5. Map to the values of $X_i$ using the inverse cumulative distribution function of the required variable.

In order to establish a reasonable number of analyses to accurately describe the results of variability on the pushover curve, different sample sizes were employed. A sample size of 10,000 was used as a benchmark for comparison and smaller samples of 250, 500, and 1000 models were investigated, again for the Iasi frame; this time considering also the correlation in between the variables. The resulting means and standard deviations of the same results considered for the sensitivity analysis are presented in Table 16.6.

The results show that comparing to the benchmark 10,000 analyses, there is no point of increasing the number of analyses above 1000 as the increase in accuracy is basically neglectable. From an engineering perspective, the results for 500 are sufficiently accurate, as well.

## 16.7   RESULTS OF THE PUSHOVER ANALYSIS

After considering the obtained results the variability of each of the employed frame structures can be captured within a satisfactory degree by using five variables ($fc, Ec, fy, bc, ac$) for each column and seven variables ($fc, Ec, fy, h_b, b_{sl}, a_b, p$) for each beam. The variables are correlated to account for building practice erection, as mentioned previously. A number of 1000 pushover analyses are carried out in SeismoStruct (Seismosoft 2020) for each of the four frames using a LHS method to determine the input sample. In Table 16.7, the input variables for Cluj and Timisoara frames are presented.

After running the analyses, for each pushover curve the base shear force diagram is extracted and the displacement at 50% maximum base shear force is extracted as a proxy for the collapse of the structure. Figure 16.3 presents the mean pushover curves for each one of the four frames.

A quadrilinear approximation of each one of the pushover curves is done, similar to the ones proposed by De Luca, Vamvatsilos and Iervolino (2013) near optimal fit. The four segments are the following:

1. An elastic segment captured by secant linear segment with the initial stiffens matching the secant stiffness of the pushover curve at 60% of peak base shear.

**TABLE 16.7**

**Statistics of the Nine Considered Random Variables for the Timisoara and Cluj Frames. Values in Parenthesis Are for the Cluj Frame**

| Random Variable | Mean Value | Coefficient of Variation | Distribution | Lower Bound | Upper Bound |
|---|---|---|---|---|---|
| $f_c$ (MPa) | 33 | 0.15 | Normal | 21 | 45 |
| $E_c$ (MPa) | 27,000 | 0.05 | Normal | 22,000 | 32,000 |
| $f_y$ (MPa) | 550 | 0.05 | Lognormal | 500 | 600 |
| $p$ (kN/m$^2$) | 7.1 | 0.4 | Normal | 2 | 13 |
| $b_{sl}$ (m) | 0.9 | 0.4 | Normal | 0.3 | 1.5 |
| $h_b$ (mm) | 451 | 0.015 | Lognormal | 420 | 480 |
| $b_c$ (mm) | 602(552) | 0.013 | Lognormal | 572(522) | 632(582) |
| $a_b$ (mm) | 25 | 0.4 | Normal | 10 | 40 |
| $a_c$ (mm) | 25 | 0.4 | Normal | 10 | 40 |

2. A hardening segment of positive stiffness is chosen to terminate at the point of maximum base shear, while minimizing the area difference in between the real pushover curve and the simplified model.
3. A softening, negative stiffness segment is defined based on the same principle of minimizing the difference in the areas.
4. A residual plateau segment is employed with a maximum displacement equal to the maximum displacement in the pushover curve and a force obtained on the same principle of minimizing the difference in the areas.

As opposed to the method provided by De Luca, Vamvatsilos and Iervolino (2013), which suggest using an elastic segment based on a tangent stiffness at 10% of the maximum base shear, a 60% value of the tangent stiffness is adopted for the elastic segment as it is implemented in codes such as FEMA (2005) and is deemed conservative both from the results of the mentioned study and also considering it is going to lead to more conservative values in terms of the corresponding

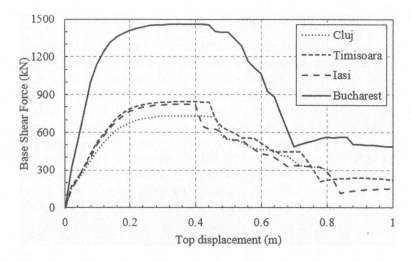

**FIGURE 16.3**    Mean pushover curves for the four analysed RC frames.

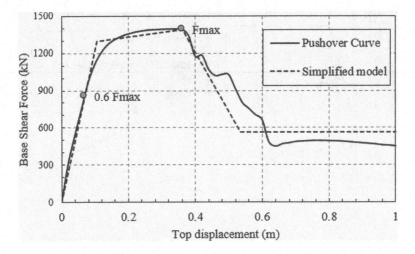

**FIGURE 16.4**   Sample pushover curve and simplified corresponding linear model.

displacement demands. A sample pushover curve and the corresponding linear fit are illustrated in Figure 16.4.

## 16.8   INCREMENTAL DYNAMIC ANALYSES

The incremental dynamic analyses (IDA) (Vamvatsikos and Cornell 2002) are conducted for each equivalent SDOF system characterized by the linear pushover curves. A total of 20 site-specific ground motion recordings are selected for each of the four analysed sites. The ground motions for Bucharest and Iasi were recorded during the Vrancea intermediate-depth earthquakes of March 4, 1977 (moment magnitude $M_W$ = 7.4 and focal depth $h$ = 94 km), August 30, 1986 (moment magnitude $M_W$ = 7.1 and focal depth $h$ = 131 km), and May 30, 1990 (moment magnitude $M_W$ = 6.9 and focal depth $h$ = 91 km). In the case of Bucharest, the same ground motion dataset as in the study of Pavel, Pricopie and Nica (2019) is employed, as well. For Cluj and Timisoara, a selection of the ground motion recordings was performed using the PEER strong ground motion database (PEER 2013).

A comparison between the mean acceleration response spectra, 16th and 84th percentiles acceleration response spectra computed for each set of 20 site-specific ground motion recordings (for Bucharest, Cluj, Iasi, and Timisoara), and the design acceleration response spectra is illustrated in Figure 16.5.

For each of the four frames, 1000 SDOF models are created from the linearized pushover curves. For each of the models an incremental dynamic analysis (IDA) is performed using the spectral acceleration at the period of the first eigenmode ($Sa(T_1)$) as an intensity measure and the mentioned set of site-specific ground records. The analyses are performed in OpenSees (McKenna and Fenves 2001), and the SDOF systems are represented by the Modified Ibarra Medina Krawinkler (Ibarra et al. 2005) deterioration model, which accounts for strength deterioration, post-capping strength deterioration, and unloading stiffness deterioration.

## 16.9   SEISMIC FRAGILITY ASSESSMENT

The fragility assessment is performed for each of the four RC frames using the results of the IDA (Vamvatsikos and Cornell 2002). The fragility functions are obtained for the serviceability limit state, ULS, and collapse limit state. Collapse is considered to occur when either the model does not converge or when the maximum displacement of the system is larger than the displacement at 50%

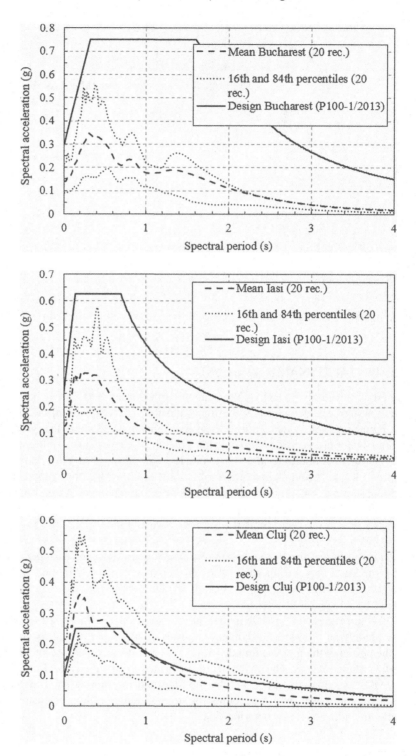

**FIGURE 16.5** Comparison between the mean acceleration response spectra, 16th and 84th percentiles acceleration response spectra computed for each set of 20 site-specific ground motion recordings and the design acceleration response spectra for: (a) Bucharest; (b) Cluj; (c) Iasi; and (d) Timisoara.

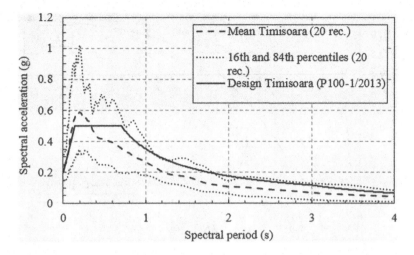

**FIGURE 16.5**  (Continued)

maximum base shear force, obtained from the pushover curve. The occurrence of the other two design limit states is considered when the maximum interstory drift surpasses 0.5% for the SLS and 2.5% for the ULS, as specified by the design code P100-1/2013 (MDRAP 2013). In order to compute the maximum interstory drift from the maximum displacement recorded from the IDAs, the coefficient of distortion as developed by Qi and Moehle (1991) is used. The procedure for deriving the fragility functions involves the fitting of a lognormal CDF on the results of the IDA. The procedure is described in the paper of Baker (2015). The fragility functions for each individual structure are illustrated in Figure 16.6.

The probability of exceedance of either of the three limit states (serviceability, ultimate, and collapse) is obtained by convolving the site-specific seismic hazard from the study of Pavel et al. (2016) for Cluj, Iasi, and Timisoara and from the study of Pavel and Vacareanu (2017) for Bucharest with the seismic fragilities derived in this study. The seismic hazard curves for the spectral acceleration corresponding to the fundamental Eigen period $SA(T_1)$ are illustrated in Figure 16.7.

The results in terms of annual probabilities of exceedance of a particular limit state are reported in Table 16.8.

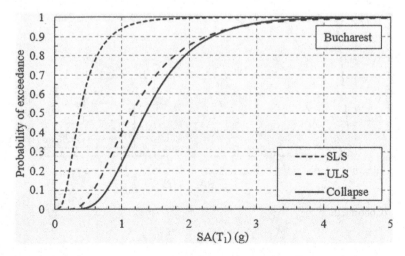

**FIGURE 16.6**  Fragility functions for serviceability limit state (SLS), ultimate limit state (ULS), and collapse limit state for the structure situated in: (a) Bucharest; (b) Cluj; (c) Iasi; and (d) Timisoara.

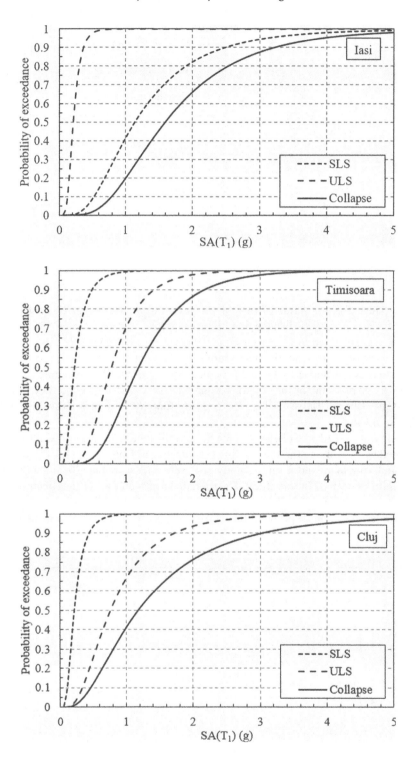

**FIGURE 16.6** (Continued)

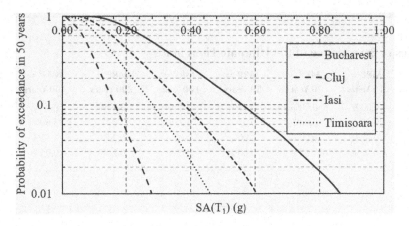

**FIGURE 16.7**   Seismic hazard curves for SA(T₁) for the four analysed sites.

**TABLE 16.8**
**Annual Probabilities of Exceedance of a Particular Limit State**

| Structure | Serviceability Limit State | Ultimate Limit State | Collapse Limit State |
|---|---|---|---|
| Bucharest | $2.63 \bullet 10^{-2}$ | $8.50 \bullet 10^{-4}$ | $1.66 \bullet 10^{-4}$ |
| Iasi | $1.83 \bullet 10^{-3}$ | $8.86 \bullet 10^{-5}$ | $1.06 \bullet 10^{-5}$ |
| Timisoara | $5.54 \bullet 10^{-3}$ | $9.53 \bullet 10^{-5}$ | $1.16 \bullet 10^{-5}$ |
| Cluj | $1.53 \bullet 10^{-3}$ | $5.19 \bullet 10^{-5}$ | $2.33 \bullet 10^{-5}$ |

The resulting annual probabilities of exceedance of the three limit states are one order of magnitude larger for Bucharest as compared to Iasi, Timisoara, and Cluj.

Next, the probabilities of exceedance of the three damage states are evaluated considering the seismic hazard levels corresponding to seven mean return periods: 10 years, 20 years, 50 years, 100 years, 200 years, 500 years, and 1000 years. The results are reported in Tables 16.9–16.11. The largest exceedance probabilities for all the three limit states are obtained for the structure situated at Bucharest, while the smallest ones are derived for the structures situated at Cluj. In addition, one can notice that the probabilities of exceedance of the three limit states obtained for the two sites which are under the influence of the Vrancea intermediate-depth seismic source (Bucharest and Iasi) are significantly larger than the ones obtained for Cluj and Timisoara. It can also be observed that the collapse rates corresponding to the ground motion with a mean return period of 1000 years are in the range 0.37–5.00% when considering that the mean return period of the design peak ground acceleration is 225 years for all the sites.

**TABLE 16.9**
**Probabilities of Exceedance of the Serviceability Limit State**

| Structure | MRP = 10 Years | MRP = 20 Years | MRP = 50 Years | MRP = 100 Years | MRP = 200 Years | MRP = 500 Years | MRP = 1000 Years |
|---|---|---|---|---|---|---|---|
| Bucharest | 0.95% | 5.58% | 22.84% | 41.41% | 58.54% | 74.41% | 82.08% |
| Iasi | 0.04% | 1.37% | 19.92% | 50.55% | 75.39% | 91.76% | 96.34% |
| Timisoara | 0.08% | 1.35% | 10.56% | 26.35% | 45.09% | 64.78% | 75.12% |
| Cluj | 0.00% | 0.02% | 1.33% | 4.93% | 15.31% | 33.54% | 44.66% |

**TABLE 16.10**
**Probabilities of Exceedance of the Ultimate Limit State**

| Structure | MRP = 10 Years | MRP = 20 Years | MRP = 50 years | MRP = 100 Years | MRP = 200 Years | MRP = 500 Years | MRP = 1000 Years |
|---|---|---|---|---|---|---|---|
| Bucharest | 0.00% | 0.00% | 0.13% | 0.87% | 3.13% | 9.05% | 15.25% |
| Iasi | 0.00% | 0.00% | 0.03% | 0.27% | 1.07% | 3.60% | 6.53% |
| Timisoara | 0.00% | 0.00% | 0.00% | 0.02% | 0.24% | 1.51% | 3.77% |
| Cluj | 0.00% | 0.00% | 0.00% | 0.03% | 0.20% | 0.95% | 1.83% |

**TABLE 16.11**
**Probabilities of Exceedance of the Collapse Limit State**

| Structure | MRP = 10 Years | MRP = 20 Years | MRP = 50 Years | MRP = 100 Years | MRP = 200 Years | MRP = 500 Years | MRP = 1000 Years |
|---|---|---|---|---|---|---|---|
| Bucharest | 0.00% | 0.00% | 0.00% | 0.05% | 0.38% | 2.13% | 5.00% |
| Iasi | 0.00% | 0.00% | 0.00% | 0.01% | 0.08% | 0.46% | 1.11% |
| Timisoara | 0.00% | 0.00% | 0.00% | 0.00% | 0.01% | 0.11% | 0.37% |
| Cluj | 0.00% | 0.00% | 0.00% | 0.02% | 0.09% | 0.41% | 0.77% |

## 16.10 CONCLUSIONS

In this study, a probabilistic seismic performance assessment is performed for four seven-story RC frames under the influence of either intermediate-depth or crustal earthquakes. In order to represent the variability of the response of the structures, 12 variables have been initially chosen. The variables are considered to be perfectly correlated within an element; however, different correlations coefficients have been used for elements which are on the same story as well as on different stories. The seismic performance is evaluated through IDA performed on equivalent SDOF systems. Fragility functions for three limit states (serviceability, ultimate, and collapse) are also derived for each individual structure. Subsequently, the annual probabilities of exceedance of each damage state are computed by convolving the site-specific seismic hazard with the previously derived fragilities. The main results of this study can be summarized as follows:

- A response sensitivity analyses carried out on the 12 input variables showed that the variability of the width of the beam and the modulus of elasticity of steel did not influence the static nonlinear response of the structure.
- A number of simulations with different numbers of samples have been considered in order to identify the number of required analysis. Around 500 pushover analyses appear to be sufficient in order to describe the variability of the pushover curves obtained from around 10,000 simulations.
- The resulting annual probabilities of exceedance of the three limit states are one order of magnitude larger for Bucharest as compared to Iasi, Timisoara, and Cluj.
- The largest exceedance probabilities for all the three limit states are obtained for the structure situated at Bucharest, while the smallest ones are derived for the structures situated at Cluj. This can be due to the significant long-period spectral accelerations that occur in Bucharest, as opposed to the other three sites.

- The probabilities of exceedance of the three limit states obtained for the two sites which are under the influence of the Vrancea intermediate-depth seismic source (Bucharest and Iasi) are significantly larger than the ones obtained for Cluj and Timisoara.
- The use of nonlinear SDOF models can model sufficiently accurate the nonlinear seismic performance of medium-height RC frames as the ones analysed in this study. Considering the reduced time necessary for running nonlinear dynamic analyses, they can provide a viable solution for performing seismic fragility assessments which require a significant number of analyses.

## REFERENCES

Baker, J. W. 2015. "Efficient analytical fragility function fitting using dynamic structural analysis." *Earthquake Spectra* 31 (1): 579–599.

CEN. 2004. Eurocode 8 – Design of structures for earthquake resistance – Part 1: General rules, seismic actions. Bruxelles, Belgium: CEN.

—. 2005. Eurocode 8 – Design of structures for earthquake resistance – Part 3: Assessment and retrofitting of buildings. Bruxelles, Belgium: CEN.

De Luca, F., D. Vamvatsilos, and I. Iervolino. 2013. "Near-optimal piecewise linear fits of static pushover capacity curves for equivalent SDOF analysis." *Earthquake Engineering & Structural Dynamics* 42 (4): 523–543.

Decanini, L., F. Mollaioli, and A. Mura. 2001. "Equivalent SDOF systems for the estimation of seismic response of multistory frame structures." *Transactions on the Built Environment* 57: 101–110.

Fajfar, P. 2000. "A nonlinear analysis method for performance-based seismic design." *Earthquake Spectra* 16 (3): 573–592.

FEMA. 2005. Improvement of nonlinear static seismic analysis procedures. Washington DC: Federal Emergency Management Agency.

Franchin, P., F. Mollaioli, and F. Noto. 2017. "RINTC project: influence of structure-related uncertainties on the risk of collapse of Italia code-conforming reinforced concrete buildings." COMPDYN 2017–6th ECCOMAS thematic conference on computational methods in structural dynamics and earthquake engineering. Rhodes Island, Greece.

Graziotti, F., A. Penna, and G. Magenes. 2016. "A nonlinear SDOF model for the simplified evaluation of the displacement demand of low-rise URM buildings." *Bulletin of Earthquake Engineering* 14: 1589–1612.

Haselton, C. B., and G. G. Deierlein. 2008. Assessing seismic collapse safety of modern reinforced concrete moment-frame buildings. Berkeley, California: Pacific Earthquake Engineering Research Center College of Engineering.

Ibarra, L. F., R. A. Medina, and H. Krawinkler. 2005. "Hysteretic models that incorporate strength and stiffness deterioration." *Earthquake Engineering and Structural Dynamics* 34 (12) 1489–1511.

Iman. 2020. lhsgeneral(pd,correlation,n. Accessed August 2020. https://www.mathworks.com/matlabcentral/fileexchange/56384-lhsgeneral-pd-correlation-n.

Iman, R. L., J. C. Helton, and J. E. Campbell. 1981. "An approach to sensitivity analysis of computer models, Part 1. Introduction, input variable selection and preliminary variable assessment." *Journal of Quality Technology* 13 (3): 174–183.

JCSS. 2000. Probabilistic Model Code. Accessed August 2020.

Kuramoto, H., M. Teshigawara, T. Okuzono, N. Koshika, M. Takayama, and T. Hori. 2000. "Predicting the earthquake response of buildings using equivalent single degree of freedom system." Proceedings of the 12th World Conference on Earthquake Engineering. Auckland. Paper no. 1039.

Mander, J. B., M. J. N. Priestley, and R. Park. 1988. "Theoretical stress-strain model for confined concrete." *Journal of Structural Engineering* 114 (8): 1804–1826.

McKenna, F., and G. Fenves. 2001. Open System for Earthquake Engineering Simulation v.3.2.1. Accessed 2020. opensees.berkeley.edu.

MDRAP. 2013. "P100-1/2013 Cod de proiectare seismică. Partea I – prevederi de proiectare pentru clădiri." Bucharest, Romania: MDRAP.

Menegotto, M., and P. E. Pinto. 1973. "Method of analysis for cyclically loaded R.C. plane frames including changes in geometry and non-elastic behaviour of elements under combined normal force and bending." IABSE Symposium on the Resistance and Ultimate Deformability of Structures Acted on by Well Defined Repeated Loads. Zurich. pp. 15–22.

Oviedo, J. A., M. Midorikawa, and T. Asari. 2010. "An equivalent SDOF system model for estimating the response of R/C building structures with proportional hysteretic dampers subjected to earthquake motions." *Earthquake Engineerign and Structural Dynamics* 40 (5): 571–589.

Pavel, F., A. Pricopie, and G. Nica. 2019. "Collapse assessment for a RC frame structure in Bucharest (Romania)." *International Journal of Civil Engineering* 17: 1373–1381.

Pavel, F., and R. Vacareanu. 2017. "Ground motion simulations for seismic stations in southern and eastern Romania and seismic hazard assessment." *Journal of Seismology* 21: 1023–1037.

Pavel, F., R. Vacareanu, J. Douglas, M. Radulian, C. Cioflan, and A. Barbat. 2016. "An updated probabilistic seismic hazard assessment for Romania and comparison with the approach and outcomes of the SHARE project." *Pure and Applied Geophysics* 173: 1881–1905.

PEER. 2013. PEER ground motion database. Accessed August 2020. https://ngawest2.berkeley.edu/.

Pianosi, F., F. Sarrazin, and T. Wagener. 2015. "A Matlab toolbox for Global Sensitivity Analysis." *Environmental Modelling & Software* 70: 80–85.

Qi, X., and Moehle, J.P. 1991. Displacement design approach for reinforced concrete structures subjected to earthquakes. Rep. No. UCB/EERC - 91/02. Earthquake Engineering Research Center.

Scott, M. H., and G. L. Fenves. 2006. "Plastic hinge integration methods for force-based beam–column elements." *ASCE Journal of Structural Engineering* 132 (2): 244–252.

Seismosoft. 2020. SeismoStruct – a computer program for static and dynamic nonlinear analysis of framed structures. Accessed August 2020. http://www.seismosoft.com.

Soong, T., and M. Grigoriu. 1993. Random vibration of mechanical and structural systems. Englewood Cliffs: Prentice Hall.

Suzuki, A., and I. Iervolino. 2019. "Seismic fragility of code-conforming Italian buildings based on SDoF approximation." *Journal of Earthquake Engineering* pp. 1–35. doi:10.1080/13632469.2019.1657989.

Vamvatsikos, D., and C. A. Cornell. 2005. "Direct estimation of seismic demand and capacity of multidegree-of-freedom systems through incremental dynamic analysis of single degree of freedom approximation." *Journal of Structural Engineering* 131 (4): 589–599.

Vamvatsikos, D., and C. A. Cornell. 2002. "Incremental dynamic analysis." *Earthquake Engineering and Structural Dynamics* 31 (3): 491–514.

Wagener, T., D. P. Boyle, M. J. Lees, H. S. Wheater, H. V. Gupta, and S. Sorooshian. 2001. "A framework for development and application of hydrological models." *Hydrology and Water Resources* 5: 13–26.

# 17 Research on Safety Assessment and Maintenance Decision for the Metal Structure of the Casting Crane

Hui Jin, Jiaming Cheng, Guoliang Chen, and Weiming Zhang

Southeast University, Nanjing, China

## CONTENTS

## 17.1 INTRODUCTION

The casting cranes are used in smelting and pouring workshops in the metallurgical industry. The frequent hoisting impact load and the harsh working environment with high temperatures and high dust lead to the performance degradation of the metal structure, which is prone to cause a crack, corrosion, and other failure modes. The troubleshoot of these failure modes and the determination of the repair cycle are relied heavily on subjectivity and experience, and are apt to over-repair or under-repair, can hardly meet the technical requirements of modern casting cranes. Therefore, contraposing the problems of high maintenance cost and low maintenance efficiency of the metal structure of casting crane, it is of great significance to research safety evaluation and maintenance decisions to achieve economical and reliable maintenance results.

In the first part, the authors establish a safety evaluation model to form a comprehensive weight using experience factors, based on the objective weighting of the entropy weight method (EWM) and the subjective weighting of the analytic hierarchy process (AHP), to integrate the objective analysis data and subjective prior knowledge. Then, a safety evaluation system with state value as the main consideration was built on the combination of knowledge and fuzzy comprehensive evaluation. Next, the performance degradation models of four indicators for the metal structures of the casting crane including crack, strength, stiffness, and corrosion are established in the second part. In the third part, the proposed performance degradation model of safety evaluation indicators based on the Gamma process is employed to establish the probability density function of four safety indicators, and a new component maintenance decision and group maintenance decision-making model is then proposed, the presented examples validate the applicability of the novel maintenance decision for the metal structure of the casting crane.

## 17.2   COMPREHENSIVE SAFETY EVALUATION OF METALLURGICAL CRANE BASED ON AHP AND ENTROPY WEIGHT METHOD

Metallurgical crane is widely used as a piece of important auxiliary machinery in modern industrial production. The safety of metallurgy cranes has received widespread attention, especially in recent years, because of the high accident rate and huge economic losses. Hence, it is vital to present a relatively effective and comprehensive safety evaluation method for the safe operation of the crane. Nevertheless, the industry lacks a common and reasonable standard for crane safety evaluation. Traditional crane safety evaluation methods such as traditional empirical methods and fault tree analysis methods heavily rely on expert experience and are unable to effectively avoid the influence of adverse subjective factors such as subjective arbitrariness, thinking uncertainty, and cognitive ambiguity. In recent years, the AHP has been widely used in fault diagnosis and fuzzy evaluation of mechanical devices and has achieved good results. Therefore, the AHP has attracted some researcher's attention and be applied for the safety assessment of the crane. Fei, Li and Fan (2018) proposed a safety performance assessment model for bridge cranes and use the AHP to determine the weights of the indicators. Sarkar and Shah (2012) employed the AHP to elicit subjective knowledge from the expert and formalized it into a set of weighted safety factors. However, the standard AHP still depends on the experience and knowledge of experts and is too subjective to affect the accuracy of the evaluation results. The EWM, a known objective weight evaluation method, is sometimes prone to given greater weight to the indicators with a small difference and then contrary to fact. The authors take advantage of two methods and find a balance between subjective consciousness and objective data, to propose a comprehensive safety evaluation method for the metallurgical cranes based on the objective weighting of the EWM and the subjective weighting of the AHP. The empirical factors are used to form a comprehensive weight model to achieve the combination of objective analysis data and subjective prior knowledge and then combined with a fuzzy comprehensive evaluation to present an effective metallurgical crane safety evaluation algorithm.

### 17.2.1   THE ANALYTIC HIERARCHY PROCESS

The AHP is a multi-quasi-measurement decision-making method that combines qualitative and quantitative analysis established by Saaty (1988). Four main steps are included in the AHP:

1. Establish a hierarchical model
   The whole system is decomposed according to the contained subsystems, and then the subsystems are decomposed into indicators to determine the evaluation value of each indicator

$$\mathbf{A} = \left\{ A_1, A_2, A_3, A_4, ..., A_n \right\} \tag{17.1}$$

where **A** is the evaluation set, $n$ is the number of the factors, and $A_n$ is a single factor evaluation value.

2. Construct a judgment matrix

According to expert experience, the relative importance of each single factor evaluation indicator can be obtained by comparing each factor in pairs, and the judgment matrix can be obtained comprehensively:

$$\mathbf{A} = \begin{pmatrix} a_{11} & a_{12} & \cdots & a_{1n} \\ a_{21} & a_{22} & & a_{2n} \\ \vdots & & \ddots & \\ a_{n1} & a_{n2} & \cdots & a_{nn} \end{pmatrix} \tag{17.2}$$

$$a_{ij} = \begin{cases} 1 & i = j\,(i, j = 1,\ldots,n) \\ a_{ji}^{-1} & i \neq j \end{cases} \tag{17.3}$$

3. Obtain weight

The weight is obtained by solving the eigenvalues of the judgment matrix

$$\mathbf{AW} = \lambda_{\max}\mathbf{W} \tag{17.4}$$

where $\lambda_{\max}$ is the maximum eigenvalue of the judgment matrix and **W** is the eigenvector.

4. Consistency judgment

A consistency test is required to avoid the logical confusion of indicators. The smaller value of $CI$ suggests the better consistency of the matrix, and $CI = 0$ means the judgment matrix is completely consistent.

$$CI = \frac{\lambda_{\max} - n}{n - 1} \tag{17.5}$$

### 17.2.2 LOGICAL HIERARCHICAL TREE FOR SAFETY EVALUATION

The indicators included strength, stiffness, weld, local deformation, and corrosion are selected as the evaluation indicator for the safety assessment of a double girder metallurgical crane based on the inspection and maintenance records. Zhang, Jin and Chen (2019) presented the detailed evaluation method for the evaluation indicators. Therefore, the logical hierarchical tree for safety evaluation can be depicted with the obtained evaluation indicators, shown in Figure 17.1.

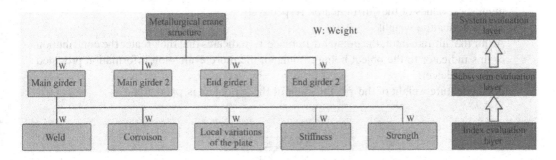

**FIGURE 17.1** Logic hierarchy tree for safety evaluation of a double girder metallurgical crane.

According to Figure 17.1, the logical hierarchy tree shows the evaluation indicators of the main girder 1. Besides, the evaluation indicators of the main girder 2, the end girder 1 and the end girder 2 are the same as the main girder 1.

### 17.2.3 ENTROPY WEIGHT METHOD

Entropy in physics is the quantity related to the starting state and the ending state. Shannon (1948) founded information theory in 1948. Shannon called the uncertainty of information source signals as information entropy. In information entropy, the increase in entropy indicates the loss of information. The greater the amount of information in the system, the higher the degree of order and the lower the entropy. Thus, the information is negative entropy, and entropy can be used to measure the amount of information. The procedure for EWM is presented as follows:

1. Construct decision matrix
   The set of the evaluation indicators and objects are defined as $\mathbf{D} = (D_1, D_2, \ldots, D_n)$ and $\mathbf{M} = (M_1, M_2, \ldots, M_m)$, respectively. The evaluation state value of each object set $M_i$ against the indicator set $D_i$ is expressed as $x_{ij}(i = 1, 2, \ldots, m; j = 1, 2, \ldots, n)$, so the decision matrix $\mathbf{X}$ can be expressed as:

$$\mathbf{X} = \begin{pmatrix} x_{11} & \cdots & x_{1n} \\ \vdots & \ddots & \vdots \\ x_{m1} & \cdots & x_{mn} \end{pmatrix} \tag{17.6}$$

2. Standardize decision matrix
   According to the different attributes of indicators, the indicators can generally be divided into two categories: Efficiency-oriented and Cost-oriented indicators. As for the efficiency-oriented indicator, the higher the state value of the indicator corresponds to the better state. The cost-oriented indicator is vice-versa, the higher state value implies the worse state.

$$v_{ij} = \begin{cases} \dfrac{x_{ij} - \min x_j}{\max x_j - \min x_i} & \textit{Efficiency – oriented} \\[4mm] \dfrac{\max x_j - x_{ij}}{\max x_j - \min x_i} & \textit{Cost – oriented} \end{cases} \tag{17.7}$$

where $v_{ij}$ is the standardized value of $x_{ij}$. $\max x_j$ and $\min x_j$ are the maximum and minimum state values of the $j$th indicator, respectively.

3. Calculate feature weight
   As for the $j$th indicator, the greater difference $v_{ij}$ indicates that the greater the contribution of this indicator to the object being evaluated, the more evaluation information provided by this indicator.
   The feature weight of the $j$th indicator of the $i$th object is $p_{ij}$, defined as

$$p_{ij} = \frac{v_{ij}}{\displaystyle\sum_{i=1}^{m} v_{ij}} \tag{17.8}$$

4. Calculate entropy

$$e_j = -\frac{1}{\ln m} \sum_{i=1}^{m} p_{ij} \ln p_{ij} \qquad (17.9)$$

If $p_{ij} = 0$ or $p_{ij} = 1$, then $p_{ij} \ln p_{ij} = 0$.

5. Calculate the coefficient of difference and weight

According to the Eq. (17.34), the information is negative entropy. The greater entropy value indicates the smaller difference among $v_{ij}$. The coefficient of difference $d_j$ is used to represent the difference in the $j$th indicator value of each object, defined as $d_j = 1 - e_j$.

6. Calculate the weight

$$w_j = \frac{d_j}{\sum_{k=1}^{n} d_k} (j = 1, 2, \ldots, n) \qquad (17.10)$$

## 17.2.4  DETERMINATION FOR COMPREHENSIVE WEIGHT

Supposing the subjective weight $\mathbf{W}_1$ is obtained by AHP, and $\mathbf{W}_2$ is the result calculated using the EWM, the comprehensive weight $\mathbf{W}$ can be obtained by the linear combination of the two weights

$$\mathbf{W} = \alpha \mathbf{W}_1 + (1 - \alpha) \mathbf{W}_2 \qquad (17.11)$$

where $\alpha (0 < \alpha < 1)$ is the scale factor. If the subjective weight is set as the main weight, it is generally taken as 0.618.

## 17.2.5  FUZZY COMPREHENSIVE EVALUATION METHOD BASED ON A COMPREHENSIVE WEIGHT

Fuzzy comprehensive evaluation method (FCEM) is featured by the ability to accurately cope with inaccurate and incomplete information. With the development of an intelligent crane safety evaluation system, the fuzzy relation matrix of a fuzzy comprehensive evaluation in hoisting machinery system is mainly obtained through expert experience, which is of low efficiency and high cost. Therefore, this paper proposes a new safety evaluation method based on a comprehensive weight and FCEM.

The standard FCEM consists of five-element sets as followings:

1. Set of evaluation indicator ($\mathbf{U}$)

The set composed by the attributes of the evaluation object can be expressed as $\mathbf{U} = [u_1, u_2, \ldots, u_n]$ and $u_i (i = 1, 2, \ldots, n)$ is the attribute of the evaluation object, respectively.

2. Set of weight ($\mathbf{W}$)

The weight set is a set calculated by a comprehensive weight model, indicating that the importance of each attribute in the evaluation layer is different, expressed as $\mathbf{W} = [w_1, w_2, \ldots, w_n]$ and $w_i (i = 1, 2, \ldots, n)$ is the membership of $u_i$ to $\mathbf{W}$. Generally, $w$ should meet the normalization and non-negative conditions $\sum_{i=1}^{n} w_i = 1 (w_i \geq 0)$.

3. Set of evaluation ($\mathbf{V}$)

The evaluation set $\mathbf{V}$ is the set composed of attribute state values of the evaluation object, expressed as: $\mathbf{V} = [v_1, v_2, \ldots, v_m]$.

4. Matrix of fuzzy relation ($\mathbf{R}$)

The fuzzy relationship between the evaluation index set $\mathbf{U}$ and the evaluation set $\mathbf{V}$ can be represented by a fuzzy relation matrix

$$\mathbf{R} = \begin{pmatrix} r_{11} & r_{12} & \cdots & r_{1n} \\ r_{21} & r_{22} & & r_{2n} \\ \vdots & & \ddots & \vdots \\ r_{m1} & r_{m2} & \cdots & r_{mn} \end{pmatrix} \tag{17.12}$$

$$r_{ij} = \mu_r(u_i, v_j) \tag{17.13}$$

$$0 \leq r_{ij} \leq 1 \tag{17.14}$$

where $r_{ij}$ denotes the degree to which the object makes judgment results $v_i$ when considering attributes $u_i$ and $\mu_r(\cdot)$ is the membership function.

5. Set of Fuzzy comprehensive evaluation index ($\mathbf{B}$)

The set of fuzzy comprehensive evaluation index would be gained with the obtained weight set $\mathbf{W}$ and fuzzy relation matrix $\mathbf{R}$, which is calculated by $\mathbf{B} = \mathbf{W} \circ \mathbf{R}$ and $\circ$ is the fuzzy operator. Generally, the weighted average operator $(\cdot, +)$ is chosen as the fuzzy operator. Consequently, the system state can be determined according to the obtained comprehensive evaluation indicator $b_j (j = 1, 2, \ldots, m)$ and maximum membership principle.

Therefore, the detailed procedure for the proposed fuzzy evaluation method is shown in Figure 17.2. The Figure 17.2 shows that the comprehensive weight obtained via the AHP and EWM is combined with the FCEM to evaluate the upper layer subsystem, and continuously, until the evaluation value of the overall structural system is obtained.

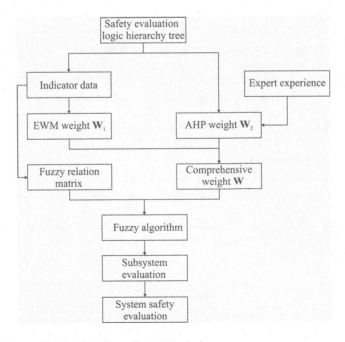

**FIGURE 17.2**  Safety evaluation algorithm of metallurgical crane.

## 17.2.6 EXAMPLE

This chapter takes a single main girder of a 200 t casting crane with four girders and four rails as an example of safety evaluation to demonstrate the validity of the proposed algorithm. The state values of the evaluation indicators are shown in Table 17.1 below.

**Step 1**. Calculate the weight using the AHP

The maximum eigenvalue of the matrix $\lambda_{max} = 5$ and eigenvector

$W_1 = (0.4, 0.2, 0.2, 0.07, 0.13)$ is obtained, and the result $CI = \dfrac{\lambda_{max} - n}{n-1} = 0$ indicates the judgment matrix is completely consistent.

**Step 2**. Obtain the weight using the EWM $W_2 = (0.0141, 0.0014, 0.0131, 0.9668, 0.0046)$.

**Step 3**. Calculate the comprehensive weight $W = (0.2526, 0.1241, 0.1286, 0.4105, 0.0842)$.

**Step 4**. Establish a fuzzy relation matrix

Chen, Liu, and G. (2018) provide a systematic evaluation membership function based on eight levels, shown as Table 17.2.

Thus, the fuzzy relation matrix of FCEM can be generated as the following:

$$
\left\{
\begin{array}{lll}
0 \le m_v < \dfrac{1}{8} & R(x,1) = 0.9 & R(x,2) = 0.1 \\[2ex]
\dfrac{1}{8} \le m_v < \dfrac{1}{4} & R(x,1) = 1 - 4m_v & R(x,2) = 4m_v \\[2ex]
\dfrac{1}{4} \le m_v < \dfrac{3}{8} & R(x,2) = 1 - 4[m_v - X(2)] & R(x,3) = 4[m_v - X(2)] \\[2ex]
\dfrac{3}{8} \le m_v < \dfrac{1}{2} & R(x,3) = 1 - 4[m_v - X(3)] & R(x,4) = 4[m_v - X(3)] \\[2ex]
\dfrac{1}{2} \le m_v < \dfrac{5}{8} & R(x,4) = 1 - 4[m_v - X(4)] & R(x,5) = 4[m_v - X(4)] \\[2ex]
\dfrac{5}{8} \le m_v < \dfrac{3}{4} & R(x,5) = 1 - 4[m_v - X(5)] & R(x,6) = 4[m_v - X(5)] \\[2ex]
\dfrac{3}{4} \le m_v < \dfrac{7}{8} & R(x,6) = 1 - 4[m_v - X(6)] & R(x,7) = 4[m_v - X(6)] \\[2ex]
m_v = \dfrac{7}{8} & R(x,7) = 0.5 & R(x,8) = 0.5 \\[2ex]
\dfrac{7}{8} \le m_v < 1 & R(x,7) = 0.1 & R(x,8) = 0.9
\end{array}
\right. \tag{17.15}
$$

where $X(2) = \dfrac{1}{8}$, $X(3) = \dfrac{1}{4}$, $X(4) = \dfrac{3}{8}$, $X(5) = \dfrac{1}{2}$, $X(6) = \dfrac{5}{8}$.

**TABLE 17.1**

**The State Values of Evaluation Indicators of Metallurgical Crane**

| Main Girder | Strength | Stiffness | Weld | Corrosion | Local Deformation |
|---|---|---|---|---|---|
| Detection 1 | 0.955 | 0.68 | 0.99 | 0.72 | 0.756 |
| Detection 2 | 0.912 | 0.685 | 0.97 | 0.71 | 0.747 |
| Detection 3 | 0.922 | 0.69 | 0.83 | 0.702 | 0.725 |
| Detection 4 | 0.91 | 0.68 | 0.65 | 0.7 | 0.724 |

**TABLE 17.2**

**Classification of the Result of Safety Evaluation**

| Status Value | Status Description |
|---|---|
| (0, 0.125] | Disposal required |
| (0.125, 0.25] | Discontinue use and overhaul required |
| (0.25, 0.375] | Overall maintenance and repair required |
| (0.375, 0.5] | Serious operation with fault and major structural parts repair required |
| (0.5, 0.625] | More serious operation with fault, pay attention to the structure and cracks, and structural parts repair required |
| (0.625, 0.75] | Belong to the faulty operating state, focus on local deformation and micro-cracks. The additional regular inspections and damaged parts repair required. |
| (0.75, 0.875] | Running in normal condition, strengthen inspection required |
| (0.875, 1.0] | Running in good condition |

$$\mathbf{R} = \begin{bmatrix} 0 & 0 & 0 & 0 & 0 & 0 & 0.1 & 0.9 \\ 0 & 0 & 0 & 0 & 0.28 & 0.72 & 0 & 0 \\ 0 & 0 & 0 & 0 & 0.104 & 0.896 & 0 & 0 \\ 0 & 0 & 0 & 0 & 0.4 & 0.6 & 0 & 0 \\ 0 & 0 & 0 & 0 & 0.2 & 0.8 & 0 & 0 \end{bmatrix} \quad (17.16)$$

**Step 5**. Obtain the set of fuzzy comprehensive evaluation indicator

$$\mathbf{B} = [0 \quad 0 \quad 0 \quad 0 \quad 0.2292 \quad 0.5182 \quad 0.0253 \quad 0.2273] \quad (17.17)$$

According to the principle of maximum membership, it is shown that the safety state of the main girder system is the third level, i.e., the safety condition in general, and it belongs to the faulty operating state. Therefore, it is necessary to pay attention to the local deformation and micro-cracks, and the damaged parts should be repaired besides regular inspection. In general, the result obtained via FCEM based on comprehensive weight is consistent with the single factor feedback results of recent detection, which indicates that the proposed method is available for the safety assessment of metallurgical crane.

## 17.3 PERFORMANCE DEGRADATION MODEL FOR SAFETY ASSESSMENT INDICATORS

The metal structure of the casting crane subjected to frequent variable impact loads and the performance of its safety evaluation indicator degraded during long-term service. According to the performance degradation characteristics of different evaluation indicators, the performance degradation models of four indicators including crack, strength, stiffness, and corrosion are established in this part. The parameters in the degradation model can be estimated via the detection and monitoring data, thereby achieving the purpose of prediction residual life through the degradation model, which lays a theoretical foundation for the maintenance decision considering the degradation process.

### 17.3.1 Strength Degradation Model

As for the metal structure of the casting crane, the metal material degradation process is relatively slow. However, under the impact load such as lifting, the metal structure will also produce crystal slip, dislocation, and other defects, which will reduce the bearing capacity of this part and affect the safety of normal use. The strength degradation described herein refers to the degradation of static strength after fatigue loading. In the process of lifting, running, and braking, the continuously generated fatigue load of the casting crane leading to the internal fatigue damage of component materials and the weakened bearing capacity. This kind of strength degradation performance can be described by microscopic cumulative damage or phenomenological residual strength. Due to the complexity of the microscopic cumulative damage model, the practical engineering application is difficult. Here, the residual strength of component materials is described from the phenomenological point of view. The residual strength refers to the residual capacity of the equipment to resist external load after a certain period of service, which can describe the degradation of materials from a macro perspective. Yao (2003) concluded that the residual strength is related to the number of load cycles $n$ and the loading stress level $S$, as shown in the following formula:

$$R(n) = f(n, S) \tag{17.18}$$

The residual strength degradation model has the following characteristics (Wu 1985):

1. The residual strength is equal to the static tensile strength of the component material when the component is not in use, i.e., $R(0) = \sigma_b$.
2. The material is destroyed when the residual strength $R(n)$ equals to peak fatigue load $\sigma_{max}$, and the component fatigue life is $N$.
3. The degradation rate of fatigue strength is slow at the initial moment, i.e., $\dfrac{dR(n)}{dn} = 0$, $n \to 0$.

Schaff and Davidson (1997) proposed a residual life model to simplify the mechanism design:

$$R(n) = R(0) - f\left(R(0), S, R(S)\right) \tag{17.19}$$

where $f\left(R(0), S, R(S)\right)$ is the loss function during loading. Shen (2009) assumes that the stress on the structure is a discrete impact stress $S$ and obeys the Poisson distribution with a strength of $\lambda$, the strength degradation process is a Gamma process. Then, a stress-strength interference model considering strength degradation is established, and the model is applied to the reliability analysis and evaluation of the offshore platform in the process of strength decline. The stress-strength interference model is

$$R(t) = \Pr\{c(t) > s(t)\} \tag{17.20}$$

where $c(t)$ is the structural strength at time $t$; $s(t)$ is the stress on the structure at time $t$; $R(t)$ is the structural reliability at time $t$. Xie and Lin (1993) used a logarithmic model to fit the strength degradation process of normalizing 45 steel, and got a better fitting result, and concluded that the logarithmic model described its residual strength degradation law is more reasonable. The logarithmic model expression is:

$$R(n) = \sigma_b + (\sigma_b - S)\frac{\ln\left[1 - \dfrac{n}{N}\right]}{\ln N} \tag{17.21}$$

Su et al. (2000), Zhang et al. (2015), and Lv, Xie and Xu (1997) analyzed the strength degradation test data of normalized 45 steel under symmetrical constant amplitude stress cycles, and then developed corresponding degradation models for residual strength as follows:

$$R(n) = \sigma_b \left\{ 1 + \frac{\ln\left(1 - \frac{n}{N+1}\right)}{\ln(N+1)} \right\} \tag{17.22}$$

$$\sigma_b = S \left( \left( \frac{R(n)}{S} \right)^{1/A} + Bn \right)^A \tag{17.23}$$

$$R^{1+q}(n) = \sigma_b^{1+q} - (1-q)S^p n \tag{17.24}$$

in which $A, B, p, q$ is the material parameter with the values of 0.2817, $1.318 \times 10^{-3}$, 8.114087, 7.357682, respectively. Li et al. (2010, 2009) carried out a tensile strength degradation test on $35CrMo$ steel and proved the feasibility of the exponential degradation model to describe the residual strength degradation law of metal materials, and the exponential degradation model is

$$R(n) = \sigma_b + \left( \frac{S - \sigma_b}{1 - \theta} \right) \left( 1 - \theta^{\frac{n}{N}} \right) \tag{17.25}$$

where $\theta$ is the material parameter, the value is $4.113 \times 10^4$.

Based on the above six models, it is found that when the stress strength interference model involves stress distribution and strength degradation distribution, the reliability function and failure probability function need to be solved by numerical integration, which is of many difficulties. Though the logarithmic model, such as Eqs. (17.21) and (17.22) is characterized by few parameters and generally only related to the static tensile strength $\sigma_b$ and the load $S$, which is beneficial to describe the residual strength degradation process under constant amplitude load. However, it is difficult to calculate the equivalent cycle times under different loads when applied for the strength degradation under variable amplitude load, which is not conducive to the construction of a degradation model under variable amplitude load. The degradation models are shown as Eqs. (17.23) and (17.24) can better describe the strength degradation process of normalized 45 steel, but there are also problems that there are many parameters, which are not conducive to engineering applications. The exponential degradation model shown in Eq. (17.25) can better describe the degradation law of the residual strength of metal materials, and the model parameters are less, so it is easier to calculate the equivalent cycle times under different loads. Therefore, the exponential degradation model is employed to describe the strength degradation process of Q345 steel for the metal structure of the casting crane. The material constant $\theta > 1$ can be given by the number of cycles and the corresponding strength recorded by historical monitoring and testing

However, the load amplitude of the casting crane is different in mid-span, end, and 1/4 span. Therefore, it is necessary to expand the strength degradation model under variable amplitude. Lv and Yao (2000) proposed the extension of the residual strength degradation model under variable amplitude load, and assuming that the component experienced $n_i$ cycles under the load $S_i$

and produces residual strength $R_i$; experienced $n_j$ cycles under the load $S_j$ and produces residual strength $R_i$. Then, the residual strength of the $i$th and the $j$th times are

$$R(n_i) = \sigma_b + \left( \frac{S_i - \sigma_b}{1 - \theta_i} \right) \left( 1 - \theta_i^{\frac{n_i}{N_i}} \right) \qquad (17.26)$$

$$R(n_j) = \sigma_b + \left( \frac{S_j - \sigma_b}{1 - \theta_j} \right) \left( 1 - \theta_j^{\frac{n_j}{N_j}} \right) \qquad (17.27)$$

Herein, it is assumed that the damage caused by $n_{ji}$ secondary cycle under the action of the load $S_j$ is equal to the damage caused by $n_i$ secondary cycle under the action of the load $S_i$, then

$$R(n_j) = \sigma_b + \left( \frac{S_j - \sigma_b}{1 - \theta_j} \right) \left( 1 - \theta_j^{\frac{n_{ji}}{N_j}} \right) \qquad (17.28)$$

$$n_{ji} = N_j \left( \frac{\ln \left( 1 - \left( 1 - \theta_i^{\frac{n_i}{N_i}} \right) \frac{(S_i - \sigma_b)(1 - \theta_j)}{(S_j - \sigma_b)(1 - \theta_i)} \right)}{\ln \theta_j} \right) \qquad (17.29)$$

and $R(n_i + n_j)$ can be obtained, defined as

$$R(n_i + n_j) = \sigma_b + \left( \frac{S_j - \sigma_b}{1 - \theta_j} \right) \left( 1 - \theta_j^{\frac{n_j + n_{ji}}{N_j}} \right) \qquad (17.30)$$

The strength degradation model under variable amplitude can be expressed as

$$R(n_j + n_i + n_{i+1} \cdots n_n) = \sigma_b - (\sigma_b - S_j) \left( \frac{n_j + n_{ji} + n_{ji+1} \cdots n_{jn}}{N_j} \right)^{\theta_j} \qquad (17.31)$$

The strength degradation model under fatigue load is:

$$R = \sigma_b + \sum_{j=1}^{m} \left( \frac{S_j - \sigma_b}{1 - \theta_j} \right) \left( 1 - \theta_j^{\frac{n_j + n_{ji} + n_{ji+1} \cdots n_{jn}}{N_j}} \right) \qquad (17.32)$$

Strength obeys a certain distribution in the process of strength degradation. Guo and Yao (2003) assume that the critical fatigue damage and the actual fatigue damage obey lognormal distribution, and describes the strength degradation process under variable amplitude load. Gu (1999) uses the discrete two-parameter Weibull distribution to describe the strength degradation process under variable amplitude load. Shen (2009) describes the strength degradation process of materials with a random process and assumes the strength degradation process obeys gamma random process.

The lognormal distribution model of strength degradation is shown as

$$F(R) = \int_0^\infty f_Y(y) \int_0^\infty f_X(x) dx dy \tag{17.33}$$

where $f_X(x)$ is the probability density function of critical fatigue damage, $f_Y(y)$ is the load transformation density function.

The strength degradation process of Weibull distribution in different circulation fraction $n/N$ is described as follows

$$f_R(r|0) = \frac{m(0)}{b(0)} \left( \frac{r}{b(0)} \right)^{m(0)-1} \exp\left[ -\left( \frac{r}{b(0)} \right)^{m(0)} \right] \tag{17.34}$$

$$f_R(r|1) = \frac{m(1)}{b(1)} \left( \frac{r}{b(1)} \right)^{m(1)-1} \exp\left[ -\left( \frac{r}{b(1)} \right)^{m(1)} \right] \tag{17.35}$$

$$f_R(r|x) = \frac{m(x)}{b(x)} \left( \frac{r}{b(x)} \right)^{m(x)-1} \exp\left[ -\left( \frac{r}{b(x)} \right)^{m(x)} \right] \tag{17.36}$$

in which $r$ is the variable. $m(0)$, $b(0)$ is the shape parameter and scale parameter of the Weibull distribution of static strength, respectively. $f_R(r|0)$ represents the distribution of the residual strength $R(n)$ of static strength $\sigma_b$ when $n$ equals to 0. $f_R(r|1)$ and $f_R(r|x)$ indicate the distribution of the residual strength $R(n)$ when $n$ equals to $N$ and $x$, respectively. Also, $m(0)$, $b(0)$, $m(1)$, $b(1)$, $m(x)$, $b(x)$ all are the function of load $S$.

Gamma distribution is a strictly regularized degradation, and its intensity degradation process density function is described as follows

$$f(R|\alpha(n), \beta) = \frac{\beta^{\alpha(n)}}{\Gamma(\alpha(n))} R^{\alpha(n)-1} e^{(-\beta R)} \qquad R \in (0, \infty) \tag{17.37}$$

where $\Gamma(\alpha(n)) = \int_0^\infty x^{\alpha(n)-1} e^{-x} dx$, $n$ is the number of cycles, $R$ is the strength degradation value, $\alpha(n)$, and $\beta$ is the shape parameter and scale parameter, respectively. Besides, the value of $\alpha(n)$ is determined by the relationship between the number of cycles and strength degradation. In stationary gamma random distribution, $\alpha(n) = cn$, $c$ is the coefficient of shape parameter function and can be obtained by parameter fitting.

By comparing the three models, it is found that when using the lognormal distribution model to describe, the lifting load value should be recorded whenever the time history load changes and transformed into a unified load transformation density function. Due to the great changes in the stress that happens during the casting lifting process, it is difficult to calculate with the lognormal distribution model. The relationship between the load $S$ and the shape parameter $m$ and the scale parameter $b$ is required to describe the strength degradation of the two-parameter Weibull distribution, and the test or sufficient strength degradation data are also needed to study. When the gamma stochastic process is used for degradation modeling, a corresponding probability density function can be formed whenever the number of cycles $n$ changes. Hence, it is not necessary to discretize the process of structural strength degradation, and only the corresponding strength data under different cycles is required to calculate the shape parameter $\alpha(n)$ and scale parameter $\beta$ by maximum natural estimation of distance estimation in engineering practice. Therefore, the gamma stochastic process is recommended to be used as a strength degradation model.

### 17.3.2 Stiffness Degradation Model

The metal structure (main beam, plate, bolt, etc.) of the casting crane will inevitably produce dislocation, sliding, and other damage under the action of continuous fatigue load. However, the structure, load-bearing, and working environment of the casting crane are relatively complex, and the internal damage of the metal structure is difficult to measure, which brings some difficulties to the cumulative damage analysis of the metal structure of the casting crane. The macroscopic appearance of stiffness material performance is closely related to the micro-damage inside the material, and it is easier to measure and monitor. Therefore, fatigue stiffness degradation is an index that can reflect the fatigue performance of the whole structure. In 1972, Salkind (1972) first proposed the use of stiffness change to measure structural fatigue damage. Later, Zong and Yao (2016), An et al. (2019), Li and Nie (2018), Guo (2017), and Zhang and Wang (2016) studied the stiffness degradation method to characterize fatigue damage. The stiffness degradation is similar to strength degradation, which is mainly related to the number of load cycles $n$, the applied load $S$, and the initial stiffness $E(0)$.

The stiffness can be obtained after dimensionless treatment as:

$$\frac{E(n)}{E(0)} = f(n,S) \tag{17.38}$$

where $E(n)$ is the residual stiffness of material after n cycle times under load $S$. $f(n,S)$ is the stiffness degradation function. When $n$ equals zero or $N$, the value of $f(n,S)$ equals unity or zero. Besides, $N$ is the number of stiffness failure cycles under load $S$.

Echtermeyer, Engh and Buene (1995) used a logarithmic model to describe the stiffness degradation of the ruptured fiber composite laminate, and it fits the test results to a good result. The logarithmic model is expressed as:

$$E(n) = E(0) - \alpha \log(n) \tag{17.39}$$

where $\alpha$ is the material parameter. Zhang and Sandor (1992) proposed the stiffness degradation model based on composite material elastic modulus $E(n)$, load $S$, and cycle number $n$ as:

$$E(n) = E(0)\left[1 - K\left(\frac{S}{E(0)}\right)^{\alpha} n^{\beta}\right] \tag{17.40}$$

where $\alpha$ and $\beta$ are the parameter of the material. Li (2013) carried out a fatigue test on modulated 45 steel and recorded the stiffness measurement under different cycles. The proposed stiffness degradation model is:

$$\frac{E(n)}{E(0)} = 1 - S^{a}\left(\frac{n}{N}\right)^{b} \tag{17.41}$$

in which, $a$ and $b$ are the parameter related to loading condition $S$. When $S = 300$ Mp, $a = -0.31$, $b = 3.27$; when $S = 320$ Mp, $a = -0.24$, $b = 2.67$.

It can be seen from the above four models that Eqs. (17.34–17.35) are composite stiffness models while Eq. (17.36) is a steel stiffness degradation model and has a good fitting effect on steel stiffness degradation. The power function degradation model represented by Eq. (17.36) is suitable for engineering applications due to its simplicity and fewer parameters. Therefore, the power function degradation model shown in Eq. (17.36) is selected to analyze the stiffness degradation of the metal structure of the casting crane.

Supposing the component undergoes $n_i$ secondary cycle under the action of the load $S_i$ and generates residual stiffness $E_i$. Under the action of the load $S_j$, $n_j$ secondary cycle is experienced and residual stiffness $E$ is generated. Then, the formula of residual stiffness for the $i$th and $j$th are as follows:

$$\frac{E(n_j)}{E(0)} = 1 - S_j^{a_j} \left( \frac{n_j}{N_j} \right)^{b_j} \tag{17.42}$$

$$\frac{E(n_i)}{E(0)} = 1 - S_i^{a_i} \left( \frac{n_i}{N_j} \right)^{b_i} \tag{17.43}$$

Assuming that the stiffness degradation caused by $n_{ji}$ secondary cycle under the action of the load $S_j$ is equal to that caused by $n_i$ secondary cycle under the action of the load $S_i$, i.e., $E(n_i) = E(n_{ji})$, then

$$\frac{E(n_{ji})}{E(0)} = 1 - S_j^{a_j} \left( \frac{n_{ji}}{N} \right)^{b_j} \tag{17.44}$$

$$n_{ji} = N_j \sqrt[b_j]{\frac{S_i^{a_i}}{S_j^{a_j}} \left( \frac{n_i}{N_I} \right)^{b_i}} \tag{17.45}$$

Therefore, the sum of stiffness degradation under two different loads is

$$\frac{E(n_j + n_i)}{E(0)} = 1 - S_j^{a_j} \left( \frac{n_j + n_{ji}}{N_j} \right)^{b_j} \tag{17.46}$$

When $i = 1 \cdots m$, that is to say, when there are $n$ fatigue loads $S_i$, stiffness degradation is:

$$\frac{E(n_j + n_i + \cdots + n_m)}{E(0)} = 1 - S_j^{a_j} \left( \frac{n_j + n_{ji} + \cdots + n_{jm}}{N_j} \right)^{b_j} \tag{17.47}$$

where $n_{ji} \sim n_{jm}$ can be obtained by Eq. (17.40).

The above is the material stiffness degradation model of the metal structure of the casting crane. Besides, the stiffness degradation of the main beam of the metal structure can be expressed by the following:

$$K(n) = \frac{F}{\omega_n} \tag{17.48}$$

where $F$ is the fixed load for measuring stiffness; $\omega_n$ is the mid-span and lower deflection under a fixed load $F$ after $n$ cycles; $K(n)$ is the structural stiffness after $n$ cycles. Thus, the stiffness degradation model of the main beam can be derived by fitting different cycles and corresponding structural stiffness.

Due to the uncertainties exist in the lifting quality, speed, starting, braking, and metal structure materials, it is necessary to conduct a reliability analysis during the stiffness degradation process. When the stiffness degradation process is strictly regularized, the *B-S* distribution can be introduced to build the probability density function of the component stiffness degradation model. The *B-S* distribution is used to approximate the failure of life distribution to the stiffness degradation to simplify the function (Kou 2019), which is expressed as follows

$$f(t,\mu,\omega) = \frac{1}{2\sqrt{2\pi}\mu\omega}\left[\left(\frac{\omega}{t}\right)^{0.5} + \left(\frac{\omega}{t}\right)^{1.5}\right]\exp\left[-\frac{1}{2\alpha^2}\left(\frac{t}{\omega} - 2 + \frac{\omega}{t}\right)\right], \quad t > 0 \qquad (17.49)$$

where $\mu = \sqrt{1/\beta E_f}$, $\omega = \sqrt{\beta E_f / \alpha(n)}$, $E_f$ is the threshold of the stiffness failure, $\alpha(n)$ and $\beta$ is the shape parameter and scale parameter of Gamma distribution, respectively.

### 17.3.3 CORROSION DEGRADATION MODEL

Metal corrosion is a process of physical and chemical properties change due to the interaction with the environment. According to the environment of corrosion, it can be divided into atmospheric corrosion, freshwater corrosion, seawater corrosion, soil corrosion, and concrete corrosion. The corrosion of the metal structure of the casting crane belongs to atmospheric corrosion. The corrosion rate is related to water, oxygen, carbon dioxide, $SO_2$, $Cl^-$ and other factors. Metal and its anti-corrosion coating contact with corrosive media in the environment, when the anti-corrosion coating is damaged by a corrosive medium, the metal structure will react with the environment to produce corrosion. The corrosion types also can be divided into overall corrosion, pitting corrosion, crevice corrosion, denudation, filiform corrosion, etc. When there is no anti-corrosion coating on the surface of the metal structure, general corrosion generally occurs. Pitting corrosion generally occurs in highly humid areas such as submarine pipelines and underground pipelines. Crevice corrosion generally occurs between metal connectors, such as bolts and welding. The volume expansion of the metal caused by corrosion causes the surface to peel off and bulge. This type of corrosion may occur on the surface of painted metal, stainless steel profiles, alloys, etc. Filamentous corrosion often occurs in humid environments and mostly occurs in coated metals. It can be seen that in addition to pitting corrosion, other types of corrosion may occur in the foundry crane factory. Kou (2019) indicates that the corrosion rate of steel is doubled when the temperature increases by 10°C; however, the relationship between humidity and corrosion is more close. The steel structure will not be corroded or the corrosion rate is very low when the relative humidity $RH \leq 60\%$, and the highest corrosion rate will be achieved when the value of $RH$ ranges from between 90% and 98%.

Two traditional methods are often used for expressing the degree of corrosion of steel structures, namely, the weight method and the depth method. The weight of the metal structure will change before and after corrosion, so the weight method is used to measure the corrosion rate. The depth method is based on the weight method, and the deep corrosion rate is used to characterize the metal corrosion rate. The expression of the weight method is

$$V_w = \frac{W}{AT} \qquad (17.50)$$

where $V_w$ is the corrosion rate of steel characterized by weight; $W$ is the reduction or loss of weight before and after corrosion; $A$ and $T$ are the steel area and corrosion time of steel, respectively.

The expression of the depth method is as follows:

$$V_d = \frac{24 \times 365 \times V_w}{100\rho} \qquad (17.51)$$

in which, $V_d$ is the corrosion rate of steel characterized by depth; $\rho$ is the density of the steel. When $V_d < 0.1$, the steel belongs to corrosion-resistant; $0.1 \ll V_d \ll 1$ indicates the steel category is available, $V_d > 1$ represents the steel category is unavailable (Ma, Wang and Sun 2009).

Many researchers have studied the depth of rust pit and achieve some results. Tang (1995) used GM (1.1), the exponential regression model $\left(Y = Ae^{Bx}\right)$ and power function regression model $\left(Y = AX^B\right)$ to simulate the data of low alloy steel under atmospheric corrosion, and found that the three models had good fitting results in 1–4 years of corrosion of low-alloy steel, but only the power exponential regression model could fit the data within 8 years. To increase the prediction accuracy of the model, Wang et al. (2004) divided the corrosion law into three stages, namely, the linear stage, the parabolic stage, and the linear stage composite model, and verified with 16 years of corrosion data. Liang and Hou (1995) carried out a large number of atmospheric corrosion data analyses, which shows that the power function has good prediction results for the depth of atmospheric corrosion rust pits. Therefore, the power function is used to predict the corrosion depth of the metal structure of the casting crane

$$D = CT^n \tag{17.52}$$

where $D$ is the corrosion depth; $T$ is the time of metal structure exposed to air; $C$ and $n$ are the parameters.

However, most metal structures have anti-corrosion coatings, and metal structures should not begin to corrode until the anti-corrosion coating fails. Pan (2009) proposed the power function prediction model for anti-corrosion coatings and compared with the measured data, it was found that the model has a better prediction effect. The model is defined as

$$S = Bt^A \tag{17.53}$$

in which, $S$ is the corrosion area of the coating (mm); $t$ is the corresponding time; $A$ and $B$ are the coefficients. Therefore, the corrosion pit depth prediction model of the metal structure is as follows

$$D = C(T - t_c)^n \tag{17.54}$$

where $t_c$ is the coating failure life, determined by Eq. (17.47), and it is considered that the coating will fail when the area ratio of coated steel plate (or the area ratio of coating opening) reaches 5%.

Nevertheless, the corrosion rate of metals may be different due to different materials and environments whether it is physical corrosion or electrochemical corrosion. It is undeniable that the depth of rust pits is increasing in the process of metal corrosion. Li and Zhang (2017) established gamma stochastic process to describe the corrosion pit deepening process of oil and gas pipeline, and proved that the reliability of the remaining life prediction model reached 95%. From the prediction model of rust pit, it can be seen that the deepening process of rust pit changes exponentially. Therefore, the shape parameter $\alpha(t) = ct^b$ and the scale parameter $\beta$ is recommended here. Then the probability density function of the corrosion process is

$$f\left(D \mid ct^b, \beta\right) = \frac{\beta^{ct^b}}{\Gamma(ct^b)} D^{ct^b - 1} e^{-\beta D}, \quad D \in (0, \infty) \tag{17.55}$$

where $\Gamma(ct^b) = \int_0^\infty x^{ct^b - 1} e^{-x} \, dx$, $D$ is the corrosion depth.

### 17.3.4 CRACK DEGRADATION MODEL

Fatigue cracks are one of the most harmful failures of the metal structure of casting cranes. The expansion of fatigue cracks often causes the failure of the beam structure and causes major safety accidents. Currently, it is generally believed that factors such as stress, strain, and energy are closely related to the fatigue life of the structure.

Therefore, the fatigue life is linked to these factors in the process of life prediction, and the life prediction model is established based on damage accumulation theory, damage mechanics theory, and fracture mechanics theory. The methods of stress fatigue life prediction ($S$-$N$ curve), strain fatigue life prediction ($\varepsilon - N$ curve), fatigue cumulative damage theory, and fracture mechanics are widely used in industrial design. The most widely used method is the fatigue life prediction method based on fracture mechanics, that is, the fatigue crack growth is described when the stress intensity factor at the crack tip reaches the threshold value, and the crack length $a$ and cycle number $N$ are used to describe the crack growth rate. Herein, to facilitate the combination of physical models and filtering algorithms, the classical Paris formula and the two-parameter exponential formula is used to predict the crack growth life. The expression of the Paris formula is as follows:

$$\frac{da}{dN} = C(\Delta K)^m \tag{17.56}$$

Where $\Delta K$ is the amplitude of stress intensity factor, which is related to crack length, stress form, and crack surface stress, $da/dN$ is the fatigue crack growth rate. The improved two-parameter exponential formula can be expressed as:

$$\frac{da}{dN} = e^\alpha e^{\frac{\beta}{\Delta K}} \tag{17.57}$$

in which, $\alpha$ and $\beta$ is the parameter of material.

The fatigue crack growth is affected by the random fluctuation of external load, geometric structure, structural material properties, and manufacturing process. The fatigue life is different even if the same type of cracks is under the same working conditions. Therefore, it is not accurate enough to predict the crack growth life without considering the reliability. The dispersion caused by the structure size and shape dispersion, material property dispersion, processing technology, installation and debugging, etc., should be considered. Uncertainties such as accidental factors such as environmental differences are taken into account (Li and Zhang 2017). The distribution models currently used in fatigue structure reliability mainly include Weibull distribution, Gamma distribution, extreme value distribution, normal distribution, lognormal distribution, etc. Considering the strict regularization of the gamma distribution and the change of the distribution characteristics over time, the corresponding reliability distribution function can be constructed according to the change of the crack length during the crack propagation process. Hence, the Gamma distribution is employed to describe the reliability of the crack propagation process. The details can be seen in Section 17.4.

## 17.4 MAINTENANCE DECISION ON THE PROBABILITY DENSITY FUNCTION OF THE SAFETY ASSESSMENT INDICATORS

The performance degradation models of safety evaluation indicators based on the Gamma process are employed to establish the probability density function of four safety indicators, and the parameter estimation of Gamma function was conducted according to the time data and state values of different components. Then, the Weibull cumulative distribution function is used to simplify the reliability and failure probability of the Gamma function. Thus, the maintenance decision model of a single component is presented based on the theory of the minimum cost ratio. The maintenance time variation penalty function is built using the out-of-production loss cost and the single-component group maintenance interval is determined according to the penalty function. A group maintenance decision is then formed to minimize the penalty function. Lastly, ten random fatigue cracks are randomly generated to verify the validity of the group maintenance decision model.

### 17.4.1 PROBABILITY DENSITY FUNCTION OF EVALUATION INDICATOR BASED ON GAMMA RANDOM PROCESS

The difficulty of the maintenance decision of casting crane is how to consider the influence of uncertain factors in the operation process and to construct the distribution function of structure life in the process of structural degradation. The performance degradation process of casting crane metal structure is non-negative, and the degradation increment is only related to the state quantity at the last moment due to the influence of the environment. The Gamma process is an independent and strictly regularized stochastic degradation process, which can describe the discontinuous problem of degradation increment caused by the impact process, so it is widely used in various kinds of performance degradation research.

Herein, the gamma stochastic process is used to describe the crack growth process of the metal structure of the casting crane. The crack length $X(t)$ is a random variable obeying gamma distribution at time $t$, $X(t) \sim Ga(\alpha(t), \beta)$, where the shape parameter $\alpha(t)$ is a non-negative, right continuous and monotone increasing path, and $\alpha(0) \equiv 0$ and $\beta > 0$ are scale parameters. Then, the distribution probability density function is as follows:

$$f(x \mid \alpha(t), \beta) = \frac{\beta^{\alpha(t)}}{\Gamma(\alpha(t))} x^{\alpha(t)-1} e^{(-\beta x)} I_{(0,\infty)}(x) \tag{17.58}$$

where

$$\Gamma(\alpha(t)) = \int_0^\infty x^{\alpha(t)-1} e^{-x} dx \tag{17.59}$$

$$I_{(0,\infty)}(x) = \begin{cases} 1, & x \in (0,\infty) \\ 0, & x \notin (0,\infty) \end{cases} \tag{17.60}$$

$X(t)$ is a continuous-time stochastic process, which has the following characteristics:

1. $X(0) = 0$ holds with probability 100%.
2. $X(t)$ has a stable independent increment.
3. For any $t \geq 0$ and $\Delta t$, $X(t + \Delta t) - X(t) \sim Ga(\alpha(t + \Delta t) - \alpha(t), \beta)$.

The mean and variance of the degradation process are as follows:

$$E(X(t)) = \frac{\alpha(t)}{\beta} \tag{17.61}$$

$$Var = \frac{\alpha(t)}{\beta^2} \tag{17.62}$$

The failure time of $X(t)$ reaching the failure threshold $a_{cr}$ for the first time is defined as the failure time, and the structure life is the length of time from the beginning of use to reaching the failure threshold $a_{i,cr}$ for the first time is $T$. The cumulative failure rate is:

$$F_T\left(p \mid X(t_i)\right) = P(t \geq T) = P\left(X(t) \geq a_{cr}\right)$$

$$= \int_{+\infty}^{a_{cr}} Ga(\alpha(t_i), \beta)\, da \qquad (17.63)$$

$$= \frac{\psi(\alpha(t_i), \beta a_{cr})}{\Gamma(\alpha(t_i))}$$

in which, $F\left(p \mid X(t_i)\right)$ is the failure probability of the fatigue structure at time $t$, $\psi\left(\alpha(t_i), \beta a_{cr}\right)$ is the incomplete gamma function, defined as

$$\psi\left(\alpha(t_i), \beta a_{cr}\right) = \int_{\beta a_{cr}}^{+\infty} y^{\alpha(t_i)-1} e^{-y}\, dy \qquad (17.64)$$

Since Eq. (17.59) is difficult to be obtained by analytical solution, the Monte Carlo sampling is used to obtain the integral result and calculate the cumulative failure rate. Thus, the reliability function is expressed as

$$R(t) = P(T \leq t) = P\{X(t) \leq a_{cr}\} = \int_0^{a_{cr}} \frac{\beta^{\alpha(t)}}{\Gamma(\alpha(t))} x^{\alpha(t)-1} e^{-\beta x}\, dx = \frac{\psi_R(\alpha(t), \beta a_{cr})}{\Gamma(\alpha(t))} \qquad (17.65)$$

Therefore, parameter estimation is then conducted for the aforementioned probability density function. According to $n$ sample records $\left(t_i, X(t_i)\right), i = 1\ldots n$, the moment estimation method, and maximum natural estimation method can be used to estimate the shape parameter $\alpha$ (T) and the scale parameter $\beta$. Moment estimation can be used when the sample parameters are large, and the maximum likelihood estimation can be used when the sample parameters are small. Since the crack growth process is exponential, $\alpha(t) = ce^{bt}$ is selected as shape parameter coefficients where $c$, $b$ is the shape parameter.

For the nonstationary gamma process, the logarithm of both sides of Eq. (17.56) can be obtained as

$$\log E(X(t)) = bt + \log(c/\beta) \qquad (17.66)$$

Then, the function is an oblique line with $t$ as the horizontal axis and $\log E\left(X(t)\right)$ as the longitudinal axis and the slope of $b$. According to the least square method, the following results are obtained

$$\hat{b} = \frac{n \sum_{i=1}^{n} t_i \log(x_i) - \left(\sum_{i=1}^{n} t_i\right)\left(\sum_{i=1}^{n} \log(x_i)\right)}{n \sum_{i=1}^{n} t_i^2 - \left(\sum_{i=1}^{n} t_i\right)^2} \qquad (17.67)$$

When the estimated value $\hat{b}$ of parameter $b$ is obtained, the non-stationary gamma process concerning time $t$ becomes a stationary process of $e^{bt}$. The moment estimation or maximum natural estimation then can be used to calculate $c$ and $\beta$. For the gamma process of known $b$, let $\Delta x_i = x_i - x_{i-1}$, $\Delta t_i = t_i - t_{i-1}$, $i = 1\ldots n$. Then $\Delta x_i / \Delta t_i$ is defined as the degradation rate $R_i$. Each degradation rate

**TABLE 17.3**

**Parameter Estimation for Gamma Stochastic Process**

| Parameter | $\hat{\beta}$ | $\hat{c}$ | $\hat{b}$ |
|---|---|---|---|
| Estimated value | 1.0147 | 4.7906 | $1.8172 \times 10^{-6}$ |

is independent of each other and obeys gamma distribution. The mean and variance of the sample degradation rate $R_i$ are

$$\bar{R} = \frac{1}{n} \sum_{i=1}^{n} \frac{\Delta x_i}{\Delta t_i} \tag{17.68}$$

$$S^2 = \frac{1}{n-1} \sum_{i=1}^{n} \left( \frac{\Delta x_i}{\Delta t_i} - \bar{R} \right)^2 \tag{17.69}$$

therefore

$$E(S^2) = \frac{1}{n-1} \sum_{i=1}^{n} E\left( \frac{\Delta x_i}{\Delta t_i} - \bar{R} \right)^2 = \frac{\alpha}{n\beta^2} \sum_{i=1}^{n} \left( \frac{1}{\Delta t_i} \right) \tag{17.70}$$

According to Eqs. (17.63–17.65), the estimated parameters are obtained as

$$\hat{c} = \frac{\bar{R}^2 \sum_{i=1}^{n} \frac{1}{\Delta t_i}}{nS^2} \tag{17.71}$$

$$\hat{\beta} = \frac{\bar{R} \sum_{i=1}^{n} \frac{1}{\Delta t_i}}{nS^2} \tag{17.72}$$

Consequently, the Gamma random process parameters of fatigue crack propagation of a structure can be fitted as shown in Table 17.3.

### 17.4.2 COMPONENT MAINTENANCE DECISION OF CASTING CRANE METAL STRUCTURE

The whole structure is deemed to have an infinite service life, and only considers the economic correlation to optimize the maintenance decision of local components or equipment. When the metal structural components of the casting crane fail, the maintenance cost includes preventive maintenance cost $C_P$, maintenance cost $C_P$, failure cost $C_g$, loss of work stoppage $C_t$ and maintenance time of different components of casting crane $d_i$. The unit time maintenance cost of $i$th component repair type maintenance is as follows:

$$C_i = \frac{C_P + (C_f + C_g) \int_{t_0}^{T_i} \lambda(t)dt}{T_i} \tag{17.73}$$

where $\lambda(t)$ is the failure probability function, defined as

$$\lambda_i(t) = \frac{f_i(t)}{R_i(t)} = \beta^{\alpha(t)} \frac{x^{\alpha(t)-1} e^{-\beta x}}{\beta a_{cr} \int_0^{} t^{\alpha(t)-1} e^{-\beta t} dt} \tag{17.74}$$

Therefore, the unit time maintenance cost of component $i$ is

$$\hat{C}_i = \frac{C_P R_i(T) + (C_f + C_g)(1 - R_i(T))}{\int_{t_0}^{T} R_i(t) dt} \tag{17.75}$$

in which, $R_i(t)$ is the component reliability function.

The optimal maintenance time $t_i$ of the $i$th fatigue component can be obtained by Eqs. (17.68–17.70). Let $X_i(t_0) = 0$, that is, there is no damage to the $i$th component at the initial moment. At this time, the optimal maintenance time of the fatigue component obtained by the $i$th component is the nominal optimal maintenance time $\overline{t}_i$. $\hat{C}_i$ represents the minimum maintenance cost of the nominal optimal maintenance time. The nominal optimal maintenance time and the nominal optimal maintenance cost will be used in the calculation of the number of single component maintenance in a maintenance cycle and the construction of the penalty function caused by the deviation of the maintenance time from the optimal maintenance time.

Aiming to reduce the computation cost, the failure rate function and a cumulative failure rate of the three-parameter Gamma distribution are approximated by the WeibullCDF function, and the time series $t=[t_1, t_2, \ldots, t_n]$ failure rate function is solved by the Monte Carlo sampling toolbox of Matlab. The solution is as follows:

$$P(x > a_{cr}) = \int_{a_{cr}}^{\infty} \lambda_i(t_j) dx = \frac{\beta^{\alpha(t_j)}}{\beta a_{cr} \int_0^{} t^{\alpha(t_j)-1} e^{-\beta t} dt} \int_{a_{cr}}^{\infty} x^{\alpha(t_j)-1} e^{-\beta x} dx, \qquad j = 1, 2, \ldots, n \tag{17.76}$$

$$R(t_j) = \frac{\beta^{\alpha(t_j)}}{\Gamma(\alpha(t_j))} \int_0^{a_{cr}} x^{\alpha(t_j)-1} e^{-\beta x} dx \tag{17.77}$$

Then, we perform WeibullCDF fitting on the obtained $n$ function values, and the fitting formula is as follows:

$$y = y_0 + A_1 \left(1 - e^{-(x/a)^b}\right) \tag{17.78}$$

where $A_1$, $a$, $b$, $y_0$ are the parameters of the WeibullCDF function, and the quality of the fitting result is judged according to $R^2$. During the fitting process, it should be noted that the value of the time series $t = [t_1, t_2, \ldots, t_n]$ should not be too large. If the time series is too large, the maintenance period will be missed and resulting in large errors of the optimal solution of the maintenance time. From Figures 17.3 and 17.4, we can see that by substituting the failure rate function and the reliability function into Eqs. (17.68) and (17.70), respectively, the optimal maintenance time of the component can be obtained, that is, the minimum value in Figures 17.3 and 17.4.

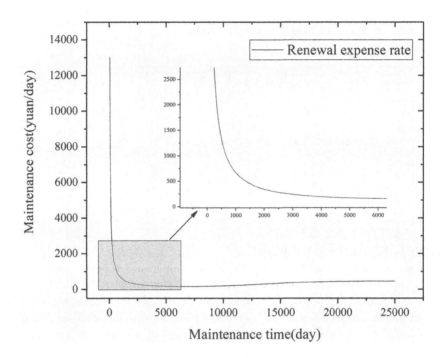

**FIGURE 17.3**    Renewal expense curve of the $i$th component.

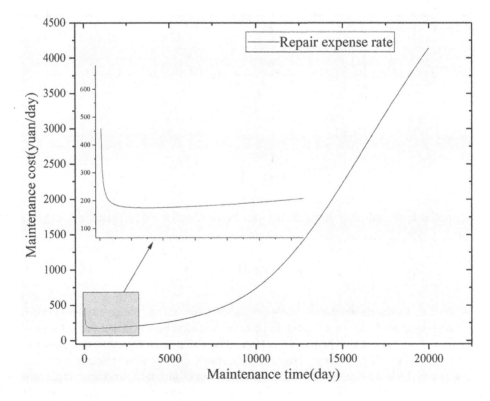

**FIGURE 17.4**    Repair expense curve of the $i$th component.

### 17.4.3 GROUP MAINTENANCE DECISION OF THE CASTING CRANE METAL STRUCTURE

The current maintenance schemes mainly focus on a single component. It is considered that the physical structure, maintenance cost, and fault of components in the system are independent of each other. However, due to the casting crane is a large and complex system, the component types and structural forms are diverse, which makes the components have certain interaction. This correlation exists in casting crane. The maintenance correlation between system components can be divided into three categories: economic correlation, fault correlation, and structural correlation. Structural correlation considers that in a structure of multiple components, one component is repaired while the other is repaired. Fault correlation considers that when one component of the system fails, it will lead to changes in the fault segments of other components. Economic correlation considers that the cost of maintaining several components together is lower than that of separate maintenance.

As for the casting cranes, the influence between system component failures, and the relationship between structures are very complicated. Therefore, to facilitate the establishment and use of the maintenance model, the group maintenance decision in this research only considers economic relevance and considers multi-component group maintenance to reduce the number of maintenance and the impact of the number of downtimes on production to achieve the purpose of reducing maintenance costs. Thus, the following assumptions are made:

1. Failures between different components do not affect each other.
2. The maintenance of a single component does not affect other components.
3. The component degradation data required in this paper (the fatigue structure degradation data is the crack length) can be obtained through simulation, inspection, or monitoring.
4. The residual life of components can be predicted by computer means such as particle filter and genetic algorithm.
5. The reliability model of the degradation process can be constructed.

One casting crane may have multiple same failure modes. The group maintenance considered in this research is the maintenance of multiple fatigue structures at the same time. Supposing that $n$ fatigue structures are involved in the system, and the optimal maintenance time for each fatigue structure is $t_i (i = 1...n)$ when no overlap exists in maintenance time, the maintenance cost of a single fatigue structure is:

$$C_i = C_P(t_i) + C_t \qquad t_i \langle RUL_i, R_i \rangle 0.995 \qquad (17.79)$$

where $C_P(t_i)$ is the preventive maintenance cost at $t_i$. Because the maintenance time of the group maintenance process may be adjusted, the maintenance cost will change to a certain extent. $C_t$ is the downtime loss cost. $RUL_i$ is the prediction in the remaining life, which is used to constrain the maintenance timeless than the remaining life of the structure to guarantee the whole safety of the structure. $R_i$ is the reliability of the component at the time of maintenance to ensure that the reliability is not less than 0.995.

Assuming that the maintenance decision in the system can be divided into $K$ maintenance groups, respectively, $Z_1,...,Z_k$. Then, the maintenance costs of the $K$ maintenance groups are:

$$C_Z = \sum_{i=1}^{K} \sum_{j=Z_i} C_j(t_{Z_i}) + \sum_{i=1}^{K} C_{t_i} \qquad (17.80)$$

where $t_{Z_i}$ is the maintenance time of the $i$th maintenance group, $\sum_{i=1}^{K}\sum_{j=Z_i} C_j(t_{Z_i})$ is the sum of the

$K$th group preventive maintenance cost, and $\sum_{i=1}^{K} C_{t_i}$ is the total cost of $K$ shutdowns. Then the cost

savings of the maintenance decision group is expressed as:

$$C_{S_i} = (\text{length}(Z_i) - 1)C_{t_i} - \sum_{j=1}^{\text{length}(Z_i)} \mu_j(t_j, \Delta t_j) \tag{17.81}$$

in which $C_{S_i}$ is the cost savings of the $i$th maintenance group, $\text{length}(Z_i)$ indicates the number of maintenance components in the $i$th maintenance group, $\mu_i(t_i, \Delta t_i)$ is the penalty function for delaying or advancing the repair time in the repair group $Z_i$, defined as

$$\mu_j(t_j, \Delta t_j) = C_j(t_j + \Delta t_j) - C_j(t_j) \tag{17.82}$$

where $|\Delta t_i| < \bar{t_i}$ indicates that the maintenance time adjustment range is less than the interval between two maintenance times for the same component, that is, less than the nominal optimal maintenance time of the component $\bar{t_i}$. When the repair type is repairable maintenance, $\mu_i(t_i, \Delta t_i)$ is expressed as (Guo 2014)

$$\mu_j(t_j, \Delta t_j) = (C_f + C_g)\int_{t_0}^{T_i}\lambda(t)dt - (C_f + C_g)\int_{t_0}^{T_{i+1}}\lambda(t)dt - \Delta t_j \widehat{C_i'}$$

$$= (C_f + C_g)\int_{T_i}^{T_{i+1}}\lambda(t)dt - \Delta t_j \widehat{C_i'} \tag{17.83}$$

in the formulation, $\widehat{C_i'}$ represents the minimum maintenance cost of the nominal optimal maintenance time, $\Delta t_j \widehat{C_j'}$ indicates the cost of changing the repair time.

Thus, the overall cost savings for group maintenance is

$$C_G = \sum_{i=1}^{K} C_{S_i} \tag{17.84}$$

from which we can conclude that when $C_Z$ reaches the minimum or $C_G$ reaches the maximum, the group maintenance saves the most.

The maintenance time adjustment range is determined by the Eq. (17.80), and $t_i^{\min}, t_i^{\max}$ are the upper and lower limits of the solution, respectively. Maintenance components can only be repaired in the maintenance interval $\left[t_i^{\min}, t_i^{\max}\right]$, especially $t_i^{\max}\langle RUL_i$ and $R(t_i^{\max})\rangle 0.995$. When the intervals of the two fatigue members are $\left[t_i^{\min}, t_i^{\max}\right]$, $\left[t_{i+1}^{\min}, t_{i+1}^{\max}\right]$ and $t_{i+1}^{\min} \ll t_i^{\max}$, then the group maintenance interval of the two components is $\left[t_{i+1}^{\min}, t_i^{\max}\right]$. The grouping principle is: according to the maintenance sequence, the next maintenance activity $i + 1th$ is compared with the previous $i$th maintenance activity, if the common maintenance interval is grouped, and the $(i + 2)\,th$ maintenance activities are compared with the $i$th and the $(i + 1)$th times until there is no common maintenance interval before. The optimal maintenance time after grouping is $t' \in \left[t_{i+1}^{\min}, t_i^{\max}\right]$, and let $\mu_i(t_i, t' - t_i) + \mu_{i+1}(t_{i+1}, t' - t_{i+1})$ reaches the minimum

$$\mu_j(t_j, \Delta t_j) - C_{t_j} = 0 \tag{17.85}$$

## 17.4.4 ILLUSTRATIVE EXAMPLE

Ten maintenance decisions for fatigue structures are taken as examples to test the effectiveness of the group maintenance method. Based on the fatigue crack growth curve, taking days as the unit, assuming that the casting crane is hoisted 40 times a day, the Gamma stochastic process is used to establish the performance degradation model of the fatigue components. Supposing that the preventive maintenance cost $C_P = 6000$ yuan/day, the sum of the failure maintenance cost and the failure loss cost $C_f + C_g = 6$ yuan/day, the shutdown loss cost $C_t = 8000$ yuan and the initial time is zero.

Firstly, we substitute $c$, $b$, and $\beta$ in Table 17.4 into Eqs. (17.60) and (17.69) to obtain the reliability and failure rate of fatigue components. Then, the optimal repair time for repair maintenance and update maintenance is calculated according to Eqs. (17.78) and (17.80). The maintenance method of fatigue components plays a role in repairing the casting crane, so the repair maintenance is adopted here. Due to the lack of historical data, the randomly generated data yielded by the Gamma random process is employed to predict the remaining life. The predicted crack length interval is from the initial crack to the critical crack length $X_i(a_{cr}) = 101.91$ mm, as shown in Table 17.5 below.

According to Eq. (17.68), the original data processing software is used to obtain the WeibullCDF function fitting parameters and its $R^2$ as shown in the following Table 17.6. Since the failure rate in the initial state is zero, the parameter of the WeibullCDF function is set as $y_0 = 0$.

According to Eq. (17.64), the optimal maintenance time of fatigue components can be determined as shown in Table 17.6, in which the nominal optimal maintenance time $\hat{t}_i'$ is the sum of the time from the minimum detection crack state to the initial detection crack and the optimal maintenance time, and the optimal maintenance time $t_i$ and the nominal optimal maintenance time $\hat{t}_i'$ are

## TABLE 17.4
### Fatigue Component Information

| $i$ | $c$ | $b(\times10^{-5})$ | $\beta$ | $d_i$ | $X_i(t_0)$ (mm) | $X_i(a_{cr})$ (mm) | $C_P$ (yuan) | $C_f + C_g$ (yuan) |
|---|---|---|---|---|---|---|---|---|
| 1 | 3.1695 | 3.338 | 0.1426 | 2 | 16 | 101.91 | 12,000 | 120,000 |
| 2 | 2.0362 | 3.396 | 0.1145 | 1 | 2 | 101.91 | 6000 | 60,000 |
| 3 | 2.8854 | 1.206 | 0.1472 | 3 | 5.6 | 101.91 | 18,000 | 180,000 |
| 4 | 2.4090 | 2.392 | 0.1488 | 2 | 4.8 | 101.91 | 12,000 | 120,000 |
| 5 | 3.0016 | 2.462 | 0.1444 | 1 | 2.2 | 101.91 | 6000 | 60,000 |
| 6 | 1.3421 | 2.744 | 0.1265 | 1 | 7.6 | 101.91 | 6000 | 60,000 |
| 7 | 2.3960 | 3.064 | 0.1468 | 1 | 6.3 | 101.91 | 6000 | 60,000 |
| 8 | 2.8277 | 2.971 | 0.1279 | 2 | 2.3 | 101.91 | 12,000 | 120,000 |
| 9 | 3.3162 | 2.129 | 0.1831 | 1 | 5.30 | 101.91 | 6000 | 60,000 |
| 10 | 3.4160 | 2.938 | 0.1444 | 1 | 2.50 | 101.91 | 6000 | 60,000 |

## TABLE 17.5
### Remaining Life Prediction Table

| Component | 1 | 2 | 3 | 4 | 5 | 6 | 7 | 8 | 9 | 10 |
|---|---|---|---|---|---|---|---|---|---|---|
| Remaining life (day) | 3972 | 3863 | 3952 | 4268 | 3634 | 3835 | 3819 | 4103 | 3688 | 3832 |

**TABLE 17.6**

**Fitting Parameters for the WeibullCDF Function**

| i | $A_1$ | a | b | $y_0$ | $R^2$ |
|---|---|---|---|---|---|
| 1 | 7503.718 | 6792.045 | 5.487 | 0 | 0.997 |
| 2 | 8104.785 | 13,298.030 | 4.376 | 0 | 0.992 |
| 3 | 1.398 | 303,105.520 | 1.140 | 0 | 0.573 |
| 4 | 0.00743 | 927.509 | 3.704 | 0 | 0.931 |
| 5 | 2564.911 | 22,123.656 | 3.723 | 0 | 0.976 |
| 6 | 14.087 | 28,965.587 | 2.410 | 0 | 0.734 |
| 7 | 2098.497 | 9366.241 | 5.046 | 0 | 0.986 |
| 8 | 5264.869 | 13,314.560 | 4.170 | 0 | 0.996 |
| 9 | 0.00288 | 872.689 | 6.033 | 0 | 0.866 |
| 10 | 12020.789 | 983.216 | 4.989 | 0 | 0.994 |

both less than the remaining life value, and the reliability is 0.995. This reliability threshold meets the new requirements of reliability and safety. The minimum detection cracks setting to 2 mm (Zuo and Huang 2016; Huang 2012), then the integral for the time from the minimum detection crack to the initial detection crack is:

$$
\begin{cases}
N_f = \dfrac{a^{1-\frac{m}{2}} - a_0^{1-\frac{m}{2}}}{\left(1 - \dfrac{m}{2}\right) C\gamma^m \pi^{m/2} (\Delta\sigma)^m} & m \neq 2 \\[4mm]
N_f = \dfrac{1}{C\gamma^m \pi^{m/2} (\Delta\sigma)^2} \ln\dfrac{a}{a_0} & m = 2
\end{cases}
\tag{17.86}
$$

where $N_f$ is the number of cycles, $C$, $m$ are parameters of material, $a_0$ is the minimum detection crack length, $\gamma$ is the geometric shape factor, $\Delta\sigma$ is the equivalent stress amplitude, $a$ is the initial detection crack length.

It should be noted from Table 17.7 that the value of the 9th component is $\infty$ when the reliability is 0.995, for which the value of the fitting parameters $A_1$ is less than 0.0028. Due to the data applied to for fitting process ranges from the 525th to the 1000th days, the optimal maintenance

**TABLE 17.7**

**Optimal Maintenance Schedule for Each Component**

| i | $R_i = 0.995$ | $\hat{t}'_i$ | $t_i$ | $\hat{C}_i$ | $\widehat{C}_i$ | $d_i$ |
|---|---|---|---|---|---|---|
| 1 | 508 | 1019 | 317 | 44.77 | 13.92 | 2 |
| 2 | 507 | 289 | 289 | 25.54 | 25.54 | 1 |
| 3 | 2170 | 800 | 326 | 103.75 | 42.27 | 3 |
| 4 | 958 | 823 | 399 | 38.31 | 18.58 | 2 |
| 5 | 649 | 386 | 326 | 23.34 | 19.70 | 1 |
| 6 | 1077 | 928 | 369 | 22.99 | 9.14 | 1 |
| 7 | 720 | 919 | 410 | 17.53 | 7.82 | 1 |
| 8 | 479 | 357 | 270 | 55.11 | 41.70 | 2 |
| 9 | $\infty$ | 1024 | 567 | 12.37 | 6.85 | 1 |
| 10 | 52 | 180 | 45 | 158.36 | 39.68 | 1 |

**TABLE 17.8**

**Optimal Maintenance Time of Internal Components in the Maintenance Cycle $T$**

| $i$ | NOR | $t_i$ | $d_i$ |
|---|---|---|---|
| 1 | 2 | 317 | 2 |
| 2 | 4 | 289 | 1 |
| 3 | 1 | 326 | 3 |
| 4 | 2 | 399 | 2 |
| 5 | 4 | 326 | 1 |
| 6 | 4 | 369 | 1 |
| 7 | 6 | 410 | 1 |
| 8 | 2 | 270 | 2 |
| 9 | 3 | 567 | 1 |
| 10 | 4 | 45, 225,405 | 1 |

*Note:* NOR: number of repairs.

time of the 9th component and the maximum reliability of the maintenance interval both are less than 99.5%.

The number of maintenance and maintenance time of each component in the maintenance period $T$ is shown in Table 17.8. It shows that a total of 12 shutdowns are required if the components are repaired separately. Frequent shutdowns and close maintenance activities will cause certain losses to the production of the enterprise. Therefore, the group maintenance mode will greatly reduce the maintenance process. Then 12 maintenance activities are arranged in ascending order, and Eq. (17.76) is used to calculate the maintenance interval, as shown in Table 17.8 below.

It is found that the interval for the first maintenance of the 10th component is [0,64]. However, the maintenance time corresponds to 52 days when the reliability is 99.5%, so the first maintenance interval of the 10th component is adjusted to [0,52]. The 12 maintenance activities are grouped using the group maintenance scheme, as shown in the following Table 17.9. The

**TABLE 17.9**

**Maintenance Sequence in Maintenance Cycle $T$**

| GO | Order | MI | SC | TS | MT |
|---|---|---|---|---|---|
| $Z_1$ | 1 | [0,52] | 0 | 0 | 45 |
| | 2 | [181,233] | 0 | 0 | 226 |
| | 2,3 | [181,233] | 4232.22 | 4232.22 | 225 |
| | 2,3,4 | [181,233] | 5207.87 | 9440.09 | 225 |
| $Z_2$ | 2,3,4,5 | [181,233] | 4681.71 | 14,103.38 | 226 |
| | 6 | [274,433] | 0 | 14,103.38 | 333 |
| | 6,7 | [274,433] | 7872.19 | 21,975.57 | 333 |
| | 6,7,8 | [274,433] | 6725.34 | 28,700.91 | 336 |
| | 6,7,8,9 | [274,433] | 4563.95 | 33,264.86 | 336 |
| | 6,7,8,9,10 | [374,426] | 5366.07 | 38,630.93 | 380 |
| | 6,7,8,9,10,11 | [374,426] | 7048.20 | 45,679.13 | 380 |
| $Z_3$ | 6,7,8,9,10,11,12 | [374,426] | 5673.58 | 51,352.71 | 384 |

*Notes:* GO, group order; MI, maintenance interval; SC, saving cost; TS, total savings; MT, maintenance time.

components that cannot be grouped are repaired according to the optimal maintenance time of a single component. The grouped components are repaired in the group maintenance interval, and the cost savings are defined as the cost saved by the $i$th additional component under the optimal group maintenance time. The total savings is the sum of all the group maintenance costs saved, which is calculated by Eq. (17.80). In Table 17.9, $Z_1$, $Z_2$, $Z_3$ represent group 1, group 2, and group 3, respectively.

It can be seen from the above table that 10 fatigue components can form 12 maintenance activities. The 12 maintenance activities also can be divided into 3 groups. In particular, the $Z_1$ group saving of single maintenance activities is zero. The group $Z_2$ is composed of the 2nd–5th component. The 6th–12th component constitute the group $Z_3$. The result shows that there is a cost-saving for each group. In detail, $Z_2$ and $Z_3$ saves costs up to 14103.38 yuan and 37249.33 yuan, respectively. The total group maintenance saving cost is up to 51352.71 yuan, and the maintenance time is within the range of 99.5% reliability and remaining life. Therefore, we can conclude that the proposed group maintenance scheme is applied successfully for the example and shows great potential for the maintenance of casting cranes in the industry.

## 17.5  CONCLUDING REMARKS

The main contribution of this chapter can be summarized as:

1. This chapter builds a logical hierarchical tree for double-girder metallurgical crane safety evaluation contains five evaluation indicators and a two-layer subsystem based on the AHP. Due to the weight factor determined by the AHP can only reflect the importance of strength indicator in the systematic evaluation, is unable to reflect the status factor of weld point, and the EWM excessively amplifies the importance of the weld point indicator and ignores the other evaluation indicators, thus we propose a comprehensive evaluation method combining with two methods. The obtained results demonstrate that the proposed comprehensive evaluation method can reasonably reflect the system state value thus can be considered as an effective safety evaluation method for the metallurgical crane.

2. As for the casting crane, the evaluation indicators of faulty components are determined according to the failure mode of important components, and performance degradation studies are conducted on each evaluation index to establish the performance degradation models of corrosion, strength, stiffness, and crack with its corresponding probability density distribution functions. Then, the framework of the optimal maintenance decision model for a single component of the casting crane is constructed according to the probability density distribution function of the evaluation index and minimum cost rate.

3. A new component and group maintenance decision for the metal structure of the casting crane is proposed. The Gamma random process is used to describe the probability density function of the crack growth process, the parameters of the Gamma function are estimated according to the known number of cycles, and the corresponding crack length. Also, the WeibullCDF function is used to simplify the expressions of the reliability and failure probability of the Gamma function. Finally, a maintenance decision-making model for fatigue crack growth of a single component is established according to the theory of minimum cost rate; a group maintenance decision-making model is formed to minimize the penalty function, and the case verifies the effectiveness of the group maintenance model.

## ACKNOWLEDGMENT

The National Key Research and Development Program of China (No. 2017YFC0805100), the Open Research Fund Program of Jiangsu Key Laboratory of Engineering Mechanics, the Priority Academic Program Development of the Jiangsu Higher Education Institutions.

# REFERENCES

B Li, GZ Nie. 2018. Research on failure of reinforced concrete columns based on stiffness degradation. *Jiangxi Building Materials* 11: 23–24+27 (in Chinese).

CC Zhang, JJ Wang. 2016. Research advances in damage of composite structures on stiffness degradation. *Materials Reports* 30: 8–13 (in Chinese).

CF Liang, WT Hou. 1995. Atmospheric corrosion of carbon steels and low rolly steels. *Corrosion Science and Protection Technology* 03: 183–186 (in Chinese).

DS Pan. 2009. *Corrosion of H-steel components bending load performance.* Master dissertation. Xi'an University of Architecture and Technology.

AT Echtermeyer, B Engh, L Buene. 1995. Lifetime and Young's modulus changes of glass/phenolic and glass/polyester composites under fatigue. *Composites* 26 (1): 10–16.

F. Wu. 1985. Structural fatigue strength. Northwestern Polytechnical University Press, (in Chinese).

S Fei, S, X Li, Y Fan. 2018. Safety evaluation for bridge crane based on FTA and AHP. MATEC Web of Conferences, 207: 03014. EDP Sciences.

FJ Zuo, HZ Huang. 2016. *Research on methods for reliability analysis and fatigue life prediction for mechanical structure.* Doctoral dissertation. University of Electronic Science and Technology of China.

HB Lv, WQ Yao. 2000. Residual strength model of elements' fatigue reliability evaluation. *Acta Aeronautica et Astronautica Sinica* 01: 75–78 (in Chinese).

HX Kou. 2019. *Research on stiffness degradation model of composite wind turbine blades.* Doctoral dissertation. Lanzhou University of Science & Technology.

J Ma, M Wang, XJ Sun. 2009. Steel structure corrosion present situation and protection development. *Building Structure* 39: 344–345 (in Chinese).

JC Shen. 2009. *Research on stress-strength interference model of strength degradation obeying gamma process.* Doctoral dissertation. Lanzhou University of Technology.

JD Zong, WX Yao. 2016. Composite model of residual stiffness degradation for FRP composites. *Acta Materiae Compositae Sinica* 33: 280–86 (in Chinese).

JR Wang, Z. Zhang, LQ Zhu, et al. 2004. Research on mathematical model of carbon steel and low alloy steel in atmosphere. *Journal of Aeronautical Materials* 24: 41–46 (in Chinese).

L. Chen, GS. Liu, KQ. G. 2018. A method for safety evaluation of hoisting machinery system based on the membership function model. *Hoisting and Conveying Machinery* 12: 87–94 (in Chinese).

L Li, LY Xie, XH He, et al. 2009. Investigation of residual strength degradation of 35CrMo steel. *China Mechanical Engineering* 20: 2890–92+97 (in Chinese).

L Li, LY Xie, XH He, et al. 2010. Strength degradation law of metallic material under fatigue loading. *Journal of Mechanical Strength* 32: 967–971 (in Chinese).

L Li, ZL Sun. 2013. *Research on the law of the stiffness degradation and strength degradation of the bolt under fatigue loads.* Master dissertation. Northeast University.

L Zhang, W Ji, W Zhou, et al. 2015. Fatigue cumulative damage models based on strength degradation. *Transactions of the Chinese Society of Agricultural Engineering* 31: 47–52 (in Chinese).

LY Xie, WQ Lin. 1993. Probabilistic linear cumulative fatigue damage rule. *Journal of Mechanical Strength* 03: 41–44 (in Chinese).

Q Guo. 2014. Group maintenance strategy based on particle swarm optimization. *Guide of Cci-tech Magazine* 14: 346–347.

Q Huang. 2012. *Crack Failure research of crane based on finite element method and fracture mechanics.* Doctoral dissertation. Wuhan University of Science and Technology.

QH Tang. 1995. Fitting and prediction of atmospheric corrosion data of low alloy steel. *Corrosion & Protection* 04: 188–191 (in Chinese).

TL Saaty. 1988. What is the analytic hierarchy process? *Mathematical models for decision support*, 109–121. Springer.

MJ Salkind. 1972. Fatigue of composites. Composite materials: testing and design (second conference). ASTM International.

D Sarkar, S Shah. 2012. Evaluating tower crane safety factors for a housing project through AHP. *NICMAR-Journal Construction Management* 27: 61–70.

JR Schaff, BD Davidson. 1997. Life prediction methodology for composite structures. Part II—Spectrum fatigue. *Journal of Composite Materials* 31 (2): 158–81.

CE Shannon. 1948. A mathematical theory of communication. *Bell System Technical Journal* 27 (3): 379–423.

SQ Guo, WX Yao. 2003. Reliability model for structural elements based on fatigue residual life. *Journal of Nanjing University of Aeronautics & Astronautics* 01: 25–29 (in Chinese).

WG Lv, LY Xie, H Xu. 1997. A nonlinear noel of strength degradation. *Journal of Mechanical Strength* 55–57+62 (in Chinese).

WX Yao. *Structural fatigue life analysis.* 2003. National Defense Industry Press, (in Chinese).

Y Gu. 1999. Tensile residual strength of composite materials and its distribution. *Journal of Nanjing University of Aeronautics & Astronautics* 02: 3–5 (in Chinese).

Y Yang, F Li, YZ Hou, et al. 2012. Opportunistic group maintenance optimization of multi-unit under system under dependence. *Computer integrated manufacturing systems* 18: 827–832 (in Chinese).

YY Li, XS Zhang. 2017. *Study on maintenance strategy of oil & gas corrosive pipeline based on survival analysis.* Master dissertation. Xi'an University of Architecture and Technology.

D Zhang, BI Sandor. 1992. Damage evaluation in composite materials using thermographic stress analysis. *Advances in fatigue lifetime predictive techniques.* ASTM International.

WM Zhang, H Jin, GL Chen. 2019. Casting Crane Safety Evaluation Method Based on AHP and Rough Set Theory.

ZM Guo, YT Zhang, J Fan, et al. 2017. Damage model of RC members based on stiffness degradation and fiber-beam elements. *Journal of Hunan University (Natural Sciences)* 44: 76–87 (in Chinese).

XG An, Q Ma, HX Kou. 2019. Stiffness degradation analysis of wind turbine blade based on fatigue test data. *Journal of Lanzhou University of Technology* 45: 33–37 (in Chinese).

ZX Su, HZ Liu, JP Wang, et al. 2000. Non-linear model of fatigue damage based on the residual strength degradation law. *Journal of Mechanical Strength* 03: 238–240 (in Chinese).

# 18 Seismic Risk-Based Design for Bridge

*Ji Dang*
Saitama University, Saitama, Japan

## CONTENTS

## 18.1 INTRODUCTION

The loads on structures have great uncertainty, and the hazards that the structures may likely experience in their lifetime are difficult to predict when engineers try to design them. The complete restoration works of the Hanshin Expressway took about 1 year and 8 months after the 1995 Kobe earthquake. This caused economical loss not only due to the damage but also due to blocked facilities. Tohoku earthquakes in 2011 and Kumamoto earthquakes in 2016 also caused great damage resulting in heavy economical loss. All these three earthquakes were commonly considered beyond the design level for the codes being practiced at that time.

Generally, seismic design levels were selected based on the expecting scenario or expected exceedance probability based on their realization of the seismic environment.

As can be seen in Figure 18.1, after experiencing a higher level of shake, i.e., unexpectedly large or beyond the most considered earthquakes (MCEs), engineers had to adjust their knowledge and design code, until the next higher one.

Thus, as there are expected earthquakes or MCEs, there is a probability of unexpectedly large earthquakes also. Accordingly, there are Black Swan and corresponding catastrophes.

It is a wakeup call for engineers to consider the probability of the earthquakes exceeding the most considered level in design calculation and the corresponding structural damage. New design concepts, such as the next-generation performance-based seismic design (Applied Technology Council, 2006, 2018), pointed out the importance of using risk as a design tool to evaluate new and existing structures. Recent concepts such as resilience (Bruneau et al., 2003) and anti-catastrophe (Honda et al., 2017) were also proposed and have been discussed intensively.

Their common concept is to take all the levels of possible earthquakes into account instead of using fixed levels, to consider the uncertainties of earthquake, and to control the downtime or other consequences due to fail, rather than using a design only to prevent fail in the most considered level. Thus, some common methods such as incremental dynamic analysis (IDA), fragility analysis and risk, and life cycle cost (LCC) analysis can provide this kind of information for decision-making or design.

In this chapter, the concepts of risk-based seismic design and LCC-based design are introduced to control the unfunctional time due to damage of key components such as rubber bearings. As an example, performance evaluation of a function-separated bridge was conducted by risk assessment.

DOI: 10.1201/9781003194613-18

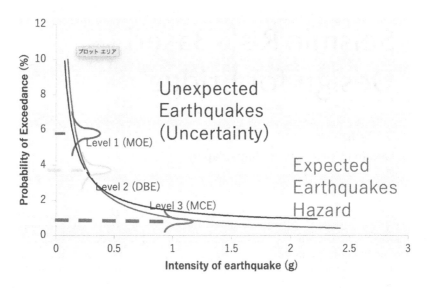

FIGURE 18.1 Why there are always unexpectedly large earthquakes.

## 18.2 RISK-BASED SEISMIC DESIGN

Similar to the performance-based seismic design, risk-based seismic design checks the seismic risk of a structure. A design code can set acceptable levels for different types of structures based on their social impact, or the stakeholders can set a higher standard based on their own risk management needs.

If the risk assessment results show that the seismic risk is higher than the security requirement, then the structure parameters should be considered to be updated for better performance, such as an increase in the cross section of members, use of higher performance material, and an increase in the numbers of dampers similar to performance-based design.

There are a number of other ways to control the risk, such as using an easier-to-repair or easier mechanism or a fail-safe system to maintain the function or shorten the downtime. These methods are difficult to be accepted in performance-based design, as the level of earthquake is fixed and the code will be focused on functional recovery beyond design events. In the risk seismic-based seismic design, a designer can use all kinds of new ideas to decrease both the possibility of damage and the consequences of damage, and they can be balanced by the risk.

Basic concept and procedure of risk-based seismic design is shown in Figure 18.2.

Based on the design conditions, some condition parameters for the structures can be decided, such as floors, structural systems, locations of members, and the weight of floors. Using the condition parameters, normal structural load-based design can be used to find the cross section of members, stiffness of floors, and so on. They are considered as design parameters to build analysis structural models, such as lumped mass spring models (multiple degrees of freedom [MDoF] models).

Instead of normal structural nonlinear seismic response simulation, IDA can be performed to find the IDA curves, the relationship of seismic intensity measure (IM) to structural damage measure (DM) of one earthquake. Using multiple ground motion record as the seed earthquakes, the uncertainty of phase character of input earthquakes can be taken into account in this IDA approach, and a ground view of structural damage or even collapse vulnerability can be summarized in the form of the relationship between the IM and the probability of some level or some scenario of damage due to some DM threshold, which is defined as the fragility curve.

Fragility curves can be drawn for any measure of damage to members or other damage scenarios, so that they can be used to evaluate monetary or time loss.The seismic hazard curves show the exceedance probability of IM for a certain time period, such as for the next 50 years. The seismic

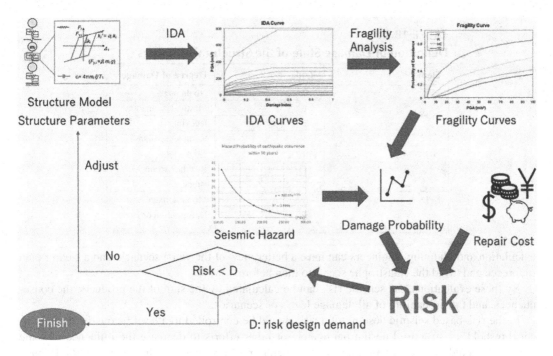

**FIGURE 18.2** The basic concept of seismic risk-based design.

hazard curves can be obtained from probability seismic hazard analysis (PSHA). With the same IM, the fragility curves and seismic hazard curves can serve together by their convolution to find the likelihood of the same level of damage due to large earthquakes in the future time domain, e.g., 50 years.

An example of seismic hazard curve is shown in Figure 18.3. The horizontal axis is the IM, which uses PGV (peak ground velocity) in the unit of kine (cm/s). The vertical axis is the exceedance probability of ground shake in this location of construction based on the survey and seismic simulation of the earthquake environment.

The definitions of damage level for structural members are listed in Table 18.1.

For each damage level or damage scenario, the consequences, such as repairing or reconstruction fee should be predicted based on some preplanning works. By this kind of consideration and

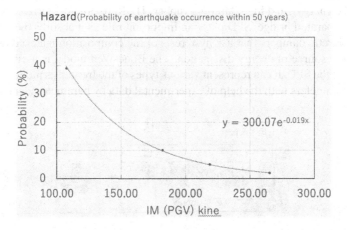

**FIGURE 18.3** Seismic hazard curves.

**TABLE 18.1**
**Definition of Damage State of the Structural Members**

| Element | Rank | Degree of Damage |
|---------|------|------------------|
| Pier | 0 | No damage |
| | 1 | Plastic deformation |
| | 2 | Buckling |
| | 3 | Complete collapse |
| HDR | 0 | No damage |
| | 1 | Residual strain |
| | 2 | Break |
| SPD | 0 | Minor damage |
| | 1 | Deformation or Break |

calculation and planning, engineers can have a better view of the worst scenario and a clearer plan to foresee and avoid the catastrophe scenario than before.

By these evaluations, the seismic risk can be calculated by the sum of the products, the consequences, and the probability of all damage levels or scenarios.

In the risk-based seismic design, total risk should be controlled to a level below the design target threshold, or structural design parameters or other efforts to decrease the influence of some scenarios should be planed based on the consideration of resilience, such as the four Rs: Robust, Redundancy, Rapidity, and Resourceful.

From the next section, the basic process of risk-based design has been explained in detail, with a case study of comparing the normal seismic resistant structure, the seismic isolation structure, and the author-proposed function-separated structure.

## 18.3   INCREMENTAL DYNAMIC ANALYSIS

Taking the resilience of bridge into consideration, a system of multiple energy dissipation and bearing can be considered. In this system, the multiple functions are dispersed into individual damper or bearing, consequently even if one device is damaged, the whole system will not lose its total functionality.

Here, as shown in Figure 18.4, the concept of function separation is brought forward with a model consisting of sliding bearing (to support vertical load), Bingham damper (to absorb deformation, such as temperature expansion and contraction and girder), shear panel damper (SPD) (to absorb earthquake energy), and high damping rubber (HDR; to provide horizontal restoring force).

In a function-separated bridge, SPD plays an important role as it absorbs the earthquake forces. SPDs are the hysteretic dampers that use hysteresis of the composition material (such as low-yield point steel) as their source of energy dissipation. The Bouc-Wen model is used for analysis of the hysteretic nature of the SPD. It can represent various types of the hysteresis model with optimization of its structural parameters with the help of experimental data by parameter identification methods.

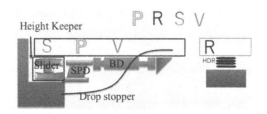

**FIGURE 18.4**   A function-separated bridge.

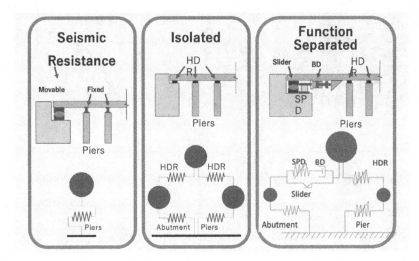

**FIGURE 18.5** Define of structural models.

Once the parameters are identified and the hysteretic data is in agreement with the experimental data with small errors from practical view, then it is considered as the good data. This method is very widely used for structural nonlinear analysis due to its flexibility.

To explain the seismic risk-based design, benchmark analysis is explained below, to evaluate the seismic resilience of function-separated bridges, and compare with convenient isolated bridges using HDR only. Here, incremental dynamic analysis was conducted to three structural models, using a set of 100 earthquakes.

The three structural models are seismic bridges, isolated bridges, and function-separated bridges, as shown in Figure 18.5.

For each structural model, the earthquake waves were scaled to a small IM interval, 10 kine in PGV, before being input into the structural models. With the increase of input wave's IM, structural nonlinear response will increase with DM, such as response ductility ratio. Until the DM reaches the ultra-limit, such as collapse or facture, the analysis will be continued. The relationship of IM to DM can be summarized as the IDA curves for each earthquake. Using a set of 100 earthquakes as input waves, a total of 100 IDA curves can be constructed.

Finally, the fragility curves, shown in Figures 18.6–18.8, can be calculated from these IDA curves to present the relationship of damage probability with the increase of IM.

## 18.4 SEISMIC RISK ANALYSIS

The seismic risk can be expressed by the following formula:

$$Risk = \sum_{i}^{n} P^i \left( C_R^i + C_T^i + C_L^i \right)$$  (18.1)

where:
  $i$: Index of damage scenario
  $P$: Probability of the damage scenario
  $C_R$: Recovery cost
  $C_T$: Travel time cost
  $C_L$: Human life cost

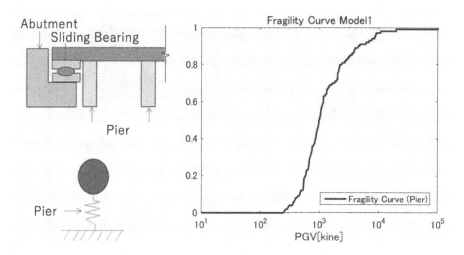

**FIGURE 18.6**  Fragility curve for a seismic resistance bridge.

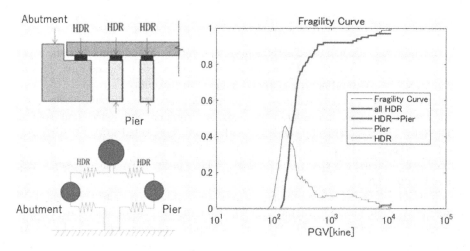

**FIGURE 18.7**  Fragility curve for a seismic isolation bridge.

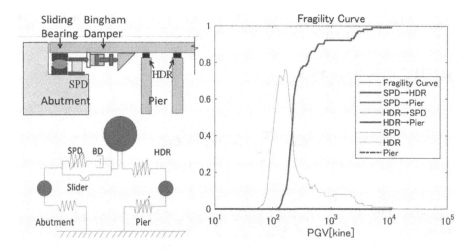

**FIGURE 18.8**  Fragility curve for a function-separated bridge.

Here the probability of damage scenario is calculated by the following equation:

$$P^i = \sum_j^m P_j\left(F_j|H_j\right)P_j\left(H_j\right)\Delta IM \tag{18.2}$$

where
j: Earthquake IM interval number
$P_j\left(F_j|H_j\right)$: The condition probability of structure failure under seismic IM in level of j
$P_j\left(H_j\right)$: Probability of seismic IM happens around an interval of j based on the mesh hazard
curve to m intervals.
$\Delta IM$ : IM interval

A simplified numerical formula to calculate the damage scenario probability is as follows:

$$P^i = 1/4 \sum_j^m \left(H_j + H_{j+1}\right)\left(F_j + F_{j+1}\right)\Delta IM \tag{18.3}$$

The repair cost $C_r$ should be evaluated based on the geological condition of the bridge and the method of repairing. In some case, very severe damage can be conveniently repaired such as damage to the piers in an easy-to-access location. The same level of damage may cost a lot if the location is different; for example, if the piers are in the river. Only the mechanical property of the damage level should not be considered, the economic calculations due to all kinds of constraints should also be taken into account.

The cost of travel time can be considered during the downtime of the bridge, from the time of earthquake to its function recovery; the cost of detour time of traffic of the bridge, according to the cost-benefit analysis manual can be calculated as follows:

$$C_t = Q \times \Delta T \times \alpha_T \tag{18.4}$$

where:
Q: Traffic volume per day (vehicle/day)
$\Delta T$ : Increment of travel time (min)
$\alpha_T$ : Original unit price of time value (yen/min/vehicle)

The calculation of travel time cost of the benchmark bridge is shown in Table 18.2.
It should be mentioned that, even when using the detour route, the travel time lost may be under-estimated by this method. If one bridge damage and the traffic of the detour route will increase, as a consequence, heavy traffic may cause severe jam in the detour route, and the actual travel time lost maybe 10 times the given calculation.

**TABLE 18.2**

**Calculation the Travel Time Cost**

| | Using Bridge | Using Detour Route | | Normal Vehicle | Large Vehicle |
|---|---|---|---|---|---|
| **Distance (km)** | 1.3 | 2.2 | $Q$ | 9122.4 | 1411.6 |
| **Speed (km/h)** | 50 | 33.6 | $\alpha_T$ | 40.1 | 64.18 |
| **Travel Time (min)** | 2 | 6 | $C_u$ | 1,463,240 | 362,374.7 |
| **Time loss (min)** | 0 | 4 | $C_T$ (thousand jpy) | 1825.6 | |

Next, the human life cost can be calculated as follows:

$$C_L = q \times L/v \times B \times I \tag{18.5}$$

where:

$q$: Traffic volume per minute
$L$: Span of bridge
$v$: Average passing by speed
$B$: Average boarding rate
$I$: Loss of personal injury

In this case, let's simplify the question by assuming the time passing by the bridge is 1 minute, and the number of cars on the bridge when earthquake assaulted is $q \times L / v = q$. According to the Ministry of Land, Infrastructure, Transport and Tourism's (MLIT) manual (2008), the loss of personal injury can be taken as $I = 245.674$ million Japanese Yen. Thus, the human life cost can be also calculated.

The earthquake risk assessments of the seismic resistance bridge, seismic isolation bridge, and the function-separated bridges were conducted and listed in Tables 18.3–18.5. The risk of earthquake damage is shown in Figure 18.9 quantitatively by the monetary value.

**TABLE 18.3**
**The Risk Evaluation of Seismic Resistant Bridge (million yen)**

| Rank | | Loss Cost | | | | |
|------|---------|-----------|---------|---------|-------|--------|
| Pier | Days to Close | $C_R^i$ | $C_T^i$ | $C_L^i$ | $P^i$ | Risk |
| 0 | 0 | 0 | 0 | 0 | 0.13 | 0 |
| 1 | 5 | 8 | 9 | 0 | 0.75 | 12.5 |
| 2 | 30 | 110 | 55 | 5320 | 0.12 | 662.5 |
| 3 | 90 | 229 | 164 | 5320 | 0 | 4.0 |
| | | | | | Sum | 678.92 |

**TABLE 18.4**
**The Risk Evaluation of Seismic Isolation Bridge (million yen)**

| Rank | | | Loss Cost | | | | |
|------|------|---------|-----------|---------|---------|-------|--------|
| HDR | Pier | Days to Close | $C_R^i$ | $C_T^i$ | $C_L^i$ | $P^i$ | Risk |
| 0 | 0 | 0 | 0 | 0 | 0 | 0.43 | 0 |
| | 1 | 5 | 8 | 9 | 0 | 0.16 | 2.6 |
| | 2 | 30 | 110 | 55 | 0 | 0.01 | 1.5 |
| | 3 | 90 | 229 | 164 | 5320 | 0 | 0 |
| 1 | 0 | 15 | 42 | 27 | 0 | 0.16 | 11.0 |
| | 1 | 20 | 50 | 37 | 0 | 0.06 | 5.0 |
| | 2 | 45 | 152 | 82 | 0 | 0 | 0.8 |
| | 3 | 105 | 271 | 192 | 5320 | 0 | 0 |
| 2 | 0 | 30 | 84 | 55 | 0 | 0.13 | 18.2 |
| | 1 | 35 | 92 | 64 | 0 | 0.05 | 7.5 |
| | 2 | 60 | 194 | 110 | 5320 | 0 | 15.8 |
| | 3 | 120 | 313 | 219 | 5320 | 0 | 0 |
| | | | | | | Sum | 62.45 |

**TABLE 18.5**

**The Risk Evaluation of Function Separated Bridge (million yen)**

| Rank | | | | Cost | | | | |
|---|---|---|---|---|---|---|---|---|
| SPD | HDR | Pier | Days to Close | $C_R^i$ | $C_T^i$ | $C_L^i$ | $P^i$ | Risk |
| 0 | 0 | 0 | 0 | 0 | 0 | 0 | 0.27 | 0 |
| | | 1 | 5 | 8 | 9 | 0 | 0.05 | 0.9 |
| | | 2 | 30 | 110 | 55 | 0 | 0 | 0 |
| | | 3 | 90 | 229 | 164 | 5320 | 0 | 0 |
| | 1 | 0 | 15 | 42 | 27 | 0 | 0.04 | 2.8 |
| | | 1 | 20 | 50 | 37 | 0 | 0.01 | 0.7 |
| | | 2 | 45 | 152 | 82 | 0 | 0 | 0 |
| | | 3 | 105 | 271 | 192 | 5320 | 0 | 0 |
| | 2 | 0 | 30 | 84 | 55 | 0 | 0.02 | 2.4 |
| | | 1 | 35 | 92 | 64 | 0 | 0 | 0.5 |
| | | 2 | 60 | 194 | 110 | 0 | 0 | 0 |
| | | 3 | 120 | 313 | 219 | 5320 | 0 | 0 |
| 1 | 0 | 0 | 0 | 20 | 0 | 0 | 0.42 | 8.4 |
| | | 1 | 5 | 28 | 9 | 0 | 0.08 | 3.1 |
| | | 2 | 30 | 130 | 55 | 0 | 0 | 0 |
| | | 3 | 90 | 249 | 164 | 5320 | 0 | 0 |
| | 1 | 0 | 15 | 62 | 27 | 0 | 0.06 | 5.6 |
| | | 1 | 20 | 70 | 37 | 0 | 0.01 | 1.3 |
| | | 2 | 45 | 172 | 82 | 0 | 0 | 0 |
| | | 3 | 105 | 291 | 192 | 5320 | 0 | 0 |
| | 2 | 0 | 30 | 104 | 55 | 0 | 0.03 | 4.3 |
| | | 1 | 35 | 112 | 64 | 0 | 0.01 | 0.9 |
| | | 2 | 60 | 214 | 110 | 5320 | 0 | 0 |
| | | 3 | 120 | 333 | 219 | 5320 | 0 | 0 |
| | | | | | | | 合計 | 30.77 |

SPD, HDR, and piers were evaluated based on cost to required performance for each degree of damage, respectively. The cost of travel time loss and human suffering loss was also calculated. Probability of earthquake damage occurrence can be calculated by the results of IDA and hazard curves of the area around Tokyo, published by Japan Seismic Hazard Information Station (J-SHIS). The earthquake risk can be calculated from the probability and cost as shown.

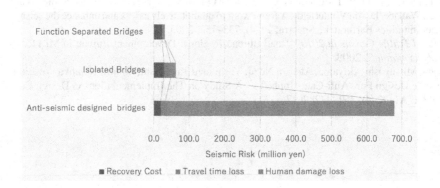

**FIGURE 18.9**   Risk evaluation for three types of bridges

In Figure 18.9, function-separated bridges are compared with the other two bridges. The first one is the anti-seismic designed bridge without a seismic isolation device. The other one is the isolated bridges having HDR. The initial cost of function-separated bridges is the same as that of the isolated bridges. But, the function-separated bridges have the least cost compared to other types.

Using inexpensive and easy-to-install devices makes it possible to recover the economic loss more quickly than using isolated bridges. It also results in less travel time loss and human suffering loss as compared with isolated bridges.

## 18.5    CONCLUSIONS

In this chapter, seismic risk-based design was introduced by using IDA and earthquake risk assessment to evaluate the seismic resilience performance of bridge structures in case of future large earthquakes.

First, the basic concept and procedure of seismic risk-based design was introduced, including IDA, fragility analysis, seismic hazard, structural seismic damage cost, and risk assessment.

By using IDA, engineers can select multiple earthquakes waveform seed and conduct seismic nonlinear response analysis at many intensity levels. Thus, the fragility curves including the uncertainty of earthquake types and phase property can be considered better than using only a design spectrum or even time history analysis with few standard waves.

The fragility curves can be obtained from multiple IDA curves.The convolution of the fragility curves and the seismic hazard of the site, provides an idea about certain expected damage level properties. Moreover, the probability of different intensity earthquakes at the construction site can be evaluated from the seismic hazard map. With the information of probable structural damage, damage scenarios, repair cost and recovery time, the seismic risk can be calculated..

A benchmark bridge seismic analysis was conducted for the function-separated bridges, it was found by IDA that its main damage scenario is as follows: the SPD of the abutment breaks before the HDR of the piers. According to earthquake risk assessment, the risk of function-separated bridges is about the half of the traditional isolated bridges. It was also found that function-separated bridges are economically and more quickly restored by a combination of low-cost and easy-to-install bearings considering the earthquake risk, and they contribute to the restoration of the damaged area at low cost.

## REFERENCES

Applied Technology Council. FEMA 445 Next-Generation Performance-Based Seismic Design Guidelines, Program Plan for New and Existing Buildings, 2006.

Applied Technology Council. FEMA P-58 Next-Generation Performance-Based Seismic Design Guidelines, Volume 1- Methodology, 2018.

M. Bruneau, S. E. Chang, R. T. Eguchi, G. C. Lee, T. D. O'Rourke, A. M. Reinhorn, M. Shinozuka, K. Tierney, W. A. Wallace, D. von Winterfeldt. A framework to quantitatively assess and enhance the seismic resilience of communities Earthquake Spectra, 19 (4), 733–752, 2003.

MLIT. "Road Traffic Census in 2010": Road Bureau/Regional Development Bureau in MLIT, "Cost-benefit analysis manual", 2008.

Riki Honda, Mitsuyoshi Akiyama, Atsushi Nozu, Yoshikazu Takahashi, Shojiro Kataoka, Yoshitaka Murono. Seismic Design For "Anti-Catastrophe" — A Study on The Implementation As Design Codes, Journal of JSCE A1, 72 (4), 459–472, 2017.

# 19 Determination of Maximum Entropy Probabilistic Distribution of Soil Properties

## A Case Study of Nipigon River Landslide in Canada

Jian Deng, Navjot Kanwar, and Umed Panu
Lakehead University, Thunder Bay, ON, Canada

## CONTENTS

## 19.1 INTRODUCTION

In geotechnical engineering, increasing interest can be found in probabilistic reliability analysis and design, which explicitly accounts for inherent probabilistic uncertainties or randomnesses ubiquitous in soil properties (Harr, 1987; Phoon and Retief, 2016). In probabilistic approaches, accurate quantification of randomnesses in soil random variables is the key first step because all subsequent reliability analyses are dependent upon this quantification (Haldar and Mahadevan, 2000; Ranganathan, 2006). These randomnesses are usually quantified by probability density functions (PDFs) or cumulative distribution functions (CDFs) which are determined from available sample knowledge. Several soil properties were discovered to follow either lognormal or normal distributions (Wang et al., 2015). In another research, site-specific characteristics of soil properties were found to tend towards lognormal, normal, or Gaussian mixture distributions (Cao et al., 2017).

An appropriate approach to convey soil sample information is sample moments. Parameters in lognormal and normal distributions can be estimated by the method of moments. Although a PDF on a finite domain is completely characterized by its full set of sample moments, the first four moments can be often used to approximate a distribution function remarkably good. Several methods have been put forward to generate a probability distribution from sample moments. For example, various orthogonal polynomials were used to establish PDFs (Kennedy and Lennox, 2000). Probability weighted moments are more accurate for small samples and thus excel to estimate quantile functions, i.e., inverse CDFs (Deng and Pandey, 2008, 2009a,b). A noteworthy method is grounded on maximum entropy principle (MEP), which can be used to estimate both discrete and continuous PDFs with the fewest assumptions about the sample data from a true distribution. Entropy is defined as a measure of uncertainty being involved in a random variable (Kapur, 1989). The maximum entropy method (MEM) is to select a PDF that maximizes Shannon's entropy constrained by the known knowledge in terms of moments. The MEM was proven to a rational method for selecting the least biased PDF that contains minimum spurious information and is consistent with available sample data (Shore and Johnson, 1980).

The MEP was adapted by Li et al. (2012) who estimated the PDF and evaluated slope stability using a four-moment procedure to conduct a reliability analysis for earth slopes. Based on the MEP, Deng and Pandey (2008) derived quantile functions constrained by probability-weighted moments. Fracture diameter distributions of rocks were estimated using the MEP by Zhu et al. (2014). The MEP was also used to develop a statistical constitutive equation of rocks (Deng and Gu, 2011). Baker (1990) presented a procedure to estimate the PDF based on information theory concepts. The procedure links Jayne's MEP with Akaike information criterion (AIC) to select the optimal order among a hierarchy of models (Akaike, 1971). It should be noted that Baker did not establish the validity of the underlying distribution based purely on sample information. More recently, the maximum entropy distribution constrained by fractional moments is used for reliability analysis (Zhang et al., 2020).

In this chapter, we carried out a series of *in-situ* field soil tests to investigate soil properties, which will play a critical role in slope stability. The objective was to determine the failure mechanism behind Nipigon River landslides in the area of northwestern Ontario, Canada. In 1990, a catastrophic landslide occurred on the east bank of the Nipigon River. The landslide mobilized approximately 300,000 m$^3$ of soil, extended a maximum width of 290 m, and destroyed almost 350 m inshore area. Since then, progressive landslides at various scales have occurred along the river banks, causing significant environmental and economic problems (Dodds et al., 1993). Because only the soil sample is known, the MEP and Akaike estimation procedures were combined to determine what and how many sample moments are needed to estimate the PDF of soils. Uncertainties in soil properties were characterized by entropy distributions for use in reliability analysis and design of slopes in the Nipigon River area.

This chapter is structured in four sections. Section 19.2 presents a distribution-free methodology to quantify the randomness of soil properties based on the combination of the MEP and the AIC. Section 19.3 introduces the Nipigon River landslide area and field vane shear testing, and presents results of entropy modeling of soil variables, along with comparative analysis and discussion. Section 19.4 concludes the chapter.

## 19.2 METHODOLOGY

First, the MEP is introduced to quantify a random variable by entropy distributions based on moments. The random variable will be the undrained shear strength of soil, which is introduced in Section 19.3. A nondimensional analysis is then conducted using a transformation of random variables to remove the effect of possible overflow in the calculation. The AIC is further used to decide the optimal order of entropy PDFs. The entire procedure, given in a flowchart, generates an unbiased and optimal PDF in accordance with the quantity and nature of the available sample data.

### 19.2.1 MAXIMUM ENTROPY PRINCIPLE

Entropy is a measure of uncertainty inherent in a random variable, which can be mathematically defined as:

$$H[f(x)] = -\int_R f(x)\ln[f(x)]\,dx, \tag{19.1}$$

where $H$ is the entropy of $X$, a continuous random variable on the domain $R$, $f(x)$ is the PDF. One can obtain the PDF by maximizing Eq. (19.1) with constraints of moments as follows:

$$\int_R f(x)\,dx = 1, \quad \int_R x^k f(x)\,dx = \mu_k, \ k = 1,\ldots,K, \tag{19.2}$$

where $\mu_k$ is the $k$th raw moment (moment about the origin), and $K$ is the highest order of moments. This PDF is able to best represent the current state of knowledge about the random variable. The first equation in Eq. (19.2) represents the normalization condition, and the second equation is the constraint of sample raw moments. The problem of maximization in Eqs. (19.1) and (19.2) can be solved by an augmented Lagrangian function:

$$\bar{H} = \int_R f(x)\ln[f(x)]dx + (\lambda_0 - 1)\int_R f(x)\,dx + \sum_{k=1}^{K} \lambda_k \left[ \int_R x^k f(x)\,dx - \mu_k \right], \tag{19.3}$$

where $\bar{H}$ is the Lagrangian and $\lambda_k$ is an unknown Lagrangian multiplier. The term $(\lambda_0 - 1)$ is used for ease of calculation instead of $\lambda_0$. The Lagrangian maximization yields:

$$\frac{\partial \bar{H}}{\partial f(x)} = 0. \tag{19.4}$$

Substitution of Eq. (19.3) into Eq. (19.4) results in:

$$f(x) \approx f_K(x|\lambda) = \exp\left[ -\sum_{k=0}^{K} \lambda_k x^k \right], \tag{19.5}$$

where $\lambda_k$, $k = 0,1,\ldots,K$ is the Lagrangian multiplier and is also called the coefficients of the PDF $f(x)$, and $f_K(x|\lambda)$ is the $k$-th order maximum entropy PDF, which is an approximation of the true PDF $f(x)$ in the random space. $f_K(x|\lambda)$ is also simply called the $K$-th order entropy distribution or maximum entropy distribution. Substitution of Eq. (19.5) into the term $\ln[f(x)]$ of Eq. (19.1) and considering the second equation of Eq. (19.2) yield an approximate of $H[f(x)]$,

$$\hat{H}[f(x)] = \sum_{k=0}^{K} (\lambda_k \mu_k). \tag{19.6}$$

It was proven that the maximization of $H[f(x)]$ in Eq. (19.1) subject to the constraints of Eq. (19.2) is equivalent to the minimum of the following function (Kapur and Kesavan, 1992):

$$\Gamma(\lambda_1,\ldots,\lambda_K) = \sum_{k=0}^{K} (\lambda_k \mu_k) + \ln\left\{ \int_R \exp\left[ -\sum_{k=0}^{K} \lambda_k x^k \right] dx \right\}. \tag{19.7}$$

This can be done by a function in MATLAB software, fminunc, which is used to find the minimum of an unconstrained multivariable function. Substituting Eq. (19.5) into the first equation of Eq. (19.2) gives the first coefficient:

$$\lambda_0 = -\ln\left\{\int_R \exp\left[-\sum_{k=0}^{K}\lambda_k x^k\right]dx\right\}. \tag{19.8}$$

Eq. (19.5) is a family of exponential PDFs generated from the MEP that is only based on the sample information in terms of raw moments. The PDF is distribution-free because no classical distributions are assumed *a priori* during the derivation. These exponential functions have been extensively studied due to their important statistical properties. Maximum likelihood estimation also results in these exponential families (Campbell, 1970). Eq. (19.5) can also be proven to be equivalent to Charlier's Type C orthogonal expansion of a PDF (e.g., the Gram-Charlier and Charlier and Edgeworth series), which is useful for approximating to arbitrary distributions in engineering analysis (Ochi, 1986). In engineering practice, the information for a random variable under consideration is usually obtained from a sample space collected on site or tested in labs. The maximum entropy distribution would be obtained from a sample of the variable with dimension. Since $x^k$ in Eq. (19.7) is involved, nondimensional analysis is necessary to reduce the risk of overflow or underflow in the calculation, which is discussed in the next subsection.

### 19.2.2   SAMPLE-BASED MAXIMUM ENTROPY DISTRIBUTION: NONDIMENSIONAL ANALYSIS

The nondimensional analysis includes three steps: (1) transformation of a random variable's original domain into [0,1] and calculation of central moments of the transformed variable; (2) transformation of the central moments to raw moments and determination of the maximum entropy distribution; (3) Transformation of domain of the maximum entropy distribution from [0,1] to the original domain.

#### 19.2.2.1   Transformation of Domain of a Random Variable to [0, 1]

Consider a sample with $N$ elements from a random variable $X$: $x_i (i = 1,2,...,N)$, where $N$ is the sample size and $x_i$ is the $i$-th sample value with dimension. The $r$-th central moment $\upsilon_{X_r}$ of the sample is defined as:

$$\upsilon_{X_r} = \frac{1}{N}\sum_{i=1}^{N}(x_i - \mu_X)^r, \ \mu_X = \frac{1}{N}\sum_{i=1}^{N}x_i, \ \upsilon_{X_2} = \frac{1}{N-1}\sum_{i=1}^{N}(x_i - \mu_X)^2 \tag{19.9}$$

where $\mu_X$ is the sample mean of $X$ and $\upsilon_{X_1}=\mu_X$.

To remove the effect of dimension and to prevent overflow or underflow in the calculation, an appropriate scale can be obtained by a linear transformation of the domain of the random variable $X$ into a variable $Y$ between 0 and 1:

$$Y = \frac{X - x_{min}}{x_d}, \ x_d = x_{max} - x_{min}, \tag{19.10}$$

where $[x_{min}, x_{max}]$ is the domain of $X$, and the domain of $Y$ is $[0,1]$. The sample central moments of the transformed random variable $Y$ are then:

$$\mu_Y = \frac{\mu_X - x_{min}}{x_d}, \ \upsilon_{Y_r} = \frac{\upsilon_{X_r}}{(x_d)^r}, \tag{19.11}$$

where $\upsilon_{Y_r}$ is the $r$-th central moment on the domain $[0,1]$, $\mu_Y$ is the mean value of $Y$ on $[0,1]$, and $\upsilon_{Y_1} = \mu_Y$.

### 19.2.2.2 Transformation of Central Moments to Raw Moments

Application of the binomial theorem gives:

$$(y - \mu_Y)^r = \sum_{i=0}^{r} \left\{ (-1)^i \frac{r!}{i!(r-i)!} (\mu_Y)^i y^{r-i} \right\}. \tag{19.12}$$

Doing the expectation of both sides yields:

$$\mathrm{E}\left[(y - \mu_Y)^r\right] = \sum_{i=0}^{r} \left\{ (-1)^i \frac{r!}{i!(r-i)!} (\mu_Y)^i \mathrm{E}\left[y^{r-i}\right] \right\}, \tag{19.13}$$

where $\mathrm{E}[\cdot]$ indicates the expected value. Thus the relationship between central moments and raw moments is:

$$\upsilon_{Y_r} = \sum_{i=0}^{r} \left\{ (-1)^i \frac{r!}{i!(r-i)!} (\mu_Y)^i (\mu_{Y_r})^{r-i} \right\}, \tag{19.14}$$

which means that central moments, $\upsilon_{Y_r}$, can be transformed to raw moments, $\mu_{Y_r}$ (Siddall, 1983).

### 19.2.2.3 Transformation of Domains of Maximum Entropy Distributions

If the sample raw moments of $Y$ are used to approximate the distribution raw moments,

$$\mu_{Y_r} \approx \int y^k f(y) dy, \tag{19.15}$$

then the maximum entropy probability density distribution of $Y$, $f_Y(y)$, can be obtained from the MEP in Section 19.2.1 as follows:

$$f(y) \approx f_N(y|\lambda) = \exp\left[ -\sum_{k=0}^{N} (\eta_k y^k) \right]. \tag{19.16}$$

Because the linear transformation $Y = (X - x_{min})/x_d$ of Eq. (19.9) is a monotonically increasing function, one has

$$P(X \le x) = P(Y \le y) \tag{19.17}$$

or

$$F_X(x) = F_Y(y), \; F_X(x) = F_X(x_{min} + y x_d), \tag{19.18}$$

where $P(X \le x) = F_X(x)$ is the CDF of the random variable $X$, and $P(Y \le y) = F_Y(y)$ is the CDF of $Y$. Differentiating both sides of Eq. (19.18) with respect to $y$ yields

$$f_Y(y) = f_X(x) \cdot \frac{d(x_{min} + y x_d)}{dy} = f_X(x) \cdot x_d, \tag{19.19}$$

where $f_Y(y)$ and $f_X(x)$ are the probability density functions (PDF) of random variables $Y$ and $X$, respectively. Thus, if the PDF of $Y$, $f_Y(y)$, and the domain distance of $X$, $x_d$, are known, the uncertainty in $X$ in terms of the probability density distribution $f_X(x)$ can be obtained from Eq. (19.19) as

$$f_X(x) = \frac{f_Y(y)}{x_d} = \frac{f_Y\left(\dfrac{x - x_{min}}{x_d}\right)}{x_d}. \tag{19.20}$$

In Section 19.2.3, AIC is used to decide the optimal order of sample moments in the MEP. Substituting Eq. (19.16) into Eq. (19.20) yields

$$f_X(x) = \frac{\exp\left[-\displaystyle\sum_{k=0}^{N} \eta_k \left(\dfrac{x - x_{min}}{x_d}\right)^k\right]}{x_d} = \exp\left[-\sum_{k=0}^{N} \lambda_k x^k\right]. \tag{19.21}$$

The calculation of $\lambda_k$ from $\eta_k$ can be completed by using four functions in MATLAB symbolic math toolbox: poly2sym, expand, simplify, and sym2poly.

### 19.2.3 AKAIKE INFORMATION CRITERION

Let $f(x)$ be the unknown but true distribution and $f_K(x|\lambda)$ be $k$-th order PDF from a sample. The difference between $f_K(x|\lambda)$ and $f(x)$ can be measured by Kullback-Leibler ($KL$) entropy:

$$KL[f(x), f_K(x, \lambda)] = \int_R f(x) \ln \frac{f(x)}{f_K(x, \lambda)} dx = C - L(\lambda, K), \tag{19.22}$$

$$C = \int_R f(x) \ln f(x) dx, \ L(\lambda, K) = \int_R f(x) \ln f_K(x, \lambda) dx. \tag{19.23}$$

Here, $C$ is not related to $f_K(x, \lambda)$, so when the $KL$ entropy is minimized with respect to $\lambda$, $C$ is regarded as a constant. $L(\lambda, K)$ is the expectation of $\ln f_K(x, \lambda)$; therefore, a natural estimate $\hat{L}(\lambda, K)$ of $L(\lambda, K)$ can be obtained from the sample, $x_i \, (i = 1, 2, ..., N)$:

$$\hat{L}(\lambda, K) = \frac{1}{N} \sum_{i=1}^{N} \ln f_K(x_i, \lambda), \ \widehat{KL}(\lambda, K) = C - \hat{L}(\lambda, K), \tag{19.24}$$

where $\widehat{KL}(\lambda, K)$ is a sample estimate of the $KL$ entropy. If $K$ is given, minimization of $\widehat{KL}(\lambda, K)$ will result in the best choice of $\lambda$,

$$\min_\lambda \left\{\widehat{KL}[\lambda, K]\right\} = C + \min_\lambda \left\{-\hat{L}(\lambda, K)\right\} = C + \min_\lambda \left\{-\frac{1}{N} \sum_{i=1}^{N} \ln f_K(x_i, \lambda)\right\}. \tag{19.25}$$

The minimization in Eq. (19.25) is equivalent to the minimization in Eq. (19.7), which can be performed by a MATLAB function, fminunc. However, AIC suggested that the term $-\hat{L}(\lambda, K)$ is a biased likelihood function (Burnham and Anderson, 2003). One of the unbiased estimates of $-\hat{L}(\lambda, K)$ is given as

$$\hat{\Gamma}(\lambda, K) = -\hat{L}(\lambda, K) + \frac{K}{N}. \tag{19.26}$$

This equation can be expanded as by considering Eqs. (19.5), (19.6), (19.24), and (19.26)

$$\hat{\Gamma}(\lambda, K) = \sum_{k=0}^{K} \lambda_k \left[ \frac{1}{N} \sum_{i=1}^{N} (x_i)^k \right] + \frac{K}{N} = \sum_{k=0}^{K} \lambda_k \mu_k + \frac{K}{N} = \hat{H}[f(x)] + \frac{K}{N}. \tag{19.27}$$

where $\hat{\Gamma}(\lambda, K)$ is called the unbiased $KL$ entropy. In Section 19.2.2, the maximum entropy PDF can be derived on the basis of sample moments with a specified order of $K$. Given a series of $K$, there must exist a number, $K$, to minimize $\hat{\Gamma}(\lambda, K)$ in Eq. (19.27), which means to minimize $\widehat{KL}[f(x), f_K(x, \lambda)]$ in Eq. (19.24). This $K$ is the optimal order of the entropy distribution.

## 19.2.4 STATISTICAL GOODNESS-OF-FIT TEST

The underlying maximum entropy distribution needs to be statistically tested to determine its goodness-of-fit to the sample data. The non-parametric Chi-square ($\chi^2$) goodness-of-fit test is used to determine if the sample value is significantly different from the expected value from an underlying probabilistic distribution, on the basis of the error between the expected and observed PDF (Haldar and Mahadevan, 2000). The main advantage of the test is that it can be applied for any probabilistic distribution for which the CDF can be determined.

The first step is to divide the domain of the sample data into $m$ intervals; an empirical relationship between $m$ and the sample size $n$ is given by

$$m = 1 + 3.3 \log 10^n. \tag{19.28}$$

Then the test statistic is defined as

$$\chi^2 = \sum_{i=1}^{m} (c_i) = \sum_{i=1}^{m} \left[ \frac{(n_i - e_i)^2}{e_i} \right], \quad c_i = \frac{(n_i - e_i)^2}{e_i}, \tag{19.29}$$

where $n_i$ and $e_i$ are the expected and the observed frequency for interval $i$ from the assumed PDF respectively. The assumed distribution will be acceptable at the significance level if

$$\chi^2 < c_{1-\alpha, f}, \ f = m - 1 - k, \tag{19.30}$$

where $\alpha$ is a specified significance level, $k$ is the number of distribution parameters, $m$ is the number of intervals, $c_{1-\alpha, f}$ is the value of the $\chi^2$ distribution with $f$ degrees of freedom at the CDF of the nonconfidence level $(1 - \alpha)$. A significance level of $\alpha = 5\%$ implies that for five out of a total of 100 different samples, the assumed distribution cannot be an acceptable model. The common significance levels lie between 1% and 10%. For the present chapter, $\chi^2$ is only used for comparison: the smaller the value of $\chi^2$, the better the probability distribution. Furthermore, $e_i$ and $n_i$ are replaced by the probability for interval $i$ due to the large sample size.

## 19.2.5 FLOWCHART OF THE METHODOLOGY

The above MEM can be summarized in a flowchart of Figure 19.1 for determination of an entropy PDF from a sample. The results provide a universal form of PDF per the implemented constraints. The entropy distributions can cover most classical distributions as special cases under certain moment constraints (Table 19.1). The entropy distribution from the MEM can be proved the most unbiased due to a well-documented maximization (using the MEM) and a well-ordered minimization (using AIC). In next section, the soil property in Nipigon River area is tested and modeled using the procedure in Figure 19.1.

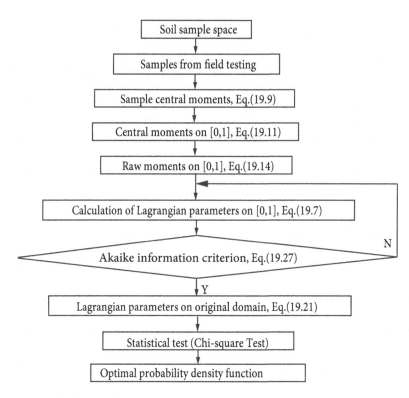

**FIGURE 19.1**    Flowchart for determination of the entropy distribution from a sample.

**TABLE 19.1**
**Maximum Entropy Probability Distributions**

| Given Constraints | Probability Distribution |
| --- | --- |
| $\int_a^b f(x)dx = 1$ | Uniform |
| $\int_a^b f(x)dx = 1$ and expected value | Exponential |
| $\int_a^b f(x)dx = 1$ and expected value, standard deviation | Normal |
| $\int_a^b f(x)dx = 1$ and expected value, standard deviation, minimum and maximum values | Beta |
| $\int_a^b f(x)dx = 1$ and mean occurrence rate between arrivals of independent events | Exponential |

## 19.3   A CASE STUDY OF SOIL IN NIPIGON RIVER LANDSLIDE AREA

### 19.3.1   NIPIGON RIVER LANDSLIDE

On April 23, 1990, a large landslide occurred on the east bank of the Nipigon River, 8 km downstream of the dam of Alexander Generating Station and 13 km upstream of the Town of Nipigon in the District of Thunder Bay in northwestern Ontario, Canada (Figure 19.2). The landslide pushed

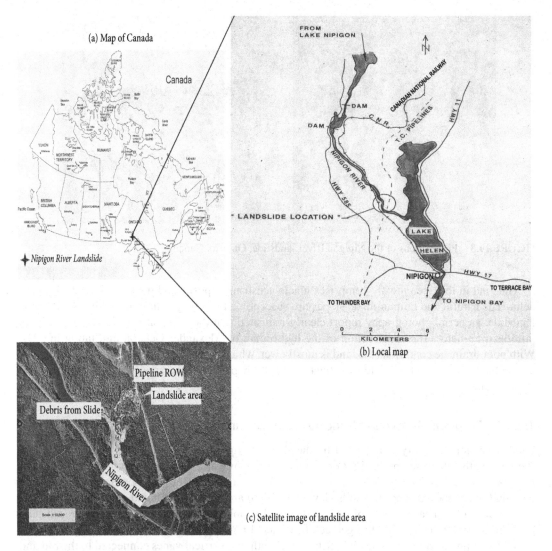

(a) Map of Canada

Canada

Nipigon River Landslide

(b) Local map

(c) Satellite image of landslide area

**FIGURE 19.2**    Location of Nipigon River landslide, Ontario, Canada, showing fiber optics cable right-of-way (ROW). (Adapted from Dodds et al., 1993.)

the soil into the Nipigon River 300 m upstream and downstream and formed several islands in the river. The main current of the Nipigon River was redirected by these newly-formed islands and enormous erosion was caused on the west bank. Events like this probably caused several landslides farther south one month later.

The landslide had significant environmental and economic impacts. The TransCanada (T.C.) pipelines carrying natural gas were bent and laterally displaced up to 8.3 m westward towards the river by the soil movement, leaving it suspended for 75 m without soil support (Dodds et al., 1993). The pipeline did not prominently rupture, but a fiber optics cable inside the pipeline was broken due to the soil movement (Figure 19.2, Adapted from Dodds et al., 1993). The landslide increased the sediment load to the Nipigon River, which caused problems for the residential water supply of the Town of Nipigon and is also suspected to have affected sensitive fish habitat and feeding behavior for several years (Dodds et al., 1993). Since April 1990, progressive landslides along the river banks have continued to occur at various scales (Figure 19.3); therefore, it is important to study the failure probability of landslides in this area. The present chapter is part of the investigation.

**FIGURE 19.3**  Photographs of the Nipigon River landslide, Ontario, Canada.

The land in this area mostly comprises glaciolacustrine deposits and pockets of sand silt in the delta. The natural and human-made slopes are susceptible to sliding failure. Immediately after the landslide, an aerial photography report clearly indicated that the Nipigon River had a lot of bank failures even before the construction of the hydro dam of Alexander Generating Station in 1931. With poor drainage conditions the land is mostly wet, which further induce potential soil instability on the banks. Our recent on-site investigation clearly depicts that the area is prone to small-scale slope failures.

### 19.3.2  FIELD SOIL TESTING IN NIPIGON RIVER LANDSLIDE AREA

Short-term slope stability is imparted by the shear strength of undrained soils; therefore, *in-situ* shear strength data were obtained for undrained conditions using the vane shear test (VST) (ASTM International, 2008). The VST is the most frequently adopted method to measure and obtain the field undrained shear strength of soils (Jay et al., 2016) and facilitates a rapid and robust computation approach for shear strength of undisturbed and remolded soils. Furthermore, VST results can be directly input into GEOSLOPE geotechnical modeling software.

VST equipment comprises a solid pushing rod with two vertical vanes connected to the rod and a vane shear reading meter connected to the top of the rod to measure the torque (Figure 19.4). The testing was conducted in a pre-drilled borehole. The rod was inserted to the specified depth

**FIGURE 19.4**  A vane shear tester equipped with a 33-mm diameter vane.

**TABLE 19.2**

**Soil Physical Properties from a Typical Laboratory Test of the Soil**

| Liquid Limit (%) | Plasticity Index (%) | Plastic Limit (%) | Liquidity Index |
|---|---|---|---|
| 45.5 | 20.8 | 24.7 | 0.33 |
| Moisture Content (%) | Silt (%) | Sand (%) | Clay (%) |
| 31.5 | 71 | 11 | 18 |

**TABLE 19.3**

**Measured Undrained Shear Strength (kPa) ($n = 121$)**

| | | | | | | | |
|---|---|---|---|---|---|---|---|
| 60.80 | 56.05 | 52.25 | 41.80 | 80.75 | 38.95 | 18.05 | 23.75 |
| 32.30 | 61.75 | 56.05 | 52.25 | 42.75 | 81.70 | 39.90 | 18.05 |
| 25.65 | 33.25 | 61.70 | 57.00 | 52.25 | 44.65 | 84.55 | 39.90 |
| 19.00 | 26.60 | 33.25 | 62.70 | 57.00 | 52.25 | 44.65 | 85.50 |
| 39.90 | 19.00 | 26.60 | 33.25 | 63.65 | 57.00 | 53.20 | 45.60 |
| 87.40 | 39.90 | 19.00 | 27.55 | 33.25 | 64.60 | 57.00 | 53.20 |
| 46.55 | 95.00 | 39.90 | 20.90 | 28.50 | 33.25 | 64.60 | 57.95 |
| 54.15 | 47.50 | 95.00 | 39.90 | 20.90 | 29.45 | 33.25 | 64.60 |
| 57.95 | 54.15 | 49.40 | 95.00 | 40.85 | 22.80 | 29.45 | 33.25 |
| 64.60 | 57.95 | 55.10 | 49.40 | 96.90 | 40.85 | 22.80 | 30.40 |
| 34.20 | 65.55 | 58.90 | 55.10 | 51.30 | 104.5 | 41.80 | 23.75 |
| 32.30 | 35.15 | 68.40 | 68.40 | 68.40 | 71.25 | 71.25 | 74.10 |
| 77.90 | 77.90 | 80.75 | 65.55 | 66.50 | 66.50 | 66.50 | 66.50 |
| 67.45 | 67.45 | 68.40 | 68.40 | 68.40 | 35.15 | 36.10 | 37.05 |
| 37.05 | 38.00 | 38.00 | 38.00 | 38.00 | 38.00 | 38.00 | 104.50 |
| 114.00 | | | | | | | |

(ASTM International, 2008). Boreholes ($n = 121$) were drilled in the firm clayey silt layer, generally 0.5–1 m deep into the soil layer using hollow-stem augers. With a single thrust, the vane was pushed to the required depth. After a few minutes, the torque was applied at a rate of 0.1°/s. At the same time, the maximum torque at failure was recorded every 15 s. The vane was rotated continuously seven to ten times. At the end of the test, the residual torque was recorded.

Given that undrained shear strength values determined by VST were influenced by anisotropy and strain rate, a correction factor is necessarily applied. For example, the average plasticity index is 20.5% from field data and 20.8% from laboratory testing (Table 19.2); therefore, the resulting correction factor is 0.986. After this correction, all 121 samples were assumed to be from the same population. The results of the VST listed in Table 19.3 shows that even though great care is taken to keep the conditions of testing as uniform as possible, the individual observations exhibit an intrinsic variability that cannot be overlooked. This means that the undrained shear strength is subject to random variations and hence call for a statistical or probabilistic approach. Table 19.3 will be used to obtain an optimal and unbiased maximum entropy probabilistic distribution.

## 19.3.3 RESULTS

Following the procedure in Section 19.2, one can derive entropy distribution functions and make a comparison to classical probabilistic distributions. First, a frequency histogram is plotted. The number of intervals, $m$, is determined and rounded using the empirical formula in Eq. (19.28). For

**TABLE 19.4**

**Coefficients of Entropy Distribution of Soil Strength of Frictional Angle (Moment Order $K = 1$–5)**

|  | $K = 1$ | $K = 2$ | $K = 3$ | $K = 4$ | $K = 5$ |
|---|---|---|---|---|---|
| $\lambda_0$ | 4.3548840 | 6.76542171 | 8.02392832 | 10.16360285 | 12.32234087 |
| $\lambda_1$ | 0.0075689875 | −0.10864651 | −0.19942748 | −0.40671402 | −0.66969328 |
| $\lambda_2$ |  | 0.0010620140 | 0.0028577423 | 0.0092029036 | 0.020292643 |
| $\lambda_3$ |  |  | $-10.28541728 \times 10^{-6}$ | $-8.595990827 \times 10^{-5}$ | $-2.07525336 \times 10^{-4}$ |
| $\lambda_4$ |  |  |  | $3.04649777 \times 10^{-7}$ | $2.46970131 \times 10^{-6}$ |
| $\lambda_5$ |  |  |  |  | $-5.56575232 \times 10^{-9}$ |
| $\hat{\Gamma}(\lambda, K)$ | −0.0247 | −0.394 | −0.409 | −0.416(optimal) | −0.412 |

the sample of $n = 121$ elements (see Section 19.3.2), the number of intervals is $m = 8$. Then, two commonly used classical PDFs are assumed to fit the soil property and the function parameters were calculated by the maximum likelihood method with a confidence interval of 95%. The two common PDFs are normal and lognormal distributions as follows, respectively,

$$f_1\left(x|\mu = 51.6847, \sigma = 21.1043\right) = \frac{1}{21.1043\sqrt{2\pi}} \exp\left[-\frac{1}{2}\left(\frac{x - 51.6847}{21.1043}\right)^2\right], \tag{19.31}$$

$$f_2\left(x|\lambda_x = 3.8582, \zeta_x = 0.4298\right) = \frac{1}{0.4298 \; x \; \sqrt{2\pi}} \exp\left[-\frac{1}{2}\left(\frac{\ln x - 3.8582}{0.4298}\right)^2\right], \tag{19.32}$$

where $f_1(x)$ is the normal PDF and $f_2(x)$ is the lognormal PDF. $X$ is the undrained shear strength at Nipigon River slope being regarded as a random variable.

To obtain entropy distributions, six central moments were calculated as: 51.68471074, 441.71152244, $5.3459286922 \times 10^3$, $5.71212819 \times 10^5$, $1.855471519 \times 10^7$, and $1.244953484 \times 10^9$. The algorithm in Section 19.2 was applied to obtain the coefficients of entropy distributions of moment orders up to five (Table 19.4). These coefficients ($\lambda_k$) are calculated with eight significant digits after the decimal point for the purpose of accuracy. Table 19.4 also lists the unbiased $KL$ entropy calculated from Eq. (19.27) for various orders of moment constraints. The fourth order entropy distribution had the minimum unbiased $KL$ entropy, $\hat{\Gamma}(\lambda, K)$, and therefore is the most unbiased, optimal probabilistic model for the soil property,

$$\begin{aligned} f_3(x) = \exp(&-10.16360285 + 0.40671402x - 0.0092029036x^2 \\ &+ 8.595990827 \times 10^{-5} x^3 - 3.04649777 \times 10^{-7} x^4), \end{aligned} \tag{19.33}$$

where $f_3(x)$ is the maximum entropy PDF for the random variable $X$, $x \in [0,134]$.

To conduct the statistical goodness-of-fit test ($\chi^2$), the 121 observations were divided into eight intervals. In each interval, the number of observations was tabulated under $n_i$ in Table 19.5. Here, $e_i$ and $n_i$ are the observed and expected frequency for interval $i$ from the assumed PDF, respectively. $c_i$ is defined in Eq. (19.29). The theoretical frequency for each interval in the case of normal, lognormal, and entropy distributions is calculated. The degrees of freedom are $f = 8 - 1 - 2 = 5$. For the significance level $\alpha = 5\%$, the corresponding $c_{0.95,5}$ is found to be 11.07 (Haldar and Mahadevan, 2000), which is smaller than the $c_i$ for the normal distribution and the entropy distribution with

**TABLE 19.5**

**Chi-Square Results for Soil Shear Strength with Normal, Lognormal, and Entropy Distributions**

| Soil Shear Strength (kPa) | Observed Frequency $n_i$ | Lognormal $e_i$ | Lognormal $c_i$ | Normal $e_i$ | Normal $c_i$ | Entropy $(K=2)$ $e_i$ | Entropy $(K=2)$ $c_i$ | Entropy $(K=3)$ $e_i$ | Entropy $(K=3)$ $c_i$ | Entropy $(K=4)$ $e_i$ | Entropy $(K=4)$ $c_i$ | Entropy $(K=5)$ $e_i$ | Entropy $(K=5)$ $c_i$ |
|---|---|---|---|---|---|---|---|---|---|---|---|---|---|
| ≤ 15 | 0 | 0.698 | 0.698 | 4.509 | 4.509 | 4.741 | 4.741 | 2.706 | 2.706 | 1.495 | 1.495 | 1.076 | 1.076 |
| 15 – 30 | 18 | 18.879 | 0.041 | 14.288 | 0.964 | 14.331 | 0.939 | 14.426 | 0.885 | 15.893 | 0.279 | 17.101 | 0.047 |
| 30 – 45 | 36 | 37.601 | 0.068 | 27.867 | 2.373 | 27.346 | 2.738 | 31.804 | 0.553 | 34.646 | 0.052 | 34.244 | 0.089 |
| 45 – 60 | 26 | 30.143 | 0.057 | 33.481 | 1.671 | 32.962 | 1.470 | 34.465 | 2.079 | 31.266 | 0.887 | 29.686 | 0.457 |
| 60 – 75 | 26 | 17.245 | 4.444 | 24.785 | 0.059 | 25.101 | 0.032 | 21.896 | 0.768 | 19.736 | 1.988 | 20.823 | 1.286 |
| 75 – 90 | 8 | 8.633 | 0.047 | 11.300 | 0.963 | 12.073 | 1.374 | 9.816 | 0.336 | 11.229 | 0.928 | 12.028 | 1.349 |
| 90 – 105 | 6 | 4.107 | 0.087 | 3.171 | 2.523 | 3.665 | 1.486 | 3.776 | 1.309 | 5.272 | 0.100 | 4.605 | 0.421 |
| > 105 | 1 | 1.926 | 0.445 | 0.546 | 0.375 | 0.701 | 0.126 | 1.532 | 0.185 | 1.362 | 0.096 | 1.160 | 0.022 |
| Total | 121 | 119.236 | 7.185 | 119.950 | 13.441 | 120.924 | 12.909 | 120.424 | 8.824 | 120.903 | 5.828 | 120.727 | 4.752 |
| $c_{0.95,5}$ | | Acceptable | | Unacceptable | | Unacceptable | | Acceptable | | Acceptable | | Acceptable | |

$K = 2$ in Table 19.5. Thus, these two distributions are not acceptable at a 5% significance level. In contrast, the lognormal and other three entropy distributions with 3–5 order moments are acceptable. Among them, the entropy distribution with the four order moments $(K = 4)$ owned one of the smallest $c_i$, which indirectly proves that the entropy method in the present chapter is correct.

The fitted distributions in Figure 19.5 show that the entropy distribution with $K = 4$ fits the histogram best. The entropy distributions constrained by more than five order moments would exhibit multiple modals, thus they are not considered here.

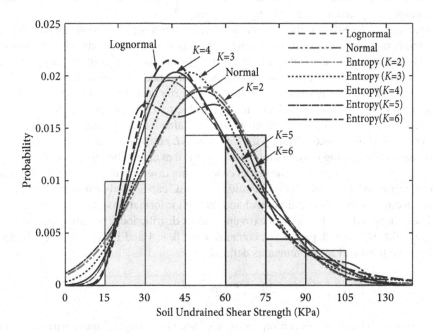

**FIGURE 19.5**   Frequency histogram and the fitted distributions.

### 19.3.4 Discussion

For the soil undrained shear strength dataset presented here, Eq. (19.33) shows that the first four moments must be used to adequately describe the probabilistic distribution of the random variable. Although a random variable is completely specified only if all its moments are known, most commonly used distributions can be adequately described by their first four raw moments or central moments. Sample raw moments and sample central moments can be converted interchangeably (Siddall, 1983). The first moment is used to describe the average value; the second moment measures dispersion; the third moment characterizes skewness or asymmetry, which is a measurement of whether the probability mass tends more to the right or the left of the average with respect to a normal distribution; the fourth moment provides kurtosis or peakedness, which represents the width of a distribution.

In the MEM, there is an issue of sample size requirements for assessing the moments to represent the population. The larger the sample size, the more accurately the sample can represent the population. In practice, the sample size can be estimated by Cochran's sample size formula (Cochran, 1977)

$$M \geq \left( \frac{z_{\alpha/2} \cdot \sigma}{d} \right)^2 , \tag{19.34}$$

where $M$ is the minimum sample size, $\sigma$ is the population standard deviation, $d$ is the desired level of precision (in the same unit of measure as $\sigma$), and $1 - \alpha$ is the desired confidence level, i.e., 95%; $z_{\alpha/2}$ is the value of the standard normal variate evaluated at the probability level of $(1 - \alpha/2)$. For $1 - \alpha = 95\%$, $z_{\alpha/2} = z_{0.025} = 1.96$, and for $1 - \alpha = 90\%$, $z_{\alpha/2} = z_{0.05} = 1.645$. For example, if $\sigma = 22$kPa, $d = 4$ kPa, then the minimum sample size $M$ necessary to obtain an estimate with an maximum absolute error of 4 kPa at a confidence level 95% can be given as $\left( \frac{z_{\alpha/2} \cdot \sigma}{d} \right)^2 = \left( \frac{1.96 \cdot 22}{4} \right)^2 = 117$.

When the number of data points is small (e.g., fewer than 20 data points), which is frequently encountered in geotechnical practice, the standard deviation will be significantly underestimated when using the standard deviation equation (Cao et al., 2017). In this case, the population standard deviation may be estimated from literature data. Another efficient method for small samples is to apply probability weighted moments (Deng and Pandey, 2008, 2009a,b).

One advantage of the MEM is that the entropy distribution's sophistication can be adjusted by the order of sample moments. Figure 19.6 illustrates that entropy distributions with too few moments (e.g., $K = 1$) cannot capture the full sample information and thus under-fit the sample uncertainty. On the other hand, too many orders of moments-constrained entropy distribution in Eq. (19.5) (e.g., $K = 7$) would over-fit the sample uncertainty. Only the maximum entropy distribution, which has a minimum value of the unbiased $KL$ entropy, is able to represent the sample information adequately and optimally. For the Nipigon River soil undrained shear strength, the optimal order of moments is $K = 4$ (Figure 19.6). Among the unbiased $KL$ entropy in Eq. (19.27), the term $K/N$ is able to prevent us from fitting too elaborate distributions that cannot be justified by the sample data, because it is directly proportional to the model order $K$ but inversely proportional to the sample size $N$ (Novi Inverardi and Tagliani, 2003). Therefore, the AIC can not only parsimoniously discard the redundant information but also retain useful and relevant information in the sample.

As pointed out by Siddall (1983), the maximum entropy distribution is presumably used to represent a probabilistic distribution with parametric domains at $0 < \beta_1 < 4$ and $1 < \beta_2 < 9$, where $\beta_1$ and $\beta_2$ are the standardized third and fourth moments defined by

$$\beta_1 = \frac{c_3^2}{c_2^3}, \quad \beta_2 = \frac{c_4}{c_2^2}, \tag{19.35}$$

in which $c_i (i = 2,3,4)$ is the $i$-th central moment of a sample. Beyond these parametric domains, the distribution would have a top sharp spike, which is difficult to represent by an exponential

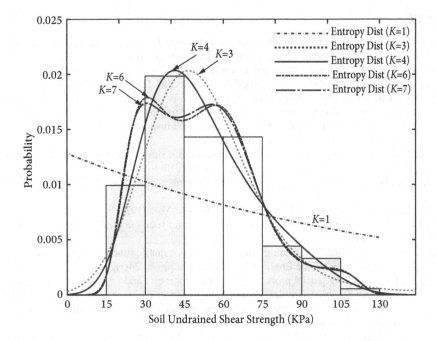

**FIGURE 19.6** Comparison of entropy distributions of various orders.

distribution function. This is one of the disadvantages for the maximum entropy distributions. However, the distributions of soil properties are observed to lie within these parametric domains. Entropy distributions can cover most classical distributions as special cases as shown in Table 19.1.

## 19.4 CONCLUSION

The chapter presented a distribution-free approach to quantify the randomness of soil properties in terms of PDFs by using a combination of MEP and AIC. The PDF is distribution-free because classical distributions are not assumed *a priori* during the derivation. The methodology was described in detail and the algorithm was implemented in MATLAB environment.

In a case study at the Nipigon River landslide area, Ontario, Canada, a sample of 121 values of the soil undrained shear strength was obtained by the VST technique and the MEM was applied to develop a probabilistic density function for the soil property. The four-order entropy distribution was the most unbiased, optimal, probabilistic model for the soil sample, which was also supported by the Chi-square test. The proposed methodology can reasonably and accurately quantify the uncertainties in soil properties by establishing a judiciously chosen maximum entropy probabilistic distribution. The entropy distribution will be implemented in reliability analysis and design of the Nipigon River slope. The methodology can be readily extended to probabilistic modelling of other random variables when sample data are available, such as the soil cohesion and frictional angle.

## ACKNOWLEDGMENT

The research was partly supported by Regional Research Fund of Lakehead University. Help from Dr. Eltayeb Mohamedelhassan at Lakehead University was greatly acknowledged on the VST equipment and field soil testing. The first two authors would like to thank Dr. Zhong-qi Quentin Yue from the University of Hong Kong for his encouragement to prepare the manuscript.

## REFERENCES

Akaike H. Information theory and an extension of the maximum likelihood principle, in Petrov BN, Csaki F (eds), The 2nd International Symposium on Information Theory, Tsahkadsor, Armenia, USSR, September 2–8, 1971, Budapest: Akademiai Kiado, pp. 267–281.

ASTM International. ASTM D2573 – 08. Standard Test Method for Field Vane Shear Test in Cohesive Soil. USA, ASTM International (ASTM), 2008.

Baker DR. Probability estimation and information principles. Structural Safety 1990; 9:97–116.

Burnham KP and Anderson DR. Model Selection and Multimodel Inference: A Practical Information-Theoretic Approach (2nd ed.), 2003, Germany, Springer.

Campbell LL. Equivalence of Gauss' principle and minimum discrimination information estimation of probabilities. The Annals of Mathematical Statistics 1970; 41: 1011–1015.

Cao Z, Wang Y, and Li D. Probabilistic Approaches for Geotechnical Site Characterization and Slope Stability Analysis, Springer-Verlag, Berlin Heidelberg, 2017.

Cochran WG. Sampling Techniques. John Wiley and Sons, New York, 1977.

Deng J and Gu D. On a statistical damage constitutive model for rock materials. Computers & Geosciences 2011; 37(2): 122–128.

Deng J. and Pandey MD. Estimation of minimum cross-entropy quantile function using fractional probability weighted moments. Probabilistic Engineering Mechanics 2009a; 24(1): 43–50.

Deng J. and Pandey MD. Using partial probability weighted moments and partial maximum entropy to estimate quantiles from censored samples. Probabilistic Engineering Mechanics 2009b; 24(3): 407–417.

Deng J. and Pandey MD. Estimation of the maximum entropy quantile function using fractional probability weighted moments. Structural Safety 2008; 30(4): 307–319.

Dodds RB, Burak JP, and Eigenbrod KD. Nipigon River Landslide. In: Proceedings of Third International Conference on Case Histories in Geotechnical Engineering, St. Louis, Missouri, June 1–4, 1993, Paper No. 2.49.

Haldar A and Mahadevan S. Probability, Reliability and Statistical Methods in Engineering Design. John Wiley, New York, 2000.

Harr ME, Reliability-Based Design in Civil Engineering. McGraw-Hill, New York, 1987.

Jay A, Nagaratnam S, and Braja D. Correlations of Soil and Rock Properties in Geotechnical Engineering. Springer, New Delhi, 2016.

Kapur JN. Maximum Entropy Models in Science and Engineering, USA, John Wiley & Sons, 1989.

Kapur JN and Kesavan HK. Entropy Optimization Principles with Applications. Academic Press, Boston, 1992.

Kennedy CA, Lennox WC. Solution to the practical problem of moments using nonclassical orthogonal polynomials, with applications for probabilistic analysis. Probabilistic Engineering Mechanics 2000; 15(4): 371–379.

Li C, Wang W, and Wang S. Maximum-entropy method for evaluating the slope stability of earth dams. Entropy 2012; 14: 1864–1876.

Novi Inverardi, PL and Tagliani A. Maximum entropy density estimation from fractional moments, Communications in Statistics – Theory and Methods 2003; 32: 2327–2345.

Ochi MK. Non-Gaussian random processes in ocean engineering. Probabilistic Engineering Mechanics 1986, 1:28–39.

Phoon KK, Retief JV. Reliability of Geotechnical Structures in ISO2394. USA, CRC Press, 2016.

Ranganathan R. Structural Reliability Analysis and Design. India, Jaico Publishing House, 2006.

Shore JE, Johnson RW. Axiomatic derivation of the principle of maximum entropy and the principle of minimum cross-entropy. IEEE Transaction on Information Theory 1980; 26(1): 26–37.

Siddall JN. Probabilistic Engineering Design. USA, CRC Press, 1983.

Wang Y, Zhao T, and Cao Z. Site-specific probability distribution of geotechnical properties. Computers and Geotechnics 2015; 70: 159–168.

Zhang X, Low YM, and Koh CG. Maximum entropy distribution with fractional moments for reliability analysis. Structural Safety 2020; 83: 101904.

Zhu H, Zuo Y, Li X, Deng J, Zhuang X. Estimation of the fracture diameter distributions using the maximum entropy principle. International Journal of Rock Mechanics and Mining Sciences 2014; 72: 127–137.

# 20 Advances in Online Decentralized Structural Identification

*Ke Huang*
Changsha University of Science and Technology, Changsha, China

*Ka-Veng Yuen*
University of Macau, Macau, China

## CONTENTS

## 20.1 INTRODUCTION

Structural health monitoring (SHM) using vibration data has received considerable attention for decades, since it provides health status assessment of the monitored structures (Yuen 2010). Conventional SHM using smart wireless sensor network (WSN) relies on centralized processing techniques, which indicates that all the observations are collected from the local sensors to a single centralized storage device and processed upon completion by a single centralized unit. Centralized identification is most commonly used for structural identification purpose, owing to its simple architecture. As a result, the majority of existing SHM methods are in this category (Yi et al. 2015; Yuen and Kuok 2016; Yuen and Huang 2018a,b; Yan et al. 2019).

However, some critical issues arise from centralized identification approaches. First, the typical sampling rate for recording the structural dynamic responses is 100 Hz or above, and consequently, a challenging problem for data transmission and processing in the network is induced. Second, in centralized-based methods, there is commonly one central unit to process the entire dataset, so the computational demand is prohibitive. Finally, centralized identification approaches often lead to a serious packet loss rate and long decision-making period, in contrast to decentralized identification approaches. As a result, the potential of decentralized identification approaches is desired to be explored for SHM. Smart wireless sensors are equipped with on-board microprocessors, which possess the function of sensing and computing. Therefore, decentralized identification approaches can take advantage of smart wireless sensors to distribute parts of the calculation to the nodes and fuse the local estimates at the base station. Thus, in contrast to the conventional centralized identification, decentralized identification methods can substantially reduce the data transfer requirements in WSN. Therefore, it can save the energy consumption of the sensor nodes and potentially extend the service life of the system.

Existing decentralized identification methods for SHM focused on offline damage metrics evaluation. The first class of methods (Lynch et al. 2003; Hackmann et al. 2012) does not require the wireless communication among the sensor nodes. Each sensor node processes its own observed data and then transmits the local results to the base station. The second class of methods (Gao et al. 2006; Wang et al. 2009) requires the wireless collaboration among the sensor nodes to perform structural damage detection algorithms. Despite the recent advanced decentralized identification methods proposed for SHM, online/real-time structural identification in decentralized manner is still rather limited.

In this chapter, a dual-rate decentralized strategy for online estimation is presented. In this method, the preliminary local estimates are calculated using the raw observations from the corresponding sensor node only. These local estimates are then condensed before transmitting to the base station for data fusion. At the base station, Bayesian fusion method is used to integrate the condensed local estimates transmitted from the sensor nodes, in order to obtain reliable global estimates. Consequently, the large uncertainty in the local estimates can be significantly reduced. In addition, two different rates for sampling and transmission are utilized to alleviate the data transmission burden in WSN, so that online updating can be achieved efficiently.

Next, based on the framework of dual-rate decentralized approach for online estimation, the treatment of asynchronism encountered in different sensor nodes is addressed for online model updating. Time synchronization of the sensor nodes for online estimation purpose is one of the most critical issues in WSN. On the one hand, it is not a trivial task to keep the clocks of the sensor nodes in the network synchronous. On the other hand, even if the synchronized clocks of the sensor nodes can be guaranteed, the sampling time errors caused by clock jitter still result in asynchronous data.

Motivated by the importance of time synchronization in WSN, time synchronization protocols were utilized to achieve clock synchronization only, such as reference broadcast synchronization (Elson and Estrin 2001), time-synchronization protocol for sensor networks (Ganeriwal et al. 2003), and flooding time synchronization protocol (Maróti et al. 2004). However, the performance of these protocols is unsatisfactory for structural identification purpose, because of the variations in temperature and humidity. There have also been investigations on quantifying the asynchronism in the observations. The first class of methods (Lei et al. 2005; Maes et al. 2016) targets to detecting and compensating the initial time shifts among the sensor nodes. Another class of methods (Li et al. 2016; Zhu and Au 2018) attempts to adjust and align the random time drift. Nonetheless, the majority of existing structural identification approaches considering the asynchronism among the sensor nodes concentrated on offline modal identification.

In this chapter, a decentralized identification approach using directly asynchronous observations is introduced for online model parameter estimation. The presented method exploits the fact that locally estimated model parameters are insensitive to asynchronism in the network, since the locally updated results are obtained using the data from the corresponding sensor node only. Moreover, the locally estimated structural parameters to be fused at the base station are not sensitive to asynchronism. The presented approach is able to handle the asynchronous issues without asynchronism modelling or time shifts quantification. Thus, it is effective for online updating of structural systems using asynchronous data directly.

The third issue to be introduced is on the hierarchical outlier detection for decentralized identification. Outliers are the data points that differ significantly from other observations. Therefore, outlier detection is to identify the data points which are inconsistent with the majority of the data, for achieving reliable estimates. Compared with conventional centralized identification methods, outlier detection for decentralized identification is substantially more difficult. Based on the scope of the data utilized for outlier detection, outliers can be classified into two categories, namely local and global outliers. In particular, a local outlier is identified considering a subset of the measurements, while a global outlier is detected with respect to the whole data set (Zhang et al. 2010).

Existing outlier detection methods for SHM include statistical-based techniques (Huang et al. 2020), nearest neighbor-based techniques (Gul and Catbas 2009), spectral decomposition-based techniques (Makki et al. 2015), clustering-based techniques (Diez et al. 2016), and classification-based techniques (Kerschen et al. 2004). However, the majority of the literature were proposed based on centralized identification approaches, so they performed less satisfactorily on detecting both local and global outliers. Besides, most of the existing outlier detection methods heavily depend on suitable threshold settings, which are difficult to be prescribed in practice.

In this chapter, based on the concept of outlier probability proposed in previous studies (Yuen and Mu 2012; Mu and Yuen 2014; Yuen and Ortiz 2017), an algorithm that performs hierarchical outlier detection is presented for online decentralized identification. The local outliers regarding the raw measurements are identified at the corresponding sensor node and then removed from further calculation. The condensed local estimates are sent back to the base station, where the transmitted local estimates are examined for global outlier detection. After excluding the global outliers, reliable fusion results can be obtained by Bayesian fusion algorithm. The presented approach provides an effective way to remove both local and global outlies without requiring the information about the outlier characteristics.

In Section 20.2, the formulation of the problem for decentralized identification is presented. In Section 20.3, a dual-rate decentralized approach for online estimation is introduced. Then, in Section 20.4, an online estimation algorithm using directly asynchronous data is presented. In Section 20.5, an algorithm that performs hierarchical outlier detection is introduced to remove local and global outliers.

## 20.2 FORMULATION OF THE PROBLEM

A structural dynamical system with $N_d$ degrees of freedom (DOFs) is typically formulated as:

$$\mathbf{M}\ddot{x}(t) + \mathbf{C}\big[\psi_{\mathrm{C}}(t)\big]\dot{x}(t) + \mathbf{K}\big[\psi_{\mathrm{K}}(t)\big]x(t) = \mathbf{D}f(t) \tag{20.1}$$

where $\ddot{x}(t) \in \mathbb{R}^{N_d}$, $\dot{x}(t) \in \mathbb{R}^{N_d}$, and $x(t) \in \mathbb{R}^{N_d}$ denote the acceleration, velocity, and displacement vectors at time $t$, respectively; $\mathbf{M} \in \mathbb{R}^{N_d \times N_d}$, $\mathbf{C} \in \mathbb{R}^{N_d \times N_d}$, and $\mathbf{K} \in \mathbb{R}^{N_d \times N_d}$ stand for the mass, damping, and stiffness matrices of the system, respectively. The damping matrix $\mathbf{C}\big[\psi_{\mathrm{C}}(t)\big]$ and stiffness matrix $\mathbf{K}\big[\psi_{\mathrm{K}}(t)\big]$ are parameterized by the uncertain model parameters in $\psi(t) \equiv \big[\psi_{\mathrm{C}}(t)^{\mathrm{T}}, \psi_{\mathrm{K}}(t)^{\mathrm{T}}\big]^{\mathrm{T}} \in \mathbb{R}^{N_\psi}$; $f(t) \in \mathbb{R}^{N_f}$ is the excitation acting on the structure at time $t$; $\mathbf{D} \in \mathbb{R}^{N_d \times N_f}$ indicates the excitation distribution matrix.

By introducing the augmented state vector $u(t) = \big[x(t)^{\mathrm{T}}, \dot{x}(t)^{\mathrm{T}}, \psi(t)^{\mathrm{T}}\big]^{\mathrm{T}} \in \mathbb{R}^{2N_d + N_\psi}$, the state-space representation of the dynamical system in Eq. (20.1) is expressed as follows:

$$\dot{u}(t) = \begin{bmatrix} \dot{x}(t) \\ \mathbf{M}^{-1}\big\{\mathbf{D}f(t) - \mathbf{C}\big[\psi_{\mathrm{C}}(t)\big]\dot{x}(t) - \mathbf{K}\big[\psi_{\mathrm{K}}(t)\big]x(t)\big\} \\ -\mu\mathbf{1}_{N_\psi \times 1} \end{bmatrix} \equiv g\big[u(t), f(t); \psi(t)\big] \tag{20.2}$$

where $\mu$ is a small positive value to prevent singularity; $\mathbf{1}_{a \times b}$ denotes the $a \times b$ all-ones matrix, and $g$ defines the nonlinear state-space function. The discretization of Eq. (20.2) at a sampling time step $\Delta t$ is given by:

$$u_i = \mathbf{A}_{i-1} u_{i-1} + \mathbf{B}_{i-1} f_{i-1} + \xi_{i-1} \tag{20.3}$$

where $u_i \equiv u(i\Delta t)$, $f_{i-1} \equiv f[(i-1)\Delta t]$, $\mathbf{A}_{i-1}$ and $\mathbf{B}_{i-1}$ denote the transitional and input-to-state matrices, respectively; $\xi_{i-1}$ represents the interception term due to the local linear approximation.

Assume that discrete-time response is observed by $N_s$ local wireless sensors and $N_c$ measurement channels for each sensor node are available. The corresponding local sampling rate is $r_s$, so the sampling time step can be given by $\Delta t = 1/r_s$. Then, the noise-corrupted measurements $y_i^{(s)} \in \mathbb{R}^{N_c}$ of the $s$th sensor node at the $i$th time step are given by:

$$y_i^{(s)} = h^{(s)}\left(u_i, f_i, w_i^{(s)}\right), \ i = 1, 2, \ldots; s = 1, 2, \ldots, N_s \tag{20.4}$$

where $y_i^{(s)} \equiv y^{(s)}(i\Delta t)$ and $h^{(s)}$ represents the measurement equation of the $s$th sensor node. The measurement noise $w_i^{(s)} \in \mathbb{R}^{N_c}$ is assumed to be a Gaussian i.i.d. (independent and identically distributed) process with zero mean and covariance matrix $\mathbf{R}^{(s)} \in \mathbb{R}^{N_c \times N_c}$.

Conventional centralized identification methods require transmission of all the measurements $\mathbb{D}_i^{(s)} = \{y_i^{(s)}\}(i = 1, 2, \ldots; s = 1, 2, \ldots, N_s)$ to the central processing unit. Consequently, the heavy data transmission and processing in the network lead to long system response time and significantly deteriorate the prompt tracking capability of the underlying system. Since the smart wireless sensors are equipped with sensing and computing devices, it is desired to process the measurements first at the sensor nodes and extract important information, which will be sent back to the base station. Then, based on the transmitted local information, fusion approach can be implemented at the base station to obtain reliable estimation results. Next, an online dual-rate decentralized identification strategy is elaborated.

## 20.3  ONLINE DUAL-RATE DECENTRALIZED IDENTIFICATION STRATEGY

### 20.3.1  LOCAL IDENTIFICATION USING EXTENDED KALMAN FILTER

At each sensor node, the extended Kalman filter (EKF) (Hoshiya and Saito 1984) is implemented for online identification using the raw observations from the corresponding sensor node only. The processing rate for the sensor nodes is $r_s$. Therefore, when the measurements $\mathbb{D}_i^{(s)}$ up to the $i$th time step are available, the updated state estimate $u_{i|i}^{(s)}$ and the covariance matrix $\mathbf{P}_{i|i}^{(s)}$ of the prediction error for $u_{i|i}^{(s)}$ can be obtained as $u_{i|i}^{(s)} \equiv E\left[u_i^{(s)}\middle|\mathbb{D}_i^{(s)}\right]$ and $\mathbf{P}_{i|i}^{(s)} \equiv E\left[\left(u_i^{(s)} - u_{i|i}^{(s)}\right)\left(u_i^{(s)} - u_{i|i}^{(s)}\right)^{\mathrm{T}}\middle|\mathbb{D}_i^{(s)}\right]$.

### 20.3.2  BAYESIAN FUSION ALGORITHM AT THE BASE STATION

In order to reduce the data transfer requirements and ensure the online tracking capability, the transmission/processing rate $r_b$ at the base station is taken as a much lower value than the sampling rate $r_s$ at the sensor nodes. Specifically, only for every $r_s/r_b$ sampling time step, the local information is sent back to the base station. The time index at the base station is denoted as $I$, i.e. $i = Ir_s/r_b$ at the same physical time. On the other hand, direct transmission of $u_{i|i}^{(s)}$ and $\mathbf{P}_{i|i}^{(s)}$ involves $\left(2N_d + N_\psi\right)\left(2N_d + N_\psi + 3\right)/2$ real numbers for individual sensor node, which may easily exceed the bandwidth limit of the wireless channel. As a result, it is desired to condense the local estimates $u_{i|i}^{(s)}$ and $\mathbf{P}_{i|i}^{(s)}$, before transmission to the base station for calculating the fusion results. Next, based on the condensed local estimates, a Bayesian fusion algorithm is presented.

Denote $\left\{z_i^{(1)}, z_i^{(2)}, \ldots, z_i^{(N_s)}\right\}$, $z_i^{(s)} \in \mathbb{R}^{N_z}$ and $\left\{\mathbf{P}_{z,i}^{(1)}, \mathbf{P}_{z,i}^{(2)}, \ldots, \mathbf{P}_{z,i}^{(N_s)}\right\}$, $\mathbf{P}_{z,i}^{(s)} \in \mathbb{R}^{N_z \times N_z}$ as the condensed local information about the state estimate and its covariance matrix, respectively, and they are assigned to be transmitted to the base station for calculating the fusion results. Then, with the transmitted data $\left\{z_i^{(1)}, z_i^{(2)}, \ldots, z_i^{(N_s)}\right\}$, the posterior probability density function (PDF) for the fused state estimate $z_i^f \in \mathbb{R}^{N_z}$ at the $l$th time step is given by the Bayes' theorem:

$$p\left(z_i^f \mid z_i^{(1)}, z_i^{(2)}, \ldots, z_i^{(N_s)}\right) = c_0 p\left(z_i^f\right) p\left(z_i^{(1)}, z_i^{(2)}, \ldots, z_i^{(N_s)} \mid z_i^f\right) \tag{20.5}$$

where $c_0$ is a constant and $p\left(z_i^f\right)$ is the prior PDF of the fused state estimate. Assume that the data from any two sensor nodes are statistically independent and thus the likelihood function $p\left(z_i^{(1)}, z_i^{(2)}, \ldots, z_i^{(N_s)} \mid z_i^f\right)$ is expressed as the product of conditional PDFs. Each of the conditional PDF is written as:

$$p\left(z_i^{(s)} \mid z_i^f\right) = (2\pi)^{-\frac{N_z}{2}} \left|\mathbf{P}_{z,i}^{(s)}\right|^{-\frac{1}{2}} \exp\left[-\frac{1}{2}\left(z_i^{(s)} - z_i^f\right)^{\mathrm{T}} \left(\mathbf{P}_{z,i}^{(s)}\right)^{-1} \left(z_i^{(s)} - z_i^f\right)\right] \tag{20.6}$$

The non-informative prior distribution $p\left(z_i^f\right)$ is utilized, so the posterior PDF of the fused state estimate can be obtained. Then, the optimal fused state estimate $z_i^f$ can be calculated by maximizing the posterior PDF and it is given by:

$$z_i^f = \left[\sum_{s=1}^{N_s}\left(\mathbf{P}_{z,i}^{(s)}\right)^{-1}\right]^{-1} \left[\sum_{s=1}^{N_s}\left(\mathbf{P}_{z,i}^{(s)}\right)^{-1} z_i^{(s)}\right] \tag{20.7}$$

In addition, the covariance matrix for $z_i^f$ is given as follows:

$$\mathbf{P}_{z,I}^f = \left[\sum_{s=1}^{N_s}\left(\mathbf{P}_{z,i}^{(s)}\right)^{-1}\right]^{-1} \tag{20.8}$$

According to Eq. (20.7), the optimal fused state estimate $z_i^f$ turns out to be a weighted average of the condensed state estimates from the sensor nodes. The posterior uncertainty of $z_i^f$ is significantly reduced by integrating the local information. In addition, it is noted that Eqs. (20.7) and (20.8) give the general formulas for the fusion results and they can be simplified with different condensed local estimation results.

Sections 20.3.1 and 20.3.2 form the framework of the dual-rate decentralized method for online estimation. Next, a data condensation and extraction algorithm is elaborated for compressing and extracting the local estimation results.

## 20.3.3 DATA CONDENSATION AND EXTRACTION

### 20.3.3.1 Condensation and Extraction of the Locally Updated State Estimate

The components of the locally updated state estimate $u_{i|i}^{(s)}$ are denoted as $x_{i|i}^{(s)}$, $\dot{x}_{i|i}^{(s)}$, and $\psi_{i|i}^{(s)}$, which indicate the estimated displacement, velocity, and model parameter vectors at the $s^{th}$ the sensor node, respectively. Use $\phi_n \in \mathbb{R}^{N_d}$, $n = 1, 2, \ldots, N_n$ to indicate the $n$th eigenvector of the nominal model of the underlying system. These eigenvectors are mass normalized and $\mathbf{M}$-orthogonal, namely $\phi_n^{\mathrm{T}} \mathbf{M} \phi_n = 1$ and $\phi_n^{\mathrm{T}} \mathbf{M} \phi_{n'} = 0$, $n \neq n'$. Next, a nominal eigen-matrix $\mathbf{\Phi}$ is introduced to consist of the eigenvectors of the nominal models $\mathbf{\Phi} = [\phi_1, \phi_2, \ldots, \phi_{N_n}] \in \mathbb{R}^{N_d \times N_n}$. In other words, only the first $N_n$ eigenvectors of the nominal models are included in the eigen-matrix $\mathbf{\Phi}$. It is noted that the eigen-matrix $\boldsymbol{\Phi}$ will be pre-calculated and pre-stored at the sensor nodes and the base station.

Then, the estimated displacement and velocity vectors $x_{i|i}^{(s)}$ and $\dot{x}_{i|i}^{(s)}$ can be projected into the subspace spanned by the nominal models:

$$x_{i|i-1}^{(s)} = \sum_{n=1}^{N_n} \boldsymbol{\phi}_n q_{n,i}^{(s)} + \boldsymbol{\varepsilon}_{q,i} = \boldsymbol{\Phi} q_i^{(s)} + \boldsymbol{\varepsilon}_{q,i} \tag{20.9}$$

$$\dot{x}_{i|i-1}^{(s)} = \sum_{n=1}^{N_n} \boldsymbol{\phi}_n \dot{q}_{n,i}^{(s)} + \boldsymbol{\varepsilon}_{\dot{q},i} = \boldsymbol{\Phi} \dot{q}_i^{(s)} + \boldsymbol{\varepsilon}_{\dot{q},i} \tag{20.10}$$

where $q_{n,i}^{(s)}$ and $\dot{q}_{n,i}^{(s)}$, $n = 1,2,\ldots,N_n$ denote the displacement and velocity coordinates, respectively. Then, the displacement and velocity coordinate vectors are formulated as $q_i^{(s)} = \left[ q_{1,i}^{(s)}, q_{2,i}^{(s)}, \ldots, q_{N_n,i}^{(s)} \right]^{\mathrm{T}} \in \mathbb{R}^{N_n}$ and $\dot{q}_i^{(s)} = \left[ \dot{q}_{1,i}^{(s)}, \dot{q}_{2,i}^{(s)}, \ldots, \dot{q}_{N_n,i}^{(s)} \right]^{\mathrm{T}} \in \mathbb{R}^{N_n}$. The terms $\boldsymbol{\varepsilon}_{q,i}$ and $\boldsymbol{\varepsilon}_{\dot{q},i}$ in Eqs. (20.9) and (20.10) are the projection errors and they satisfy $\boldsymbol{\phi}_n^{\mathrm{T}} \mathbf{M} \boldsymbol{\varepsilon}_{q,i} = 0$ and $\boldsymbol{\phi}_n^{\mathrm{T}} \mathbf{M} \boldsymbol{\varepsilon}_{\dot{q},i} = 0$, $n = 1,2,\ldots,N_n$. Since the eigenvectors are $\mathbf{M}$-orthogonal (Yuen 2012), the displacement and velocity coordinate vectors can be calculated as follows:

$$q_i^{(s)} = \left( \boldsymbol{\Phi}^{\mathrm{T}} \mathbf{M} \right) x_{i|i}^{(s)} \tag{20.11}$$

$$\dot{q}_i^{(s)} = \left( \boldsymbol{\Phi}^{\mathrm{T}} \mathbf{M} \right) \dot{x}_{i|i}^{(s)} \tag{20.12}$$

It is noted that $\boldsymbol{\Phi}^{\mathrm{T}} \mathbf{M}$ can be pre-calculated and pre-stored in all sensor nodes to condense the locally estimated displacement and velocity vectors.

As a result, in this section, the condensed local state estimate can be assigned as $z_i^{(s)} = \left[ q_i^{(s)\mathrm{T}}, \dot{q}_i^{(s)\mathrm{T}}, \boldsymbol{\psi}_i^{(s)\mathrm{T}} \right]^{\mathrm{T}} \in \mathbb{R}^{N_z}$, where $N_z = 2N_n + N_\psi$. Since the number of significant modes is much less than the number of DOFs, namely, $N_n \ll N_d$, transmitting $z_i^{(s)}$ requires only $2N_n + N_\psi$ real numbers, compared with the locally estimated state estimate $u_{i|i}^{(s)}$ for $2N_d + N_\psi$ real numbers.

Therefore, the locally estimated displacement and velocity vectors $x_{i|i}^{(s)}$ and $\dot{x}_{i|i}^{(s)}$ can be condensed to the displacement and velocity coordinate vectors $q_i^{(s)}$ and $\dot{q}_i^{(s)}$, respectively, to form the condensed local state estimate $z_i^{(s)}$. Contrarily, after fusion at the base station, $x_{i|i}^{(s)}$ and $\dot{x}_{i|i}^{(s)}$ are extracted from the fused state estimate. The fused state estimate is denoted as $z_I^f = \left[ q_I^{f\mathrm{T}}, \dot{q}_I^{f\mathrm{T}}, \boldsymbol{\psi}_I^{f\mathrm{T}} \right]^{\mathrm{T}}$, which includes the fused displacement coordinate vector $q_I^f$, the fused velocity coordinate vector $\dot{q}_I^f$ and the fused model parameter vector $\boldsymbol{\psi}_I^f$. The extraction of the locally estimated displacement and velocity vectors is given by:

$$x_{i|i}^{(s)} = \boldsymbol{\Phi} q_I^f \tag{20.13}$$

$$\dot{x}_{i|i}^{(s)} = \boldsymbol{\Phi} \dot{q}_I^f \tag{20.14}$$

### 20.3.3.2  Condensation and Extraction of the Covariance Matrix for the Locally Updated State Estimate

According to Eqs. (20.11) and (20.12), the covariance matrices of the displacement and velocity coordinate vectors are calculated by:

$$\mathbf{P}_{q,i}^{(s)} = \boldsymbol{\Phi}^{\mathrm{T}} \mathbf{M} \mathbf{P}_{x,i|i}^{(s)} \mathbf{M} \boldsymbol{\Phi} \tag{20.15}$$

$$P_{\hat{q},i}^{(s)} = \Phi^T M P_{\dot{x},i|i}^{(s)} M \Phi \tag{20.16}$$

where $P_{x,i|i}^{(s)}$ and $P_{\dot{x},i|i}^{(s)}$ are the submatrices in $P_{i|i}^{(s)}$ and they correspond to the covariance matrices of $x_{i|i}^{(s)}$ and $\dot{x}_{i|i}^{(s)}$, respectively.

In the literature (Sun 2004; Sun and Deng 2004), it has been suggested to use only the diagonal elements of the covariance matrix to relieve the heavy data transmission pressure. As a result, only the variances in the diagonal of $P_{\hat{q},i|i}^{(s)}$, $P_{\dot{\hat{q}},i|i}^{(s)}$, and $P_{\psi,i|i}^{(s)}$ are sent back to the base station. Herein, $P_{\psi,i|i}^{(s)}$ indicates the submatrix in $P_{i|i}^{(s)}$ and it corresponds to the covariance matrix of $\psi_{i|i}^{(s)}$. Then, the condensed variance vector is defined as $\tau_i^{(s)} \equiv \left[ \overline{P}_{\hat{q},i|i}^{(s)}, \overline{P}_{\dot{\hat{q}},i|i}^{(s)}, \overline{P}_{\psi,i|i}^{(s)} \right]^T \in \mathbb{R}^{2N_n+N_\psi}$, where $\overline{A}$ indicates a row vector of the diagonal values in matrix $A$. In this way, the covariance matrix $P_{i|i}^{(s)}$ can be condensed to a variance vector $\tau_i^{(s)}$, which contains only $2N_n + N_\psi$ real numbers. On the other hand, after fusing the local information, the fused variance vector $\tau_i^f$ can be obtained and the covariance matrix $P_{i|i}^{(s)}$ can be extracted from the transmitted fused variance vector $\tau_i^f$:

$$P_{i|i}^{(s)} = \begin{bmatrix} \Phi & 0 & 0 \\ 0 & \Phi & 0 \\ 0 & 0 & I \end{bmatrix} \Lambda\left(\tau_i^f\right) \begin{bmatrix} \Phi^T & 0 & 0 \\ 0 & \Phi^T & 0 \\ 0 & 0 & I \end{bmatrix} \tag{20.17}$$

where $\Lambda\left(\tau_i^f\right)$ refers to a diagonal matrix with the elements in $\tau_i^f$ on the diagonal.

In conclusion, the locally updated state estimate $u_{i|i}^{(s)}$ and its covariance matrix $P_{i|i}^{(s)}$ will be condensed as $z_i^{(s)}$ and $\tau_i^{(s)}$, respectively. Then, the condensed results are transmitted to the base station, where the fusion results are calculated. Finally, the locally updated results will be extracted from the transmitted fusion results by using Eqs. (20.13), (20.14), and (20.17).

Furthermore, given the condensed local estimation results $z_i^{(s)}$ and $\tau_i^{(s)}$, the calculation of the fusion results in Eqs. (20.7) and (20.8) can be simplified. Specifically, the components of the fused state estimate in Eq. (20.7) are expressed as:

$$z_{j,I}^f = \frac{\sum_{s=1}^{N_s} \dfrac{z_{j,i}^{(s)}}{\tau_{j,i}^{(s)}}}{\sum_{s=1}^{N_s} \dfrac{1}{\tau_{j,i}^{(s)}}}, j = 1,2,\ldots,2N_n + N_\psi \tag{20.18}$$

where $z_{j,i}^{(s)}$, $\tau_{j,i}^{(s)}$ and $z_{j,I}^f$ indicate the $j$th component of $z_i^{(s)}$, $\tau_i^{(s)}$, and $z_i^f$, respectively.

In addition, the components of the fused variance vector are:

$$\tau_{j,I}^f = \frac{1}{\sum_{s=1}^{N_s} \dfrac{1}{\tau_{j,i}^{(s)}}}, j = 1,2,\ldots,2N_n + N_\psi \tag{20.19}$$

where $\tau_{j,I}^f$ indicates the $j$th component of $\tau_i^f$.

Validated examples can be found in Huang and Yuen (2019). In the next two sections, two novel methods are introduced based on the framework of the online dual-rate decentralized identification.

## 20.4  ONLINE DECENTRALIZED IDENTIFICATION USING POSSIBLY ASYNCHRONOUS DATA

Conventionally, the transmitted local estimates are required to be synchronous in order to obtain reliable fusion results. In particular, it is essential to keep the displacement and velocity components in the transmitted local estimates synchronized. However, in practice, it is inevitable for the raw measurements to suffer from asynchronism caused by random errors, and it will lead to asynchronous local estimation results. Therefore, in this section, only the model parameter vector and the condensed information about its covariance matrix are sent back to the base station in order to tackle the estimation error due to asynchronism. Another advantage for transmitting only the local estimation results about the model parameters is that it can significantly reduce the data transfer requirements in WSN and the calculation burden for the base station. Next, a data condensation technique is presented for compressing the covariance matrix of the locally estimated model parameters.

Define the asynchronous data set of the $s$th sensor node up to the $i$th time step as $\mathbb{D}_{i+o(t)^{(s)}}^{(s)} = \left[ y_{1+o(t)^{(s)}}^{(s)}, y_{2+o(t)^{(s)}}^{(s)}, \ldots, y_{i+o(t)^{(s)}}^{(s)} \right]$, where $o(t)^{(s)}$ indicates the unknown time shift in the data recording at the $s$th sensor node. Then, the augmented state vector $u_i$ in Eq. (20.4) and its prediction-error covariance matrix can be updated recursively by using EKF.

It is to be recalled that $\mathbf{P}_{\psi,i|i}^{(s)}$ indicates the prediction-error covariance matrix of $\psi_{i|i}^{(s)}$. Based on the fact that the covariance matrix $\mathbf{P}_{\psi,i|i}^{(s)}$ is real and symmetric, the matrix $\mathbf{P}_{\psi,i|i}^{(s)}$ can be approximated by:

$$\mathbf{P}_{\psi,i|i}^{(s)} \approx \tilde{\mathbf{P}}_{\psi,i|i}^{(s)} = \sum_{m=1}^{N_\psi} \rho_{m,i}^{(s)} \mathbf{V}_m \tag{20.20}$$

where $\mathbf{V}_m$ refers to the $m$th basis matrix and $\rho_{m,i}^{(s)}$ represents the condensation coefficient for the basis matrix $\mathbf{V}_m$. A coefficient vector $\rho_i^{(s)}$ is then defined to include all the condensation coefficients: $\rho_i^{(s)} = \left[ \rho_{1,i}^{(s)}, \rho_{2,i}^{(s)}, \ldots, \rho_{N_\psi,i}^{(s)} \right]^{\mathrm{T}} \in \mathbb{R}^{N_\psi}$ for the $s$th sensor node at the $i$th time step. The approximate expression in Eq. (20.20) is interpreted below. The dimension of a symmetric $N_\psi \times N_\psi$ real matrix is larger than $N_\psi$. However, considering the heavy data transfer requirements, the covariance matrices are desired to be condensed to certain extent. As a result, $N_\psi$ terms are suggested to be used in Eq. (20.20), and in such a way the variances of different model parameters can be discriminated. In addition, there is no requirement to match the approximated matrices with the actual ones. An alternative expression for $\mathbf{V}_m$ is:

$$\mathbf{V}_m \equiv v_m \left( v_m \right)^{\mathrm{T}} \tag{20.21}$$

where $\left\{ v_1, v_2, \ldots, v_{N_\psi} \right\}$ is an orthonormal basis, which satisfies $\left( v_m \right)^{\mathrm{T}} v_{m'} = 0$, $m \neq m'$ and $\left( v_m \right)^{\mathrm{T}} v_m = 1$. Therefore, the basis matrices are orthogonal, that is, $\mathbf{V}_m \left( \mathbf{V}_{m'} \right)^{\mathrm{T}} = \mathbf{0}_{N_\psi \times N_\psi}$, $m \neq m'$. It is noteworthy that the basis matrices $\mathbf{V}_m$, $m = 1, 2, \ldots, N_\psi$, in Eqs. (20.20) and (20.21) are precalculated and pre-stored in the network and do not need transmission in the monitoring duration. On the other hand, the coefficient vector $\rho_i^{(s)}$ is time-dependent in order to describe the variation of the covariance matrix $\mathbf{P}_{\psi,i|i}^{(s)}$. Because of the orthogonality of the basis matrices, the components of $\rho_i^{(s)}$ are calculated as follows:

$$\rho_{m,i}^{(s)} = \left\langle \mathbf{P}_{\psi,i|i}^{(s)}, \mathbf{V}_m \right\rangle_F \tag{20.22}$$

where $\langle \mathbf{X}, \mathbf{Y} \rangle_F$ refers to the Frobenius inner product of matrices $\mathbf{X}$ and $\mathbf{Y}$.

Therefore, based on the framework of dual-rate decentralized approach for online estimation presented previously, only the locally estimated model parameter vector $\psi_{i|i}^{(s)} \in \mathbb{R}^{N_\psi}$ and the coefficient vector $\rho_i^{(s)} \in \mathbb{R}^{N_\psi}$ are required to be sent back to the base station for calculating the fused results. Therefore, in this section, the transmitted local estimates are assigned as $\psi_{i|i}^{(s)}$ and $\rho_i^{(s)}$.

On the other hand, the inverse of $\tilde{\mathbf{P}}_{\psi,i|i}^{(s)}$ in Eq. (20.20) can be calculated efficiently by:

$$\left(\tilde{\mathbf{P}}_{\psi,i|i}^{(s)}\right)^{-1} = \sum_{m=1}^{N_\psi} \frac{1}{\rho_{m,i}^{(s)}} \mathbf{V}_m \tag{20.23}$$

Therefore, the fused model parameter vector also can be computed efficiently by:

$$\psi_I^f = \sum_{m=1}^{N_\psi} \sum_{\tilde{s}=1}^{N_s} \frac{1}{\rho_{m,i}^{(\tilde{s})} \sum_{s=1}^{N_s} \frac{1}{\rho_{m,i}^{(s)}}} \mathbf{V}_m \psi_{i|i}^{(\tilde{s})} \tag{20.24}$$

Moreover, the covariance matrix $\mathbf{P}_{\psi,I}^f$ for $\psi_I^f$ is:

$$\mathbf{P}_{\psi,I}^f = \sum_{m=1}^{N_\psi} \left( \frac{1}{\sum_{s=1}^{N_s} \frac{1}{\rho_{m,i}^{(s)}}} \right) \mathbf{V}_m \tag{20.25}$$

It is noted that Eq. (20.25) is in the same fashion as Eq. (20.20). Thus, only the fused model parameter vector $\psi_I^f \in \mathbb{R}^{N_\psi}$ and the fused coefficients $\rho_I^f \in \mathbb{R}^{N_\psi}$ defined in Eq. (20.26) are required to be sent back to the sensor nodes for updating the local estimates.

$$\rho_I^f = \left[ \frac{1}{\sum_{s=1}^{N_s} \frac{1}{\rho_{1,i}^{(s)}}}, \frac{1}{\sum_{s=1}^{N_s} \frac{1}{\rho_{2,i}^{(s)}}}, \ldots, \frac{1}{\sum_{s=1}^{N_s} \frac{1}{\rho_{N_\psi,i}^{(s)}}} \right]^T \tag{20.26}$$

Validated examples can be found in Huang and Yuen (2020b).

## 20.5 HIERARCHICAL OUTLIER DETECTION FOR ONLINE DECENTRALIZED IDENTIFICATION

In addition to the data asynchronism encountered in WSNs, it is also unavoidable for WSNs to have abnormal observations, which can be regarded as outliers. In particular, isolated outliers exhibit extraordinarily large error and these outliers result from multiple disturbances of the monitoring environment. Compared with those isolated erroneous data points, segmental outliers due to events usually last for a certain period of time and significantly alter the historical pattern of the observations. Furthermore, faulty sensors also generate similar long segmental outliers as events (Zhang et al. 2010). As a result, the segmental outliers cannot be detected by considering only the data observed from one sensor node. In this section, based on the framework of dual-rate decentralized approach for online estimation presented previously, an efficient algorithm for hierarchical outlier detection is introduced in order to remove both isolated and segmental outliers in the data set.

### 20.5.1 LOCAL OUTLIER DETECTION AT THE SENSOR NODES

First, the algorithm in Mu and Yuen (2014) is utilized to detect the local outliers and it is briefly introduced below. At the $s$th sensor node, the $c$th component of the measurement $y_i^{(s)}$ is denoted as $y_{c,i}^{(s)}$. Then, the (local) outlier probability $P_o^l\left(y_{c,i}^{(s)}\right)$ for $y_{c,i}^{(s)}$ is given as follows (Mu and Yuen 2014):

$$P_o^l\left(y_{c,i}^{(s)}\right) \approx \sum_{\ell=0}^{\ell_{c,i}^{(s)}}\binom{\ell_{reg}+1}{\ell}\left[2\mathcal{F}\left(-\left|e_{c,i}^{(s)}\right|\right)\right]^\ell\left[1-2\mathcal{F}\left(-\left|e_{c,i}^{(s)}\right|\right)\right]^{\ell_{reg}+1-\ell} \tag{20.27}$$

where $\begin{pmatrix} b \\ a \end{pmatrix}=b!/\left[a!(b-a)!\right]$ indicates a binomial coefficient; $e_{c,i}^{(s)}$ is the normalized residual for $y_{c,i}^{(s)}$ and $\mathcal{F}(\cdot)$ represents the cumulative distribution function of the standard normal variable. In order to effectively evaluate the outlierness of $y_{c,i}^{(s)}$, a moving time window is considered and it is defined to include no more than $\ell_{reg}$ regular data points. $\ell_{c,i}^{(s)}$ indicates the number of data points with absolute normalized residuals not less than $\left|e_{c,i}^{(s)}\right|$ in the regular data set. It is suggested to utilize a value of $\ell_{reg}$ to cover roughly 50 fundamental periods of the underlying system.

Therefore, according to Eq. (20.27), a data point $y_{c,i}^{(s)}$ is recognized as a local outlier if $P_o^l\left(y_{c,i}^{(s)}\right)$ is greater than or equal to 0.5. Therefore, it will be removed from the identification process.

With the measurements excluding the local outliers, the locally updated state estimate $u_{i|i}^{(s)}$ and its covariance matrix $\mathbf{P}_{i|i}^{(s)}$ can be obtained recursively using EKF. Moreover, only the locally estimated model parameter vector $\psi_{i|i}^{(s)}$ and the condensed information about its covariance matrix are sent back to the base station for further detection of global outliers. In this section, the condensed information about the covariance matrix of $\psi_{i|i}^{(s)}$ is taken as the diagonal elements in its covariance matrix and it is noted as $\vartheta_i^{(s)} \equiv \overline{\mathbf{P}}_{\psi,i|i}^{(s)\ \mathrm{T}} \in \mathbb{R}^{N_\psi}$. Therefore, the fusion results for the model parameter vector $\psi_I^f$ and the variance vector $\vartheta_I^f$ can be calculated by using Eqs. (20.7) and (20.8), respectively. The components of $\psi_I^f$ and $\vartheta_I^f$ are given as follows:

$$\psi_{m,I}^f = \frac{\displaystyle\sum_{s=1}^{N_s}\frac{\psi_{m,i|i}^{(s)}}{\vartheta_{m,i}^{(s)}}}{\displaystyle\sum_{s=1}^{N_s}\frac{1}{\vartheta_{m,i}^{(s)}}}, m=1,2,\ldots,N_\psi \tag{20.28}$$

$$\vartheta_{m,I}^f = \frac{1}{\displaystyle\sum_{s=1}^{N_s}\frac{1}{\vartheta_{m,i}^{(s)}}}, m=1,2,\ldots,N_\psi \tag{20.29}$$

where $\psi_{m,i|i}^{(s)}$, $\vartheta_{m,i}^{(s)}$, $\psi_{m,I}^f$, and $\vartheta_{m,I}^f$ indicate the $m$th component of $\psi_{i|i}^{(s)}$, $\vartheta_i^{(s)}$, $\psi_I^f$ and $\vartheta_I^f$, respectively.

### 20.5.2 GLOBAL OUTLIER DETECTION AT THE BASE STATION

In real applications, it is prone for raw measurements to suffer from segmental outliers, e.g., due to biased sensors. In this circumstance, the segmental outliers in the sensor node level are not detectable, because the local measurements have the same level of error. As a result, global outlier detection is necessary for examining the transmitted local estimates. It should be noted that the global outlier detection problem inherently differs from the previous one, which focuses on handling the observations directly. The data points in global outlier detection are the transmitted local estimates denoted as $\mathfrak{D}_I=\left\{\mathfrak{D}_I^{(s)},s=1,2,\ldots,N_s\right\}$, where $\mathfrak{D}_I^{(s)}\equiv\left\{\psi_{i|i}^{(s)},\vartheta_i^{(s)}\right\}$ and $i=Ir_s/r_b$.

First, a suspicious dataset $\mathfrak{D}_{sus,I}$ which includes all possible global outliers is formulated and thus the rest of the data points are grouped as the initial regular dataset $\mathfrak{D}_{R0,I}$. The suspicious data points in $\mathfrak{D}_I$ are identified using a modified Mahalanobis distance (MD). With the transmitted local estimation results $\mathfrak{D}_I$ at the $I$th time step, the modified MD for $\mathfrak{D}_I^{(s)}$ is defined by:

$$d\left(\mathfrak{D}_I^{(s)}\right) \equiv \sqrt{\sum_{m=1}^{N_\psi} \frac{\left(\psi_{m,I}^{f\backslash s} - \psi_{m,i|i}^{(s)}\right)^2}{\vartheta_{m,I}^{f\backslash s} + \vartheta_{m,i}^{(s)}}}, \quad s = 1,2,\ldots,N_s \tag{20.30}$$

where $\psi_I^{f\backslash s}$ indicates the fused model parameter vector without the local estimates from the $s$th sensor node and it is obtained by Eq. (20.28); $\psi_{m,I}^{f\backslash s}$ is the $m$th component of $\psi_I^{f\backslash s}$; $\vartheta_{m,I}^{f\backslash s}$ is the associated variance of $\psi_{m,I}^{f\backslash s}$ by using Eq. (20.29). Next, the transmitted local estimates with modified MD value larger than $\sqrt{\mathcal{F}_{\chi_{N_\psi}^2}^{-1}(0.95)}$ are grouped as suspicious dataset, where $\mathcal{F}_{\chi_{N_\psi}^2}^{-1}(\cdot)$ is the quantile function of the Chi-square distribution with $N_\psi$ DOF. Thus, the suspicious dataset $\mathfrak{D}_{sus,I}$ can be obtained by:

$$\mathfrak{D}_{sus,I} = \left\{ \mathfrak{D}_I^{(s)} = \left[\psi_{i|i}^{(s)}, \vartheta_I^{(s)}\right] : d\left(\mathfrak{D}_I^{(s)}\right) > \sqrt{\mathcal{F}_{\chi_{N_\psi}^2}^{-1}(0.95)}, s = 1,2,\ldots,N_s \right\} \tag{20.31}$$

The initial regular dataset $\mathfrak{D}_{R0,I}$ is then readily obtained by excluding all suspicious data points and $N_{R0,I}$ is defined to indicate the number of data points in $\mathfrak{D}_{R0,I}$. As a result, the initial fusion results $\psi_I^{f0}$ and $\vartheta_I^{f0}$ can be calculated by:

$$\psi_{m,I}^{f0} = \frac{\sum_{\mathfrak{D}_I^{(s)} \in \mathfrak{D}_{R0,I}} \frac{\psi_{m,i|i}^{(s)}}{\vartheta_{m,i}^{(s)}}}{\sum_{\mathfrak{D}_I^{(s)} \in \mathfrak{D}_{R0,I}} \frac{1}{\vartheta_{m,i}^{(s)}}}, m = 1,2,\ldots,N_\psi \tag{20.32}$$

$$\vartheta_{m,I}^{f0} = \frac{1}{\sum_{\mathfrak{D}_I^{(s)} \in \mathfrak{D}_{R0,I}} \frac{1}{\vartheta_{m,i}^{(s)}}}, m = 1,2,\ldots,N_\psi \tag{20.33}$$

where $\psi_{m,I}^{f0}$ and $\vartheta_{m,I}^{f0}$ indicate the $m$th component of $\psi_I^{f0}$ and $\vartheta_I^{f0}$, respectively. It is noted that Eqs. (20.32) and (20.33) are formulated in the same fashion as Eqs. (20.28) and (20.29), respectively, except that $\mathfrak{D}_{sus,I}$ is excluded.

After obtaining the initial fusion results, the outlier probability of each suspicious data point $\mathfrak{D}_{sus,I}^{(s^*)} = \left\{ \psi_{i|i}^{(s^*)}, \vartheta_I^{(s^*)} \right\} \in \mathfrak{D}_{sus,I}$, $s^* = 1,\ldots,N_s - N_{R0,I}$ can be evaluated. Define the residual vector $\delta_I^{(s^*)} \equiv \psi_{i|i}^{(s^*)} - \psi_I^{f0}$. It can be observed that $\delta_I^{(s^*)}$ follows an $N_\psi$-variate Gaussian distribution with zero mean and the diagonal covariance matrix has variances from the elements in vector $\vartheta_i^{(s^*)} + \vartheta_I^{f0}$. Therefore, the contour of the residuals with the same probability density as $\delta_I^{(s^*)}$ is governed by:

$$\sum_{m=1}^{N_\psi} \frac{\left(\tilde{\delta}_{m,I}^{(s^*)}\right)^2}{\vartheta_{m,i}^{(s^*)} + \vartheta_{m,I}^{f0}} = \sum_{m=1}^{N_\psi} \frac{\left(\delta_{m,I}^{(s^*)}\right)^2}{\vartheta_{m,i}^{(s^*)} + \vartheta_{m,I}^{f0}} = C_I^{(s^*)}, s^* = 1,\ldots,N_s - N_{R0,I} \tag{20.34}$$

where $\tilde{\delta}_I^{(s^*)}$ indicates the points on the contour in the residual space; $\delta_{m,I}^{(s^*)}$ is the $m$th component of $\tilde{\delta}_I^{(s^*)}$. It is noted that Eq. (20.34) describes the surface of a hyper-ellipsoid and the residuals within the hyper-ellipsoid are associated with higher probability density values than outside.

As a result, the probability of a data point falling inside the hyper-ellipsoid is utilized to assess the outlierness of $\mathfrak{D}_{sus,I}^{(s^*)}$. It is seen that $C_I^{(s^*)}$ in Eq. (20.34) is the sum of squares of $N_\psi$ independent normal variables, so it follows the Chi-square distribution with $N_\psi$ DOF, i.e., $C_I^{(s^*)} \sim \chi_{N_\psi}^2$. Therefore, the (global) outlier probability $P_o^g\left(\mathfrak{D}_{sus,I}^{(s^*)}\right)$ for a suspicious data point $\mathfrak{D}_{sus,I}^{(s^*)}$ can be given as follows:

$$P_o^g\left(\mathfrak{D}_{sus,I}^{(s^*)}\right) \equiv \mathcal{F}_{\chi_{N_\psi}^2}\left(C_I^{(s^*)}\right) \tag{20.35}$$

where $\mathcal{F}_{\chi_{N_\psi}^2}(.)$ is the cumulative distribution function of the Chi-square distribution with $N_\psi$ DOF.

Therefore, according to Eq. (20.35), the data points with $P_o^g\left(\mathfrak{D}_{sus,I}^{(s^*)}\right) < 0.5$ are recognized as regular data points and then reallocated to form an updated regular dataset $\mathfrak{D}_{R,I}$. With the data points in $\mathfrak{D}_{R,I}$, the updated fusion results $\psi_I^f$ and $\vartheta_I^f$ are calculated by Eqs. (20.32) and (20.33).

Validated examples can be found in Huang and Yuen (2020a).

## 20.6   SUMMARY

This chapter presented several recent advances in online decentralized structural identification. First, a dual-rate decentralized method for online estimation is introduced. At the sensor nodes, the preliminary local estimates are calculated by EKF using the observations from the corresponding sensor node only. These local estimates will be condensed and then sent back to the base station for data fusion. Moreover, it is proposed to utilize two different rates for sampling and transmission/fusion. In this way, the data transfer requirements in WSNs and the calculation burden for the base station are reduced, so that efficient online updating can be achieved. Then, a novel approach for online updating of the structural parameters using asynchronous data is presented, based on the framework of online dual-rate decentralized identification. The presented approach is able to handle the asynchronism directly without extra calculation to model the asynchronism or determine the time shifts. Finally, a hierarchical outlier detection method is introduced to remove the abnormal data in the decentralized identification. The presented hierarchical outlier detection method is capable for detecting both isolated outliers and segmental outliers, so reliable estimates can be obtained. Furthermore, it does not require characteristic information about the outliers and ad-hoc threshold setting for the outlier detection criterion. Through these recent developments, it is expected to achieve efficient and reliable online decentralized identification for structural systems.

## REFERENCES

Diez A, Khoa NLD, Alamdari MM, Wang Y, Chen F, Runcie P. 2016. A clustering approach for structural health monitoring on bridges. *J Civ Struct Health Monit* 6(3):429–445.

Elson J, Estrin D. 2001. Time synchronization for wireless sensor network. Proceedings of the 15th parallel and distributed processing symposium, pp. 1965–1970.

Ganeriwal S, Kumar R, Srivastava MB. 2003. Timing-sync protocol for sensor networks. *Proceedings of the 1st international conference on embedded networked sensor systems*, pp. 138–149.

Gao Y, Spencer BF Jr, Ruiz-Sandoval M. 2006. Distributed computing strategy for structural health monitoring. *Struct Control Hlth* 13(1):488–507.

Gul M, Catbas FN. 2009. Statistical pattern recognition for structural health monitoring using time series modeling: Theory and experimental verifications. *Mech Syst Signal Pr* 23(7):2192–2204.

Hackmann G, Sun F, Castaneda N, Lu C, Dyke S. 2012. A holistic approach to decentralized structural damage localization using wireless sensor networks. *Comput Commun* 36(1):29–41.

Hoshiya M, Saito E. 1984. Structural identification by extended Kalman filter. *J Eng Mech* 110(12):1757–1770.

Huang B, Yi TH, Li HN. 2020. Anomaly identification of structural health monitoring data using dynamic independent component analysis. *J Comput Civil Eng* 34(5):04020025.

Huang K, Yuen KV. 2019. Online dual-rate decentralized structural identification for wireless sensor networks. *Struct Control Hlth* 26(11):e2453.

Huang K, Yuen KV. 2020a. Hierarchical outlier detection approach for online distributed structural identification. *Struct Control Hlth* 27(11):e2623.

Huang K, Yuen KV. 2020b. Online decentralized parameter estimation of structural systems using asynchronous data. *Mech Syst Signal Pr* 145:106933.

Kerschen G, De Boe P, Golinval JC, Worden K. 2004. Sensor validation using principal component analysis. *Smart Mater Struct* 14(1):36–42.

Lei Y, Kiremidjian AS, Nair KK, Lynch JP, Law KH. 2005. Algorithms for time synchronization of wireless structural monitoring sensors. *Earthq Eng Struct D* 34:555–573.

Li J, Mechitov KA, Kim RE, Spencer Jr BF. 2016. Efficient time synchronization for structural health monitoring using wireless smart sensor networks. *Struct Control Hlth* 23(3):470–486.

Lynch JP, Sundararajan A, Law KH, Kiremidjian AS, Kenny T, Carryer E. 2003. Embedment of structural monitoring algorithms in a wireless sensing unit. *Struct Eng Mech* 15(3):285–297.

Maes K, Reynders E, Rezayat A, De Roeck G, Lombaert G. 2016. Offline synchronization of data acquisition systems using system identification. *J Sound Vib* 381:264–272.

Makki Alamdari M, Samali B, Li J, Kalhori H, Mustapha S. 2015. Spectral-based damage identification in structures under ambient vibration. *J Comput Civil Eng* 30(4):04015062.

Maróti M, Kusy B, Simon G, Lédeczi Á. 2004. The flooding time synchronization protocol. *Proceedings of the 2nd international conference on embedded networked sensor systems*, pp. 39–49.

Mu HQ, Yuen KV. 2014. Novel outlier-resistant extended Kalman filter for robust online structural identification. *J Eng Mech* 141:04014100.

Sun S. 2004. Multi-sensor optimal information fusion Kalman filters with applications. *Aerosp Sci Technol* 8(1):57–62.

Sun SL, Deng ZL. 2004. Multi-sensor optimal information fusion Kalman filter. *Automatica* 40(6):1017–1023.

Wang M, Cao J, Liu M, Chen B, Xu Y, Li J. 2009. Design and implementation of distributed algorithms for WSN-based structural health monitoring. *Int J Sens Netw* 5(1):11–21.

Yan WJ, Zhao MY, Sun Q, Ren WX. 2019. Transmissibility-based system identification for structural health monitoring: Fundamentals, approaches, and applications. *Mech Syst Signal Pr* 117:453–482.

Yi TH, Li HN, Song G, Zhang XD. 2015. Optimal sensor placement for health monitoring of high-rise structure using adaptive monkey algorithm. *Struct Control Hlth* 22(4):667–681.

Yuen KV. 2010. *Bayesian Methods for Structural Dynamics and Civil Engineering*. New York: Wiley.

Yuen KV. 2012. Updating large models for mechanical systems using incomplete modal measurement. *Mech Syst Signal Pr* 28:297–308.

Yuen KV, Huang K. 2018a. Identifiability-enhanced Bayesian frequency-domain substructure identification. *Comput-Aided Civ Inf* 33(9):800–812.

Yuen KV, Huang K. 2018b. Real-time substructural identification by boundary force modeling. *Struct Control Hlth* 25(5):e2151.

Yuen KV, Kuok SC. 2016. Online updating and uncertainty quantification using nonstationary output-only measurement. *Mech Syst Signal Pr* 66-67:62–77.

Yuen KV, Mu HQ. 2012. A novel probabilistic method for robust parametric identification and outlier detection. *Probabilist Eng Mech* 30:48–59.

Yuen KV, Ortiz GA. 2017. Outlier detection and robust regression for correlated data. *Comput Method Appl M* 313:632–646.

Zhang Y, Meratnia N, Havinga PJM. 2010. Outlier detection techniques for wireless sensor networks: A survey. *IEEE Commun Surv Tut* 12(2):159–170.

Zhu YC, Au SK. 2018. Bayesian operational modal analysis with asynchronous data, part I: Most probable value. *Mech Syst Signal Pr* 98:652–666.

# 21 Stochastic Optimization Methods in Machine Learning

*Loc Vu-Quoc*
University of Illinois at Urbana-Champaign, Urbana, Illinois, USA


*Alexander Humer*
Johannes Kepler University, Linz, Austria

## CONTENTS

## 21.1  INTRODUCTION

Classical optimization methods have been developed further to tackle new problems, such as noisy gradients, encountered in deep-learning training with random minibatches. To avoid confusion, we will use the terminology "full batch" (instead of just "batch") when the entire training set is used for training. A minibatch is a small subset of the training set. There is much room for new research on learning rate since: *"The learning rate may be chosen by trial and error. This is more of an art than a science, and most guidance on this subject should be regarded with some skepticism."* (Goodfellow, Bengio, & Courville, 2016, p. 287). At the time of this writing, we are aware of two review (Bottou, Curtis, & Nocedal, 2018) papers on optimization algorithms for machine learning, and in particular deep learning (Sun, Cao, Zhu, & Zhao, 2020), aiming particularly at experts in the field: Bottou, Curtis, and Nocedal (2018), as mentioned above, and (Sun, Cao, Zhu, & Zhao, 2020) our contribution complements these two review papers. We are aiming at bringing first-time learners up to speed to benefit from, and even to hopefully enjoy, reading these and others related papers. To this end, we deliberately avoid the dense mathematical-programming language, not familiar to readers outside the field, as used in Bottou, Curtis, and Nocedal (2018), while providing more details on algorithms that have proved important in deep learning than Sun, Cao, Zhu, and Zhao (2020).

## 21.2   STANDARD SGD, MINIBATCH, FIXED LEARNING-RATE SCHEDULE

The stochastic gradient descent (SGD) algorithm, originally introduced by Robbins and Monro (1951a) (another classic) according to many sources has been playing an important role in training deep-learning networks: *"Nearly all of deep learning is powered by one very important algorithm: stochastic gradient descent (SGD). Stochastic gradient descent is an extension of the gradient descent algorithm."* (Goodfellow, Bengio, & Courville, 2016, p. 147). The number M of examples in a training set $\mathbb{X}$ could be very large, rendering prohibitively expensive to evaluate the cost function and to compute the gradient of the cost function with respect to the number of parameters, which by itself could also be very large. At iteration k within a training session $\tau$, let $\mathbb{I}_k^{|m|}$ be a randomly selected set of m indices, which are elements of the training-set indices $[M] = \{1,\ldots,M\}$. Typically, m is much smaller than M: *"The minibatch size m is typically chosen to be a relatively small number of examples, ranging from one to a few hundred. Crucially, m is usually held fixed as the training set size M grows. We may fit a training set with billions of examples using updates computed on only a hundred examples."* (Goodfellow, Bengio, & Courville, 2016, p. 148). Generated as in Eq. (21.2), the random-index sets $\mathbb{I}_k^{|m|}$, for $k = 1,2,\ldots$, are non-overlapping such that after $k_{max} = M/m$ iterations, all examples in the training set are covered, and a training session, or training epoch, is completed. At iteration $k$ of a training epoch $\tau$, the random minibatch $\mathbb{B}_k^{|m|}$ is a set of $m$ examples pulled out from the much larger training set $\mathbb{X}$ using the random indices in $\mathbb{I}_k^{|m|}$ with the corresponding targets in the set $\mathrm{T}_k^{|m|}$:

$$M_1 = 1,\ldots,M =: [M], k_{max} = M/m, \tag{21.1}$$

$$\mathbb{I}_k^{|m|} = \left\{i_{1,k},\ldots,i_{m,k}\right\} \subseteq M_k, M_{k+1} = M_k - \mathbb{I}_k^{|m|} \text{ for } k = 1,\ldots,k_{max}, \tag{21.2}$$

$$\mathbb{B}_k^{|m|} = \left\{x^{|i_{1,k}|},\cdots,x^{|i_{m,k}|}\right\} \subseteq \mathbb{X} = \left\{x^{|1|},\cdots,x^{|M|}\right\}, \tag{21.3}$$

$$\mathrm{T}_k^{|m|} = \left\{y^{|i_{1,k}|},\cdots,y^{|i_{m,k}|}\right\} \subseteq \mathrm{Y} = \left\{y^{|1|},\cdots,y^{|M|}\right\}. \tag{21.4}$$

Note that once the random index set $\mathbb{I}_k^{|m|}$ had been selected, it was deleted from its superset $M_k$ to form $M_{k+1}$ so the next random set $\mathbb{I}_{j+1}^{|m|}$ would not contain indices already selected in $\mathbb{I}_k^{|m|}$. Unlike the iteration counter $k$ within a training epoch $\tau$, the global iteration counter $j$ is not reset to one, the beginning of a new training epoch $\tau + 1$, but continues to increment for each new minibatch. The cost-function estimate is the average of the cost functions, each of which is the cost function of an example $x^{|i_k|}$ in the minibatch for iteration $k$ in epoch $\tau$:

$$\tilde{J}(\theta) = \frac{1}{m} \sum_{a=1}^{a=m \leq M} J_{i_a}(\theta), \text{ with } J_{i_a}(\theta) = J\left(f\left(x^{|i_a|},\theta\right),y^{|i_a|}\right), \text{ and } x^{|i_a|} \in \mathbb{B}_k^{|m|}, y^{|i_a|} \in \mathrm{T}_k^{|m|}, \tag{21.5}$$

where we wrote the random index as $i_a$ instead of $i_{a,k}$ as in Eq. (21.2) to alleviate the notation. The corresponding gradient estimate is:

$$\tilde{g}(\theta) = \frac{\partial \tilde{J}(\theta)}{\partial \theta} = \frac{1}{m} \sum_{a=1}^{a=m \leq M} \frac{\partial J_{i_a}(\theta)}{\partial \theta} = \frac{1}{m} \sum_{a=1}^{a=m \leq M} g_{i_a}. \tag{21.6}$$

The epoch stopping criterion is usually determined by a computation "budget", i.e., the maximum number of epochs allowed. For example, Bergou, Diouane, Kungurtsev, and Royer (2018) set a budget of 1600 epochs maximum in their numerical examples. There are several known problems with SGD: *"Despite the prevalent use of SGD, it has known challenges and inefficiencies. First, the direction may not represent a descent direction, and second, the method is sensitive to the step-size (learning rate) which is often poorly overestimated."* (Paquette & Scheinberg, 2018). For the above reasons, it may not be appropriate to use the norm of the gradient estimate being small as stationarity condition, i.e., where the local minimizer or saddle point is located; See Bergou, Diouane, Kungurtsev, and Royer (2018).

## 21.3 MOMENTUM AND FAST (ACCELERATED) GRADIENT

The standard update of the whole set of $P_T$ network parameters $\theta$ for gradient descent,

$$\theta \leftarrow \theta - \epsilon \, \partial J / \partial \theta = \theta - \epsilon g, \text{ with } g := \partial J / \partial \theta \in \mathbb{R}^{P_T \times 1}, \tag{21.7}$$

where $J$ denotes the cost function and $\epsilon$ is the learning rate, would be slow when encountering deep and narrow valley, as shown in Figure 21.1, and can be replaced by the general update with momentum as follows:

$$\tilde{\theta}_{k+1} = \tilde{\theta}_k - \epsilon_k \tilde{g}\left(\tilde{\theta}_k + \gamma_k\left(\tilde{\theta}_k - \tilde{\theta}_{k-1}\right)\right) + \zeta_k\left(\tilde{\theta}_k - \tilde{\theta}_{k-1}\right) \tag{21.8}$$

from which the following methods are obtained:

- Standard SGD update Eq. (7) with $\gamma_k = \zeta_k = 0$, (Robbins & Monro, 1951b)
- SGD with classical momentum: $\gamma_k = 0$ and $\zeta_k \in (0,1)$ *("small heavy sphere"* or heavy point mass) (Polyak, 1964)
- SGD with fast (accelerated) gradient: $\gamma_k = \zeta_k \in (0,1)$, (Nesterov, 1983, 2018)

The continuous counterpart of the parameter update Eq. (21.8) with classical momentum, i.e., when $\gamma_k \in (0,1)$ and $\zeta_k = 0$, is the equation of motion of a heavy point mass (thus no rotatory inertia) under viscous friction at slow motion (proportional to velocity) and applied force $-\tilde{g}$ given below with

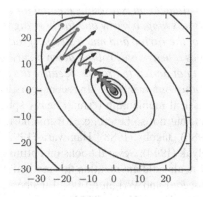

**FIGURE 21.1** SGD with momentum, small heavy sphere: The descent direction (black arrows) bounces back and forth between the steep slopes of a deep and narrow valley. The small-heavy-sphere method (SGD with momentum) follows a faster descent (red path) toward the bottom of the valley. (Source: Goodfellow, Bengio, and Courville (2016, p. 289), with permission.)

its discretization by finite difference in time, where $h_k$ and $h_{k-1}$ are the time-step sizes, (Goudou & Munier, 2009; Polyak, 1964)

$$\frac{d^2\tilde{\theta}}{(dt)^2} + \nu\frac{d\tilde{\theta}}{dt} = -\tilde{g} \Rightarrow \frac{\dfrac{\tilde{\theta}_{k+1} - \tilde{\theta}_k}{h_k} - \dfrac{\tilde{\theta}_k - \tilde{\theta}_{k-1}}{h_{k-1}}}{h_k} + \nu\frac{\tilde{\theta}_{k+1} - \tilde{\theta}_k}{h_k} = -\tilde{g}_k, \tag{21.9}$$

$$\tilde{\theta}_{k+1} - \tilde{\theta}_k - \zeta_k\left(\tilde{\theta}_k - \tilde{\theta}_{k-1}\right) = -\epsilon_k\tilde{g}_k, \text{ with } \zeta_k = \frac{h_{k-1}}{h_k}\frac{1}{1+\nu h_k} \text{ and } \epsilon_k = \left(h_k\right)^2\frac{1}{1+\nu h_k}, \tag{21.10}$$

which is the same as the update Eq. (21.8) with $\gamma_k = 0$. The term $\zeta_k\left(\theta_k - \theta_{k-1}\right)$ is often called the *"momentum"* term since it is proportional to (discretized) velocity. Polyak (1964) on the other hand explained the term $\zeta_k\left(\theta_k - \theta_{k-1}\right)$ as *"giving inertia to the motion, [leading] to motion along the "essential" direction, i.e., along 'the bottom of the trough'"*, and recommended to select $\zeta_k \in (0.8, 0.99)$, i.e., close to one, without explanation. The reason is to have low friction, i.e., $\nu$ small, but not zero friction ($\nu = 0$), since friction is important to slow down the motion of the sphere up and down the valley sides (like skateboarding from side to side in a half-pipe), thus accelerate convergence toward the trough of the valley; from Eq. (21.10), we have

$$h_k = h_{k-1} = h \text{ and } \nu \in [0, +\infty) \Rightarrow \zeta_k \in (0,1], \text{ with } \nu = 0 \Rightarrow \zeta_k = 1. \tag{21.11}$$

For more insight into the update Eq. (21.10), consider the case of constant coefficients $\zeta_k = \zeta$ and $\epsilon_k = \epsilon$, and rewrite this recursive relation in the form:

$$\tilde{\theta}_{k+1} - \tilde{\theta}_k = -\epsilon\sum_{i=0}^{k}\zeta^i g_{k-i}, \text{ using } \tilde{\theta}_1 - \tilde{\theta}_0 = -\epsilon\tilde{g}_0 \tag{21.12}$$

that is, without momentum for the first term. So the effective gradient is the sum of all gradients from the beginning $i = 0$ until the present $i = k$ weighted by the exponential function $\zeta^i$ so there is a fading memory effect, i.e., gradients that are farther back in time have less influence than those closer to the present time. The summation term in Eq. (21.8) also provides an explanation of how the "inertia" (or momentum) term work: (1) Two successive opposite gradients would cancel each other, whereas; (2) Two successive gradients in the same direction (toward the trough of the valley) would reinforce each other. See Bertsekas and Tsitsiklis (1995, pp. 104–105) and Hinton (2012) who provided a similar explanation: *"Momentum is a simple method for increasing the speed of learning when the objective function contains long, narrow and fairly straight ravines with a gentle but consistent gradient along the floor of the ravine and much steeper gradients up the sides of the ravine. The momentum method simulates a heavy ball rolling down a surface. The ball builds up velocity along the floor of the ravine, but not across the ravine because the opposing gradients on opposite sides of the ravine cancel each other out over time."* In recent years, Polyak (1964) (English version) has often been cited for the classical momentum ("small heavy sphere") method to accelerate the convergence in gradient descent, but not so before, e.g., Rumelhart, Hinton, and Williams (1986), Plaut, Nowlan, and Hinton (1986), Jacobs (1988), Hagiwara (1992), and Hinton (2012) used the same method without citing Polyak (1964). Several books on optimization not related to neural networks, many of them well-known, also did not mention this method: Polak (1971), Gill, Murray, and Wright (1981), Polak (1997), Nocedal and Wright (2006), Luenberger and Ye (2016), and Snyman and Wilke (2018). The book by Priddy and Keller (2005) on neural networks did cite both the original Russian version and the English translated version Polyak (1964) (whose name was spelled as "Poljak" before 1990), and referred to another neural-network book (Bertsekas & Tsitsiklis, 1995), in which the formulation was discussed.

## 21.4  INITIAL-STEP-LENGTH TUNING

The initial step length $\epsilon_0$, or learning-rate initial value, is one of the two most influential hyperparameters to tune, i.e., to find the best performing values. During tuning, the step length $\epsilon$ is kept constant at $\epsilon_0$ in the parameter update Eq. (21.7) throughout the optimization process, i.e., a fixed step length is used, without decay as in Eqs. (21.14–21.16) in Section 21.5.

Wilson, Roelofs, Stern, Srebro, and Recht (2018) proposed the following simple tuning method:

*"To tune the step sizes, we evaluated a logarithmically-spaced grid of five step sizes. If the best performance was ever at one of the extremes of the grid, we would try new grid points so that the best performance was contained in the middle of the parameters. For example, if we initially tried step sizes 2, 1, 0.5, 0.25, and 0.125 and found that 2 was the best performing, we would have tried the step size 4 to see if performance was improved. If performance improved, we would have tried 8 and so on."* The above logarithmically-spaced grid was given by $2^k$, with $k = 1, 0, -1, -2, -3$. This tuning method appears effective as shown in Figure 21.2 on the CIFAR-10 dataset mentioned above, for which the following values for $\epsilon_0$ had been tried for different optimizers, even though the values did not always belong to the sequence $\{a^k\}$, but could include close, rounded values:

- SGD (Section 21.2): 2, 1, 0.5 (best), 0.25, 0.05, 0.01144
- SGD with momentum (Section 21.3): 2, 1, 0.5 (best), 0.25, 0.05, 0.01
- AdaGrad: 0.1, 0.05, 0.01 (best, default), 0.0075, 0.005
- RMSProp (Section 21.9.3): 0.005, 0.001, 0.0005, 0.0003 (best), 0.0001
- Adam (Section 21.9.5): 0.005, 0.001 (default), 0.0005, 0.0003 (best), 0.0001, 0.00005

## 21.5  STEP-LENGTH DECAY, ANNEALING, AND CYCLIC ANNEALING

In the update of the parameter $\theta$ as in Eq. (21.7), the learning rate (step length) $\epsilon$ has to be reduced gradually as a function of either the epoch counter $\tau$ or of the global iteration counter $j$. Let $\dagger$ represents either $\tau$ or $j$, depending on user's choice. If the learning rate $\epsilon$ is a function of epoch $\tau$, then $\epsilon$ is held constant in all iterations $k = 1,\ldots,k_{max}$ within epoch $\tau$, and we have:

$$j = (\tau - 1) * k_{max} + k. \tag{21.13}$$

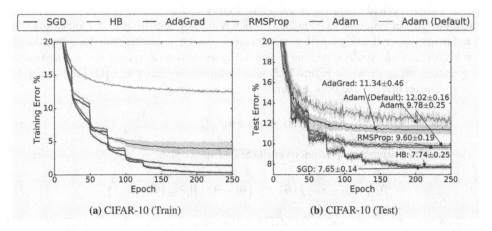

**(a)** CIFAR-10 (Train)                           **(b)** CIFAR-10 (Test)

**FIGURE 21.2**  Standard SGD and SGD with momentum vs AdaGrad, RMSProp, Adam on CIFAR-10 dataset. From Wilson, Roelofs, Stern, Srebro, and Recht (2018), who proposed a method for step-size tuning and step-size decaying to achieve lowest training error and generalization (test) error for both Standard SGD and SGD with momentum ("Heavy Ball" or better yet "Small Heavy Sphere" method) compared to adaptive methods such as AdaGrad, RMSProp, Adam. (Source: Wilson, Roelofs, Stern, Srebro, and Recht (2018), with permission.)

The following learning-rate scheduling, linear with respect to $\dagger$, is one option:

$$\epsilon(\dagger) = \begin{cases} \left(1 - \dfrac{\dagger}{\dagger_c}\right)\epsilon_0 + \dfrac{\dagger}{\dagger_c}\epsilon_c & \text{for } 0 \leq \dagger \leq \dagger_c \\[2ex] \epsilon_{\dagger_c} & \text{for } \dagger_c \leq \dagger \end{cases} \quad , \tag{21.14}$$

$$\dagger = \text{epoch } \tau \text{ or global iteration } j$$

where $\epsilon_0$ is the learning-rate initial value, and $\epsilon_{\dagger_c}$ the constant learning-rate value when $t \geq \dagger_c$. Other possible learning-rate schedules are:

$$\epsilon(\dagger) = \frac{\epsilon_0}{\sqrt{\dagger}} \to 0 \text{ as } \dagger \to \infty, \tag{21.15}$$

$$\epsilon(\dagger) = \frac{\epsilon_0}{\dagger} \to 0 \text{ as } \dagger \to \infty, \tag{21.6}$$

with $\dagger$ defined as in Eq. (21.14), even though authors, such as Reddi, Kale, and Kumar (2019) used Eqs. (21.15) and (21.16) with $\dagger = j$ as global iteration counter. Another step-length decay method proposed by Wilson, Roelofs, Stern, Srebro, and Recht (2018) is to reduce the step length $\epsilon(\tau)$ for the current epoch $\tau$ by a factor $\varpi \in (0,1)$ when the cost estimate $\tilde{J}_{\tau-1}$ at the end of the last epoch $(\tau - 1)$ is greater than the lowest cost in all previous global iterations, with $\tilde{J}_j$ denoting the cost estimate at global iteration $j$, and $k_{max}(\tau - 1)$ the global iteration number at the end of epoch $(\tau - 1)$:

$$\epsilon(\tau) = \begin{cases} \varpi\epsilon(\tau - 1) \text{ if } \tilde{J}_{\tau-1} > \min_j\{\tilde{J}_j, j = 1, \ldots, k_{max}(\tau - 1)\} \\ \epsilon(\tau - 1) \text{ Otherwise} \end{cases}. \tag{21.17}$$

Recall, $k_{max}$ is the number of non-overlapping minibatches that cover the training set, as defined in Eq. (1.1). Wilson, Roelofs, Stern, Srebro, and Recht (2018) set the step-length decay parameter $\varpi = 0.9$ in their numerical examples, in particular Figure 21.2.

In additional to decaying the step length $\epsilon$, which is already annealing, cyclic annealing is introduced to further reduce the step length down to zero ("cooling"), quicker than decaying, then bring the step length back up rapidly ("heating"), and doing so for several cycles. The cosine function is typically used, such as shown in Figure 21.3, as a multiplicative factor $a_k \in [0,1]$ to the step length $\epsilon_k$ in the parameter update, and thus the name "cosine annealing":

$$\tilde{\theta}_{k+1} = \tilde{\theta}_k - a_k \epsilon_k \tilde{g}_k \tag{21.18}$$

as an add-on to the parameter update for vanilla SGD Eq. (21.7), or

$$\tilde{\theta}_{k+1} = \tilde{\theta}_k - a_k \epsilon_k \tilde{g}\left(\tilde{\theta}_k + \gamma_k\left(\tilde{\theta}_k - \tilde{\theta}_{k-1}\right)\right) + \zeta_k\left(\tilde{\theta}_k - \tilde{\theta}_{k-1}\right) \tag{21.19}$$

as an add-on to the parameter update for SGD with momentum and accelerated gradient Eq. (21.8). The cosine annealing factor can take the form (Loshchilov & Hutter, 2019):

$$a_k = 0.5 + 0.5 \cos\left(\pi T_{cur}/T_p\right) \in [0,1], \text{ with } T_{cur} := j - \sum_{q=1}^{q=p-1} T_q, \tag{21.20}$$

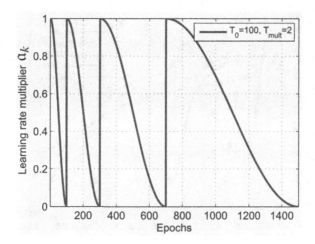

**FIGURE 21.3** Annealing factor $a_k$ as a function of epoch number. Four annealing cycles $p = 1,...,4$, with the following schedule for $T_p$ in Eq. (21.20): (1) Cycle 1, $T_1 = 100$ epochs, epoch 0 to epoch 100, (2) Cycle 2, $T_2 = 200$ epochs, epoch 101 to epoch 300, (3) Cycle 3, $T_3 = 300$ epochs, epoch 301 to epoch 700, (4) Cycle 4, $T_4 = 800$ epochs, epoch 701 to epoch 1500. (Source: Loshchilov and Hutter (2019), with permission.)

where $T_{cur}$ is the number of epochs from the start of the last warm restart at the end of epoch $\sum_{q=1}^{q=p-1} T_q$, where $a_k = 1$ ("maximum heating"), $j$ the current global iteration counter, $T_p$ the maximum number of epochs allowed for the current $p$th annealing cycle, during which $T_{cur}$ (Loshchilov & Hutter, 2019) would go from 0 to $T_p$, when $a_k = 0$ ("complete cooling"). Figure 21.3 shows four annealing cycles, which helped reduce dramatically the number of epochs needed to achieve the same lower cost as obtained without annealing. Figure 21.4 shows the effectiveness of cosine annealing in bringing down the cost rapidly in the early stage, but there is a diminishing return, as the cost reduction decreases with the number of annealing cycles. Up to a point, it is no longer as effective as SGD with weight decay in Section 21.7.

The sufficient conditions for convergence, for convex functions, are

$$\sum_{j=1}^{\infty} \epsilon_j^2 < \infty, \text{ and } \sum_{j=1}^{\infty} \epsilon_j = \infty. \tag{21.21}$$

The inequality on the left of Eq. (21.21), i.e., the sum of the squared of the step lengths being finite, ensures that the step length would decay quickly to reach the minimum, but is valid only when the minibatch size is fixed. The equation on the right of Eq. (21.21) ensures convergence, no matter how far the initial guess was from the minimum.

## 21.6 MINIBATCH-SIZE INCREASE, FIXED STEP LENGTH

The minibatch parameter update from Eq. (21.8), without momentum and accelerated gradient, which becomes Eq. (21.7), can be rewritten to introduce the error due to the use of the minibatch gradient estimate $\tilde{g}$ instead of the full-batch gradient $g$ as follows:

$$\tilde{\theta}_{k+1} = \tilde{\theta}_k - \epsilon_k \tilde{g}_k = \tilde{\theta}_k - \epsilon_k \left[ g_k + (g_k - g_k) \right]$$

$$\Rightarrow \frac{\Delta \tilde{\theta}_k}{\epsilon_k} = \frac{\tilde{\theta}_{k+1} - \tilde{\theta}_k}{\epsilon_k} = -g_k + (g_k - g_k) = -g_k + e_k, \tag{21.22}$$

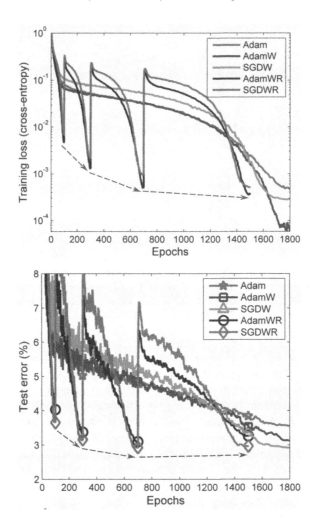

**FIGURE 21.4** AdamW vs Adam, SGD, and variants on CIFAR-10 dataset: While AdamW achieved lowest training loss (error) after 1800 epochs, the results showed that SGD with weight decay (SGDW) and with warm restart (SGDWR) achieved lower test (generalization) errors than Adam (including its variants AdamW, AdamWR). See Figure 21.3 for the scheduling of the annealing multiplier $a_k$, for which the epoch numbers 100, 300, 700, 1500 for complete cooling ($a_k = 0$) coincided with the same epoch numbers for the sharp minima. There was, however, a diminishing return beyond the 4th cycle as indicated by the dotted arrows, for both training error and test error, which actually increased at the end of the 4th cycle (right subfigure, red arrow). (Adapted from Loshchilov and Hutter (2019), with permission.)

where $g_k = g(\theta_k)$ and $\tilde{g}_k = \tilde{g}_b(\theta_k)$, with $b = k$, and $\tilde{g}_b(\cdot)$ is the gradient estimate function using minibatch $b = 1, \ldots, k_{max}$. To show that the gradient error has zero mean (average), based on the linearity of the expectation function $\mathbb{E}(\cdot) = \langle \cdot \rangle$, which is defined as

$$\mathbb{E}_{x \sim P} g(x) = \sum_x g(x) P(x) = \langle g(x) \rangle, \quad \mathbb{E}_{x \sim p} g(x) = \int g(x) p(x) dx = \langle g(x) \rangle, \qquad (21.23)$$

where $g(x = x)$ is a function of a random variable x, having a probability distribution $P(x = x)$, and the probability distribution density $p(x = x)$ for the continuous case, i.e.,

$$\langle \alpha u + \beta v \rangle = \alpha \langle u \rangle + \beta \langle u \rangle, \qquad (21.24)$$

$$\langle e_k \rangle = \langle g_k - \tilde{g}_k \rangle = \langle g_k \rangle - \langle \tilde{g}_k \rangle = g_k - \langle \tilde{g}_k \rangle, \tag{21.25}$$

from Eqs. (21.1–21.3) on the definition of minibatches and Eqs. (21.5–21.6) on the definition of the cost and gradient estimates (without omitting the iteration counter $k$), we have

$$g(\theta_k) = \frac{1}{k_{max}} \sum_{b=1}^{b=k_{max}} \frac{1}{m} \sum_{a=1}^{a=m} g_{i_{a,b}} \Rightarrow \langle g_k \rangle = g_k = \frac{1}{k_{max}} \sum_{b=1}^{b=k_{max}} \frac{1}{m} \sum_{a=1}^{a=m} \langle g_{i_{a,b}} \rangle = \langle g_{i_{a,b}} \rangle, \tag{21.26}$$

$$\tilde{g}_k = \frac{1}{m} \sum_{a=1}^{a=m \le M} g_{i_{a,k}} \Rightarrow \langle \tilde{g}_k \rangle = \frac{1}{k_{max}} \sum_{k=1}^{k=k_{max}} \langle g_{i_{a,k}} \rangle = \frac{1}{k_{max}} \sum_{k=1}^{k=k_{max}} \langle g_k \rangle = g_k \Rightarrow \langle e_k \rangle = 0. \tag{21.27}$$

Or alternatively, the same result can be obtained with:

$$g_k = \frac{1}{k_{max}} \sum_{b=1}^{b=k_{max}} \tilde{g}_b(\theta_k) \Rightarrow g_k = \langle \tilde{g}_b(\theta_k) \rangle = \langle \tilde{g}_k(\theta_k) \rangle = \langle \tilde{g}_k \rangle \Rightarrow \langle e_k \rangle = 0. \tag{21.28}$$

The mean value of the "square" of the gradient error $\langle e^T e \rangle$, in which we omitted the iteration counter subscript $k$ to alleviate the notation, relies on some identities related to the covariance matrix $\langle e, e \rangle$. The mean of the square matrix $x_i^T x_j$, where $\{x_i, x_j\}$ are two random row matrices, is the sum of the product of the mean values and the covariance matrix of these matrices

$$\langle x_i^T x_j \rangle = \langle x_i \rangle^T \langle x_j \rangle + \langle x_i, x_j \rangle, \text{ or } \langle x_i, x_j \rangle = \langle x_i \rangle^T x_j - \langle x_i \rangle^T x_j, \tag{21.29}$$

where $\langle x_i, x_j \rangle$ is the covariance matrix of $x_i$ and $x_j$, and thus the covariance operator $\langle \cdot, \cdot \rangle$ is bilinear due to the linearity of the mean (expectation) operator $\langle \cdot \rangle$ in Eq. (21.24):

$$\left\langle \sum_i \alpha_i u_i, \sum_j \beta_j v_j \right\rangle = \sum_i \alpha_i \beta_j \langle u_i, v_j \rangle, \forall \alpha_i, \beta_j \in \mathbb{R} \text{ and } \forall u_i, v_j \in \mathbb{R}^n \text{ random.} \tag{21.30}$$

Eq. (21.29) is the key relation to derive an expression for the square of the gradient error $\langle e^T e \rangle$, which can be rewritten as the sum of four covariance matrices upon using Eq. (21.29) and either Eq. (21.27) or Eq. (21.28), i.e., $\langle \tilde{g}_k \rangle = \langle g_k \rangle = g_k$, as the four terms $g_k^T g_k$ cancel each other out:

$$\langle e^T e \rangle = \langle (\tilde{g} - g)^T (\tilde{g} - g) \rangle = \langle \tilde{g}, \tilde{g} \rangle - \langle \tilde{g}, g \rangle - \langle g, \tilde{g} \rangle + \langle g, g \rangle, \tag{21.31}$$

where the iteration counter $k$ had been omitted to alleviate the notation. Moreover, to simplify the notation further, the gradient related to an example is simply denoted by $g_a$ or $g_b$, with $a, b = 1, \ldots, m$ for a minibatch, and $a, b = 1, \ldots, M$ for the full batch:

$$\tilde{g} = \tilde{g}_k = \frac{1}{m} \sum_{a=1}^m g_{i_{a,k}} = \frac{1}{m} \sum_{a=1}^m g_{i_a} = \frac{1}{m} \sum_{a=1}^m g_a, \tag{21.32}$$

$$g = \frac{1}{k_{max}} \sum_{k=1}^{k_{max}} \frac{1}{m} \sum_{a=1}^m g_{i_{a,k}} = \frac{1}{M} \sum_{b=1}^M g_b. \tag{21.33}$$

Now assume the covariance matrix of any pair of single-example gradients $g_a$ and $g_b$ depends only on parameters $\theta$, and is of the form:

$$\langle g_a, g_b \rangle = \mathfrak{C}(\theta) \delta_{ab}, \forall a, b \in 1, \ldots, M, \tag{21.34}$$

where $\delta_{ab}$ is the Kronecker delta. Using Eqs. (21.32–21.33) and Eq. (21.34) in Eq. (21.31), we obtain a simple expression for $\langle e^T e \rangle$:

$$\langle e^T e \rangle = \left( \frac{1}{m} - \frac{1}{M} \right) \mathfrak{c}. \tag{21.35}$$

## 21.7 WEIGHT DECAY, AVOIDING OVERFIT

Reducing, or decaying, the network parameter $\theta$ (which include the weights and the biases) is one method to avoid overfitting by adding a parameter-decay term to the update equation:

$$\tilde{\theta}_{k+1} = \tilde{\theta}_k + \epsilon_k \, \tilde{d}_k - \eth \tilde{\theta}_k, \tag{21.36}$$

where $\eth \in (0,1)$ is the decay parameter, and there the name "weight decay," which is equivalent to SGD with $L_2$ regularization, by adding an extra penalty term in the cost function. *"Regularization is any modification we make to a learning algorithm that is intended to reduce its generalization error but not its training error"* (Goodfellow et al., 2016, p. 117). Weight decay is only one among other forms of regularization, such as large learning rates, small batch sizes and dropout. The effects of the weight-decay parameter $\eth$ in avoiding network model overfit is shown in Figure 21.5.

$$\tilde{\theta}_{k+1} = \tilde{\theta}_k + a_k \left( \epsilon_k \tilde{d}_k - \eth \tilde{\theta}_k \right) = \tilde{\theta}_k - a_k \left( \epsilon_k \tilde{g}_k + \eth \tilde{\theta}_k \right) \tag{21.37}$$

Rögnvaldsson (1998) wrote: *"In the neural network community the two most common methods to avoid overfitting are early stopping and weight decay* (Plaut, Nowlan, & Hinton, 1986). *Early stopping has the advantage of being quick, since it shortens the training time, but the disadvantage of being poorly defined and not making full use of the available data. Weight decay, on the other hand, has the advantage of being well defined, but the disadvantage of being quite time consuming"* (because of tuning). Examples of tuning the weight decay parameter $\eth$, which is of the order of $10^{-3}$, see Rögnvaldsson (1998) and Loshchilov and Hutter (2019). In the case of weight decay with cyclic annealing, both the step length $\epsilon_k$ and the weight decay parameter $\eth$ are scaled by the annealing multiplier $a_k$ in the parameter update (Loshchilov & Hutter, 2019):

$$\tilde{\theta}_{k+1} = \tilde{\theta}_k + a_k \left( \epsilon_k \tilde{d}_k - \eth \tilde{\theta}_k \right) = \tilde{\theta}_k - a_k \left( \epsilon_k \tilde{g}_k + \eth \tilde{\theta}_k \right) \tag{21.38}$$

The effectiveness of SGD with weight decay, with and without cyclic annealing, is presented in Figure 21.4.

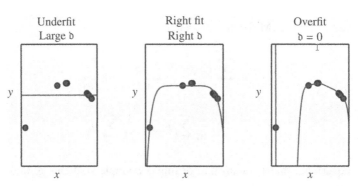

FIGURE 21.5 Weight decay: Effects of magnitude of weight-decay parameter $\eth$. (Adapted from Goodfellow, Bengio, and Courville (2016, p. 116), with permission).

## 21.8   COMBINING ALL ADD-ON TRICKS

To have a general parameter-update equation that combines all of the above add-on tricks, start with the parameter update with momentum and accelerated gradient Eq. (21.8)

$$\tilde{\theta}_{k+1} = \tilde{\theta}_k - \epsilon_k \tilde{g}\left(\tilde{\theta}_k + \gamma_k\left(\tilde{\theta}_k - \tilde{\theta}_{k-1}\right)\right) + \zeta_k\left(\tilde{\theta}_k - \tilde{\theta}_{k-1}\right) \tag{21.39}$$

and add the weight-decay term $\partial\tilde{\theta}_k$ from Eq. (21.36), then scale both the weight-decay term and the gradient-descent term $\left(-\epsilon_k \tilde{g}_k\right)$ by the cyclic annealing multiplier $a_k$ in Eq. (21.37), leaving the momentum term $\zeta_k\left(\tilde{\theta}_k - \tilde{\theta}_{k-1}\right)$ alone, to obtain:

$$\tilde{\theta}_{k+1} = \tilde{\theta}_k - a_k\left[\epsilon_k \tilde{g}\left(\tilde{\theta}_k + \gamma_k\left(\tilde{\theta}_k - \tilde{\theta}_{k-1}\right)\right) + \partial_k\tilde{\theta}_k\right] + \zeta_k\left(\tilde{\theta}_k - \tilde{\theta}_{k-1}\right). \tag{21.40}$$

## 21.9   ADAPTIVE METHODS: ADAM, VARIANTS, CRITICISM

The Adam algorithm was introduced by Kingma and Ba (2017) and has been *"immensely successful in development of several state-of-the-art solutions for a wide range of problems,"* as stated by Reddi, Kale, and Kumar (2019). According to Bock et al. (2018), *"In the area of neural networks, the ADAM-Optimizer is one of the most popular adaptive step size methods* (Kingma & Ba, 2017; Huang, Wang, & Dong, 2019). *The 5865 citations in only three years shows additionally the importance of the given paper."* Huang, Wang, and Dong (2019) concurred: *"Adam is widely used in both academia and industry. However, it is also one of the least well-understood algorithms. In recent years, some remarkable works provided us with better understanding of the algorithm, and proposed different variants of it."*

### 21.9.1   Unified Adaptive Learning-Rate Pseudocode

Reddi, Kale, and Kumar (2019) suggested a unified pseudocode that included not only the standard SGD, but also a number of successful adaptive learning-rate methods: AdaGrad, RMSProp, AdaDelta, Adam. Four new quantities are introduced for iteration $k$ in SGD: (1) $m_k$ at SGD iteration $k$, as the first moment, (2) its correction $\hat{m}_k$,

$$m_k = \phi_k\left(\tilde{g}_1, \ldots, \tilde{g}_k\right), \hat{m}_k = \chi_{\phi_k}\left(m_k\right) \tag{21.41}$$

(3) the second moment (variance) $V_k$, and (4) its correction $\hat{V}_k$.

$$V_k = \psi_k\left(\tilde{g}_1, \ldots, \tilde{g}_k\right), \hat{V}_k = \chi_{\psi_k}\left(V_k\right) \tag{21.42}$$

The descent direction estimate $\tilde{d}_k$ at SGD iteration $k$ for each training epoch is

$$\tilde{d}_k = -\hat{m}_k \tag{21.43}$$

The adaptive learning rate $\epsilon_k$ is obtained from rescaling the fixed learning-rate schedule $\epsilon(k)$, also called the "global" learning rate particularly when it is a constant, using the 2nd moment $\hat{V}_k$ as follows:

$$\epsilon_k = \frac{\epsilon(k)}{\sqrt{\hat{V}_k} + \delta}, \text{ or } \epsilon_k = \frac{\epsilon(k)}{\sqrt{\hat{V}_k + \delta}} \text{ (element-wise operations)} \tag{21.44}$$

where $\epsilon(k)$ can be either Eq. (21.14) (which includes $\epsilon(k) = \epsilon_0$) or Eq. (21.15); $\delta$ is a small number to avoid division by zero; the operations (square root, addition, division) are element-wise, with both $\epsilon_0$ and $\delta = O(10^{-6})$ to $O(10^{-8})$ (depending on the algorithm) being constants.

### 21.9.2 AdaGrad: Adaptive Gradient

Starting the line of research on adaptive learning-rate algorithms, Duchi, Hazan, and Singer (2011) selected the following functions:

$$\phi_k(\tilde{g}_1,\ldots,\tilde{g}_k) = \tilde{g}_k, \text{ with } \chi_{\phi_k} = I \text{ (Identity)} \Rightarrow m_k = \tilde{g}_k = \hat{m}_k \tag{21.45}$$

$$\psi_k(\tilde{g}_1,\ldots,\tilde{g}_k) = \sum_{i=1}^{k} \tilde{g}_i \odot \tilde{g}_i = \sum_{i=1}^{k} (\tilde{g}_i)^2 \text{ (element-wise square)} \tag{21.46}$$

$$\chi_{\psi_k} = I \text{ and } \delta = 10^{-7} \Rightarrow \hat{V}_k = \sum_{i=1}^{k} (\tilde{g}_i)^2 \text{ and } \epsilon_k = \frac{\epsilon(k)}{\sqrt{\hat{V}_k} + \delta} \text{ (element-wise operations)} \tag{21.47}$$

leading to an update with adaptive scaling of the learning rate

$$\tilde{\theta}_{k+1} = \tilde{\theta}_k - \frac{\epsilon(k)}{\sqrt{\hat{V}_k} + \delta} g_k \text{ (element-wise operations)} \tag{21.48}$$

in which each parameter in $\tilde{\theta}_k$ is updated with its own learning rate. For a given network parameter, say, $\theta_{pq}$, its learning rate $\epsilon_{k,pq}$ is essentially $\epsilon(k)$ scaled with the inverse of the square root of the sum of all historical values of the corresponding gradient component $(pq)$, i.e., $(\delta + \sum_{i=1}^{k} \tilde{g}_{i,pq}^2)^{-1/2}$, with $\delta = 10^{-7}$ being very small. A consequence of such scaling is that a larger gradient component would have a smaller learning rate and a smaller per-iteration decrease in the learning rate, whereas a smaller gradient component would have a larger learning rate and a higher per-iteration decrease in the learning rate, even though the relative decrease is about the same. Thus, progress along different directions with large difference in gradient amplitudes is evened out as the number of iterations increases.

### 21.9.3 RMSProp: Root Mean Square Propagation

Since *"AdaGrad shrinks the learning rate according to the entire history of the squared gradient and may have made the learning rate too small before arriving at such a convex structure,"* Tieleman and Hinton (2012) fixed the problem of continuing decay of the learning rate by introducing RMSProp with the following functions:

$$\phi_k(\tilde{g}_1,\ldots,\tilde{g}_k) = \tilde{g}_k, \text{ with } \chi_{\phi_k} = I(\text{Identity}) \Rightarrow m_k = \tilde{g}_k = \hat{m}_k \tag{21.49}$$

$$V_k = \beta V_{k-1} + (1-\beta)(\tilde{g}_k)^2 \text{ with } \beta \in [0,1) \text{ (element-wise square)} \tag{21.50}$$

$$V_k = (1-\beta) \sum_{i=1}^{k} \beta^{k-i} (\tilde{g}_i)^2 = \psi_k(\tilde{g}_1,\ldots,\tilde{g}_k) \tag{21.51}$$

$$\chi_{\psi_k} = I \text{ and } \delta = 10^{-6} \Rightarrow \hat{V}_k = V_k \text{ and } \epsilon_k = \frac{\epsilon(k)}{\sqrt{\hat{V}_k} + \delta} \text{ (element-wise operations)} \tag{21.52}$$

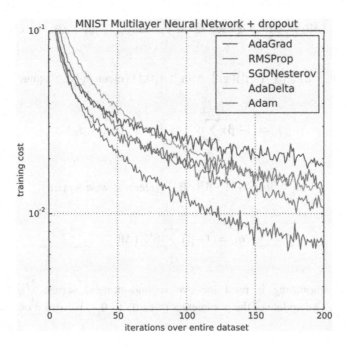

**FIGURE 21.6**   Kingma and Ba (2017): Convergence of adaptive learning-rate algorithms: AdaGrad, RMSProp, SGDNesterov, AdaDelta, Adam. (Source: Kingma and Ba (2017), with permission.)

where the running average of the squared gradients is given in Eq. (21.50) for efficient coding, and in Eq. (21.51) in fully expanded form as a series with exponential coefficients $\beta^{k-i}$, for $i = 1, \ldots, k$. Figure 21.6 shows the convergence of some adaptive learning-rate algorithms: AdaGrad, RMSProp, SGDNesterov, AdaDelta, Adam. RMSProp still depends on a global learning rate $\epsilon(k) = \epsilon_0$ constant, a tuning hyperparameter. Even though RMSProp was one of the go-to algorithms for machine learning, the pitfalls of RMSProp, along with other adaptive learning-rate algorithms, were revealed in Wilson, Roelofs, Stern, Srebro, and Recht (2018).

### 21.9.4   AdaDelta: Adaptive Delta (parameter increment)

The name "AdaDelta" comes from the adaptive parameter increment $\Delta\theta$ in Eq. (21.57). In parallel and independently, Zeiler (2012) proposed AdaDelta that not only fixed the problem of continuing decaying learning rate of AdaGrad, but also removed the need for a global learning rate $\epsilon_0$, which RMSProp still used. By accumulating the squares of the parameter increments, i.e., $(\Delta\theta_k)^2$, AdaDelta would fit in the unified framework if the symbol $m_k$ in Eq. (21.41) were interpreted as the accumulated 2nd moment $(\Delta\theta_k)^2$. Zeiler (2012) made the following observation on weaknesses of AdaGrad: *"Since the magnitudes of gradients are factored out in AdaGrad, this method can be sensitive to initial conditions of the parameters and the corresponding gradients. If the initial gradients are large, the learning rates will be low for the remainder of training. This can be combatted by increasing the global learning rate, making the AdaGrad method sensitive to the choice of learning rate. Also, due to the continual accumulation of squared gradients in the denominator, the learning rate will continue to decrease throughout training, eventually decreasing to zero and stopping training completely."* Zeiler (2012) then introduced AdaDelta, as an improvement over AdaGrad with two goals in mind: (1) to avoid the continuing decay of the learning rate, and (2) to avoid having to specify $\epsilon(k)$, called the "global learning rate," a constant. Instead of summing past squared gradients over a finite-size window, which is not efficient in coding, Zeiler (2012) suggested to use exponential smoothing for both the squared gradients $(\tilde{g}_k)^2$ and for the squared increments $(\Delta\theta_k)^2$, which is used in the update $\theta_{k+1} = \theta_k + \Delta\theta_k$, by choosing the following functions:

$$V_k = \beta V_{k-1} + (1-\beta)(\tilde{g}_k)^2 \text{ with } \beta \in [0,1) \text{ (element-wise square)} \tag{21.53}$$

$$V_k = (1-\beta)\sum_{i=1}^{k}\beta^{k-i}(\tilde{g}_i)^2 = \psi_k(\tilde{g}_1,...,g_k) \tag{21.54}$$

$$m_k = \beta m_{k-1} + (1-\beta)(\Delta\theta_k)^2 \text{ (element-wise square)} \tag{21.55}$$

$$m_k = (1-\beta)\sum_{i=1}^{k}\beta^{k-i}(\Delta\theta_i)^2 \tag{21.56}$$

Thus, exponential smoothing is used for two second-moment series: $\{(\tilde{g}_i)^2, i=1,2,...\}$ and $\{(\Delta\theta_i)^2, i=1,2,...\}$. The update of the parameters from $\theta_k$ to $\theta_{k+1}$ is carried out as follows:

$$\Delta\theta_k = -\frac{1}{\sqrt{\hat{V}_k}+\delta}\hat{m}_k, \text{ with } \hat{m}_k = \left[\sqrt{m_{k-1}}+\delta\right]\tilde{g}_k \text{ and } \theta_{k+1} = \theta_k + \Delta\theta_k \tag{21.57}$$

where $\epsilon(k)$ in Eq. (21.52) is fixed to one in Eq. (21.57), eliminating the hyperparameter $\epsilon(k)$. Another nice feature of AdaDelta is the consistency of units (physical dimensions), in the sense that the fraction factor of the gradient $\tilde{g}_k$ in Eq. (21.57) has the unit of step length (learning rate):

$$\left[\frac{\sqrt{m_{k-1}}+\delta}{\sqrt{\hat{V}_k}+\delta}\right] = \left[\frac{\Delta\theta}{\tilde{g}}\right] = [\epsilon] \tag{21.58}$$

where the enclosing square brackets denote units (physical dimensions), but that was not the case in Eq. (21.52) of RMSProp:

$$[\epsilon_k] = \left[\frac{\epsilon(k)}{\sqrt{\hat{V}_k}+\delta}\right] \neq [\epsilon] \tag{21.59}$$

Figure 21.6 shows the convergence of some adaptive learning-rate algorithms: AdaGrad, RMSProp, SGDNesterov, AdaDelta, Adam. Despite this progress, AdaDelta and RMSProp, along with other adaptive learning-rate algorithms, shared the same pitfalls as revealed in Wilson, Roelofs, Stern, Srebro, and Recht (2018).

## 21.9.5  ADAM: ADAPTIVE MOMENTS

Both 1st moment Eq. (21.60) and 2nd moment Eq. (21.62) are adaptive. To avoid possible large step sizes and non-convergence of RMSProp, Kingma and Ba (2017) selected the following functions:

$$m_k = \beta_1 m_{k-1} + (1-\beta_1)\tilde{g}_k, \text{ with } \beta_1 \in [0,1) \text{ and } m_0 = 0 \tag{21.60}$$

$$\hat{m}_k = \frac{1}{1-(\beta_1)^k} m_k \text{ (bias correction)} \tag{21.61}$$

$$V_k = \beta_2 V_{k-1} + (1-\beta_2)(\tilde{g}_k)^2 \text{ (element-wise square), with } \beta_2 \in [0,1) \text{ and } V_0 = 0 \tag{21.62}$$

$$\hat{V}_k = \frac{1}{1-(\beta_2)^k} V_k \text{ (bias correction)} \tag{21.63}$$

$$\epsilon_k = \frac{\epsilon(k)}{\sqrt{\hat{V}_k} + \delta} \text{ (element-wise operations)} \tag{21.64}$$

with the following recommended values of the parameters:

$$\beta_1 = 0.9, \beta_2 = 0.999, \epsilon_0 = 0.001, \delta = 10^{-8} \tag{21.65}$$

The recurrence relation for gradients in Eq. (21.60) leads to the following series:

$$m_k = (1-\beta_1) \sum_{i=1}^{k} \beta^{k-i} \tilde{g}_i \tag{21.66}$$

since $m_0 = 0$. Taking the expectation, as defined in Eq. (21.23), on both sides of Eq. (21.66) yields

$$\mathbb{E}[m_k] = (1-\beta_1) \sum_{i=1}^{k} \beta^{k-i} \mathbb{E}[\tilde{g}_i] = \mathbb{E}[\tilde{g}_i] \cdot (1-\beta_1) \sum_{i=1}^{k} \beta^{k-i} + D = \mathbb{E}[\tilde{g}_i] \cdot \left[1-(\beta_1)^2\right] + D \tag{21.67}$$

where D is the drift from the expected value, with $D = 0$ for stationary random processes. For non-stationary processes, Kingma and Ba (2017) suggested to keep D small by choosing small $\beta_1$ so only past gradients close to the present iteration $k$ would contribute, so to keep any change in the mean and standard deviation in subsequent iterations small. By dividing both sides by $\left[1-(\beta_1)^2\right]$, the bias-corrected 1st moment $\hat{m}_k$ shown in Eq. (21.61) is obtained, showing that the expected value of $\hat{m}_k$ is the same as the expected value of the gradient $\tilde{g}_i$ plus a small number, which could be zero for stationary processes:

$$\mathbb{E}[\hat{m}_k] = \mathbb{E}\left[\frac{m_k}{1-(\beta_1)^2}\right] = \mathbb{E}[\tilde{g}_i] + \frac{D}{1-(\beta_1)^2} \tag{21.68}$$

The argument to obtain the bias-corrected 2nd moment $\hat{V}_k$ in Eq. (21.63) is of course the same. Kingma and Ba (2017) pointed out the lack of bias correction in RMSProp leading to "*very large step sizes and often divergence*," and provided numerical experiment results to support their point. Figure 21.6 shows the convergence of some adaptive learning-rate algorithms: AdaGrad, RMSProp, SGDNesterov, AdaDelta, Adam. Their results show the superior performance of Adam compared to other adaptive learning-rate algorithms.

### 21.9.6 Criticism of Adaptive Methods, Resurgence of SGD

Yet, despite the claim that RMSProp is *"currently one of the go-to optimization methods being employed routinely by deep learning practitioners,"* and that *"currently, the most popular optimization algorithms actively in use include SGD, SGD with momentum, RMSProp, RMSProp with momentum, AdaDelta, and Adam,"* Wilson, Roelofs, Stern, Srebro, and Recht (2018) showed, through their numerical experiments, that adaptivity can overfit, and that standard SGD with step-size tuning performed better than adaptive learning-rate algorithms such as AdaGrad, RMSProp, and Adam. The total number of parameters $P_T$ in deep networks could easily exceed 25 times the number of output targets $m$, i.e., $P_T \geq 25\ m$, making it prone to overfit without employing special techniques, such as regularization or weight decay. Wilson, Roelofs, Stern, Srebro, and Recht (2018) observed that adaptive methods tended to have larger generalization (test) errors compared to SGD: *"We observe that the solutions found by adaptive methods generalize worse (often significantly worse) than SGD, even when these solutions have better training performance,"* (see Figure 21.2) and concluded in their paper: *"Despite the fact that our experimental evidence demonstrates that adaptive methods are not advantageous for machine learning, the Adam algorithm remains incredibly popular. We are not sure exactly as to why, but hope that our step-size tuning suggestions make it easier for practitioners to use standard stochastic gradient methods in their research."* The work of Wilson, Roelofs, Stern, Srebro, and Recht (2018) has encouraged researchers who were enthusiastic with adaptive methods to take a fresh look at SGD again to tease something more out of this classic method.

## 21.10 CONCLUSION

We provided an introduction to stochastic optimization methods for training neural networks. In particular, the focus was laid on diverse add-on tricks, which improve the convergence of classic SGD, and adaptive methods, which have gained great popularity recently. Understanding of both add-on tricks, adaptive methods as, e.g., Adam, and the hyperparameters involved is essential, not least to support transparency and to enable reproducibility in applications of neural networks, see, e.g., the discussion by Haibe-Kains et al. (2020) on cancer screening based on medical imaging. We provided a unified framework for adaptive methods, which includes Adam and its variants, which are widely adopted today.

## ACKNOWLEDGMENTS

A. Humer acknowledges the support by the COMET-K2 Center of the Linz Center of Mechatronics funded by the Austrian federal government and the state of Upper Austria.

## REFERENCES

Bergou, E., Diouane, Y., Kungurtsev, V., & Royer, C. W. (2018). A Subsampling Line-Search Method with Second-Order Results. https://arxiv.org/abs/1810.07211v2.
Bertsekas, D., & Tsitsiklis, J. (1995). *Neuro-Dynamic Programming*. Athena Scientific.
Bock, S., Goppold, J., & Weiß, M. (2018). An Improvement of the Convergence Proof of the ADAM-Optimizer. https://arxiv.org/abs/1804.10587.
Bottou, L., Curtis, F. E., & Nocedal, J. (2018). Optimization Methods for Large-Scale Machine Learning. *SIAM Review, 60*(2), pp. 223–311.
Duchi, J., Hazan, E., & Singer, Y. (2011). Adaptive Subgradient Methods for Online Learning and Stochastic Optimization. *Journal of Machine Learning Research, 12*, pp. 2121–2159.
Gill, P., Murray, W., & Wright, M. (1981). *Practical Optimization*. Academic Press.
Goodfellow, I., Bengio, Y., & Courville, A. (2016). *Deep Learning*. Cambridge, MA: The MIT Press.
Goudou, X., & Munier, J. (2009). The Gradient and Heavy Ball with Friction Dynamical Systems: The Quasiconvex Case. *Mathematical Programming, 116*(1–2), pp. 173–191.

Hagiwara, M. (1992). Theoretical derivation of momentum term in back-propagation. In IEEE (Ed.), *Proceedings of the International Joint Conference on Neural Networks (IJCNN'92)*, 1, pp. 682–686. Piscataway, NJ.

Haibe-Kains, B., Adam, G. A., Hosny, A., Khodakarami, F., Shraddha, T., Kusko, R., ...Dopaz. (2020). Transparency and Reproducibility in Artificial Intelligence. *Nature, 586*(7829), pp. E14–E16.

Hinton, G. (2012). A Practical Guide to Training Restricted Boltzmann Machines. In G. Montavon, G. Orr, & K. Muller (Eds.), *Neural Networks: Tricks of the Trade* (pp. 599–620). Springer.

Huang, H., Wang, C., & Dong, B. (2019). Nostalgic Adam: Weighting More of the Past Gradients When Designing the Adaptive Learning Rate. https://arxiv.org/abs/1805.07557v2.

Jacobs, R. (1988). Increased Rates of Convergence Through Learning Rate Adaptation. *Neural Networks, 1*(4), pp. 295–307.

Kingma, D. P., & Ba, J. (2017). Adam: A Method for Stochastic Optimization. https://arxiv.org/abs/1412.6980v9.

Loshchilov, I., & Hutter, F. (2019). Decoupled Weight Decay Regularization. https://arxiv.org/abs/1711.05101v3.

Luenberger, D., & Ye, Y. (2016). *Linear and Nonlinear Programming* (4th ed.). Springer.

Nesterov, I. (1983). A Method of the Solution of the Convex-Programming Problem with a Speed of Convergence O(1/k2). *Doklady Akademii Nauk SSSR, 269*3), pp. 543–547.

Nesterov, Y. (2018). *Lecture on Convex Optimization* (2nd ed.). Switzerland, Springer Nature.

Nocedal, J., & Wright, S. (2006). *Numerical Optimization* (2nd ed.). Springer.

Paquette, C., & Scheinberg, K. (2018). A Stochastic Line Search Method with Convergence Rate Analysis. https://arxiv.org/abs/1807.07994v1.

Plaut, D. C., Nowlan, S. J., & Hinton, G. E. (1986). Experiments on Learning by Back Propagation. Technical Report.

Polak, E. (1971). *Computational Methods in Optimization: A Unified Approach*. Academic Press.

Polak, E. (1997). *Optimization: Algorithms and Consistent Approximations*. Springer Verlag.

Polyak, B. (1964). Some Methods of Speeding Up the Convergence of Iteration Methods. *USSR Computational Mathematics and Mathematical Physics, 4*(5), pp. 1–17.

Priddy, K., & Keller, P. (2005). *Artificial Neural Network: An Introduction*. SPIE.

Reddi, S. J., Kale, S., & Kumar, S. (2019). On the Convergence of Adam and Beyond. https://arxiv.org/abs/1904.09237.

Robbins, H., & Monro, S. (1951a). A Stochastic Approximation Method. *Annals of Mathematical Statistics, 22*(3), pp. 400–407.

Robbins, H., & Monro, S. (1951b). Stochastic Approximation*Annals of Mathematical Statistics, 22*(2), p. 316.

Rögnvaldsson, T. S. (1998). A Simple Trick for Estimating the Weight Decay Parameter. In G. Orr, & K. Muller (Eds.), *Neural Networds: Tricks of the Trade* (pp. 68–89). Springer.

Rumelhart, D., Hinton, G., & Williams, R. (1986). Learning Representations by Back-Propagating Errors. *Nature, 323*(6088), pp. 533–536.

Snyman, J., & Wilke, D.2018. *Practical Mathematical Optimization: Basic Optimization Theory and Gradient-Based Algorithms*. Springer.

Sun, S., Cao, Z., Zhu, H., & Zhao, J. (2020). A Survey of Optimization Methods From a Machine Learning Perspective. *IEEE Transactions on Cybernetics, 50*(8), pp. 3668–3681.

Tieleman, T., & Hinton, G. (2012). Lecture 6e, RMSProp: Divide the Gradient by a Running Average of its Recent Magnitude. https://www.cs.toronto.edu/~tijmen/csc321/slides/lecture_slides_lec6.pdf.

Wilson, A. C., Roelofs, R., Stern, M., Srebro, N., & Recht, B. (2018). The Marginal Value of Adaptive Gradient Methods in Machine Learning. https://arxiv.org/abs/1705.08292v2.

Zeiler, M. D. (2012). ADADELTA: An Adaptive Learning Rate Method. https://arxiv.org/abs/1212.5701.

# 22 Period-Height Relationships for the Design, Assessment, and Reliability Analysis of Code-Compliant Reinforced Concrete Frames

*Luís Martins and Vítor Silva*
Global Earthquake Model Foundation, Pavia, Italy

*Humberto Varum*
Faculdade de Engenharia, Porto, Portugal

## CONTENTS

## 22.1  INTRODUCTION

Evaluating the period of vibration of a structure is a crucial step for the calculation of the seismic demand in the design process. While numerical modeling can be a suitable alternative, for estimating the period of vibration for single structures, when assessing a region containing a large number of buildings, this is simply not feasible. For these portfolio scale assessments, it is common to apply simplified relationships to predict the expected period of vibration (e.g., Bal *et al.* 2010a). The most common approach of these models is to use the height of the structure as the main structural parameter to estimate the dynamic properties of the structure. Despite not being a state-of-the-art method, the ease of use of the height-dependent empirical relationships contributed greatly to their popularity in the earthquake engineering community.

Such relationships have usually been developed from either monitoring the dynamic properties of real buildings (e.g., Guler *et al.* 2008, Masi and Vona 2009, Oliveira and Navarro 2010) or from modeling (e.g., Crowley and Pinho 2004, Kose 2009, Ricci *et al.* 2011, and Perrone *et al.* 2016). The former allows a more realistic estimation of the periods of vibration, however, it can be too structure specific and usually limited to solely the elastic period. Furthermore, the experimental measurement of the dynamic properties of a statistically sufficient sample of structures can be considerably

DOI: 10.1201/9781003194613-22

expensive and time-consuming. The latter approach enables considering a wider variety of structures, and the estimation of the period of vibration at various stages of damage (from elastic to yielding or even near collapse). Numerical modeling permits also the investigation of the impact of multiple variables. As an example, Verderame *et al.* (2010), Ricci *et al.* (2011), and Perrone *et al.* (2016) have developed estimates considering not only the height but also the influence of plan asymmetry, the percentage of openings in the infill panels or the Young's modulus of the masonry units.

The majority of the proposed models to estimate the fundamental period of vibration are similar to Eq. (22.1), where $H$ is the building's height and $C_t$ and $\alpha$ are constants. A special note to the work of Verderame *et al.* (2010), where the authors propose a modified version of the equation by adding a variable $S$ equal to the product of the two horizontal dimensions of the building. A summary of previously published mathematical models applicable for Europe is provided in Table 22.1.

$$T = C_t \cdot H^{\alpha} \tag{22.1}$$

Common shortcomings of existing period-height relationships are, for example, the reduced number of structures and/or the need to consider more stories. Additionally, structures designed and built according to different seismic regulations and to distinct levels of ground shaking are often analyzed together in order to attain a statistically significant sample. This study aims to evaluate in a consistent manner the variations that different design requirement can induce on the period estimates within the scope portfolio seismic risk assessment. Furthermore, structures designed according to the Eurocodes have yet been the target of limited research since the current version of Eurocode 8 (CEN 2010c) uses a mathematical formula developed in 1970s for Californian buildings. This study offers the opportunity to revise its applicability. The present study improves upon former ones by providing updated mathematical models for predicting the expected period of vibration of reinforced concrete structures

## TABLE 22.1
### Period-Height Relationships for European Structures

| Region | Elastic Period ($T_{el}$) | | Yielding Period ($T_y$) | | References |
| | Bare Frame | Infilled | Bare Frame | Infilled | |
|---|---|---|---|---|---|
| Spain | | $T = 0.051\ N$ | | | Kobayashi *et al.* (1996) |
| Spain | | $T = 0.049\ N$ | | | Navarro *et al.* (2002) |
| France | | $T = 0.015\ H$ | | | Dunand *et al.* (2002) |
| Southern Europe | | | $T = 0.1\ H^{a}$ | | Crowley and Pinho (2004) |
| Southern Europe | | | | $T = 0.055\ H$ | Crowley and Pinho (2006) |
| Southern Europe | | | $T = 0.07\ H^{b}$ | | Crowley *et al.* (2008) |
| Turkey | | $T = 0.026\ H^{0.9}$ | | | Guler *et al.* (2008) |
| Italy | | $T = 0.016\ H$ | | | Gallipoli *et al.* (2009) |
| Turkey | | | $T = 0.083\ H$ | | Bal *et al.* (2010b) |
| Portugal | $T = 0.022\ H$ | $T = 0.013\ H$ | | | Oliveira and Navarro (2010) |
| Italy | $T = 0.009\ H^{0.93}S^{0.39}$ | | | | Verderame *et al.* (2010) |
| | $T = 0.044\ H^{0.67}S^{0.14\ a}$ | | | | |
| Italy | $T = 0.039\ H^{0.70}S^{0.19}$ | | | | |
| | $T = 0.068\ H^{0.65}S^{0.10\ b}$ | | | | |
| Italy | | $T = 0.016\ H$ | | $T = 0.041\ H$ | Ricci *et al.* (2011) |
| Italy | $T = 0.055\ H$ | | | | Perrone *et al.* (2016) |

*Note:*  N – Number of stories; H – Height; S – Is product of the two principal plan dimensions of the building $Lx$ and $Ly$.

[a]  Non-ductile structures.

[b]  Ductile structures.

designed according to the most recent European regulations, and considering different design ground motion (from low to moderate-high seismic hazard), with and without infill panels. For this study, the first mode elastic and inelastic periods of vibration of reinforced concrete moment frame buildings were estimated from a significant number of finite element models. The total number of floors of these structures ranged between three and ten, and the influence of infill panels has also been considered. The elastic period was computed from eigenvalue analysis while the inelastic period was estimated from pushover analysis. This study expands on previous endeavors by: (i) including the seismic design level rather than just the total height as an input parameter for estimating the period of vibration, (ii) considering both the elastic and inelastic (at the yielding point) periods of vibration, (iii) considering the influence of infill panels and the number of bays, (iv) propagation of the building-to-building variability in the period estimates, and (v) evaluation of the impact in the calculation of and derivation of fragility models and on the final risk metrics.

## 22.2   CASE STUDY STRUCTURES

The global building database developed by the PAGER group (Jaiswal *et al.* 2010) reports that, on average, reinforced concrete structures host over 50% of the population in Southern Europe. The 14th General Census of the Population and Dwellings in Italy indicated that reinforced concrete structures comprise nearly a quarter of the total number of buildings. Additionally, Crowley *et al.* (2008) assumed all of the reinforced concrete buildings in Italy be moment-resisting frames in the assessment of the national seismic risk. For the Portuguese case, Silva *et al.* (2014) indicated that reinforced concrete construction accounts approximately half of the Portuguese building stock while also and hosting the majority of the national population. The vast majority of these structures are composed by moment frames with masonry infills, and 51% have been designed following either to latest Portuguese design regulation or the Eurocodes. It is clear that reinforced concrete construction has been increasing significantly in the last decades in Southern Europe. For example, reinforced concrete was relatively rare in Italy, Portugal, and Greece in first quarter of the 20th century, but in the last decade, it accounted for nearly 70% of the new construction. The growing acceptance and enforcement of the Eurocodes suggest also an expected increase on the number of *Eurocode-compliant* structures.

The calculation of the elastic or inelastic period of vibration using analytical approaches is more often than not performed using 2D models or equivalent single-degree-of-freedom (SDOF) systems, and less frequently 3D structures. Considering the structures covered in this study (Crowley and Pinho 2004; Silva *et al.* 2014; Ulrich *et al.* 2014) used 2D moment-resisting frames with one to eight stories and one to seven bays (on average, 3.5 bays). Similarly, Silva *et al.* (2014) assessed the dynamic response of 2D frames with a number of stories ranging from one to eight, with three bays. Bal *et al.* (2010b) and Ulrich *et al.* (2014) instead modeled 3D structures with two to eight stories and two to four bays (on average, 3.1 bays). The present study considered groups of three, five, and ten stories moment-resisting frame reinforced concrete structures designed according to the most recent European standards (CEN 2009, 2010a, 2010b, 2010c) and for increasing levels of ground shaking. Most case study structures are three-bay moment-resisting frames. However, for the taller buildings, an additional bay was considered, as the footprint of taller buildings tends to be larger (e.g., Oliveira and Navarro 2010). The structures were designed to be both regular in height and plan, as recommended by modern design regulations. The chosen concrete class has a characteristic compressive strength of 25 MPa, while the characteristic yield strength of the rebar steel is 500 MPa. A permanent vertical load of 6.25 kNm$^{-2}$ has been considered in the design. An additional live load of 2.80 kNm$^{-2}$ has also been included. Following the recommendations in Eurocode 1-1 (CEN 2009), the absolute value of the live load has been lowered at the roof level to 0.40 kNm$^{-2}$. In addition to the vertical loads, all structures have been designed to withstand the horizontal loading due to the wind excitation, considering a wind velocity of 25 ms$^{-1}$ and a Class II terrain, according to the Eurocode 1-4 (CEN 2010a). Five levels of ground motion intensity were considered with design peak ground acceleration (PGA) ranging from 0.05 g (i.e., low seismic hazard) to 0.40 g (i.e., high seismic hazard).

To avoid excessive deformations under static loading, the minimum height for the beams was set to 1/12 of the respective span while the minimum cross section for the columns was 0.25 × 0.25 m². For the ten-story structures, a vertical element has been added to account for the increase in stiffness of a lift shaft. For each intensity level, 25 structures were generated and both the span length and story height have been randomly sampled from the probability distributions found in Silva *et al.* (2014). For the sake of simplicity, the same beam length was used for each bay within the same structure, which is a common assumption in other similar studies (e.g., Ulrich *et al.* 2014, Martins *et al.* 2018).

Brick infill panels were considered on the building's facade. These have been modeled with a diagonal struts system with the strength and stiffness properties computed according to the recommendations of Crisafulli (1997) and Smyrou *et al.* (2006, 2011).

The numerical models have been constructed using force-based beam elements with distributed plasticity using the structural analysis software OpenSees (McKenna *et al.* 2000).

## 22.3  COMPUTING ELASTIC PERIOD

Although not likely to find any structure responding with its uncracked stiffness, period estimates based on gross stiffness provide a suitable lower threshold for the *true* mean value of the expected fundamental period of vibration.

Figure 22.1 depicts the fundamental periods of vibration computed from the 3D finite element models for both the bare frame and infilled structures and as expected, the average period increases with the number of stories and decreases with the seismic design level.

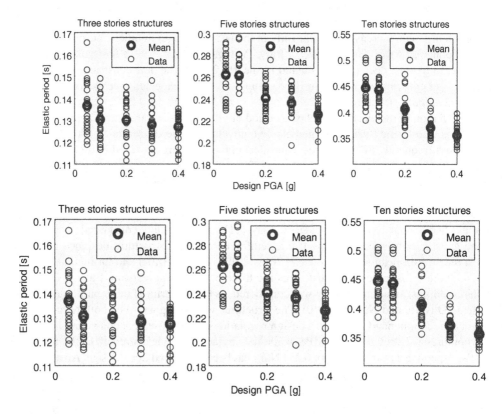

**FIGURE 22.1**  First mode elastic period *vs* design ground motion. Top: bare frame structures; bottom: infilled structures.

The results presented in Figure 22.1 illustrate a decrease of around 20% on the average period with increased ground motion level, due to the need for elements with a larger cross section (and consequently greater inertia and stiffness). These differences are obviously non-negligible and are expected to have a great influence in the development of fragility functions (e.g., Silva *et al.* 2013) or employment of displacement-based earthquake loss assessment methodologies (e.g., Bal *et al.* 2010a), further supporting the claim for including the seismic design level on the period estimates.

Moreover, comparing between the top and bottom plots indicates a decrease of more than 50% in the period of vibration. Silva *et al.* (2014) in a former study have also found a reduction in the elastic period due to the introduction of the infill panels in the same order of magnitude.

Fitting unconstrained power laws to the complete datasets led to the curves presented in Figure 22.2. The accuracy of the fits was quantified by the standard error of estimate ($S_{est}$) that provides an indication of the error introduced when predicting the period of vibration from the proposed models for any possible value for the building's height (Crowley and Pinho 2004). This parameter is calculated using Eq. (22.2), where $r_{Tpred,Tobs}$ represents the correlation between the observed periods ($T_{obs}$) and their predicted counterparts ($T_{pred}$), with a sample with size $n$. The

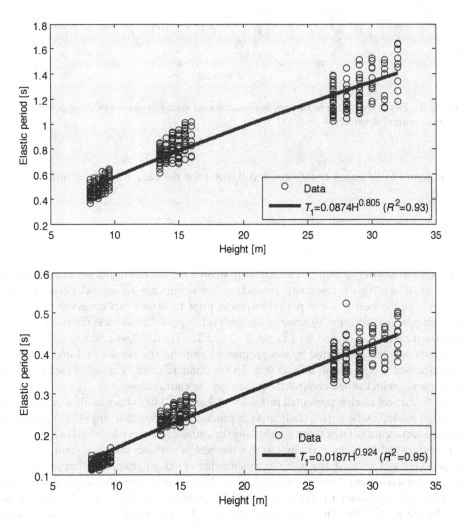

**FIGURE 22.2** Power law fitting to first mode elastic period. Top: bare frame structures; bottom: infilled structures.

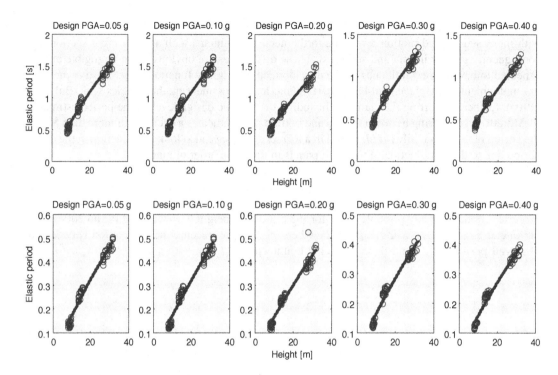

**FIGURE 22.3**  Power law fitting to first mode elastic period sorted by design PGA. Top: bare frame structures; bottom: infilled structures.

error was found to be equal to 0.094 s and 0.048 s for the bare frame and infilled structures, respectively.

$$S_{est} = S_{T_{pred}} \cdot \sqrt{1 - r^2{}_{T_{pred}, T_{obs}}} \cdot \sqrt{\frac{n}{n-2}}.$$

(22.2)

Although the models in Figure 22.2 yielded fairly small standard errors and remarkably good correlation factors ($R^2 > 0.90$), the dispersion around the best-fit curve is still considerable. Consequently, in order to have the best possible period estimate, prior to fit any mathematical model, the data should ideally be firstly sorted by seismic design level. Figure 22.3 depicts the results of fitting an unconstrained power law ($C_t$ and $\alpha$ in Tables 22.2 and 22.3) to the elastic periods of vibration given by the eigenvalue analysis sorted by design ground motion. The results in Tables 22.2 and 22.3 show, as expected, a decrease of at least 40% on the standard error of estimate for the new models when comparing with the one computed considering the entire dataset.

The mathematical models presented in Tables 22.2 and 22.3 have been calibrated to discrete values of design ground motion. This study aims to generalize the results to any design ground motion. Therefore, a surface model was fitted considering the entire dataset as illustrated in Figure 22.4. Due to the physics of the problem, it is obvious that the best-fit surfaces must not yield negative values for the period of vibration for any realistic combination of design ground motion and height within the region of interest, and therefore any best-fit surface must contain the origin. Consequently, if one assumes a two-parameter polynomial function to represent the best-fit surface, this constraint was met by setting the constant of the function to zero. The degree of the polynomial function was defined as the maximum degree without parameters with non-meaningful coefficients. The best-fit surface that complied with these conditions was found to be the polynomial function depicted in

**TABLE 22.2**

**Models Parameters (Elastic Period-Bare Frame Structures)**

| Design PGA [g] | $C_t$ | $\alpha$ | $S_{est}$ [s] |
|---|---|---|---|
| 0.05 | 0.0986 | 0.7972 | 0.0645 |
| 0.10 | 0.0989 | 0.7934 | 0.0671 |
| 0.20 | 0.0995 | 0.7641 | 0.0526 |
| 0.30 | 0.1015 | 0.7327 | 0.0518 |
| 0.40 | 0.0945 | 0.7349 | 0.0368 |

*Note:* $C_t$ and $\alpha$ – Model parameters; $S_{est}$ – Standard error of estimation.

**TABLE 22.3**

**Models Parameters (Elastic Period-Infilled Structures)**

| Design PGA [g] | $C_t$ | $\alpha$ | $S_{est}$ [s] |
|---|---|---|---|
| 0.05 | 0.0223 | 0.8896 | 0.0200 |
| 0.10 | 0.0214 | 0.8997 | 0.0220 |
| 0.20 | 0.0221 | 0.8696 | 0.0219 |
| 0.30 | 0.0259 | 0.7962 | 0.0164 |
| 0.40 | 0.0257 | 0.7838 | 0.0137 |

*Note:* $C_t$ and $\alpha$ – Model parameters; $S_{est}$ – Standard error of estimation.

Eqs. (22.3) and (22.4). For this case, the computed standard errors of estimate were 0.054 s and 0.018 s for the bare frame and infilled structures, respectively.

$$T_1(PGA, H) = -0.06 \cdot PGA + 0.07 \cdot H + 0.17 \cdot (PGA)^2 - 0.03 \cdot PGA \cdot H \qquad (22.3)$$

$$T_1(PGA, H) = -0.11 \cdot PGA + 0.02 \cdot H + 0.25 \cdot (PGA)^2 - 0.01 \cdot PGA \cdot H \qquad (22.4)$$

## 22.4   COMPUTING INELASTIC PERIOD

While period estimates, in the previous section, provide a suitable lower boundary for the true value of the period of vibration, for displacement-based assessment, the expected yield rather than the elastic period is often necessary (Crowley and Pinho 2004). Furthermore, in reinforced concrete buildings, cracking in critical elements (e.g., beams) is expected to occur under permanent loading alone. Even in the unlikely cases where cracking has not occurred, it is unreasonable to believe that a structure would respond with its gross stiffness under an earthquake, since cracking will occur at the very early stages of the dynamic response. Moreover, the derivation of fragility models (Silva *et al.* 2013) or the direct assessment of earthquake losses (Bal *et al.* 2010a) may require the calculation of the inelastic (at the yielding point) period of vibration.

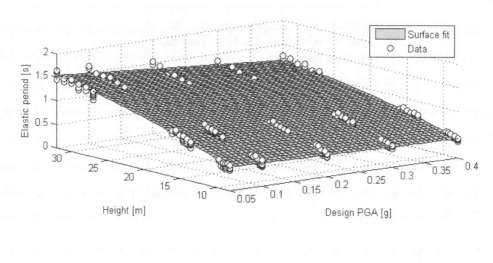

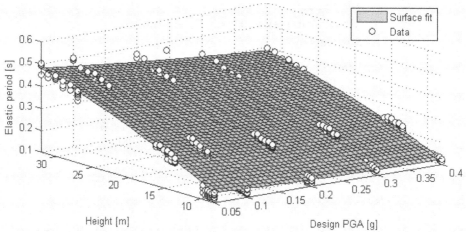

**FIGURE 22.4**   Surface fitting to first mode elastic period. Top: bare frame structures; bottom: infilled structures.

All the case study structures are regular in height and plan, and as a consequence, the influence of the higher modes may be considered negligible. Therefore, pushover algorithms have been considered an acceptable alternative to the more accurate, but also more computational demanding, nonlinear time-history analysis for estimating the inelastic period. In this study, the linearization of the capacity curve using the N2 method (Fajfar and Gašperšič 1996) has been used to compute the yielding displacement. This displacement and the associated deformed shape were then used to compute the inelastic period, as described by Eq. (22.5), where $M_{eff}$ is the effective mass of the equivalent system calculated through Eq. (22.6), $V_y$ is the base shear at the yielding point, and $d_{y,eff}$ is the correspondent displacement measured at the height of the center of the seismic forces ($h_{eff}$). Figure 22.5 depicts the inelastic periods computed from the capacity curves for bare frames and infilled structures.

$$T_y = 2\pi \sqrt{\frac{M_{eff}}{V_y / d_{y,eff}}} \qquad (22.5)$$

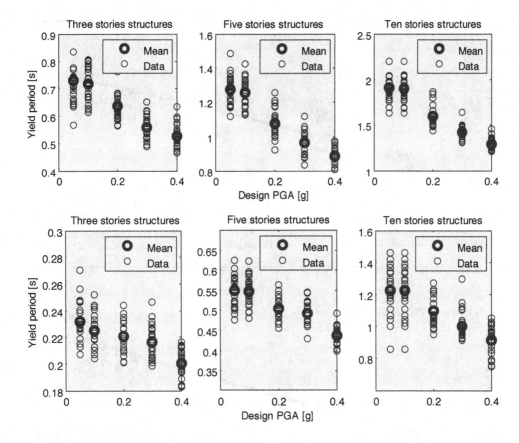

**FIGURE 22.5** Inelastic period *vs* design ground motion. Top: bare frame structures; bottom: infilled structures.

$$M_{eff} = \frac{\left( \sum\limits_{i=1}^{n} m_i \delta_i \right)^2}{\left( \sum\limits_{i=1}^{n} m_i \delta_i^2 \right)} \tag{22.6}$$

$$h_{eff} = \frac{\left( \sum\limits_{i=1}^{n} m_i \delta_i h_i \right)}{\left( \sum\limits_{i=1}^{n} m_i \delta_i \right)} \tag{22.7}$$

It should be noted that, in this context, $T_y$ could also be calculated following a mechanical approach, in which $d_{y,eff}$ is estimated following the formulae for the calculation of the displacement (or curvature) at the yielding point (e.g., Sullivan and Calvi 2012).

The results from Figure 22.6 indicate a decrease in the yield period of vibration with the design ground shaking in the order of 39% for bare frames and 24% for infilled frames. This decrease is also more pronounced for taller structures, as also observed in the assessment of the elastic period.

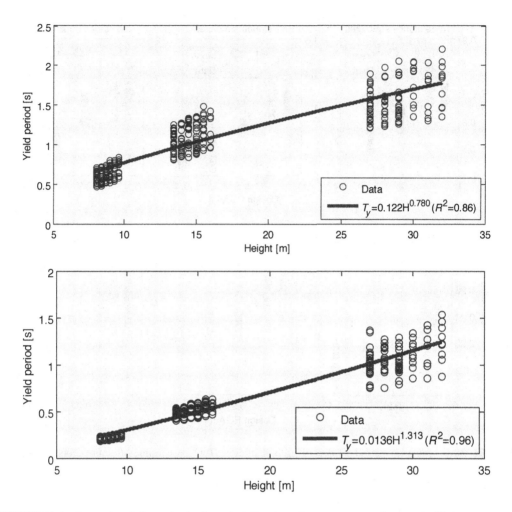

**FIGURE 22.6** Power law fitting to inelastic period. Top: bare frame structures; bottom: infilled structures.

When unconstrained power laws were fitted to the analytical results (see Figure 22.6), the standard error of estimate for the inelastic period were found to be equal to 0.201 s and 0.105 s for the bare and infilled frames, respectively, which is about double the error computed for the equivalent elastic period. This was to be expected since the nonlinear behavior of the structure inherently introduces additional variability on the response. However, it should be noted that the standard error of estimate found is still relatively low.

It has been decided to fit different mathematical laws according to the considered seismic design levels (see Figure 22.7). The standard error described in Tables 22.4 and 22.5 indicates a significant decrease in comparison with the one computed from the data prior to be sorted by seismic design level.

The same metric presented in the previous section was employed to estimate the best-fit surface was employed to model the inelastic period for the bare frame and infilled structures, as shown in Eqs. (22.8) and (22.9), and illustrated in Figure 22.8. For these surfaces, the computed standard error of estimate was 0.104 s and 0.077 s for bare frames and infilled structures, respectively. As previously discussed, increasing the design ground motion level naturally decreases the structure's period of vibration. It should be noted that this decrease is more apparent for the inelastic period

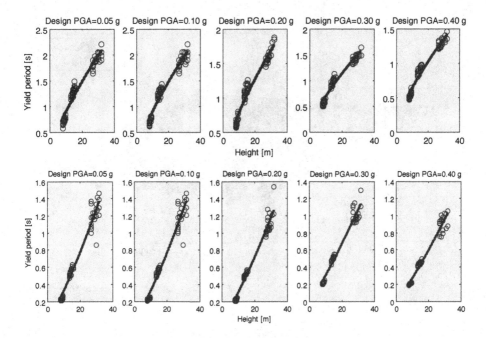

**FIGURE 22.7** Power law fitting to inelastic period sorted by design PGA. Top: bare frame structures; bottom: infilled structures.

**TABLE 22.4**

**Models Parameters (Inelastic Period-Bare Frame Structures)**

| Design PGA [g] | $C_t$ | $\alpha$ | $S_{est}$ [s] |
|---|---|---|---|
| 0.05 | 0.1647 | 0.7302 | 0.1032 |
| 0.10 | 0.1657 | 0.7267 | 0.0984 |
| 0.20 | 0.1529 | 0.7054 | 0.0771 |
| 0.30 | 0.1352 | 0.7068 | 0.0690 |
| 0.40 | 0.1320 | 0.6829 | 0.0571 |

*Note:* $C_t$ and $\alpha$ – Model parameters; $S_{est}$ – Standard error of estimation.

**TABLE 22.5**

**Models Parameters (Inelastic Period-Infilled Structures)**

| Design PGA [g] | $C_t$ | $\alpha$ | $S_{est}$ [s] |
|---|---|---|---|
| 0.05 | 0.0188 | 1.2367 | 0.0784 |
| 0.10 | 0.0179 | 1.2513 | 0.0798 |
| 0.20 | 0.0179 | 1.2296 | 0.0570 |
| 0.30 | 0.0217 | 1.1427 | 0.0565 |
| 0.40 | 0.0183 | 1.1632 | 0.0511 |

*Note:* $C_t$ and $\alpha$ – Model parameters; $S_{est}$ – Standard error of estimation.

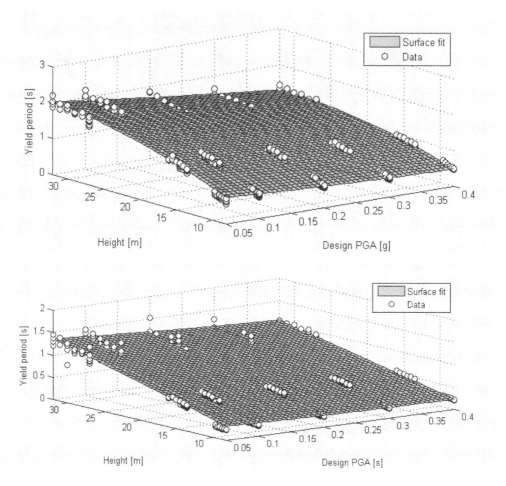

**FIGURE 22.8**   Surface fitting to yield period. Top: bare frame structures; bottom: infilled structures.

(see Figure 22.8) since structures planned for less demanding levels of ground shaking are expected to develop a mechanism at a lower load level.

$$T_y(PGA, H) = -0.74 \cdot PGA + 0.11 \cdot H + 0.98 \cdot (PGA)^2 - 0.05 \cdot PGA \cdot H \qquad (22.8)$$

$$T_y(PGA, H) = -0.16 \cdot PGA + 0.03 \cdot H + 0.30 \cdot (PGA)^2 - 0.03 \cdot PGA \cdot H \qquad (22.9)$$

## 22.5   COMPARISON WITH PREVIOUS MODELS

This section presents a comparison with the models derived herein and previous proposals. As previously mentioned, existing studies on this subject are often specific to a given region. The case study for this study comprises buildings designed according to European regulations, therefore, it was decided to limit this comparative study only to models intended for European construction. For the models listed in Table 22.1, where the number of floors rather than the height of the building was used as the input parameter an average story height of 3 m was considered.

Most of the previous models were developed from results coming from mostly structures without (or very limited) seismic design, therefore, in this section and for comparison purposes, only the

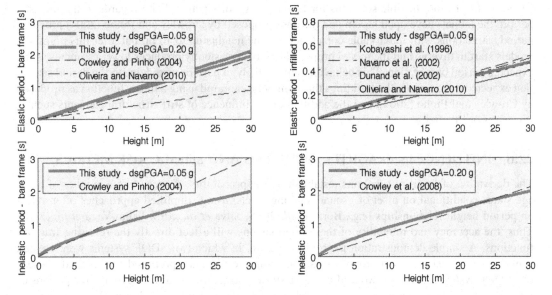

**FIGURE 22.9** Model comparison.

models developed for the lowest ground motion level have been considered. Oliveira and Navarro (2010) contain structures built in Portugal (mainly in the region of Lisbon) between 1940 and late 2000s with the majority being post-1980s buildings, therefore probably built according to the national design code introduced in 1986. For this reason, when comparing the period estimates proposed by this study with the models from Oliveira and Navarro, only functions developed for PGA equal to 0.20 g (i.e., the estimated ground shaking for the 475-year return period in Lisbon) have been used. Most of the models listed in Table 22.1 have been developed from data collected using monitoring systems installed on buildings under low-amplitude motion. It is therefore reasonable to assume that these models are most likely best suited for estimating the elastic period of vibration. Only Crowley and Pinho (2004) and Crowley *et al.* (2008) propose different models for the inelastic period of vibration. Figure 22.9 summarizes the main findings of this comparative study.

As previously mentioned, the differences in the period of vibration introduced by the seismic design level can on average reach about 20% (see Figures 22.1 and 22.5). This is consistent with the results shown on the top-left plot of Figure 22.9 where the differences between both curves become more discernible with increasing heights. Another comment on the results depicted in Figure 22.9 is that the period predictions obtained with the model from Oliveira and Navarro (2010) for bare frame structures produce consistently lower estimates than the other models. One of the reasons for this discrepancy could be the poor correlation between the periods of vibration and height indicated by Oliveira and Navarro (2010), due to the limited number of analyzed structures. When analyzing the results for the infilled frames, a stronger correlation between all the models for heights up to 10 m (i.e., around three stories) was found. For taller structures, the models start to deviate considerably. When considering the elastic period, the relationships proposed herein generally yielded slightly more flexible solutions than the remaining studies. This could result from the fact that some building collections considered in previous studies have also included buildings with a greater number of bays than the ones considered in this study (more commonly observed in taller buildings, e.g., Oliveira and Navarro 2010). However, as highlighted in previous section, the total number of bays considered herein was selected based on existing literature and is a commonly seen building typology. An additional factor to possibly have contributed for this observation was the fact that the buildings analyzed in this study have been designed according to modern design codes that generally favor more flexible solutions for better energy dissipation. With regards to the inelastic period, the models developed herein and the ones proposed by Crowley and Pinho (2004) provided period estimations with remarkable correlation for the heights up to 10 m, but after this point, the models start to diverge and the model proposed herein consistently produces lower estimates of the expected period of vibration. This behavior is most likely to have occurred due to the period saturation expected to occur in taller buildings that cannot be captured using a linear function as proposed by Crowley and Pinho (2004), and the addition of the influence of stiff structural elements such as staircases or lift shafts.

## 22.6   INFLUENCE IN FRAGILITY MODELING AND SEISMIC RISK METRICS

The derivation of fragility functions (establishing the probability of exceeding a number of damage states conditional on a set of ground shaking levels) using simplified approaches often relies on period-height relationships (e.g., Borzi *et al.* 2008, Silva *et al.* 2013, Villar-Vega *et al.* 2017). Thus, the accuracy and reliability of these relationships will affect directly the resulting fragility functions. A simple demonstration is provided herein, in which two SDOF systems were created considering the expected periods of vibration, using the formulae proposed in this study. These two SDOF systems represent two mid-rise (five stories) structures, and will have distinct periods of vibration due to the different design ground shaking (0.20 and 0.40 g – see Figure 22.6). In order to appraise the differences in the structural response only due to the distinct periods of vibration, the same yielding (1%) and ultimate (3%) global drifts were considered. However, it is acknowledged that distinct designs would certainly lead to different displacement capacities. The resulting capacity curves are presented in Figure 22.10. These two SDOF systems have been tested (using nonlinear time history analysis) against the European database of strong motion (Akkar *et al.* 2014) in order to derive a fragility function for the yielding and collapse damage states, as illustrated in Figure 22.11. Additional information about this methodology can be found in Martins and Silva (2020). Finally, these fragility functions were used to calculate the average annual probability of exceeding yielding (AAPY) or reaching collapse (AAPC), considering a hazard curve for the city of Istanbul (www.efehr.org), as described in Table 22.6.

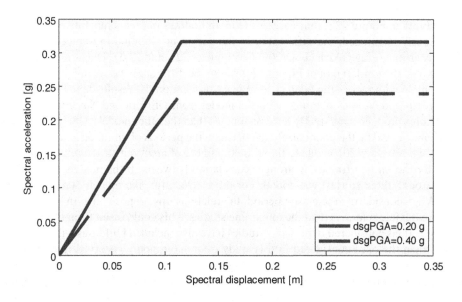

**FIGURE 22.10**   Example differences introduced in the capacity curves.

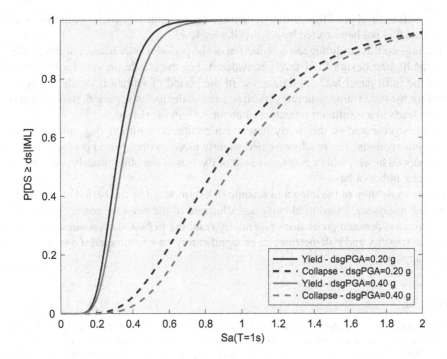

**FIGURE 22.11** Fragility curves for five stories building.

**TABLE 22.6**

**Annual Average Probability of Exceeding Yield and Reaching Collapse**

| Average Annual Probability of Yield | | Average Annual Probability of Collapse | |
|---|---|---|---|
| dsgPGA = 0.20 g | dsgPGA = 0.40 g | dsgPGA = 0.20 g | dsgPGA = 0.40 g |
| $2.36 \times 10^{-3}$ | $2.03 \times 10^{-3}$ | $4.37 \times 10^{-4}$ | $3.319 \times 10^{-4}$ |

These results shown in Figure 22.11 and Table 22.6 demonstrate that the variability in the fragility and risk metrics would have been underestimated should a single period-height relationship had been used for the derivation of the SDOF systems, as opposed to account for the effects due to the consideration of distinct design ground shaking.

## 22.7 FINAL REMARKS

Period of vibration versus height relationships has been used for several decades for the design of new structures or the assessment of large portfolios of buildings, where individualized numerical analyses are not practical. One of the major limitations of the existing height-period relationships is that often they have limited applicability outside of the region where they have been developed or calibrated, and no distinction is made regarding the design acceleration.

In this study, sets of reinforced concrete frames (bare and infilled) were created and their period of vibration was estimated using an analytical approach. These structures have been designed to be compliant with the most up-to-date European regulation and different levels of design ground motion. Tridimensional finite elemental models were created and the elastic and inelastic periods were evaluated by the means of eigenvalue and pushover analysis, respectively.

The results presented herein allowed developing functions that consider both the building's height and the design ground motion as input parameters. Considering these two parameters together has led to a reduction of about 50% in the standard error of estimate when comparing with models where the data had not been sorted by seismic design level.

The findings on this study suggest a reduction of the period of vibration of around 20% from the lowest to the highest design PGA levels considered. For the elastic period of vibration, the introduction of the infill panel leads to a decrease in the period of vibration for about one-third of its equivalent for the bare frame structures. With regards to the inelastic period, the introduction of the infill panels leads to a minimum reduction of about 30% in the period.

The models presented in this study have been compared with existing studies developed for European constructions. The results revealed slightly more flexible period predictions given by the models developed herein, which may be a result of the fact of the other models contained structures with a higher number of bays.

Finally, an evaluation of the impact in seismic risk introduced by the period of vibration has been included. The analysis allowed to identify and illustrate (i) the need for correct period estimations to avoid erroneous demand predictions potentially resulting in biased assessment, and (ii) the variability on the fragility and risk metrics can be significantly underestimated if a single period-height equation is applied.

## REFERENCES

Akkar, S., Sandıkkaya, M. A., Şenyurt, M., Azari Sisi, A., Ay, B. Ö., Traversa, P., Douglas, J., Cotton, F., Luzi, L., Hernandez, B. and Godey, S. (2014) Reference database for seismic ground-motion in Europe (RESORCE). Bulletin of Earthquake Engineering, 12(1): pp. 311–339. DOI: 10.1007/s10518-013-9506-8.

Bal, I. E., Bommer, J.J., Stafford, P. J., Crowley, H. and Pinho, R. (2010a) The influence of geographical resolution of urban exposure data in an earthquake loss model for Istanbul. Earthquake Spectra, 26(3): pp. 619–634. DOI: 10.1193/1.3459127.

Bal, İ. E., Crowley, H. and Pinho, R. (2010b) Displacement-Based Earthquake Loss Assessment: Method Development and Application to Turkish Building stock, Research Report No. ROSE-2010-02, Instituto Universitario di Studi Superiori di Pavia.

Borzi, B., Pinho, R. and Crowley, H. (2008) Simplified pushover-based vulnerability analysis for large-scale assessment of RC buildings. Engineering Structures, 30(3): pp. 804–820. DOI: 10.1016/j.engstruct. 2007.05.021.

CEN (2009) Eurocode 1 – Action on structures Part1-1: General actions densities, self-weight, imposed loads for buildings, EN1991-1-1, European Committee for Standardization, Brussels.

CEN (2010a) Eurocode 1 – Action on structures Part1-4: General actions wind actions, EN1991-1-4, European Committee for Standardization, Brussels.

CEN (2010b) Eurocode 2 – Design of concrete structures Part1-1: General rules and rules for buildings, EN1992-1-1, European Committee for Standardization, Brussels.

CEN (2010c) Eurocode 8 – Design of structures for earthquake resistance Part1: General rules, seismic actions and rules for buildings, EN1998-1, European Committee for Standardization, Brussels.

Crisafulli, F.J. (1997) Seismic behaviour of reinforced concrete structures with masonry infills. University of Canterbury.

Crowley, H., Borzi, B., Pinho, R., Colombi, M. and Onida, M. (2008) Comparison of two mechanics-based methods for simplified structural analysis in vulnerability assessment. Advances in Civil Engineering, 2008: p. 19. DOI: 10.1155/2008/438379.

Crowley, H. and Pinho, R. (2004) Period-height relationship for existing European reinforced concrete buildings. Journal of Earthquake Engineering, 8(sup001): pp. 93–119. DOI: 10.1080/13632460409350522.

Crowley, H. and Pinho, R. (2006) Simplified equations for estimating the period of vibration of existing buildings. In Proceedings of the 1st European Conference on Earthquake Engineering and Seismology.

Dunand, F., Bard, P., Chatelain, J., Guéguen, P., Vassail, T. and Farsi, M. (2002) Damping and frequency from Randomdec method applied to in situ measurements of ambient vibrations. Evidence for effective soil structure interaction. In 12th European Conference on Earthquake Engineering, London.

Fajfar, P. and Gašperšič, P. (1996) The N2 method for the seismic damage analysis of RC buildings. Earthquake Engineering and Structural Dynamics, 25(1): pp. 31–46.

Gallipoli, M. R., Mucciarelli, M. and Vona, M. (2009) Empirical estimate of fundamental frequencies and damping for Italian buildings. Earthquake Engineering & Structural Dynamics, 38(8): pp. 973–988.

Guler, K., Yuksel, E. and Kocak, A. (2008) Estimation of the fundamental vibration period of existing RC buildings in Turkey utilizing ambient vibration records. Journal of Earthquake Engineering, 12(sup2): pp. 140–150. DOI: 10.1080/13632460802013909.

Jaiswal, K., Wald, D. and Porter, K. (2010) A global building inventory for earthquake loss estimation and risk management. Earthquake Spectra, 26(3): pp. 731–748. DOI: 10.1193/1.3450316.

Kobayashi, H., Vidal, F., Feriche, D., Samano, T. and Alguacil, G. (1996) Evaluation of dynamic behaviour of building structures with microtremors for seismic microzonation mapping. In 11th WCEE. Acapulco, México.

Kose, M. M. (2009) Parameters affecting the fundamental period of RC buildings with infill walls. Engineering Structures, 31(1): pp. 93–102. DOI: 10.1016/j.engstruct.2008.07.017.

Martins, L. and Silva, V. (2020) Development of a fragility and vulnerability model for global seismic risk analyses. Bulletin of Earthquake Engineering. DOI: 10.1007/s10518-020-00885-1.

Martins, L., Silva, V., Bazzurro, P. and Marques, M. (2018) Advances in the derivation of fragility functions for the development of risk-targeted hazard maps. Engineering Structures, 173: pp. 669–680. DOI: 10.1016/j.engstruct.2018.07.028.

Masi, A. and Vona, M. (2009) Estimation of the period of vibration of existing RC building types based on experimental data and numerical results. In Increasing Seismic Safety by Combining Engineering Technologies and Seismological Data. Springer, Dordrecht, Netherlands.

McKenna, F., Fenves, G., Scott, M. and Jeremic, B. (2000) Open system for earthquake engineering simulation (OpenSees). Pacific Earthquake Engineering Research Center. University of California, Berkeley, CA.

Navarro, M., Sánchez, F., Feriche, M., Vidal, F., Enomoto, T., Iwatate, T., Matsuda, I. and Maeda, T. (2002) Statistical estimation for dynamic characteristics of existing buildings in Granada, Spain, using microtremors. In Eurodyn. Munich, Germany.

Oliveira, C. S. and Navarro, M. (2010) Fundamental periods of vibration of RC buildings in Portugal from in-situ experimental and numerical techniques. Bulletin of Earthquake Engineering, 8(3): pp. 609–642. DOI: 10.1007/s10518-009-9162-1.

Perrone, D., Leone, M. and Aiello, M. A. (2016) Evaluation of the infill influence on the elastic period of existing RC frames. Engineering Structures, 123: pp. 419–433. DOI: 10.1016/j.engstruct.2016.05.050.

Ricci, P., Verderame, G.M. and Manfredi, G. (2011) Analytical investigation of elastic period of infilled RC MRF buildings. Engineering Structures, 33(2): pp. 308–319. DOI: 10.1016/j.engstruct.2010.10.009.

Silva, V., Crowley, H., Pinho, R. and Varum, H. (2013) Extending displacement-based earthquake loss assessment (DBELA) for the computation of fragility curves. Engineering Structures, 56: pp. 343–356. DOI: 10.1016/j.engstruct.2013.04.023.

Silva, V., Crowley, H., Varum, H., Pinho, R. and Sousa, L. (2014) Investigation of the characteristics of Portuguese regular moment-frame RC buildings and development of a vulnerability model. Bulletin of Earthquake Engineering: pp. 1–36. DOI: 10.1007/s10518-014-9669-y.

Smyrou, E. (2006) Implementation and verification of a masonry panel model for nonlinear dynamic analysis of infilled RC frames. Universitá degli Studi di Pavia.

Smyrou, E., Blandon, C., Antoniou, S., Pinho, R. and Crisafulli, F. (2011) Implementation and verification of a masonry panel model for nonlinear dynamic analysis of infilled RC frames. Bulletin of Earthquake Engineering, 9(5): pp. 1519–1534. DOI: 10.1007/s10518-011-9262-6.

Sullivan, T. J. and Calvi, G. M. (2012) A model code for the displacement-based seismic design of structures DBD12. Instituto Universitario di Studi Superiori di Pavia

Ulrich, T., Negulescu, C. and Douglas, J. (2014) Fragility curves for risk-targeted seismic design maps. Bulletin of Earthquake Engineering, 12(4): pp. 1479–1491. DOI: 10.1007/s10518-013-9572-y.

Verderame, G.M., Iervolino, I. and Manfredi, G. (2010) Elastic period of sub-standard reinforced concrete moment resisting frame buildings. Bulletin of Earthquake Engineering, 8(4): pp. 955–972. DOI: 10.1007/s10518-010-9176-8.

Villar-Vega, M., Silva, V., Crowley, H., Yepes, C., Tarque, N., Acevedo, A. B., Hube, M. A., Gustavo, C. D. and María, H. S. (2017) Development of a fragility model for the residential building stock in South America. Earthquake Spectra, 33(2): pp. 581–604. DOI: 10.1193/010716eqs005m.

# 23 Reliability of the Rehabilitation of the Monumental Buildings in Seismic Regions

*Moshe Danieli*
Ariel University, Ariel, Israel

## CONTENTS

## 23.1 INTRODUCTION

Many ancient buildings, while not as large as the Pantheon or the Hagia Sophia, may also be of historical, architectural, and engineering value. Many of these buildings also have utilitarian value. The evaluation of a building in these respects determines the significance of its preservation. Ancient buildings can serve as museums, libraries, or sites of religious heritage (e.g., mosques, temples, synagogues, etc.), which require additional maintenance costs and help to improve the condition of these buildings. Heritage buildings that were built decades or even centuries in the past often require rehabilitation, especially in seismic regions. Preserving architectural-historical monuments in their original state is one of the responsibilities of a civilized society. The basis of the preservation of architectural-historical monuments is scientific research. The main aim of any work on architectural monuments is to extend their useful lifetime as a highly valuable

DOI: 10.1201/9781003194613-23

structure. Their active involvement in modern life and reliability will extend the lifetime of such monuments.

## 23.2 PRINCIPLES AND METHODS TO ENSURE RELIABILITY OF BUILDINGS

### 23.2.1 Principles

When preservation is considered, buildings are typically rehabilitated. Rehabilitation is designed to ensure the building's reliability, which is the ability of a structure or structural member to fulfill its specified requirements during the working life for which it has been designed. Reliability encompasses the safety, serviceability, and durability of a structure (ISO 2394 2015). Rehabilitation of historic heritage buildings in seismic regions is based on three approaches (Podiapolski et al. 1988, Rabin 2000, Danieli et al. 2002). Although different, these groups may include similar techniques. *Conservation* is aimed at preventing damage by maintaining and improving loading capacity of structural members to meet a required standard. Thus, conservation is based on proper structural safety in each individual case, and one should distinguish between temporary and permanent conservation. *Restoration* is primary aimed at architectural aspects of the building, including the appearance of its exterior and interior, and the functional roles of internal spaces, courtyards, etc. Restoration is most important also for historical aspects of rehabilitation. *Maintenance* refers mostly to the utilization of buildings. It entails a broad range of inspections and repair works that are conducted periodically, based on a detailed program. Maintenance may also include the monitoring of a structural member or system. The major purpose of maintenance is to prevent malfunctioning of engineering systems, such as water supply, roofing, drainpipes, yard coatings, perimeter walks, fence structures, etc., and to prevent the loss of structural integrity and ensure durability. Historic heritage buildings have long lifetimes that may extend to 150 years or more. To ensure the reliability of such buildings, conservation, restoration or repair materials should not only have adequate stiffness and strength properties, but also meet durability, corrosion-proof, and fire-proof standards. From the perspective of reliability, all building structures are complicated systems (Kapur & Lamberson 1977). Each element of such a structure has a lifetime of its own. Therefore, a long-term reliability program for such a complex system requires timely inspection and repair.

### 23.2.2 Methods

At the end of the nineteenth century, new materials, including reinforced concrete, with good compression and tension resistance capacity, began to replace stone constructions. It must be noted that concrete retains its properties for extended periods of time. The dome of the Pantheon of Rome, for example, was constructed of concrete 2000 years ago (Krautheimer and Ćurčić 1986). The use of reinforced concrete provides wide possibilities for strengthening historic masonry buildings as reinforced concrete is well compatible with stone masonry. At the same time, in many cases, the use of reinforced concrete for strengthening constructions for conservation makes it possible to produce nearly invisible elements that do not distort the monument's appearance. World conferences on building structures have devoted special sessions to challenges of rehabilitating historic heritage buildings (e.g., 13th WCEE Vancouver 2004; 14th WCEE Beijing 2008; 15th WCEE Lisbon 2012 and world conferences PROHITECH 2009, Roma; PROHITECH 2017, Lisbon), and many research projects concerning conservation and restoration of masonry buildings are underway. For the rehabilitation of historic heritage buildings, both traditional and modern methods and principles can be used (e.g., Rabin 2000, Danieli 2015, and other). Several technologies are used for their reinforcement, including metal strengthening rings, straining beams, doubling structures, carbon fiber cords, reinforcement systems consisting of carbon fiber tapes and epoxy resins (carbon structural reinforcement systems), masonry injection using cement or polymer solutions, polymer grids, concrete spraying, reinforced concrete jackets, and reinforced concrete one-or two-sided thin coatings are traditionally applied. Nontraditional components and innovative methods and materials are also used (Paret et al. 2006). All the above methods of reinforcement have advantages and disadvantages. Successful application of these methods depends on several factors, including the importance of the historical monument, its engineering condition, safety demands, long-term reliability, available technology to address these issues at the specific site, and specific considerations of engineers, owners of monuments, and so on. Besides, ensuring structural safety is supported by a long-term program of monitoring of the structures before and after reinforcement works (Bednarz et al. 2014), studies and tests of the materials used for strengthening, large-scale experimental model research, computer simulations and finite element method (FEM) analysis (e.g., Danieli et al. 2004, 2014, and other), including appropriate consideration of the materials' non-linear properties, and an estimation of the structures' load-bearing capacity in accordance with limit analysis theory.

### 23.2.3 ANALYSIS OF EARTHQUAKE RESISTANCE

According to existing building design codes (e.g., Eurocode-8), the basic method for earthquake resistance analysis is modal response spectrum analysis. When determining the design factor of seismic action, a building's behavior factor $q$ should also be taken into account. In the existing coding system of the spectrum analysis of seismic resistance, existing historic heritage buildings do not comprise a separate class. According to Eurocode-8, a building's behavior factor q is used for design purposes to reduce the forces obtained from a linear analysis in order to account for the structure's non-linear response associated with the material, the structural system, and the design procedures. For example, for masonry construction q = 1.5–3.0; for concrete buildings q = 1.5–4.5; for non-structural elements q = 1–2. A building's behavior factor $q$ is a reduction factor of seismic action, which gains significance in certain cases. If we take into account: (a) the historical significance and long-term reliability requirement for heritage buildings; and (b) some inaccuracies in determining the strength of materials (mortar, masonry), the integrity of masonry (e.g., cracking), corrosion in the concrete structure fittings, etc., for existing masonry and concrete buildings, then behavior factor $q$ as an actual seismic action reduction factor is recommended to be q = 1, including all non-structural elements.

An expert method for evaluating seismic resistance of buildings (Danielashvili & Tchatchava 1999) should also be applied to evaluate the seismic resistance of heritage buildings. An expert method evaluation shows the most vulnerable elements in terms of seismic resistance, and may be useful in determining the correct strategy for developing the strengthening project and ensuring the building's reliability.

## 23.3 EXAMPLES OF BUILDING REHABILITATION AND RELIABILITY

This chapter presents examples of actual post-rehabilitation reliability of the following buildings:

- Temple in Nikorcminda and a synagogue in Oni, in Nord Georgia, built in the eleventh and nineteenth centuries, respectively, which were damaged by an earthquake on April 30, 1991;
- Museum and synagogue in Akhaltsikhe, in South Georgia, built in the eighteenth and nineteenth centuries, respectively;
- Sports Palace in Tbilisi, built in the twentieth century.

### 23.3.1 EVALUATION OF THE SEISMIC SITUATION OF THE BUILDINGS' LOCATION

On April 29, 1991, an earthquake occurred in Racha, northern Georgia. Its magnitude on the Richter scale, according to the accepted estimations, was M = 6.9–7 (Engineering...Ed. Gabrictidze 1996). This magnitude corresponds to the intensity of $I_0 = 9.5$ on the MSK-64 12-step scale. Several thousands of aftershocks were registered in the following four-month period, with magnitudes between M = 6.2 and 5.3, which were sometimes almost as powerful as the primary shock.

### 23.3.1.1 The Temple in Nikorcminda

The distance to the earthquake epicenter was about 45 km. The estimated intensity of the primary shock was $I_0 = 7$ on the seismic scale MSK-64 (measured ground acceleration was a = 0.5 ÷ 1.0 m/sec² at the foundation.)

### 23.3.1.2 The Synagogue in Oni

The distance to the earthquake epicenter was 25–30 km. Measured ground acceleration was a = 2 m/sec².

### 23.3.1.3 The Buildings in Akhaltsikhe

The expected Richter magnitude for the Akhaltsikhe area is M = 7. The museum in Akhaltsikhe and synagogue were damaged by a series of major earthquakes: (a) in Akhalkalaki, Georgia, in 1899, at the distance of 45 km, M = 6.3; (b) in Gori, Georgia, in 1920, at the distance of 85 km, M = 6.2; (c) in Tabackuri, Georgia in 1940, at the distance of 45 km, M = 6.2; (d) in Spitak, Armenia, in 1988, at a distance of 125 km, M = 6.9–7; and (e) in Racha, Georgia, on April 29, 1991, at a distance of about 125 km from the earthquake epicenter, M = 6.9–7.

### 23.3.1.4 The Building in Tbilisi

The expected Richter magnitude for the Tbilisi area is M = 6.0–6.2.

The Sports Palace in Tbilisi suffered a series of major earthquakes in the past: (a) in Dmanisi, Georgia, in 1978, M = 5.3, in epicenter $I_0 = 7–8$; at a distance of 65 km; (b) in Spitak, Armenia, in 1988, M = 6.9–7, at a distance of 170 km; (c) in Racha, Georgia, on April 29, 1991, M = 6.9–7, at a distance of 145 km; (d) In Tbilisi, 2002, M = 4.5, in epicenter $I_0 = 7$, at a distance of 12–15 km from the earthquake epicenter.

### 23.3.2 Nikorcminda Temple

The Nikorcminda temple is a masterpiece of national Georgian architecture. It was built in the eleventh century, between 1010 and 1014. In this period, great attention was paid to the appearance of building facades. In this regard, the appearance of Nikorcminda is a great achievement of Georgian monumental plastic arts. The monument has not undergone any major changes over the years, which is very rare. Nikorcminda was rehabilitated in the sixteenth century, and the dome was preserved unchanged. The temple has a central-dome structure with a drum and a brick dome (Figure 23.1.A). The overall height of the structure is approximately 26 m. The dome span is 6.4 m, its rise is 2.8 m,

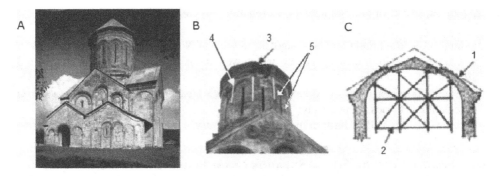

**FIGURE 23.1** The temple in Nikorcminda. A – General view; B, C – Damage to the drum and to the dome, and repair works: 1– dome; 2 – scaffolding; 3 – ring bundle of 10-mm diameter steel wires; 4 – vertical timber boards; 5 – external reinforcing hoops contain a number of 10-mm diameter wires.

and its thickness is 0.6–0.8 m. The drum has three-layer walls with an overall thickness 1.0–1.5 m, although weakened by windows. The primary shock of the earthquake caused cracking of the drum, and actually divided it into a group of vertically separated segments (Figure 23.1.B). In the dome, this generated a series of radial cracks and a closed-loop horizontal crack at the shell top. The central disk of the shell thus moved downwards by 60–100 mm but was wedged by surrounding parts of the dome, and so did not collapse (Figure 23.1.C) (Engineering…, Ed. Gabrictidze 1996, Danieli et al. 2017). Temporary conservation of the temple in Nikorcminda was carried out immediately after the main shock of the earthquake to prevent its collapse by possible aftershocks. The following measures were taken: (a) external reinforcement of the dome ring beam at the top of the drum. This included a ring, designed to withstand a tensile force of 150 kN. Horizontal force in the ring beam due to the dome weight was determined according (Timoshenko and Woinovsky 1959). A safety coefficient of 2.0 was applied to calculate the reinforcement; (b) external reinforcing hoops, mounted atop vertical timber boards. One hoop was designed to withstand a tensile force of 200 kN (Figure 23.1.B). (c) internal scaffolding, designed to support damaged areas of the structure, including the drum and the dome (Figure 23.1.C). Detailed description of the structure of the building, and damage to the building after the earthquake, and permanent conservation project of the building). Temporary conservation ensured the aftershock reliability of the temple, and the temple suffered no further damage. The temporary conservation project was planned by structural engineers Dr G. Gabrichidze and Dr M. Danielashvili (Danieli). The Georgian State Authority for Preservation of Monuments executed the temporary conservation project of the temple in Nikorcminda led by Dr M. Danielashvili (Danieli) and architect L. Medzmariashvili.

The permanent rehabilitation was completed 2 years later (Figure 23.1.A). The original shape of dome was strengthened by an external reinforced concrete shell (Danieli et al. 2002). Therefore, the internal appearance of the dome was not changed, which allowed further restoration of ancient paintings. After rehabilitation, the Nikorcminda temple is in a reliable state. The temple was included in the list of United Nations Educational, Scientific and Cultural Organization (UNESCO) World Heritage Sites in October 24, 2007.

### 23.3.3 SYNAGOGUE IN ONI

The building of the synagogue was completed in 1895. A student of a religious college (Yeshiva) in Poland brought the plans for the synagogue built in Oni. A rectangular symmetrical structure (18.5 × 14.9 m) was built of local stone on cement-lime mortar. Its maximum height is 15 m. The dome, with a span of 6.7 m and a rise of 3.0 m, is situated in the center of the structure, atop a drum supported by arches. There is a stele on the parapet of the frontal façade, and sculptural forms are placed in all four corners of the building. The portico of the building is made of stone columns covered by stone vaults. The earthquake caused substantial damage to the building, which did not collapse due to its symmetrical structure, rigid walls, steel ties, and small arch spans. Overall views of the building after earthquake are presented in Figure 23.2.A. The architectural elements were severely damaged, and fragments of a stele collapsed (Figures 23.2.A and B). Cracks developed in all bearing elements, such as arches, shell, external walls, and in the drum (Figure 23.2.C).

A method of expert evaluation of structural seismic resistance was applied to this building (Danieli and Bloch 2006). The assessment indicated the most vulnerable elements, such as the stone sculptures, and a stele made of the local stone. Rehabilitation of the building was completed 2 years later and was primarily aimed at strengthening the bearing walls, arches, corner vaults, portico, stele, sculptures, drum, etc. Details of reinforcement are shown in Figures 23.2 and 23.3. Reinforcement increased the load capacity of the above structural elements, and seismic resistance became higher than before the earthquake. This ensures the reliability of buildings. For a detailed description of the structure the building, and the damage caused to the building by the earthquake, and the conservation project of the building, see Danieli (1996, 2015).

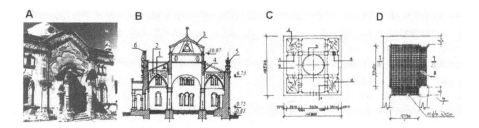

**FIGURE 23.2** The synagogue in Oni: A – view after the earthquake; B – cross-section: 1 – arch; 2 – ceiling; 3 – dome; 4 – drum; 5 – collapsed sculpture; 6 – collapsed façade element; 7 – collapsed portico; C – plafond plan with the cracks shown (dimensions in mm): 1 – arch; 2 – surmounted vaults ceiling; 3 – dome; 4 – cracks; 5 – wooden ceiling; D – details of arches and vaults reinforcement: 8 – steel mesh.

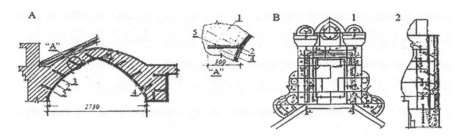

**FIGURE 23.3** Details of reinforcement (dimensions in mm): A, C – details of arches and vaults reinforcement: 1 – arch; 2 – reinforced plastering layer; 3 – steel mesh; 4 – drilling; 5– steel anchor. B – strengthening of the frontal stele: 1 – frontal façade; 2 – lateral façade.

To investigate the behavior of the building during an earthquake, a detailed modal analysis of seismic resistance was performed using FEM (Danieli & Aronchik 2014). Some results of structural analysis of the building under seismic forces are presented in Figure 23.4. In Figure 23.4, we can see that the analyses reflect the behavior of the building during an earthquake. For a detailed description the behavior of the building during an earthquake, see Danieli and Aronchik (2014). The results of the analysis correspond to the building's behavior during the earthquake, and support the recommendation to perform modal analysis of seismic resistance using FEM.

After rehabilitation of the synagogue building (Figure 23.5), the building is in a reliable state. The Georgian State Authority for Preservation of Monuments was responsible for the rehabilitation of the synagogue complex, and the authors of the rehabilitation project are the late architect Prof. Sh. Bostanashvili and structural engineer Dr. M. Danieli (Danielashvili). At present, this building has the status of an architectural and historical monument of Georgia.

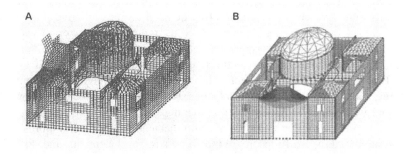

**FIGURE 23.4** A – Second mode shape in Y direction; B – bending moments in Y direction. *Y direction is perpendicular to the frontal façade.

**FIGURE 23.5**   The synagogue in Oni after strengthening and rehabilitation. A – general view; B – interior.

### 23.3.4   THE MUSEUM IN AKHALTSIKHE

The monumental building of the Museum (the former mosque known as Pasha Ahmed Dgakeli) is located in the town of Akhaltsikhe and was constructed in 1752 (Figure 23.6). In 1829–1830, it was rebuilt to serve as a Christian church. Today, the building accommodates a museum and forms part of a tourist site. The inner diameter of the supporting contour of the dome is about 16 m, inner height (rise) is about 8 m, and the thickness of the walls is 0.6–0.8 m. The dome was constructed from thin clay bricks on lime–clay mortar (Figure 23.6.B). The dome has typical cracks originating from the supporting zone in the meridian direction, and opening crack width on the lower surface of 1.5–2.0 cm (Figures 23.7.A and B). Dome conservation and reinforcement is clearly required to ensure reliability of the building, taking into account the history and architecture of this important building, the existence of developed cracks, and the need to save it from seismic loads during future severe earthquakes in order to prevent significant damage. An original reinforced structure

**FIGURE 23.6**   Dome (former mosque) in Akhaltsikhe: A – image from 1985; B – historical image without roof; C– modern image (2012).

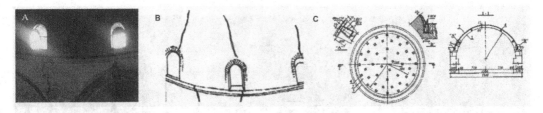

**FIGURE 23.7**   Dome in Akhaltsikhe. A, B – Inner surface cracks: A – picture; B – scheme; C – Reinforced construction (dimensions in cm): 1 – stone dome; 2 – reinforced concrete shells; 3 – supporting ring; 4 – connection elements.

was proposed (Danieli et al. 2002) (Figure 23.7.C). It is proposed to carry out the reinforcement of the existing dome from its outer surface to preserve the appearance of the dome's original interior surface (Figures 23.7.A and B). The reinforced structure consists of a thin-walled reinforced concrete shell, cast on top of the existing stone dome, and a supporting ring at its bottom (Figure 23.7.C). Thickness of 6–12 cm is recommended for the reinforced concrete shell in most stone dome spans. The necessary connection to provide interaction of the stone dome and the reinforced concrete shell is achieved by means of reinforced concrete connection elements (Figure 23.7.C). An additional linkage is the adherence force between the neighboring surfaces of the stone dome and reinforced concrete shell. The upper surface of the stone dome may be roughened to increase this force. In certain cases, this adherence force may become the principal mode of connection. Thus, the interconnected stone-reinforced concrete shell is achieved. The stresses in a stone dome may be decreased significantly as a result of strengthening by the proposed method; it could significantly raise the earthquake resistance of a stone dome; the tensile forces are perceived by the reinforced concrete ring and so the dome-supporting structures are relieved of effects from horizontal forces. For a detailed description of the structure, the damage and the conservation project of the building, see Danieli (2015). To study the stress–strain state characteristics of the strengthened structure, and to estimate the efficiency of the proposed strengthening method, a series of analyses were performed by FEM (Danieli et al. 2004). The results provide evidence of the efficiency of the proposed reinforced.

The museum building was rehabilitated in 2011–2012 (the authorities in charge of restoration were presumably unaware of the reinforcement program proposed in this paper). Existing roofing metal sheets of the dome (Figure 23.6.A) were replaced with gold-colored roofing metal sheets (see Figure 23.6.C). The reliability of the building was significantly increased as a result of these rehabilitation efforts.

### 23.3.5 Synagogue in Akhaltsikhe

The building of the synagogue was constructed in 1862–1863, 1.5–1.8 km from the former Pasha Ahmed Dgakeli mosque in Akhaltsikhe. The building is located on a hillside, in an old district of the town. There is a stone-walled yard adjacent to the façade and the side of the building, with a beautiful metal railing (Figure 23.8.A). The architecture of the building is non-eclectic, simple, and attractive. This is a perfect example of a classic architecture echoing old traditions that call for synagogue facades to comply with "strict simplicity." A beautiful wooden women's gallery is arranged inside.

#### 23.3.5.1 Structure (Construction) of the Buildings

This is a one-story building of a rectangular plan. The external dimensions of the building are $16 \times 19$ m in the plan, the height with the outer walls is 5–6 m, inside the building with the walls 5.1 m, and the height in the middle of the hall is 7.1 m. An elevated area in the central part of the building was created by elevating the bottom chord of a truss of the wood ceiling (Figure 23.8.C). The plan,

**FIGURE 23.8**  Synagogue in Akhaltsikhe: A – general view (1977); B – cracks in walls; C – interior.

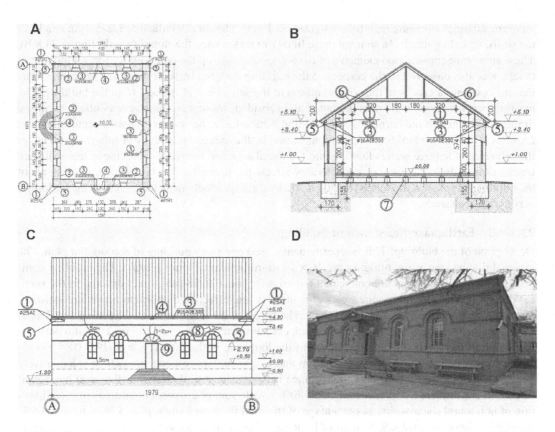

**FIGURE 23.9** Synagogue in Akhaltsikhe. Schema of strengthening: A – plan; B – cross-section; C– façade; 1– steel tie-bars; 2 – steel mesh; 3– support bar; 4 – threaded sleeve; 5 – thrust asher; 6 – lathing of floor; 7 – stone tub; 8 – typical cracks; 9 – lowered stones; D – general view after restoration (2012).

cross-section, and façade of the building are shown in Figures 23.9.A, 23.9.B, and 23.9.C, respectively. The building is a masonry construction, and its stone walls are built of limestone on lime–clay mortar, with a cornice. The thickness of the walls is 1.1 m. The interior walls are plastered with lime mortar, and the exterior is covered with 13 cm thick cut basalt stone. These stones are an integral part of the wall masonry. The wall masonry is relatively durable. Window and door openings have flat arches of a length of 1.3–1.45 m. Stone mortar masonry foundations are made at a depth of approximately 1.5 m. Foundation base is 1.70 m wide. The roof is made of wood constructions. Wooden trusses are 15.5 m long and 4.0 m high. The bottom chord has a 2 m elevation in the middle part at 10 m length. As a result, the ceiling has a polygonal form (Figure 23.8.C). There is a very impressive, 5 cm thick plank of the lathing under the lower chord. Roofing is made of galvanized roofing sheets. Drainpipes are placed in the corners. Parquet floors cover a strong, rigid timber floor. The timber floor is made of wood beams laid on stone stubs. The building has double timber doors and windows, all in satisfactory condition. Semicircular entrance staircases are made of hewn basalt stone and located on the main façade and at the side of the building. A hewn basalt stone perimeter side-walk is located along both sides of the building that have an entrance staircase. There is no perimeter walk along the remaining sides, including those facing the hillside. The existing roofing and parquet were constructed in the 1980s and are in satisfactory condition.

### 23.3.5.2 Damage to the Buildings

Considering deformations of the building of synagogue, and aiming to assess its engineering condition, the author this paper performed a survey of its building structures in 1995 and developed recommendations to ensure its reliability (Danieli 1995). The survey indicated 1.0–5.0 cm cracks in the walls, including lintels. In some of these lintels, cracks caused the stones to sink (Figure 23.8.B). These stones are construction elements of the arches that take up the compressive force. Opening of cracks was also observed in the corners of the building. Judging by their direction, we can assume that the opening of cracks in the corners attests to the separation of corners from the building. The existing perimeter side-walk was damaged and subsided. Moreover, damage was observed in the drainpipes located in the corners of the building. Cracking in the walls and lintels occurred due to an uneven yield of the building's formation, caused by the reduced efficiency of subgrade soil under the influence of surface waters flowing into the foundation soil for years. Cracking in the walls and lintels, especially in the corners, considerably affects the strength of certain structural elements of the building (especially, the lintels) and their spatial rigidity, which, in its turn, reduces the building's earthquake resistance.

### 23.3.5.3 Earthquake Resistance of Building

Description of the building: This is a permanent closed one-story building of rectangular plan. The external dimensions of the building are $16 \times 19$ m in the plan (Figure 23.9.A). The building comprises floors and walls that are connected in two orthogonal horizontal directions and in the vertical direction. The general requirements of continuity and effective diaphragm action are satisfied. Shear walls are in two orthogonal directions. The height of the walls between the ceiling and floor is $h = 5.6$ m; the thickness, including 13 cm thick cut basalt stone, $t = 1.1 > 0.35$ m. The ratio between the length of the small side and the length of the long side in the plan is $\lambda = 16/19 = 0.84 > 0.25$; Distance between shear walls is $L_{max} = 19$ m and $L_{min} = 16.0$ m, respectively (which is a relatively large value). However, the ratio of the distance between shear walls in two orthogonal horizontal directions is $L_{min}/L_{max} = 16/20 \times 100\% = 80\% > 75\%$; sum of cross-sections areas in each direction of horizontal shear walls, as percentage of the total floor area story, $p_A = 5.7\% > p_{Amin} = 3.5\%$. Ratio $h/t = 5.6/1.1 = 5.09 < 9$. System of building — regularity, construction of building — unreinforced masonry. Assessment of earthquake resistance was based on the lateral force method of analysis (used in the rehabilitation project in 1995), and modal spectrum method of analysis (2020) (Eurocode 8). Calculations were made as follows: design ground acceleration $a = 2$ m/sec$^2$; type of ground – C; Importance factor – $I = 1.2$; behavior factor $q = 1.5$; fundamental vibration period – $T_1 = 0.191$ s; tensile strength of masonry – 0.12 Mpa: module of elasticity –1200 Mpa. Walls sections table, mode shape №1 and model for analysis are shown in Figures 23.10.A, 23.10.B, and 23.10.C, respectively. Results of the calculations: Maximum shearing stress in the walls of the building can be as high as 0.08–0.15 Mpa, and 0.3–0.5 Mpa in the pillars and lintels. Maximum horizontal displacement may reach 8–10 mm, with the displacement/height of the building ratio being 1/510. Shearing stress values are relatively high and can be a cause of extensive cracking, and sometimes even masonry breakdown in pillars and lintels. Moreover, cracking in the walls, especially in the corners and pillars, considerably reduces the strength of certain structural elements of the building (walls and pillars)

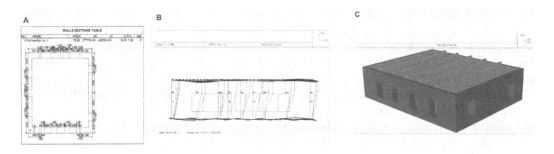

**FIGURE 23.10** Synagogue in Akhaltsikhe: A – walls sections table; B – mode shape №1; C – model for analysis.

and its spatial rigidity, which, in its turn, reduces the earthquake resistance of the building. Considering the above, rehabilitation should be performed with due regard to earthquake resistance requirements.

### 23.3.5.4   Rehabilitation of the Building

In 1975, the author provided his recommendations for rehabilitation and reliability of the synagogue, and a conservation project based thereon (the project was developed together with engineer V. Botasat). A survey of the building indicated that the existing deformation does not preclude current use of the building. However, the following measures should be taken to ensure its reliability: (a) wall reinforcement with closed mated steel tie-bars (Figures 23.9.A, B, and C) both from the inside and outside, at floor level. With this action, the main principle of building resistance in earthquakes is achieved, which is "tying of a building" (Richter 1958); (b) steel lathing reinforcement of the inner surface of corners, pillars and lintels followed by coating these surfaces with 3 cm thick M-10 cement mortar (strength – 10 Mpa) (Figure 23.9.A); (c) thorough cleaning and filling of wall and pillar cracks with M-10 cement mortar under pressure, preceded by the repair of deformed (lowered) stones; (d) re-laying of the existing perimeter side-walk; (e) laying of a new perimeter side-walk along the two remaining sides of the building, including a decorative basalt stone border; (f) repair and restoration of the drainpipes; (g) finishing, retaining the existing décor. This rehabilitation project to ensure reliability was partially performed in 1995–1996. In 2012, rehabilitation was completed, but, unfortunately did not include installation of steel tie-bars or reinforcement of the inner wall surface with steel lathing. The works that were completed provide relative reliability (and relatively safe use) of the Akhaltsikhe synagogue under a static effect, but fail to ensure adequate seismic resistance in the event of design earthquake. In 2011, this building was awarded the status of a cultural heritage monument of Georgia.

### 23.3.6   Sports Palace in Tbilisi

The Sports Palace in Tbilisi was built in 1957–1961 (architects - V. Aleksi-Meschishvili, Y. Kasradze, Th. Dgapharidze; design engineer - D. Kadgaia) (Figure 23.11.B). This is a square building with an outer perimeter of 352 m (88 × 4), inner perimeter of 304 m (76 × 4), and main arena of 44 × 27.4 m. The stands can accommodate 9305 spectators. The unique roof of the palace (included in the International Catalog of Individual Constructions) is made of dome structures embedded in a monolithic concrete octagonal supporting contour ring. The dome structures are made of precast reinforced concrete plates (span $2R = 76$ m, rise $f = 13.5$ m, effective thickness $h = 0.15$ m. Construction work was performed without suspending scaffolding (Figure 23.11.C). Precast elements were joined together using welding steel embedded parts end with concrete placed in the joints. In the course of work on the monolithic concrete octagonal contour, 2-5 mm cracks developed due to a high tension internal force. Furthermore, due to shell temperature deformation, the dome's soft roofing along the joints between the precast elements was ruptured. The penetration of atmospheric precipitation may cause

**FIGURE 23.11**   Sports Palace in Tbilisi: A, B – general view. A – after metal sheet roofing; B – soft-coated roofs; C – erection works in progress.

destruction of concrete in the joints and corrosion of metal embedded parts. Destruction of concrete in the joints is dangerous, since the main part of the shell is in a compressed state, and corrosion of embedded metal parts is dangerous, as these elements take up the tension internal force. Related to this, to protect the shells: (a) cracks in monolithic concrete octagonal contour were cleaned with a water jet and then filled under pressure with M-10 (Strength –10 Mpa) polymer-mortar solution; (b) the soft roof covering with a warmth-keeping jacket was removed (800 tons), which corresponds to 150 kg/m$^2$ (recommendations of Dr N. Achvlediani and Dr M. Danielashvili 1976). Thermal isolation was reinstalled using polystyrene foam panels (2.5 kg/m$^2$) and the soft roof covering was replaced with metal roofing sheets (7.0 kg/m$^2$) (Figure 23.11.A). By virtue of these actions, the covering of the Sports Palace was substantially unloaded and exposure to atmospheric precipitation was prevented. Breakdown of concrete and corrosion of the metal elements was also prevented. The repair significantly improved the building's reliability. It should be noted that the metal roof covering used in the Tbilisi Sport Palace was also applied to roofing of a $40 \times 40$ m$^2$ square in plan precast –a monolithic spherical reinforce-concrete shell in Sukhumi (Danieli 2017). It should be noted that cracking of the a monolithic concrete octagonal supporting contour ring shows that the use of prestressed reinforced concrete is a preferred option of a construction supporting ring for large-span concrete domes.

## 23.4 EXPERIENCE OF EARTHQUAKE RESISTANCE OF BUILDINGS

On September 8, 2009, an earthquake occurred in northern Georgia. The magnitude in the epicenter was 6.2. The epicenter was 12 km from Oni, which was the most significantly damaged location. The distance from the epicenter to Nikorcminda is about 40–45 km. The buildings of the synagogue in Oni and the temple in Nikorcminda survived this earthquake without any substantial damage, which proves the effectiveness of the strengthening and rehabilitation works, and, accordingly, the improved reliability of these buildings performed by us. On April 25, 2002, an earthquake occurred in Tbilisi. The epicenter was 12–15 km from Sport Palace. The magnitude in the epicenter was 4.5, intensity – $I_0 = 7$. The Sport Palace is a permanent closed-form reinforced concrete building, which was designed with due regard to the seismic norms and regulations that existed in 1955. The building suffered no apparent damage as a result of the 2002 earthquake or past seismic exposures.

## 23.5 CONCLUSIONS

1. In terms of reliability, a building system is a complicated structure; different system elements, whether load bearing or not, have different lifetimes. Any failure, damage, or breakdown of these elements can severely affect the long-term reliability of the system. Therefore, proper functioning of non-constructive elements such as drainpipes, roofing, and perimeter side-walks are also is crucial to ensure long-term reliability of the system. This can be achieved by regular inspections and timely repair of the system's elements.
2. In the existing coding system of the spectrum analysis of seismic resistance, existing historic heritage buildings are not considered a separate class. To analyze their seismic resistance, the recommended behavior factor q of stone and reinforced concrete buildings, including all non-structural elements, is q = 1.
3. As a result of the earthquake on April 29, 1991, the temple in Nikorcminda and the synagogue in Oni were damaged. The primary shock of the earthquake in the temple in Nikorcminda caused cracking of the drum, and actually divided it into a group of vertically separated segments. In the dome, this generated a series of radial cracks and a closed-loop horizontal crack at the shell top. Some structures of the synagogue in Oni collapsed. Temporary conservation of the temple in Nikorcminda was carried out immediately after the main shock of the earthquake to prevent its collapse by possible aftershocks and to ensure its temporary reliability. Temporary conservation, if performed in time after

earthquake, as with the Nikorcminda temple, can prevent the collapse of damaged structures and enable subsequent ensure the post-rehabilitation reliability. The rehabilitation of the synagogue in Oni was completed 2 years later. Detailed spectrum modal analysis of seismic resistance of the synagogue in Oni was performed using FEM. The results of such an analysis correspond rather closely to the building's behavior during the earthquake, which makes it possible to recommend the use of spectrum modal analysis using FEM.

4. An original reinforced stone dome structure is proposed. It consists of a new thin-walled reinforced concrete shell with a supporting ring, cast on top of the existing stone dome. This makes it possible to preserve the existing ancient appearance of the inner surface of the dome. The Akhaltsikhe stone dome can serve as an example of the efficiency of the proposed method as indicated by the results of the numerical FEM-based analysis. The design principles applied in the strengthening project of the dome in Akhaltsikhe can provide substantial contribution to successful conservation of other stone domes.

5. Reinforcement of the synagogue in Akhaltsikhe shows that long-term reliability is not guaranteed when taking into account expected ground acceleration.

6. To protect precast or precast-monolithic large-span shells from penetration of atmospheric precipitation, which may cause destruction of concrete in the joints and corrosion of metal embedded parts, roof covering should use materials that ensure protection from penetration of atmospheric water regardless of temperature action and shell deformation, such as metal roofing sheets.

7. Judging by the experience of the Sport Palace, the use of prestressed reinforced concrete is a preferred option for the construction of a supporting ring for large-span concrete domes.

8. After rehabilitation, the temple in Nikorcminda and the synagogue in Oni survived the earthquake in this region on September 8, 2009 (6.2 of the Richter scale), without any substantial damage. This proves that the principles and methods used to ensure reliability of the above heritage buildings may be recommended for the rehabilitation and long-term reliability of other heritage buildings.

# REFERENCES

Bednarz, Ł., Jasieńko, J., Rutkowski, M., Nowak, T. Strengthening and long-term monitoring of the structure of an historical church presbytery. *Engineering Structures*. vol. 81, pp. 62–75, 2014.

Danielashvili, M. A., Tchatchava, T. N. A method for quantitative estimation of the earthquake resistance of buildings. Earthquake Engineering, 1, pp. 14–16, 1999.

Danieli (Danielashvili), M. and Bloch, J., Evaluation of earthquake resistance and the strengthening of buildings damaged by earthquake. *Proceedings of the 1st European Conference on Earthquake Engineering and Seismology*, Geneva, Switzerland, Paper No: 673 (SD-R), 2006.

Danieli (Danielashvili), M., Gabrichidze, G., Goldman, A. & Sulaberidse, O., Experience in restoration and strengthening of stone made ancient domes in seismic regions. *Proc. 7th US NCEE*, Boston, MA, USA, Vol II: pp. 1167–1175, 2002.

Danieli (Danielsvilli), M., Aronchik, A., Bloch, J. Seismic safety of an ancient stone dome strengthened by an original method. *13th World Conference on Earthquake Engineering*. Vancouver, B.C., Canada, (SD-R) paper No 2789, 2004.

Danieli and A. Aronchik. Case study: The strengthening and seismic safety of the Oni synagogue in Georgia. In: *Proceedings of the 13th International Conference on Structures under Shock and Impact (SUSI XIII)*. WIT Press, 2014, pp. 456–466.

Danieli, M. Bloch, J. Gabrichidze G., Temporary conservation of a temple in Nikortzminda, Georgia, built in the 11th century. In: *Proceedings of the 3rd International Conference on Protection of Historical Constructions. PROHITECH 2017*, Lisbon (Lisboa), Portugal, 12–15 July, 2017.

Danieli, M. Construction of precast concrete shells in Georgia. *11th International Conference on Earthquake Resistant Engineering Structures*, July 5–7, 2017 ERES, Alicante, Spain July 5–7, 2017.

Danieli, M. Securing the safety of heritage buildings in active seismic regions. *International Journal of Safety and Security Engineering*, Vol. 5, No. 4 (2015) 304–321.

Danieli, M. Technical Opinion on the State of Synagogue Building in Akhaltsikhe and its Strengthening. Project. Tbilisi, 1995.

Danieli, M., Aronchik, A., Bloch, J. An Original Method for Strengthening Ancient Stone Domes in Seismic Regions and Solving Corresponding Problems of Stress-Strain State Analysis. *IJRET: International Journal of Research in Engineering and Technology*, Vol. 03, Issue 10, Oct-2014, pp.1–15, available http://www.ijret.org

Eurocode 8- Design Provisions for Earthquake Resistance of Structures, 1998;

ISO 2394: 2015. General principles on reliability for structures.

Kapur K. C., Lamberson L. R. *Reliability in Engineering Design.* John Wiley & Sons. 1977.

Krautheimer, R. & Ćurčić, S., *Early Christian and Byzantine architecture*, Penguin Books Ltd: Harmondsworth, Middlesex, England, 1986.

Krstevska, l., Tashkov, Lj., Gramatikov, K., Mazozolani, F. M., Landofo, R. Shaking table test of Mustafa Pasha Mosque model in reduced scale. PROHITECH 09. *Proc. of the international conference on protection of historical buildings*, Rome, Italy, June 21–24, 2009, vol. 2, ed. Federico M. Mazzolani: CRC Press, pp. 1633–1639, 2009.

Paret, T., Freeman, S., Searer, G., Hachem, M., Gilmarin, U., Seismic evaluation and strengthening of an historic synagogue using traditional and innovative methods and materials. *Proc. 1st European Conference on Earthquake Engineering and Seismology:* Geneva, Switzerland, (SD-R) paper No 701, 2006.

Podiapolski S. S., G. B. Bessonov et al., *Restoration of Architectural Monuments*, Stroyizdat, Moscow, 1988.

Rabin, I., *Structural Analysis of Historic Buildings: Restoration, Preservation and Adaptive Application for Architects and Engineers*, Wiley & Sons: New-York, 2000.

Richter Ch. F., Elementary Seismology, *California institute of technology*, W. H. Freeman and Company, San Francisco, 1958

Timoshenko S. P. and S. Woinovsky. *Kriger, Theory of Plates and Shells*, 2nd edition, McGraw-Hill, New York, Toronto, London, 1959.

# 24 Reliability-Based Analyses and Design of Pipelines' Underground Movements during Earthquakes

*Selcuk Toprak and Omer Yasar Cirmiktili*
Gebze Technical University, Kocaeli, Turkey

## CONTENTS

## 24.1 INTRODUCTION

Pipelines, which have been an important component for urban settlements in history, have become indispensable elements for today's societies. The old methods, such as carving waterways into stones and making pipes from terracotta to transfer water and waste water constitute the basis of today's infrastructure systems. These methods evolved into pipelines, an important part of lifeline systems, transferring many substances that are important for the society and individual life. Some of the modern uses are oil and natural gas transportation, drainage channels, culverts, water and wastewater distribution systems, telephone, internet and power lines, heat distribution, etc. Any interruption in the services of these pipelines can disrupt the life significantly and any damage, like leakage in oil pipelines, can have serious environmental consequences. Observed pipeline damages in the past earthquakes indicate that there is a strong correlation between pipeline damages and ground movements. By using reliability analyses, utility companies may obtain predicted damage levels from future earthquakes, perform risk management, determine risk levels for different hazards and consequently improve their systems, or design their systems accordingly at planning stages. This chapter presents and discusses the probabilistic approach and performance function-based reliability analysis methods namely first-order second-moment (FOSM) and Monte Carlo

DOI: 10.1201/9781003194613-24

(MC) simulation method applications on buried pipelines. Sensitivity analyses involving the key parameters and the respective uncertainties of these parameters required in the reliability analyses are discussed. The sensitivity analyses provide the classification of key seismic, pipe and ground variables quantitatively according to their importance and will be useful for reducing the risks of damage to pipelines due to the earthquakes. Case study from past earthquakes are presented to illustrate the application and limitations of reliability and sensitivity analyses to assess the effects of ground movements during earthquakes on buried pipelines.

## 24.2   PIPELINES AND THEIR CHARACTERISTICS IN RELIABILITY ANALYSES

It is very important and critical that pipelines continue to operate or get back to their operation as soon as possible even after they are exposed to significant events like earthquakes. The characteristics of pipelines which have primary effect on their seismic performance and reliability are presented in this section.

### 24.2.1   SUBCATEGORIES OF PIPELINE SYSTEMS

Pipelines in various distribution systems can be divided into different subcategories according to their function and properties. This type of categorization can be important because the size, material, and other properties of pipelines in different categories may differ significantly, influencing their response to earthquakes and consequently their reliability (e.g., O'Rourke and Toprak 1997; ALA 2005). The following is the general division in various systems: transmission, sub-transmission, distribution, and service pipelines. For example, Toprak (1998) assessed the Los Angeles water pipeline damage following the 1994 Northridge earthquake (Figure 24.1a) and Davis (2009) presents the scenario response and restoration of Los Angeles water system to a magnitude of 7.8 San Andreas fault earthquake by using similar categories. Generally, diameter size and internal pressure of the pipelines decrease from transmission to service pipelines. O'Rourke et al. (2014), Bouziou and O'Rourke (2017), and Toprak et al. (2019) analyzed the performance of water distribution pipelines in Christchurch, New Zealand during the February 22, 2011 earthquake by categorizing the pipelines. For the water supply, their study focuses on damage to watermains, which are pipelines with diameters typically between 75 and 600 mm, conveying the largest flows in the system (see Figure 24.1b), excluding repairs to smaller diameter submains and customer service laterals.

**FIGURE 24.1**   Pipeline categories in distribution systems (a) Los Angeles, United States (Toprak 1998) (b) Christchurch, New Zealand. (Toprak et al. 2019.)

## 24.2.2   PIPE MATERIAL

Observations in past earthquakes showed that the type of material pipelines are composed of affects their seismic behavior (e.g., Ballantyne et al. 1990; Eguchi 1991; Kitaura and Miyajima 1996; Toprak and Taskin 2007; Kurtulus 2011; O'Rourke et al. 2014; Toprak et al. 2019). Some characteristic information about the most common pipelines of different materials are presented herein. The most common pipe materials in the existing systems in the world are ductile iron (DI), asbestos cement (AC), cast iron (CI), steel, polyvinyl chloride (PVC), polyethylene (PE), and glass reinforced plastic (GRP) pipe. Various pipeline systems may consist of pipelines primarily made of certain materials. For example, gas pipelines in whole Turkey are either steel or PE.

AC pipes has been widely used in the past because it does not rust and is more economical than other pipes. However, they are no longer used in new lines due to its low bending strength compared to other existing pipes and the asbestos component in its structure threatening human health. Depending on the replacement program of the utility companies, significant portion of the distribution systems may still consist of AC pipelines. For example, at the time of February 22, 2011 earthquake, almost half of 1713 km water distribution pipelines of the city of Christchurch was AC pipelines. The Christchurch distribution system required repairs at 1502 locations. Sixty-nine percent of the distribution line repairs were in AC pipelines (O'Rourke et al. 2012). AC pipelines are one of the most vulnerable pipelines against earthquakes (e.g., Kuwata et al. 2018; Toprak et al. 2019).

CI pipes are generally the oldest pipelines in distribution systems. For example, at the time of 1994 Northridge earthquake, 76% and 11% of the 10.750 km distribution lines in the city of Los Angeles, United States were composed of CI and steel, respectively (Toprak et al. 1999). The Los Angeles Department of Water and Power (LADWP) distribution system required repairs at 1013 locations. Seventy-one percent of the distribution line repairs were in CI pipelines. They were preferred in the past because of their relatively strong structure against corrosion and abrasion. As CI pipes are very heavy and brittle and there are other pipe material alternatives they are used much less nowadays. DI pipes, which are similar to CI pipes in terms of appearance and some properties, are stronger and more durable against applied forces and impacts compared to cast iron pipes. DI pipes have replaced generally cast iron pipes today. Steel pipes, which were produced as rivets in the early days, gave way to welded steel pipes in time. In addition to being very strong, they are ductile and can change shape in the desired direction without breaking. In general, their resistance to tensile forces is stronger than pressure forces. Corrosion problem can be their weakness but coating can be applied to the inner and outer surfaces of the steel pipes to prevent corrosion. PVC pipe is one of the preferred pipe types due to its flexibility and lightness. It was used extensively in the past particularly to replace AC pipes (ALA 2005; Toprak et al. 2009). In some urban areas, PVC pipes comprise the majority of the water supply pipelines. Shih and Chang (2006) reported that PVC pipes made up 86% of water delivery pipelines in 11 townships and cities in the disastrous area affected by the 1999 Chi–Chi Taiwan earthquake.

PE pipes are generally used in conditions where flexibility is important. PE pipelines with high cracking and impact resistance and high flexibility provide great convenience in installation and performance during earthquakes. Lines formed with fusion welded joints are among the most preferred pipe types because of their durability, flexibility, and cost. They have been in use substantially in gas distribution systems and getting common in water systems as well. GRP pipe produced in a durable way by combining glass fiber and resin. It is very light compared to steel and concrete pipes, easy to transport and install, can be used in environments where PH conditions are unsuitable, highly resistant to corrosion and does not require internal and external coating. Glass fiber reinforcement plays an important role in making the pipe more resistant to internal and external pressures.

### 24.2.3 Pipeline Installation and Joints

Pipeline installments can be aboveground or underground. The depth of burial for underground pipelines depends on different calculations controlled by the safety, security, environmental factors, or aesthetics. In design calculations of pipelines before placement into the ground, parameters, such as the pipe geometry, the properties of the pipe material, the properties of the pipe welding and joints, the properties of the ground where the pipe will be laid, the pressure force of the soil on the pipe, other sources of pressures and loadings, the freezing level and environmental factors are taken into consideration. The design of the trench and requirements of the backfill material around the pipe are also critical in the pipeline installation and pipeline response when earthquakes occur (Mattiozzi and Strom 2008; Vitali and Mattiozzi 2014; Toprak et al. 2015).

In addition to laying the pipe, the way the pipes forming the pipeline are joined to each other is also an important issue for their performance during the earthquakes. Pipelines are classified as continuous or segmented according to the way they are joined. Continuous pipelines are joined to each other by welding from the inner and outer parts of the pipe wall. This welding process can be named as butt, single lap, or double lap welded (Indian Institute of Technology Kanpur 2007). It is known that the weld joints in steel or PE pipelines generally do not differ from the main body of the pipe in terms of strength. However, there are exceptions to that such as welded slip joints used in many high pressure water transmission and trunk pipelines. These joints are vulnerable to axial compressive loads generated by ground movements during earthquakes. A welded slip joint composed of a belled, or flared pipe end into which a straight cylindrical, or spigot, section of the pipe fits. The bell and spigot are joined by means of a circumferential weld. Because the slip joint introduces a local eccentricity and change in curvature, its axial load carrying capacity is decreased (Tawfik and O'Rourke 1985; Toprak 1998). Tawfik and O'Rourke (1985) analyzed the compressive failure mode associated with plastic flow in the curvilinear, belled ends of joints. The segmented pipelines can be formed via different type of joints such as bell mouth, mechanical coupling, etc. (Eidinger 1998; ALA 2005). For example, pipelines formed by interlacing (bellmouth) consist of cast iron, DI, and PVC for water transmission, and pipes made of concrete or PVC for sewage and waste water. Although these segmented pipelines designed and constructed to carry matters properly, the weakest points of the line against external and internal factors are usually these joints. To overcome this problem, many different types of seismic joints were used and have been under development as shown in Figure 24.2 (e.g., Toprak et al. 2015; Oda et al. 2019).

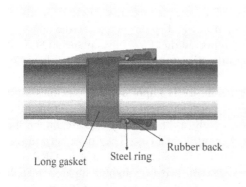

Samsun Makina Sanayi Inc. earthquake resistant type connection

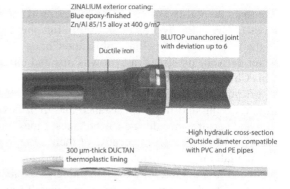

The Saint-Gobain PAM BLUTOP jointing

**FIGURE 24.2** Seismic joints used in Denizli, Turkey water pipelines replacement program. (Toprak et al. 2015.)

### 24.2.4 PIPELINE COATINGS

Coatings are materials that are applied to the surface of pipes to protect them against internal and external elements and delay the deformation period. The most common external coating materials that provide insulation against factors that adversely affect and damage the pipe from the outside for various reasons: polyethylene, bitumen, coal tar-based enamel, and epoxy. The most common interior coating materials that provide insulation against internal factors are epoxy, bitumen, and cement mortar. Internal coating is used generally to provide an efficient solution to both the corrosion, roughness, and the precipitation and deposition problems that may affect pipe life expectancy and efficiency in liquid/gas transmission. External coating protects pipelines from corrosion caused by soil and other materials contained in backfill. In addition, the external coating material affects the interaction between the pipeline and soil primarily by changing the frictional resistance at the pipe-soil interface. Because forces applied to pipeline by soil during the seismic actions transferred via the interaction at the interface, the response of the pipeline to earthquake is influenced inevitably by the coating.

## 24.3 SEISMIC EFFECTS ON PIPELINES

Ground movements triggered by earthquakes cause significant damage to infrastructure systems and results in serious loss of property and lives. The sources of seismic damages on buried pipelines can be categorized into two groups: transient ground deformations (TGD) and permanent ground deformations (PGD). TGD occur during the passage of earthquake waves in the soil. These deformations do not have much effect on continuous pipelines but can cause some damages on segmented pipelines especially if there are some weaknesses in the pipes or joints. PGDs can cause substantial damage to pipelines when they exceed some threshold values. Some examples to such PGDs are fault displacements, landslides, lateral spreads, and ground settlements. Damages that may occur in the pipeline as a result of these deformations depend on the geological structure of the area, the magnitude of the ground deformations, the extent of the ground deformation zone, the position of the pipe in relation to the displacement direction, and pipeline properties. The relative magnitudes of TGD and PGD determine which one will have predominant influence on pipeline response. In general, TGD affects larger area compared to PGD but the PGD displacement values may be much higher than the ground displacements caused by TGD. Therefore, although they occur in more localized areas, damages caused by PGD tend to be much more significant in terms of pipeline damages and service interruption effects. With the advancement in remote sensing capabilities and instrumentation, it has been possible to differentiate the extent of pipelines damages caused by PGD and TGD in recent earthquakes (Rathje et al. 2005; Stewart et al. 2009; Toprak et al. 2018). Such an example is the water pipeline damages observed in Christchurch during the February 22, 2011 earthquake (Bray et al. 2013; O'Rourke et al. 2014). Assessment results showed that the average pipeline repair rate for all pipelines in liquefaction areas where PGD observed was about six times higher than the average pipeline repair rate in TGD zones.

As shown in Figure 24.3, the position of the pipeline in relation to the PGD can be in an oblique (random), transverse (perpendicular), or longitudinal (parallel) direction, resulting in compression, tension, and bending in the pipeline. While evaluating the pipelines in such areas in terms of earthquake risks or when the pipelines planned to be installed in such areas are designed to resist earthquake effects, the seismic loads transferred from the ground to the pipelines and the resistance of the pipelines against these loads should be taken into account. Accordingly, pipelines should be able to resist the additional axial and bending stresses caused by PGDs. A pipe run oriented parallel and perpendicular to the soil movement is defined as experiencing longitudinal and transverse PGD, respectively. In most cases, the performance of buried pipelines is much more seriously impacted due to PGDs along the longitudinal direction of the pipe than transverse to the pipeline. This is in part due to the fact that a continuous pipe is inherently more flexible or compliant when subject to

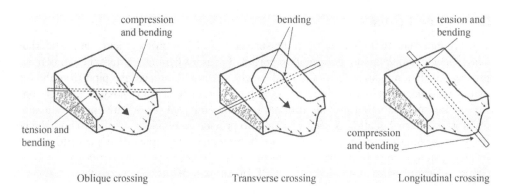

**FIGURE 24.3**   Typical pipeline-PGD interactions.

bending (transverse PGD) then when subject to axial tension or compression (longitudinal PGD) (ALA 2005).

PGD effects on pipelines can be analyzed using numerical or analytical approaches or both. There are many existing reliability methods which can be used together with either of these methods. In the following sections, selected analytical solutions developed for the longitudinal and transverse PGD effects on pipelines are presented. Then, reliability concepts and their application to these solutions as well as to general cases are described. It should be noted that analytical methods are applicable only for the boundary, soil and pipe conditions they were developed for and beyond that numerical approaches must be utilized. Because of the special importance of pipeline systems for urban life, there is huge interest in development of new methodologies both in analysis and reliability methods.

### 24.3.1   Longitudinal Permanent Ground Deformations

Longitudinal PGDs cause primarily axial stresses on the pipe (Figure 24.4a). It is possible to examine these deformations in two cases with the sliding block model proposed by O'Rourke and Nordberg (1992) and O'Rourke and Liu (2012). In the first case, if the length (L) of the PGD zone is relatively small compared to pipe length, there is slippage at the soil pipeline interface over the whole length of the PGD zone but the maximum pipe displacement at the center of the zone is less than the ground movement, δ (Figure 24.5a). The induced deformations and stresses will be controlled by the length of the PGD zone affecting the pipe axis. In the second case, if the length (L) of the PGD zone is too large compared to the pipe length, L is of sufficient size to develop zero-slip between the pipe and the ground around the pipe in the PGD zone. A slip zone of length $L_e$ exists at either side of the deformation zone. In this case, the pipe displacement matches ground displacement over

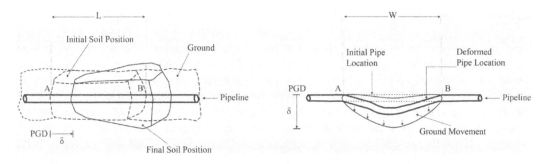

**FIGURE 24.4**   Schematic representation of ground action and pipe loading due to longitudinal (left) and transverse (right) PGD.

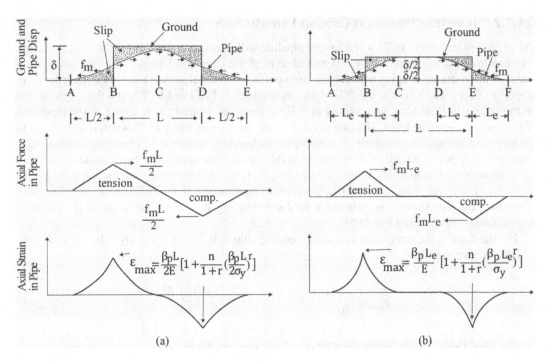

**FIGURE 24.5** Sliding block model for longitudinal PGD effects. (Redrawn after O'Rourke and Nordberg 1992.)

a region of length $L - 2L_e$ within the PGD zone and the deformations and stresses in the pipe are controlled by $\delta$ (Figure 24.5b). The peak friction force per unit length of pipe at soil pipe interface, $f_m$ can be calculated as:

$$f_m = \left[ \alpha c + \mu \gamma H \left( \frac{1+K}{2} \right) \right] \pi D \tag{24.1}$$

where $c$ = represents soil cohesion of the backfill, $\alpha$ = adhesion factor, $\gamma$ = effective unit weight of soil, $H$ = burial depth to pipe centerline, $K$ = lateral stress ratio, $D$ = outside diameter of pipe, $\mu$ = interface friction for pipe and soil; calculated as $\tan k\phi$, $\phi$ = internal friction angle of the soil, $k$ = coating dependent factor relating the internal friction angle of the soil to the friction angle at the soil-pipe interface. Over the segment AB in both cases, the force in the pipe is linearly proportional to the distance from Point A. Using a Ramberg-Osgood model for the pipe material considering inelastic pipe behavior, the pipe strain for the continuous pipelines, can be expressed as shown in Figures 24.5a and 24.5b, the following equations are obtained for Case I and II, respectively:

$$\varepsilon_{max} = \frac{\beta_p L}{2E} \left\{ 1 + \frac{n}{1+r} \left( \frac{\beta_p L}{2\sigma_y} \right)^r \right\} \qquad\qquad \varepsilon_{max} = \frac{\beta_p L_e}{E} \left\{ 1 + \frac{n}{1+r} \left( \frac{\beta_p L_e}{\sigma_y} \right)^r \right\} \tag{24.2}$$

where $n$ and $r$ = Ramberg–Osgood parameters, $E$ = elasticity modulus for the pipe material, $\sigma_y$ = yield stress for the pipe material, $\beta_p$ = pipe burial parameter. $\beta_p$ is obtained as; $\beta_p = \dfrac{\mu \gamma H (1+K)}{2t}$ for sand soil and $\beta_p = \dfrac{\alpha S_u}{t}$ for clay where $S_u$ = undrained shear strength of surrounding soil around the pipe, $t$ = pipe wall thickness.

## 24.3.2 Transverse Permanent Ground Deformations

As a result of transverse PGD, a continuous pipeline will stretch and bend under the effects of axial and bending strains (Figure 24.4b). Type and level of the damages depend on the relative magnitudes of these strains, ranging between the ruptures from tensile effects to local buckling from compressive effects. The extent of the PGD zone, represented as PGD width (W), and the magnitude of deformations in PGD area, represented as PGD amount ($\delta$), control the response of the pipelines. Two cases can be considered. The first of these is the situation in which PGD width is considered to be large and the pipeline is relatively flexible, hence lateral displacement of pipeline is assumed to closely match that of the soil. The pipe strain in this case occurs mainly due to the ground curvature (i.e., displacement controlled). In the second case, PGD width is relatively narrow and the pipeline is relatively stiff, resulting in lateral displacement of pipeline substantially less than that of the soil. Therefore, the pipe strain was assumed to be due to loading at the soil-pipe interface (i.e., loading controlled) (O'Rourke and Liu 1999).

For the Case I, the maximum and axial bending strains in the pipe are given by the following equations, respectively:

$$\varepsilon_{max} = \frac{\pi^2 \delta D}{W^2} \qquad\qquad \varepsilon_a = \frac{\left(\pi/2\right)^2 \delta^2}{W^2} \qquad\qquad (24.3)$$

for the Case II, the maximum bending strains in the pipe are given by:

$$\varepsilon_{max} = \frac{P_u W^2}{3\pi EtD^2} \qquad\qquad (24.4)$$

where $P_u$ = the maximum lateral force per unit length and is calculated as:

$$P_u = N_{ch}cD + N_{qh}\gamma HD \qquad\qquad (24.5)$$

where $N_{ch}$ = horizontal bearing capacity factor for clay and $N_{qh}$ = horizontal bearing capacity factor.

## 24.4 RELIABILITY METHODS AND THEIR APPLICATION TO PIPELINES

When designing and constructing the pipelines, engineers aim to ensure that the pipelines remain safe and stable against the factors it will be exposed to throughout its service life. Reliability analyses address the relationship between the loads that the system must carry and its capacity to carry these loads and evaluates the failure probability of the pipelines by determining whether the performance and limit situation has been exceeded (Thoft-Christensen and Baker 1982; Choi et al. 2006; Modarres et al. 2016). However, constructed pipelines face and live with several uncertainties during their lifetime in the loading conditions (e.g., earthquake) and the resistance of the system (e.g., material deterioration, and many environmental effects). Therefore, their interactions may also contain uncertainties (Baecher and Christian 2003). In order to cope with the uncertainties, the critical factors can be determined with reliability analysis approaches during the design phase, hence potential problems can be envisaged and prevented at an earlier stage. In addition, reliability analyses help to take precautions for existing pipelines to maintain their function and against uncertainties that were not considered at the time of design regarding various hazards such as possible ground movements during and after earthquakes. Analyses help engineers make a comprehensive decision about the structure, taking into account the uncertainties.

## 24.4.1 PERFORMANCE AND LIMIT STATE FUNCTION

In order to safely design a system, it is necessary to determine the critical points in the system and pay attention to how the detected points affect the system. For this, the limit states of the system can be interpreted, evaluated, and calculated with mathematical models (Ang and Tang 1975; Ditlevsen and Madsen 1996). In reliability-based design, this state is decided by performance function which is the most critical condition identified in keeping our design safe. If we define the performance function as a function of basic random variables and consider it represented by Z as:

$$Z = G(X_1, X_2, X_3, \ldots, X_m) \tag{24.6}$$

here $X = (X_1, X_2, X_3, \ldots, X_m)$ is the input vector of random variables.

The performance function for normal and log-normal distribution over the loads on the system and the resistance of the system can be determined as;

$$Z = G(X_i) = \ln(R) - \ln(Q) = \ln\left(\frac{R}{Q}\right) \tag{24.7}$$

$$Z = G(X_i) = R - Q \tag{24.8}$$

here R is the resistance and Q is the load. In order for the design to remain in the safe zone, the loads affecting the system must be less than the maximum allowable load for the safety of the system. As a result, if $Z > 0$, we come to the conclusion that the structure is in the safe zone, if $Z < 0$ it poses a risk, that is, it is in the unsafe zone, and if $Z = 0$, it is at the boundary of the safe-unsafe zone. This is called limit state situation (Figure 24.6).

If the loading on the system increases over time for various reasons to the levels that cannot met by resistance of the system, the risk factor increases, the limit state is exceeded and the system moves to the unsafe zone. This causes undesirable conditions and loss of reliability. Therefore, designs always aim to provide the $Z > 0$ condition to keep the structure in the safe zone. Reliability values may range from minimum zero to maximum one in reliability analyses. The reliability of the system increases as the result approaches to the upper limit, and the risk of system collapse increases as it approaches the zero limit.

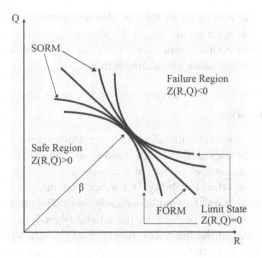

**FIGURE 24.6** Convex & concave representation of limit state, safe and failure region.

### 24.4.2 Failure Probability and Reliability Index

Reliability can be defined as the probability of the structure to fully fulfill the functions designed for on an element or system basis under certain service conditions in a certain period of time (e.g., Thoft-Christensen and Baker 1982; Modarres 1992; Bayazit 2007; Modarres et al. 2016). After this probability is determined, instead of specifying the structure just as safe or unsafe, it is necessary to determine the reliability value if it is safe, and the risk value and weak points if it is unsafe by using probability correlations (Lewis 1996). Unlike reliability, risk factors may occur in systems arising from undesirable situations that may arise due to the effect of uncertainties. This situation is called the probability of failure and can be expressed as follows:

$$P_f = \int_{G(X)\leq 0} fx(X)dX \tag{24.9}$$

where f(X) is the probability function of X. Since probability of failure and reliability are probabilities of complementary conditions, it is expressed as $P_f = 1 - P_s$. The reliability of a structure can always be improved, but the probability of failure will never be zero. For this reason, the reliability of the system must be at a level that can meet the allowable probability of failure (risk) at minimum cost. While ensuring system reliability for this purpose, optimization of the demands regarding economy and safety should be taken into consideration (Nowak and Kevin 2000). The reliability index, β can be calculated as:

$$\beta = \frac{\mu_Z}{\sigma_Z} = \frac{\mu_R - \mu_Q}{\sqrt{\sigma_R^2 + \sigma_Q^2 - 2\rho_{RQ}\sigma_R\sigma_Q}} \tag{24.10}$$

where $\mu_Z$ = mean value of limit state function, $\mu_R$ and $\mu_Q$ = means of the loads and resistance in the system, respectively. $\sigma_Z$ = standard deviation for the limit state function, $\rho_{RQ}$ = correlation coefficient between R and Q, $\sigma_R$ and $\sigma_Q$ = standard deviations of the R and Q, respectively. When R and Q random variables are independent for normal distribution, it is stated that there is a relationship between reliability index and failure probability:

$$\beta = -\phi^{-1}(P_f) \text{ or } P_f = \phi(-\beta) \tag{24.11}$$

The reliability index found as a result of the calculations serves to determine the best distance between $\mu_Z$ and the limit state for reliability. At the best distance, it is the shortest distance to the origin that provides the limit boundary state line. The relationship between β and $P_f$ ($P_f$ decreases as β increases) is very important when calculating reliability.

### 24.4.3 Reliability Methods

Reliability analyses enable systematic modeling of uncertainties arising from the type of structure and materials used, the processes in the design, production and construction stages, loading conditions and environmental factors, allowing engineers to understand how the system responds to variations in parameters that affect reliability. In essence, reliability analyses treat the uncertainty factors in the system or among the data mathematically and calculate their probabilities regarding service continuity or failure. While performing the reliability analysis, there exists various differences in the calculation steps of the limit state functions of the structure designed. The existence of different approaches in calculation steps creates the variety of methods (first-order reliability method [FORM], second-order reliability method [SORM], etc.) that form the basis of reliability analysis algorithms (Choi et al. 2006; Benjamin and Cornell 2014; Huang et al. 2016). While the limit state function is approached with first order Taylor expansion in the FORM, this calculation is made with second order Taylor expansion in SORM. As a result, the two methods use the similar approach of the boundary state function in the failure probability calculation.

A comprehensive literature review of the use of reliability methods in pipelines shows that there are two most preferred methods: FOSM method and MC simulations. Some examples are Shao et al. (2010), Nahal and Khelif (2013), An et al. (2016), Ni et al. (2018), Ebenuwa and Tee (2019), Athmani et al. (2019), Guillal (2019), Cirmiktili (2019), and Nagao and Lu (2020). FOSM is divided into two groups as Mean Value First-Order Second-Moment (MVFOSM) and Advanced First-Order Second-Moment (AFOSM). The method, which was first applied as MVFOSM, was developed as AFOSM with the studies and improvements made.

MC simulation method performs calculations based on simulating the parameters numerically with randomly constructed samples for uncertain variables and generating the necessary number of observational data without any real experience. In order for the method to be applied, the performance and limit state function should be defined properly. The correct distribution, mean value and standard deviation of the variables are important (Choi et al. 2006). In this method, it is of great importance that the functions of the variables can be taken in the analysis to make the right decision. When generating variables that affect the limit state function, special attention should be given whether the parameters for analysis are interdependent or independent (Rubinstein 1981; Abramson et al. 2002; Baecher and Christian 2003). If the dependent or independent parameters cannot be determined properly, the results will be deceiving. Parameters affecting the system are used for performance function calculation for simulation in line with certain field conditions. Many computer programs can be used to generate random values of the parameters. While these programs generate random data, they distribute numbers according to their own algorithms. Because some programs may distribute more consistently than others in terms of random number generation, attention should be paid to the program used when generating random data. Simulations are performed in high numbers to try different values of parameters to include uncertainties. The results that are evaluated as successful (satisfying the performance criteria) and unsuccessful (not satisfying the performance criteria) are used in calculating the average probability of failure. The ratio of the unsuccessful simulations to total simulations provide the average probability of failure. Approaches called inconsistency reduction methods have been developed for MC simulations to reduce the statistical errors in the method, increase the efficiency of the calculation and calculate the minimum required number of analyzes (Fenton and Griffiths 2008).

## 24.5   ILLUSTRATIVE EXAMPLE

The example reliability analyses here uses the longitudinal PGD on Balboa Boulevard and adjacent McLennon Avenue in Los Angeles which affected four gas transmission, one oil transmission, and two water trunk pipelines during the 1994 Northridge earthquake (Toprak 1998). The illustrative reliability analyses herein involved both FOSM and MC simulations but because of the similarity of the results only MC simulation results are presented. The Northridge earthquake occurred as a result of the rupture of the San Andreas fault on January 17, 1994, in the Northridge region of California, located 30 km northwest of Los Angeles. The water transmission and distribution system was severely damaged, including 15 transmission line, 74 main line (nominal pipe diameter $\geq$ 600 mm), and 1013 distribution line (diameter < 600 mm) repairs (O'Rourke and Toprak 1997). The PGD in Balboa Blvd. occurred, because of the liquefaction at deeper layers, as sliding of a relatively coherent soil mass parallel to the longitudinal axes of the pipelines. The length and magnitude of PGD were about 275 m and 0.5 m, respectively. Three out of the seven pipelines (two water and one natural gas line) at this location were severely damaged (Figure 24.7). The damaged pipelines were Granada, Rinaldi and Old Line 120. The damage in the gas transmission line (Old Line 120) caused gas leaks resulting in explosions and fires. With the rupture of the Granada and Rinaldi water trunk

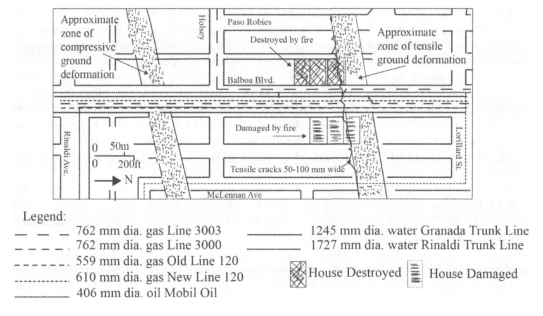

**FIGURE 24.7**  Map of Pipelines Affected by PGD. (Redrawn after O'Rourke and Palmer 1994).

lines, flooding was observed in the area and approximately 15% of the people had problems in accessing water. However, the other three natural gas transmission lines (Line 3000, Line 3003, and New Line 120) and one oil transmission (Mobil oil) pipeline in the same area were not damaged by the earthquake, and they continued to provide services during and after the earthquake. In the analysis herein, a total of four lines, two of the damaged pipelines, two of the undamaged pipelines, are presented.

The characteristic properties of the pipes in terms of mean and coefficient of variation (COV) values used in the analysis are shown in Table 24.1. As described in the Section 24.2.3, another important parameter affecting the reliability analysis of Granada and Rinaldi water pipelines is their joint characteristics. While the tensile strength of Granada and Rinaldi pipelines stays the same as the original pipe material, the compressive strengths were recalculated using reduction factors of the joints as can be seen in Table 24.1. In addition, the Granada trunk line had a mechanical coupling at the north end of the PGD zone. The New Line 120 and Mobil Oil line have full penetration girth welds by shielded electric arc practices. The coating of Granada and Rinaldi pipelines are cement mortar whereas New Line 120 and Mobil Oil line have fusion bonded epoxy and polyethylene, respectively. As shown in Figure 24.7, there are bends in the Mobil Oil line just north of the PGD zone and in the New Line 120 about 40 m south of the PGD zone. They were treated analytically as an anchor and the analytical procedure was modified to account for the asymmetric distribution of pipeline displacement and strain. The critical strain values for the pipelines so that they can safely transmit the matters without rupture under the tensile and compression loadings are taken as 0.03 and 0.01, respectively. These limits can be defined slightly different in various codes or guidelines, therefore, the designers should apply the proper code requirements for their own projects.

The probabilities of failure for the two damaged pipelines and the two undamaged pipelines were calculated using 50.000 MC simulations for each case and shown in Figure 24.8. When the probability of failure values of damaged and undamaged pipelines are examined under the varying PGD lengths, it appears that there is a very clear difference in their response. Because of their weakness at joints in compression, Granada and Rinaldi trunk lines reached the maximum probability of failure value at relatively small PGD lengths. In the pipelines that were not damaged, the probability of failure values were much lower even at PGD values higher than the observed PGD length of 275 m,

**TABLE 24.1**

**Parameter Values Used in Case Study Analysis**

| Parameter | Definition | Pipeline | Mean Value | COV (%) |
|---|---|---|---|---|
| k | Friction Reduction Coefficient | Granada & Rinaldi | 0.96 | 2 |
| | | Mobil Oil & New Line 120 | 0.52 | 2 |
| Φ | Angel of Internal Friction | All pipe | 37° | 15 |
| γ | Unit Weight of Ground | All pipe | 19.65 kN/m³ | 9 |
| H | Burial Depth | Granada | 1.8 m | 1 |
| | | Rinaldi | 2.7 m | 1 |
| | | Mobil Oil & New Line 120 | 1.5 m | 1 |
| n,r | Ramberg-Osgood Parameters | Granada & Rinaldi | n = 10 r = 11 | D* |
| | | Mobil Oil | n = 8 r = 11 | D* |
| | | New Line 120 | n = 10 r = 12 | D* |
| t | Wall Thickness | Granada & New Line 120 | 0.0064 m | 5 |
| | | Rinaldi & Mobil Oil | 0.0095 m | 5 |
| E | Modulus of Elasticity for Pipe | All pipe | 200.000 MPa | 3.3 |
| σy | Yield Stress | Granada – stress | 249 MPa | 5 |
| | | Granada – strain | 92 MPa | 5 |
| | | Rinaldi – stress | 249 MPa | 5 |
| | | Rinaldi – strain | 87 MPa | 5 |
| | | Mobil Oil | 360 MPa | 5 |
| | | New Line 120 | 415 MPa | 5 |

*Note:* D* = Deterministik COV Value.

indicating why these pipelines survived the earthquake. The results here are in harmony with the field observations. With reliability analyses, however, it is possible to quantify the level of risks pipelines are exposed to at different PGD levels. Some additional examples with various ground movements can be found in Ebenuwa and Tee (2019) and Ni et al. (2018). The utility companies use reliability analyses more often today to evaluate their system against various hazards and mitigate

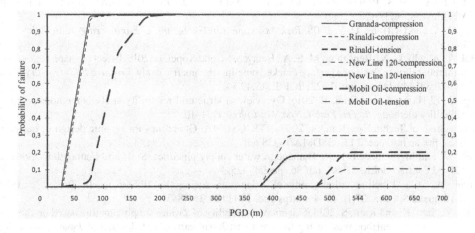

**FIGURE 24.8** Reliability analyses results for the longitudinal PGD affecting pipelines.

them. Reliability analyses can offer important options to decision makers by giving the opportunity to determine the total risk by combining all risks from different sources.

## REFERENCES

Abramson, L. W., Lee, T. S., Sharma, S. and Boyce, G. M. 2002. *Slope stability and stabilization methods.* John Wiley & Sons, Inc.

ALA (American Lifelines Alliance). 2005. Seismic guidelines for water pipelines. http://www.americanlifelinesalliance.org.

An, X., Gosling, P. D. and Zhou, X. 2016. Analytical structural reliability analysis of a suspended cable. Structural Safety 58 (January): 20–30. https://doi:10.1016/j.strusafe.2015.08.001.

Ang, A. H. and Tang, W. H. 1975. *Probability concepts in engineering planning and design.* John Wiley & Sons, New York.

Athmani, A., Khemis, A., Chaouche, A. H., et al. 2019. Buckling uncertainty analysis for steel pipelines buried in elastic soil using FOSM and MCS methods. International Journal of Steel Structures 19 (2): 381. https://doi:10.1007/s13296-018-0126-7.

Baecher, G. B. and Christian, J. T. 2003. *Reliability and statistics in geotechnical engineering.* John Wiley & Sons Inc, England.

Ballantyne, D. B., Berg, E., Kennedy, J., et al. 1990. Earthquake Loss Estimation Modeling of the Seattle Water System. K/J/C Report No: 886005.00, Kennedy/Jenks/Chilton, Federal Way, Washington, D. C.

Bayazit, M. 2007. *Reliability and risk analysis in engineering.* Birsen Press, Istanbul.

Benjamin, J. R. and Cornell, C. A. 2014. *Probability, statistics, and decision for civil engineers.* Dover Publications, Inc, New York, USA.

Bouziou, D. and O'Rourke, T. D. 2017. Response of the Christchurch water distribution system to the 22 February 2011 earthquake. Soil Dynamics and Earthquake Engineering 97:14–24.

Bray, J. D., O'Rourke, T. D., Cubrinovski, M., et al. 2013. Liquefaction impact on critical infrastructure in Christchurch. *USGS Technical Report*, March 22, 2013.

Choi, S. K., Grandhi, R. and Canfield, R. A. 2006. *Reliability based structural design.* Springer Science & Business Media.

Cirmiktili, O. Y. 2019. *Reliability analysis of pipelines.* Master's Thesis, Pamukkale University, Denizli, Turkey.

Davis, C. A. 2009. Scenario response and restoration of Los Angeles water system to a magnitude 7.8 San Andreas fault earthquake. *Proc. 7th U.S. Conf. on Lifeline Earthquake Engineering*, ASCE, Oakland, June 28–July 1, paper 232.

Ditlevsen, O. and Madsen, H. O. 1996. *Structural reliability methods.* New York: Wiley.

Ebenuwa, A. U. and Tee, K. F. 2019. Reliability estimation of buried steel pipes subjected to seismic effect. Transportation Geotechnics 20. https://doi:10.1016/j.trgeo.2019.100242.

Eguchi, R. T. 1991. Seismic hazard input for lifeline systems. *Structural Safety*, E. H. Varyrnarcke, Ed., Elsevier Science Publishers, Vol. 10, May, pp. 193–198.

Eidinger, J. 1998. Water distribution system–The Loma Prieta, Californa, Earthquake of October17, 1989–Lifelines, USGS, Professional Paper 1552-A, Schiff, A. J. (Ed.) U.S. Government Printing Office, Washington, A63–A78.

Fenton, G. A. and Griffiths, D. V. 2008. *Risk Assessment in Geotechnical Engineering.* John Wiley & Sons, Inc.

Guillal, A., Albelbaki, N., Bensghier, M. E. A., Betayeb, M. and Kopei, B. 2019. Effect of shape factor on structural reliability analysis of a surface cracked pipeline-parametric study. *Frattura Ed IntegritaStrutturale* 13 (49):341–49. https://doi:10.3221/IGF-ESIS.49.34.

Huang, C., El Hami, A. and Radi, B. 2016. Overview of structural reliability analysis methods-part I: local reliability methods. *Incert. Fiabil. Syst. Multiphys*, 17, 1–10.

Indian Institute of Technology Kanpur. 2007. IITK-GSDMA. Guidelines for seismic design of buried pipelines. iitk.ac.in/nicee/IITK-GSDMA/EQ28.pdf.

Kitaura, M. and Miyajima, M. 1996. Damage to water supply pipelines. Soils and Foundations, 36(special), 325–333. https://doi:10.3208/sandf.36.special_325.

Kurtulus, A. 2011. Pipeline vulnerability of Adapazarı during the 1999 Kocaeli, Turkey, earthquake. Earthquake Spectra, 27(1), 45–46. https://doi:10.1193/1.3541843.

Kuwata, Y., Sato, K. and Kato, S. 2018. Seismic vulnerability of a water supply pipeline based on the damage analysis of two earthquakes during the great east Japan earthquake. Journal of Japan Association for Earthquake Engineering, 18(3), 3_91–3_103. https://doi:10.5610/jaee.18.3_91.

Lewis, E. E. 1996. *Introduction to reliability engineering.* John Willey & Sons Inc, New York.

Mattiozzi, P. and Strom, A. 2008. Crossing active faults on the Sakhalin II Onshore pipeline route:pipeline design and risk analysis. *AIP Conf Proc* 1020:1004.

Modarres, M. 1992. *What every engineer should know about reliability and risk analysis* (Vol. 30). CRC Press.

Modarres, M. Kaminskiy, M. P. and Krivtsov, V. 2016. *Reliability engineering and risk analysis: a practical guide.* CRC press.

Nagao, T. and Lu. P. 2020. A Simplified reliability estimation method for pile-supported wharf on the residual displacement by earthquake. Soil Dynamics and Earthquake Engineering 129 (February). https://doi:10.1016/j.soildyn.2019.105904.

Nahal, M. and Khelif, R. 2013. Mechanical reliability analysis of tubes intended for hydrocarbons. Journal of Mechanical Science and Technology 27 (2): 367. https://doi:10.1007/s12206-012-1265-y.

Ni, P. Y., Yi. and Mangalathu, S. 2018. Fragility analysis of continuous pipelines subjected to transverse permanent ground deformation. Soils and Foundations 58 (6): 1400–1413. https://doi:10.1016/jsandf.2018.08.002.

Nowak, A. S. and Kevin, R. C. 2000. *Reliability of structures.* Boston: McGraw-Hill.

O'Rourke, M. J. and Nordberg, C. 1992. *Longitudinal permanent ground deformation effects on buried continuous pipelines.* NCEER-92-0014, Buffalo, NY.

O'Rourke, M. J. and Liu, X. 1999. *Response of buried pipelines subject to earthquake effects.* NCEER, Buffalo, NY.

O'Rourke, M. J. and Liu, X. 2012. Seismic Design of Buried and Offshore Pipelines. MCEER-12-MN04. Multidisciplinary Center for Earthquake Engineering, Buffalo, NY.

O'Rourke, T. D. and Palmer, M. C. 1994. The Northridge, California earthquake of january 17, 1994: Perfomance of gas transmission pipelines. In *The Northridge, California earthquake of january 17, 1994: Perfomance of gas transmission pipelines* (pp. 79–79).

O'Rourke, T. D. and Toprak, S. 1997. GIS assessment of water supply damage from the Northridge earthquake. *Frost, J. D. (Ed.) Geotechnical Special Publication,* ASCE, New York, pp. 117–131.

O'Rourke, T. D., Jeon, S. S., Toprak, S., Cubrinovski, M. and Jung, J. K. 2012. Lifeline system performance during the Canterbury earthquake sequence. Invited Lecture Paper, *15th World Conference on Earthquake Engineering (15WCEE),* September 24–28, 2012, Lisbon, Portugal.

O'Rourke, T. D., Jeon, S. S., Toprak, S., et al. 2014. Earthquake response of underground pipeline networks in Christchurch, NZ Earthquake Spectra, 30(1), 183–204.

Oda, K., Tanaka, R. and Kishi, S. 2019. Development of Large Diameter Earthquake Resistant Ductile Iron Pipe. Pipelines 2019. https://doi:10.1061/9780784482483.036.

Rathje, E. M. Woo, K. Crawford, M. and Neuenschwander A. 2005. Earthquake Damage Identification Using Multitemporal High-Resolution Optical Satellite Imaginery. In: *International geoscience and remote sensingsymposium,* IEEE, Seoul, South Korea, July (CD-ROM).

Rubinstein, R. Y. 1981. *Simulation and the Monte Carlo method.* John Willey & Sons, Inc.

Shao, Y., Yu, J. Z. and Yu, T. C. 2010. Mechanical model and probability analysis of buried pipelines failure under uniform corrosion. *Zhejiang DaxueXuebao (Gongxue Ban)/(Journal of Zhejiang University (Engineering Science)* 44 (6): 1225–30. Accessed November 7. https://doi:10.3785/j.issn.1008-973X.2010.06.032.

Shih, B. J. and Chang, C. H. 2006. Damage survey of water supply systems and fragility curve of PVC water pipelines in the Chi-Chi Taiwan earthquake. *Natural Hazards: Journal of the International Society for the Prevention and Mitigation of Natural Hazards,* 37(1-2), 71. https://doi:10.1007/s11069-005-4657-9.

Stewart, J. P., Hu, J., Kayen, R. E., et al. 2009. Use of airbone and terrestrial LiDAR to detect ground displacement hazards to water systems. *Journal of Surveying Engineering,* 135(3):113–124.

Tawfik, M. S. I. and O'Rourke, T. D. 1985. Load carrying capacity of welding slip joints. *Journal of Pressure Vessel Technology* 107(1), 36–43.

Thoft-Christensen, P. and Baker, M. J. 1982. Structural Reliability Theory and Its Applications. https://doi:10.1007/978-3-642-68697-9.

Toprak, S. 1998. *Earthquake effects on buried lifeline systems.* Ph.D. Thesis, Cornell University, Ithaca, NY, USA.

Toprak, S. and Taskin, F. 2007. Estimation of earthquake damage to buried pipelines caused by ground shaking. Nature Hazards, 40, 1–24. https://doi:10.1007/s11069-006-0002-1.

Toprak, S., Nacaroglu, E., Koc, A. C., et al. 2018. Comparison of horizontal ground displacements in avonside area, Christchurch from air photo, LiDAR and satellite measurements regarding pipeline damage assessment. Bulletin of Earthquake Engineering, https://doi.org/10.1007/s10518-018-0317-9.

Toprak, S., Nacaroglu, E., Van Ballegooy, S., et al. 2019. Segmented pipeline damage predictions using lique-
    faction vulnerability parameters, Soil Dynamics and Earthquake Engineering 125, 105758, ISSN 0267-
    7261, https://doi.org/10.1016/j.soildyn.2019.105758.
Toprak, S., Nacaroglu, E. and Koc, A. C. 2015. Seismic response of underground lifeline systems. In:
    *Perspectives on European Earthquake Engineering and Seismology*, Atilla Ansal, Ed.Geotechnical,
    Geological and Earthquake Engineering (Volume 39), Vol. 2, Springer.
Toprak, S., O'Rourke, T. D. and Tutuncu, I. 1999. GIS characterization of spatially distributed lifeline dam-
    age. Optimizing Post-Earthquake Lifeline System Reliability, Proceedings, Fifth U.S. Conference on
    Lifeline Earthquake Engineering, W. M. Elliott and P. McDonough, Eds., Seattle, WA, August, ASCE,
    pp. 110–119.
Toprak, S. Taskin, F. and Koc, A. C. 2009. Prediction of earthquake damage to urban water distribution
    systems: a case study for Denizli, Turkey. The Bulletin of Engineering Geology and the Environment,
    68:499–510.
Vitali, L. and Mattiozzi, P. 2014. Crossing active faults on the Sakhalin II Onshore pipeline route:analysis
    methodology and basic design. AIP Conf Proc 1020:1014.

# 25 Optimal Design of Hysteretic Damper with Stopper Mechanism for Tall Buildings under Earthquake Ground Motions of Extremely Large Amplitude

*Shoki Hashizume and Izuru Takewaki*
Kyoto University, Kyoto, Japan

## CONTENTS

## 25.1 INTRODUCTION

Natural hazards are changing their properties and the measures of resilience of structures for those disturbances are requested to upgrade (Bruneau et al. 2003, Cimellaro et al. 2010, Takewaki et al. 2011, Noroozinejad et al. 2019, Takewaki 2020). The resistance ("plan and absorb," Linkov and Trump 2019) to disturbances and the recovery ("recover and adapt," Linkov and Trump 2019) from damages are two constituents in the resilience. While the countermeasures for the resistance have been tackled extensively in the structural engineering community, the approaches to the measures for the recovery have never been investigated sufficiently. This may result from the fact that the recovery is related to many complicated factors in various multidisciplinary fields.

DOI: 10.1201/9781003194613-25

In the 20th century, building structures were designed based on the codes dealing with the resistance by the stiffness and strength of their members. However, after the unprecedented catastrophic damages induced by unpredictable natural hazards, the design principle was altered to rely on passive control devices for maintaining functions of buildings without disruption. The extensive reviews on research for passive control can be found in many papers (for example, Hanson 1993, Aiken et al. 1993, Nakashima et al. 1996, Soong and Dargush 1997, Hanson and Soong 2001, Takewaki 2009, Lagaros et al. 2013, Fukumoto and Takewaki 2017, Tani et al. 2017, Hayashi et al. 2018, Makita et al. 2018, Domenico et al. 2019, Kondo and Takewaki 2019, Kawai et al. 2020). Although the passive control is a powerful means, another difficulty was found in the 2016 Kumamoto earthquake (Japan). During two days, consecutive severe shakings (the highest level in Japan) occurred. In these earthquakes, another type of ground motions, called long-period pulse-type ground motions, of unprecedented large-amplitude were recorded whose velocity amplitude was over 2.0 m/s compared to 0.5 m/s for severe earthquake ground motions for tall buildings in Japan. The reduction of plastic deformations is strongly desired in view of the earthquake resilience (Kojima and Takewaki 2016, Ogawa et al. 2017).

Efficient and optimal use of passive dampers is getting a great concern in the field of structural control and has been tackled extensively (see, for example, Xia and Hanson 1992, Inoue and Kuwahara 1998, Uetani et al. 2003, Aydin et al. 2007, Takewaki 2009, Lavan and Levy 2010, Adachi et al. 2013a,b, Lagaros et al. 2013, Domenico et al. 2019, Akehashi and Takewaki 2019). For viscous dampers, various approaches have been proposed (Domenico et al. 2019). Since viscous dampers possess a cost and maintenance problem (Murakami et al. 2013a), hysteretic dampers have often been installed in earthquake prone countries. In addition, the combination of various types of dampers has been investigated (Uetani et al. 2003, Murakami et al. 2013a,b).

As for hysteretic dampers, Inoue and Kuwahara (1998) investigated a single-degree-of-freedom (SDOF) model and proposed a rule on the optimal quantity of hysteretic dampers by using the concept of the equivalent viscous damping (Jacobsen 1960, Caughey 1960). Since the hysteretic dampers exhibit a peculiar characteristic for random recorded ground motions, Murakami et al. (2013b) introduced a general and stable sensitivity-based approach for hysteretic dampers.

Random response characteristics observed in building structures including hysteretic dampers require a huge amount of computational load to clarify intrinsic properties of the optimal damper design (location and quantity). For overcoming such difficulty, Shiomi et al. (2016) proposed an efficient design method for hysteretic dampers by using a double impulse as a representative of pulse-type ground motions. Then, Shiomi et al. (2018) provided an innovative passive-type control system which was named a "dual hysteretic damper (DHD)" system. Afterward, Shiomi et al. (2018) proposed a method based on response sensitivity for the optimal damper placement in multi-degree-of-freedom (MDOF) systems. Recently, Hashizume and Takewaki (2020a) developed another innovative passive vibration control system named a "hysteretic-viscous hybrid (HVH)" damper system. In this system, the DSA (short-range hysteretic damper) in the DHD system was replaced by a viscous damper. They investigated the response reduction performance for the DHD and the HVH for SDOF systems and extended the research to MDOF systems (Hashizume and Takewaki 2020b).

In this chapter, the double impulse as a representative of long-period, pulse-type ground motions is input into the MDOF systems including hysteretic dampers with gap mechanism (part of the HVH system and the DHD system). A sensitivity-based method for the optimal damper design (stiffness distribution and gap displacement distribution) is then proposed. It should be pointed out that only the critical resonant double impulse is input and the intrinsic property of the optimal damper placement is made clear. Finally, the influence of the stiffness of hysteretic dampers with gap mechanism on the response suppression is investigated for a recorded ground motion of quite large amplitude.

## 25.2 DOUBLE IMPULSE REPRESENTING MAIN PART OF PULSE-TYPE GROUND MOTION

It was clarified that the double impulse of two single impulses with opposite signs captures the essential property of the main part (simulated as a one-cycle sine wave) of a near-fault pulse-type ground motion (Kojima and Takewaki 2015). The key idea is that, while the analysis of response to the forced input needs the combination of free-vibration component and forced-vibration component even in the elastic linear responses, the double impulse produces only a free-vibration component. This is helpful in avoiding the solution of the transcendental equation appearing in resonance curves. At the same time, this treatment facilitates the use of the energy balance law for obtaining the maximum response without tedious time-history response analysis. The principal part of a pulse-type near-fault ground motion is first simulated by a one-cycle sine wave $\ddot{u}_{g\sin}(t)$ as shown in Eq. (25.1) (see Figure 25.1[a]) and then modeled into a double impulse $\ddot{u}_g(t)$ described by Eq. (25.2) (see Figure 25.1[b]).

$$\ddot{u}_{g\sin}(t) = A_p \sin \omega_p t \tag{25.1}$$

$$\ddot{u}_g(t) = V\delta(t) - V\delta(t - t_0) \tag{25.2}$$

In Eqs. (25.1) and (25.2), the parameters $A_p$, $\omega_p$, $V$, $t_0$ are the acceleration amplitude and circular frequency of the one-cycle sine wave, the velocity amplitude of the double impulse and the time interval of the two impulses in the double impulse, respectively. In addition, $\delta(t)$ indicates the Dirac delta function. Kojima and Takewaki (2015) introduced a criterion on the equivalence of the maximum Fourier amplitudes in this transformation to guarantee the approximate influences on their structural responses.

The ratio $a$ of $A_p$ to $V$ as an important index of this transformation is introduced by

$$A_p = aV \tag{25.3}$$

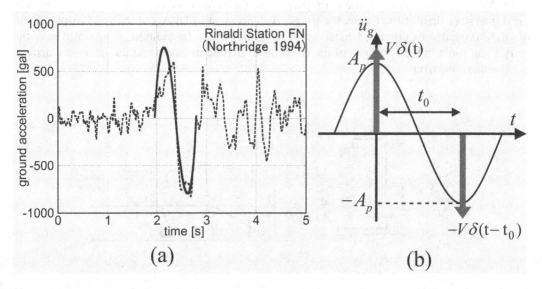

**FIGURE 25.1** Modeling of main part of ground motion into double impulse, (a) Transformation of principal part of Rinaldi station FN motion (Northridge 1994) into one-cycle sinusoidal wave, (b) Re-transformation into double impulse.

The coefficient $a$ as a function of $t_0 = \pi/\omega_p$ can be analyzed as follows (Hashizume and Takewaki 2020b).

$$a(t_0) = \frac{A_p}{V} = \frac{\max\left|\sqrt{2 - 2\cos(\omega t_0)}\right|}{\max\left|\dfrac{2\pi t_0}{\pi^2 - (\omega t_0)^2}\sin(\omega t_0)\right|} \tag{25.4}$$

The maximum velocity $V_p$ of the one-cycle sine wave can be expressed by

$$V_p = 2A_p/\omega_p \tag{25.5}$$

## 25.3   HYSTERETIC DAMPER SYSTEM WITH GAP MECHANISM

An innovative HVH damper system was introduced in the recent papers (Hashizume and Takewaki 2020a,b). A gap mechanism is installed in series to a hysteretic damper (called DLA) and a viscous damper is added in parallel. The gap mechanism plays a role as a trigger for the hysteretic damper to delay its working. As a result, this hysteretic damper with gap mechanism possesses a large-stroke performance and a function as a stopper. This large-stroke performance is beneficial for hysteretic dampers with the limit on the accumulated plastic deformation capacity.

The MDOF building structure with the HVH system is shown in Figure 25.2. The building structure is modeled by a shear-type MDOF system and the hysteretic damper is assumed to have the elastic-perfectly plastic restoring-force characteristics. In this figure, $K_{Fi}, k_L, c_i$ indicate the main structure (frame) stiffness in the $i$-th story, the stiffness ratio of the hysteretic damper (DLA) to the frame (assumed constant through all stories) and the damping coefficient of the viscous damper in the $i$-th story. Let $\mathbf{k}_d = \{k_{dj}\}$ denote the set of initial stiffnesses of hysteretic dampers. The fundamental natural period of the bare frame of 15 stories treated here is 1.60(s) and the story stiffness distribution of the bare frame is trapezoidal (the top to bottom story stiffness ratio = 0.5). Consider here the model without viscous dampers because the investigation of DLA as a stopper system is the main purpose of the investigation. Let $d_y, d_{gh}, d_{Ly}$ denote the yield interstory drift of the frame, the trigger displacement of DLA and the yield displacement of DLA, respectively, (assumed constant through all stories). In addition, $f_i$ and $\delta_i$ denote the story shear force in the $i$-th story of the frame or hysteretic dampers and the interstory drift of the frame, respectively.

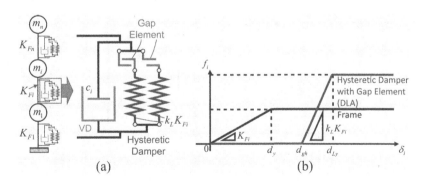

**FIGURE 25.2** Building structure including hysteretic damper with gap mechanism, (a) Hysteretic-viscous hybrid (HVH) damper system, (b) Force-interstory relation of frame and hysteretic damper system with gap mechanism (Hashizume and Takewaki 2020b).

## 25.4 OPTIMIZATION ALGORITHM

### 25.4.1 PROBLEM OF OPTIMAL DAMPER PLACEMENT

The optimal design problem of hysteretic dampers may be stated as follows.

**[Problem]** Find the stiffnesses $\mathbf{k}_d = \{k_{dj}\}$ of hysteretic dampers so as to minimize the specific seismic response $F$ subject to the sum of damper stiffnesses

$$\sum_{j=1}^{N} k_{dj} = \bar{C}_d \left( \text{or} \sum_{j=1}^{N} k_{dj} \leq \bar{C}_d \right) \tag{25.6}$$

In this problem, $\bar{C}_d$ is the specified sum of hysteretic damper stiffnesses. It can be demonstrated that the initial stiffness of hysteretic dampers is strongly related to the quantity of hysteretic dampers which corresponds to the production cost. The maximum interstory drift $D_{max}$ (or the maximum interstory-drift ductility) of the fame is employed here as $F$.

Since hysteretic dampers with gap mechanism have nonlinear restoring-force characteristics with sudden, large stiffness changes at multiple stages and input earthquake ground motions possess the nature of large randomness (because of propagation in random media), the seismic response of a building with hysteretic dampers with gap mechanism deviates greatly depending on the installed quantity and location of hysteretic dampers. The timing of damper trigger, fast-speed plastic flow and random process of input may be the main sources of the response randomness. This characteristic seems to disturb a reliable formulation of the optimal damper placement and causes a complicated situation different from other dampers (Takewaki 2009, Takewaki et al. 2011, Adachi et al. 2013a,b). To overcome this difficulty, the critical double impulse is used.

### 25.4.2 OPTIMIZATION ALGORITHM

Figure 25.3 presents a conceptual diagram of the proposed sensitivity-based optimal design algorithm. Candidate designs in Figure 25.3 are produced by reducing the damper stiffness slightly in respective story and the responses of candidate designs are obtained by time-history response analysis. The selection from the candidate designs is conducted by minimizing the maximum interstory drift. The aim of this procedure is to find the most inactive damper and reduce the quantity of such damper.

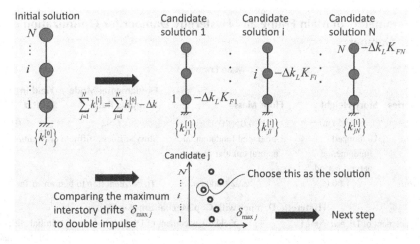

**FIGURE 25.3** Optimization algorithm of stiffness of hysteretic damper by decreasing the most sensitive stiffness.

A practical and simple procedure for optimal oil damper (often called nonlinear fluid viscous damper) design without time-consuming mathematical programming techniques has been proposed (Adachi et al. 2013a). A similar design algorithm can be devised for hysteretic dampers. The essential feature of such design algorithm may be summarized as follows:

[Step 1] The sum of hysteretic damper stiffnesses for all stories is determined (as the stiffness ratio to the main frame stiffness).

[Step 2] Produce $N$ candidates in such a way that damper stiffnesses $\Delta k_L K_{Fi}$ (in case of $i$-th story) are reduced from the present state in each story. Compute the objective function for each model constructed in Step 2 by using nonlinear time-history response analysis. Although the decrement of damper stiffness is different story by story resulting from the different values of $K_{Fi}$, this does not cause any problem (Shiomi et al. 2018).

[Step 3] Choose the design with the minimum interstory drift as the candidate design in each story.

[Step 4] Pick up the best candidate with the minimum objective function (interstory drift change) from the candidates produced in Step 3.

[Step 5] Decrease the damper stiffness in the story selected in Step 4 and return to Step 2.

## 25.5 NUMERICAL EXAMPLE OF OPTIMAL STIFFNESS DISTRIBUTION OF HYSTERETIC DAMPER WITH GAP MECHANISM

Nonlinear time-history response analysis using a Newmark-$\beta$ method (constant acceleration method) was conducted. The accuracy of the present analysis program was checked through the comparison with the general-purpose computer program, "SNAP" (SNAP 2015).

Consider a 15-story building model. Table 25.1 shows the structural parameters of the main frame and hysteretic dampers for the optimization of stiffness of hysteretic dampers. The critical timing of the second impulse in the double impulse is obtained by conducting time-history response analysis under the first impulse. The time attaining the zero value of the sum of restoring force and damping force of the damper in the first story is the critical timing of the second impulse in view

---

### TABLE 25.1

### Structural Parameters of Main Frame and Hysteretic Damper for Optimization of Stiffness of Hysteretic Damper

| | | **Main Frame** | | |
|---|---|---|---|---|
| Number of Stories | Story Height | Floor Mass | Fundamental-Mode Damping Ratio | Yield Interstory Drift $d_y$ |
| 15 | 3.5 (m) | $4.0 \times 10^5$ (kg) | 0.02 | 0.0233 (m) |
| Story stiffness in first story $K_{F1}$ | Undamped fundamental natural period $T_1$ | Undamped fundamental natural circular frequency | Story stiffness stiffness distribution | |
| $7.2 \times 10^8$ (N/m) | 1.60 (s) | 3.92 (rad/s) | Trapezoidal (top to bottom stiffness ratio = 0.5) | |

| **Hysteretic Damper with Gap Mechanism (DLA)** | | |
|---|---|---|
| Trigger displacement of DLA $d_{gh}$ $(=d_y)$ | Yield displacement of DLA $d_{Ly}$ | Initial stiffness ratio of DLA to main frame $k_L$ |
| 0.0233 (m) | 0.0467 (m) | 1, 3, 5 |

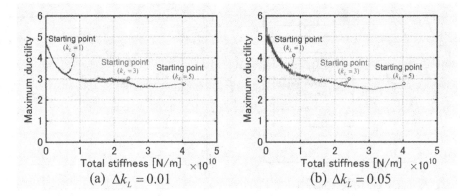

**FIGURE 25.4** Total stiffness of hysteretic damper with gap mechanism vs maximum ductility from three initial values of total stiffness ($k_L = 1,3,5$): Case of two stiffness decrements ($\Delta k_L = 0.01,0.05$).

of the maximum input of the energy by the second impulse for the constant velocity amplitude (Akehashi and Takewaki 2019).

Figure 25.4 presents the total stiffness of hysteretic dampers with gap mechanism vs the maximum ductility from three initial values of total stiffness ($k_L = 1,3,5$) in two cases of stiffness decrements ($\Delta k_L = 0.01,0.05$). It can be observed that, even if the initial values of total stiffness are the same, the paths for decreasing total stiffness are different. In addition, even if the initial values of total stiffness are different, the paths attain the same one finally.

Figure 25.5 shows the optimal stiffness distribution of hysteretic damper with gap mechanism and the corresponding interstory-drift ductilities for two stiffness decrements ($\Delta k_L = 0.01,0.05$) in three initial designs ($k_L = 1,3,5$). For reference, the damper stiffnesses and interstory-drift ductilities of the initial designs are also plotted. It can be observed that the optimal damper stiffnesses for $\Delta k_L = 0.01$ exhibit smaller values of stiffness compared to $\Delta k_L = 0.05$. As for interstory-drift ductilities, it can be assured that the maximum interstory-drift ductility is certainly reduced by the optimization. It is also understood that, while the case of $\Delta k_L = 0.01$ exhibits smaller values of interstory-drift ductilities in the initial design $k_L = 1$, $\Delta k_L = 0.05$ shows smaller values in the initial designs $k_L = 3,5$. It can also be assured that the hysteretic dampers went into the plastic range (interstory-drift ductility of two is the yield limit of hysteretic dampers in this case).

## 25.6 NUMERICAL EXAMPLE OF OPTIMAL DISTRIBUTION OF TRIGGER DISPLACEMENT

In this section, a numerical example of optimal distribution of the trigger displacement is presented. Table 25.2 shows the structural parameters of the main frame and hysteretic dampers for the optimization of the trigger displacement distribution of hysteretic dampers.

Figure 25.6 shows the optimization algorithm of trigger displacements of hysteretic dampers with gap mechanism using a sensitivity-based approach. The design problem is to find the trigger displacements of hysteretic dampers so as to minimize the maximum interstory drift under the constraint on the total quantity of trigger displacements. It is assumed here that the quantity of trigger displacements is related to the cost. The algorithm is similar to the algorithm for damper stiffness optimization (Figure 25.3). The design parameters are changed from the damper stiffnesses to the trigger displacements of hysteretic dampers.

In the numerical analysis, the case of $d_{gh} = 2d_y$ and $d_{Ly} = 3d_y$ is treated as initial values. This is because the distribution of trigger displacements around $d_{gh} = d_y$ is expected and the algorithm of decrement of trigger displacements is used. The decrement of trigger displacements is $\Delta d_{gh} = 0.02d_y$.

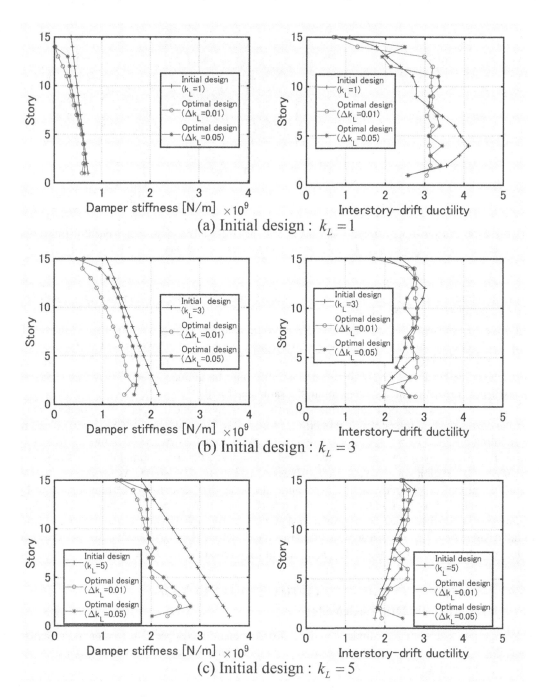

**FIGURE 25.5** Optimal stiffness distribution of hysteretic damper with gap mechanism and interstory-drift ductility of optimal design for two stiffness decrements ($\Delta k_L = 0.01, 0.05$).

Figure 25.7 illustrates the optimal trigger displacement distribution of the hysteretic damper system and the maximum interstory-drift ductility with respect to step number. The initial design for $k_L = 1$ is used as the stiffness distribution of the hysteretic dampers. It can be observed that the optimal trigger displacements distribute around $d_{gh} = d_y$ as expected and the interstory-drift ductilities corresponding to the optimal distribution of trigger displacements exhibit almost uniform

**TABLE 25.2**

**Structural Parameters of Main Frame and Hysteretic Damper for Optimization of Trigger Displacement of Hysteretic Damper**

| | | Main Frame | | |
|---|---|---|---|---|
| **Number of Stories** | **Story Height** | **Floor Mass** | **Fundamental-mode Damping Ratio** | **Yield Interstory Drift $d_y$** |
| 15 | 3.5 (m) | $4.0 \times 10^5$ (kg) | 0.02 | 0.0233 (m) |
| Story stiffness in first story $K_{F1}$ | Undamped fundamental natural period $T_1$ | Undamped fundamental natural circular frequency | Story stiffness stiffness distribution | |
| $7.2 \times 10^8$ (N/m) | 1.60 (s) | 3.92 (rad/s) | Trapezoidal (top to bottom stiffness ratio = 0.5) | |

| Hysteretic Damper with Gap Mechanism (DLA) | |
|---|---|
| (Yield displacement of DLA $d_{Ly}$) – (Trigger displacement of DLA $d_{gh}$) | Stiffness ratio of DLA to main frame $k_L$ |
| 0.0233 (m) | 1 |

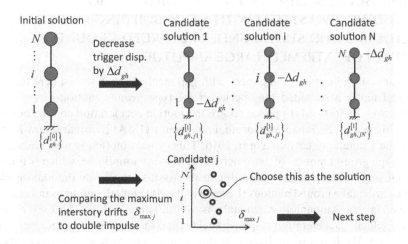

**FIGURE 25.6** Optimization algorithm of trigger displacement of hysteretic damper with gap mechanism using sensitivity-based approach.

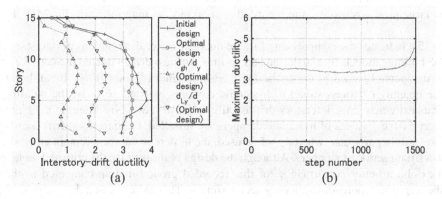

**FIGURE 25.7** Optimal trigger displacement distribution of hysteretic damper system and maximum interstory-drift ductility with respect to step number (Stiffness: Initial design for $k_L = 1$).

one in the lower two-thirds of the building model. In addition, as the step proceeds, the maximum interstory-drift ductility decreases first and increases later.

## 25.7 SYSTEM RELIABILITY

Figure 25.5 shows that, if the stiffness ratio of hysteretic dampers to the main frame is increased as $k_L = 1$, 3, 5, the interstory-drift ductility can be reduced effectively with almost uniform distribution along height. This means that the system reliability of the MDOF shear building structures against story collapse can be upgraded by the installation of hysteretic dampers with gap mechanism. If the gap mechanism is not installed, the total system of the main frame and the hysteretic dampers becomes too stiff. This may induce a large story shear in the total system. Furthermore, the effective amplitude range of hysteretic dampers is not so wide because hysteretic dampers have a limitation on the accumulated plastic deformation capacity due to fatigue. In view of these points, the hysteretic damper with gap mechanism has a strong advantage for the upgrade of system reliability of the total system.

## 25.8 RESPONSE COMPARISON OF MDOF BUILDING MODEL INCLUDING HVH SYSTEM WITH MDOF BUILDING MODEL INCLUDING DHD SYSTEM UNDER RECORDED GROUND MOTION OF EXTREMELY LARGE AMPLITUDE

The effectiveness of the hysteretic dampers with gap mechanism is shown here for the MDOF building model under a recorded long-period pulse-type ground motion of extremely large amplitude. Figures 25.8(a) and (b) present a ground motion acceleration and the corresponding velocity time history of the Japan Meteorological Agency (JMA) Nishiharamura-Komori(EW) wave during the Kumamoto earthquake in 2016. This ground motion is well known as a long-period pulse-type ground motion of extremely large velocity amplitude, which is often used for the analysis of important structures attaining at the limit state. While the maximum velocity level of severe earthquake ground motions (Level 2) in the code for tall buildings in Japan is 0.5 m/s, the recorded one in the Kumamoto earthquake is over 2.0 m/s. Figures 25.8(c)–(f) illustrate the displacement, velocity, acceleration response spectra and the input energy spectrum introduced by Ordaz et al. (2003). It can be realized that this ground motion has a large velocity response around the natural period of 0.7, 3.0(s). It should be remarked that this recorded ground motion is not necessarily critical to the present MDOF system (the fundamental natural period = 1.6 s) including hysteretic dampers with gap mechanism. However, once the MDOF system experiences the large plastic response, the recorded ground motion may influence the MDOF system greatly.

Figure 25.9 indicates the comparison of frame ductility factor distributions of interstory drift in the MDOF building models (initial design and optimal stiffness design treated in Section 25.5) with hysteretic dampers ($k_L = 1,3,5$) under the JMA Nishiharamura-Komori(EW). It can be observed that, if the quantity (stiffness value) of hysteretic dampers is small ($k_L = 1$), the building structure exhibits extremely large interstory-drift ductilities for the JMA Nishiharamura-Komori(EW). When larger stiffness values of hysteretic dampers are installed, the interstory-drift ductilities are well suppressed. In particular, when $k_L = 5$ is used, the hysteretic dampers remain an elastic range and show a strong action as stoppers. Although the design optimized for the critical double impulse does not exhibit a better performance for this recorded ground motion compared to the initial design, the response suppression tendency seems similar. This indicates that the response check by the critical double impulse is useful for the evaluation for recorded ground motions with extremely large amplitude.

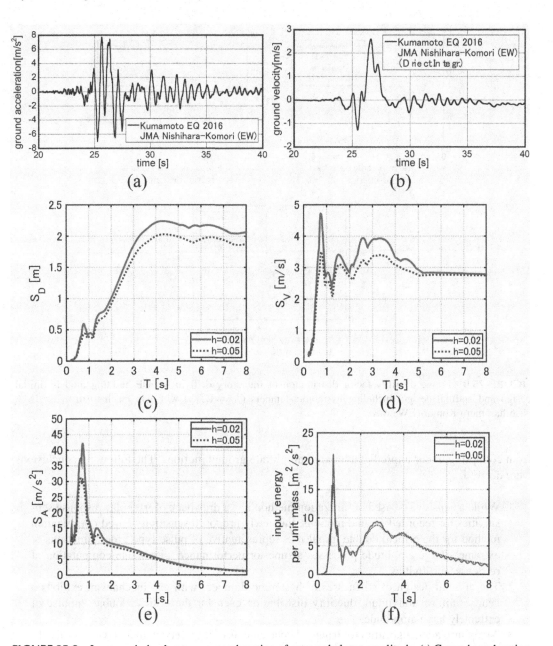

**FIGURE 25.8** Long-period pulse-type ground motion of extremely large amplitude, (a) Ground acceleration of JMA Nishiharamura-Komori(EW) wave, (b) Ground velocity of JMA Nishiharamura-Komori(EW) wave, (c) Displacement response spectrum, (d) Velocity response spectrum, (e) Acceleration response spectrum, (f) Input energy spectrum (Hashizume and Takewaki 2020b).

## 25.9 CONCLUSIONS

The optimal distribution of stiffnesses of hysteretic dampers with gap mechanism in elastic-plastic building structures was investigated for pulse-type earthquake ground motions of extremely large amplitude. Then, the optimal distribution of trigger displacements of such hysteretic dampers was also analyzed. The hysteretic dampers with gap mechanism are expected to play a role as stoppers for excessive elastic-plastic interstory deformations and prevent building structures

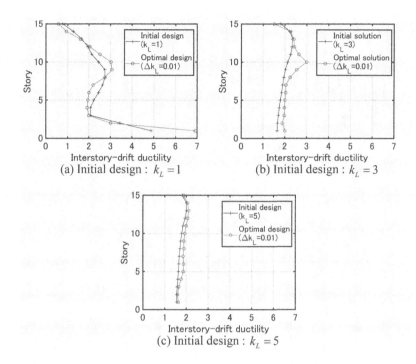

**FIGURE 25.9** Frame ductility factor distribution of interstory drift in MDOF building models (initial design and optimal design) including hysteretic dampers ($k_L = 1,3,5$) with gap mechanism under JMA Nishiharamura-Komori(EW) wave.

from collapsing under unpredictable large-amplitude ground motions. The following conclusions were derived.

1. While a sensitivity-based optimization method has a drawback of irregular response sensitivities for recorded ground motions, the newly proposed sensitivity-based optimization method for the critical double impulse as representative of pulse-type ground motions of extremely large amplitude possesses a superior performance with stable computation of response sensitivities.
2. The optimal design of stiffnesses of hysteretic dampers with gap mechanism exhibits a nearly uniform maximum ductility distribution even for the critical double impulse of extremely large amplitude.
3. An optimization algorithm for trigger displacements of hysteretic dampers with gap mechanism was also developed by using a sensitivity-based approach. It can be observed that the optimal trigger displacements distribute around $d_{gh} = d_y$ as expected.
4. It was demonstrated that, if a sufficient quantity of stiffness of hysteretic dampers is given, the MDOF building models with the initial stiffness design of hysteretic dampers and the stiffness design optimized for the critical double impulse exhibit a preferable elastic-plastic response to a recorded ground motion of extremely large amplitude.

## ACKNOWLEDGMENT

Part of the present work is supported by the Grant-in-Aid for Scientific Research (KAKENHI) of Japan Society for the Promotion of Science (No.18H01584). This support is greatly appreciated.

## REFERENCES

Adachi, F., Fujita, K., Tsuji, M. and Takewaki, I. (2013b). Importance of interstory velocity on optimal along height allocation of viscous oil dampers in super high-rise buildings, Eng Struct, 56, 489–500.

Adachi, F., Yoshitomi, S., Tsuji, M. and Takewaki, I. (2013a). Nonlinear optimal oil damper design in seismically controlled multi-story building frame, Soil Dyn Earthq Eng, 44(1), 1–13.

Aiken, I. D., Nims, D. K., Whittaker, A. S. and Kelly, J. M. (1993). Testing of passive energy dissipation systems, Earthq Spectra, 9, 335–370.

Akehashi, H. and Takewaki, I. (2019). Optimal viscous damper placement for elastic-plastic MDOF structures under critical double impulse, Front Built Environ, 5, 20.

Aydin, E., Boduroglub, M. H. and Guney, D. (2007). Optimal damper distribution for seismic rehabilitation of planar building structures, Eng Struct, 29, 176–185.

Bruneau, M., Chang, S. E., Eguchi, R. T., Lee, G. C., O'Rourke, T. D., Reinhorn, A. M., Shinozuka, M., Tierney, K., Wallace, W. A. and von Winterfeldt, D. (2003). A framework to quantitatively assess and enhance the seismic resilience of communities, Earthq Spectra, 19(4), 733–752.

Caughey, T. K. (1960). Sinusoidal excitation of a system with bilinear hysteresis, J Appl Mech, 27, 640–643.

Cimellaro, G., Reinhorn, A. and Bruneau. M. (2010). Framework for analytical quantification of disaster resilience. Eng Struct, 32(11), 3639–3649.

Domenico, D. D., Ricciardi, G., and Takewaki, I. (2019). Design strategies of viscous dampers for seismic protection of building structures: A review, Soil Dyn Earthq Eng, 118, 144–165.

Fukumoto, Y. and Takewaki, I. (2017). Dual control high-rise building for robuster earthquake performance, Front Built Environ, 3, 12.

Hanson, R. D. and Soong, T. T. (2001). Seismic design with supplemental energy dissipation devices, EERI, Oakland, CA.

Hanson, R. D. (1993). Supplemental damping for improved seismic performance, Earthq Spectra, 9, 319–334.

Hashizume, S. and Takewaki, I. (2020a). Hysteretic-viscous hybrid damper system for long-period pulse-type earthquake ground motions of large amplitude, Front Built Environ, 6, 62.

Hashizume, S. and Takewaki, I. (2020b). Hysteretic-viscous hybrid damper system with stopper mechanism for tall buildings under earthquake ground motions of extremely large amplitude, Front Built Environ, 6, 583543.

Hayashi, K., Fujita, K., Tsuji, M. and Takewaki, I. (2018). A simple response evaluation method for base-isolation building-connection hybrid structural system under long-period and long-duration ground motion, Front Built Environ, 4, 2.

Inoue, K. and Kuwahara, S. (1998). Optimum strength ratio of hysteretic damper, Earthq Eng Struct Dyn, 27, 577–588.

Jacobsen, L. S. (1960). Damping in composite structures, Proc 2nd WCEE, Tokyo, pp. 1029–1044.

Kawai, A., Maeda, T. and Takewaki, I. (2020). Smart seismic control system for high-rise buildings using large-stroke viscous dampers through connection to strong-back core frame, Front Built Environ, 6, 29.

Kojima, K. and Takewaki, I. (2015). Critical earthquake response of elastic-plastic structures under near-fault ground motions (Part 1: Fling-step input), Front Built Environ, 1, 12.

Kojima, K. and Takewaki, I. (2016). A simple evaluation method of seismic resistance of residential house under two consecutive severe ground motions with intensity 7, Front Built Environ, 2, 15.

Kondo, K. and Takewaki, I. (2019). Simultaneous approach to critical fault rupture slip distribution and optimal damper placement for resilient building design, Front Built Environ, 5, 126.

Lagaros, N., Plevris, V. and Mitropoulou, C. C. (Eds.) (2013). Design optimization of active and passive structural control systems, Information Science Reference, Hershey, PA, USA.

Lavan, O. and Levy, R. (2010). Performance based optimal seismic retrofitting of yielding plane frames using added viscous damping, Earthq Struct, 1(3), 307–326.

Makita, K., Murase, M., Kondo, K. and Takewaki, I. (2018). Robustness evaluation of base-isolation building-connection hybrid controlled building structures considering uncertainties in deep ground, Front Built Environ, 4, 16.

Murakami, Y., Noshi, K., Fujita, K., Tsuji, M. and Takewaki, I. (2013a). Simultaneous optimal damper placement using oil, hysteretic and inertial mass dampers, Earthq Struct, 5(3), 261–276.

Murakami, Y., Noshi, K., Fujita, K., Tsuji, M. and Takewaki, I. (2013b). Optimal placement of hysteretic dampers via adaptive smoothing algorithm, ICEAS13 in ASEM13, Jeju, South Korea, pp. 1821–1835.

Nakashima, M., Saburi, K. and Tsuji, B. (1996). Energy input and dissipation behaviour of structures with hysteretic dampers, Earthq Eng Struct Dyn, 25, 483–496.

Noroozinejad, E., Takewaki, I., Yang, T. Y., Astaneh-Asl, A., Gardoni, P. (Eds.) (2019). Resilient structures and infrastructures, Springer, Singapore.

Ogawa, Y., Kojima, K. and Takewaki, I. (2017). General evaluation method of seismic resistance of residential house under multiple consecutive severe ground motions with high intensity, Int J Earthq Impact Eng, 2(2), 158–174.

Ordaz, M., Huerta, B. and Reinoso, E. (2003). Exact computation of input-energy spectra from Fourier amplitude spectra, Earthq Eng Struct Dyn, 32, 597–605.

Shiomi, T., Fujita, K., Tsuji, M. and Takewaki, I. (2016). Explicit optimal hysteretic damper design in elastic-plastic structure under double impulse as representative of near-fault ground motion, Int J Earthq Impact Eng, 1(1/2), 5–19.

Shiomi, T., Fujita, K., Tsuji, M. and Takewaki, I. (2018). Dual hysteretic damper system effective for broader class of earthquake ground motions, Int J Earthq Impact Eng, 2(3), 175–202.

SNAP. (2015). An elastic-plastic analysis program for arbitrary-shape three-dimensional frame structures, Ver.6.1., Kozo System, Inc, Tokyo.

Soong, T. T. and Dargush, G. F. (1997). Passive energy dissipation systems in structural engineering, JohnWiley & Sons, Chichester.

Takewaki, I. (2009). Building control with passive dampers: Optimal performance-based design for earthquakes, John Wiley & Sons Ltd., Asia, Singapore.

Takewaki, I. (2020). New architectural viewpoint for enhancing society's resilience for multiple risks including emerging COVID-19, Frontiers in Built Environment, 6, 143.

Takewaki, I., Fujita, K., Yamamoto, K. and Takabatake, H. (2011). Smart passive damper control for greater building earthquake resilience in sustainable cities, Sustainable Cities Society, 1(1), 3–15.

Tani, T., Maseki, R. and Takewaki, I. (2017). Innovative seismic response controlled system with shear wall and concentrated dampers in lower stories, Frontiers in Built Environment, 3, 57.

Uetani, K., Tsuji, M. and Takewaki, I. (2003). Application of optimum design method to practical building frames with viscous dampers and hysteretic dampers, Eng Struct, 25, 579–592.

Xia, C. and Hanson, R. D. (1992). Influence of ADAS element parameters on building seismic response, J Struct Eng, 118(7), 1903–1918.

# 26 Reliability Analysis and Resilience Assessment of Substation Systems Using the State Tree Method
## *A Framework and Application*

*Jichao Li, Qingxue Shang, and Tao Wang*
Institute of Engineering Mechanics,
China Earthquake Administration, Harbin, China

## CONTENTS

## 26.1 INTRODUCTION

Electricity has significant effects on post-earthquake rescue and relief responses, especially in terms of saving lives in emergencies. Among the vital components of power networks, substations play a key role to transform and distribute power. Such a functionality is achieved by various interconnected electrical equipment, such as transformers, disconnecting switches, and circuit breakers, leading to the high degree of redundancy and functional interactions of the substations. As a result, it is challenging to quantify the seismic reliability and resilience of substations.

Substations were generally treated as single nodes in the studies focusing on power system reliability (Vanzi, 1996), In Federal Emergency Management Agency multi-hazard loss estimation methodology (FEMA, 2003), the damage states of the substation were divided into five categories, i.e., none, slight/minor, moderate, extensive, and complete, according to the proportion of damaged subcomponents. Recent years, the seismic resilience, which is defined as the ability of a system to resist, restore, and adapt to an earthquake impact (Bruneau et al., 2003), of power systems has been investigated from many aspects, including the frameworks, the optimization of recovery process, and the strategy of enhancement (Espinoza et al., 2020). In contrast, studies focused on the seismic resilience of substations were barely found in the literature. Liu, Shu, and Hu (2019) quantified the resilience of substations based on damage indices of different components, such as buildings, transformers, and other high-voltage electrical equipment. Nevertheless, the dynamic recovery process of the damaged system was ignored. Li et al. (2019) proposed a probability-based method, namely, the state tree method to evaluate complex systems. The state tree method explicitly considers the interrelations between components by integrating the merits of the success path method and the fault tree method. The operational state of a system can be directly determined based on that of each component in the system. It enables the use of a Monte Carlo simulation to calculate the reliability of the system, and to consider the dynamic recovery processes of the damaged system.

In this study, the state tree model-based probabilistic framework was proposed to assess the seismic performance of substations, quantitatively. The seismic reliability of the substation at different functionality levels was computed. Key components are identified considering their effects on the system fragility with sensitive analysis. The seismic resilience of the substation was quantified in terms of recovery curves and indices. The effects on the substation's resilience due to different recovery strategies were discussed.

## 26.2 A QUANTITATIVE FRAMEWORK FOR SEISMIC RELIABILITY AND RESILIENCE ASSESSMENT

The proposed framework includes four steps: (1) definition of system functionality; (2) analysis of the system using the state tree method demonstrating the system's network topology; (3) quantification of system reliability through a Monte Carlo simulation; (4) assessment of system resilience considering various recovery priorities, as shown in Figure 26.1. The components' fragility functions and recovery parameters have to be obtained prior to the calculation.

## 26.3 FUNCTIONALITY DEFINITION AND SYSTEM MODELLING OF A CASE SUBSTATION

A typical 220/110/10 kV substation, as suggested by State Grid Corporation of China (2011), is used as an example. The substation is located in a low seismic area. The maximum considered earthquake (MCE) is 0.2 g. The layout of the substation is shown in Figure 26.2. The electric power enters the substation from the 220 kV portion and exists from the 110 kV portion, passing through the transformer. There are 12 and 8 lines in the 110 kV portion and the 220 kV portion, respectively. The lines and transformers work independently and are connected to the bus, which is used to gather and distribute power. Two buses are used in the 110 kV portion and the 220 kV portion. To further improve the system's reliability, one of the two buses in the 220 kV portion is separated into two segments. To guarantee the functionality of the substation, the control house must be operational.

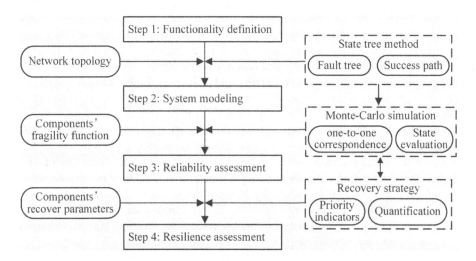

**FIGURE 26.1** Quantitative framework for seismic reliability and resilience assessment.

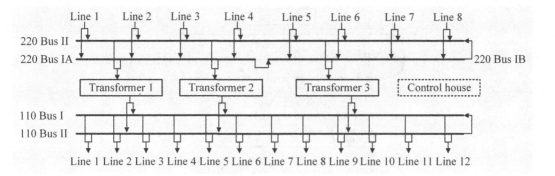

**FIGURE 26.2**  Layout of a typical 220-kV substation.

The substation's functionality, $F_s$, represents the power accessibility of the 12 outgoing lines. Therefore, it is defined as the ratio of the number of available line-out units and the total amount. The $F_s$ values range from zero to one with an increment of 1/12. Note that the components that are not directly related to the outputs are implicitly considered in the state tree model with the success paths. Due to economic issues, the transformers are not operating with full loads (GB-T 13462-2008, 2008). Thus, two transformers are assumed to be sufficient to sustain the power supply. It is further assumed that two line-out units can be supplied by one line-in unit because substations are redundantly designed. Therefore, the functionality of the substation is determined by Eq. (26.1).

$$F_s = \min\left(N_{line-out}, 6N_{transformer}, 2N_{line-in}\right)/12 \qquad (26.1)$$

Substations are always parallelly designed, therefore, power can be delivered in several paths. These paths are defined as the success paths because they determine the potential directions of power flow in the substation. All components are classified into five functional units: the line-out unit, the bus-110 unit, the transformer unit, the bus-220 unit, and the line-in unit. Power is inaccessible from the line-out unit unless all five functional units operate normally. The fault trees of the five functional units are first established. Then they are assembled together according to the success paths. The state tree model of the substation is shown in Figure 26.3. More details on this approach have been described in a previous study (Li et al., 2019).

## 26.4  FRAGILITY FUNCTIONS AND RECOVERY TIME OF COMPONENTS

A double lognormal model (Electric Power Research Institute, 1994) was used to define the equipment's fragilities, as shown in Eqs. (26.2) and (26.3), respectively. Both the aleatory randomness $\beta_r$ and the epistemic uncertainty $\beta_u$ are considered in such a model to reflect the large discrepancy in the seismic capacities of electrical equipment.

$$F_{ds}(edp) = \Phi\left(\frac{\ln(edp/\bar{x}_m)}{\beta_r}\right) \qquad (26.2)$$

$$\bar{x}_m = \hat{x}_m \cdot e^{-\Phi^{-1}(Q)\cdot\beta_u} \qquad (26.3)$$

where $\bar{x}_m$ is the median of the capacity of the component, $\hat{x}_m$ is the median of the median capacity of the component, $Q$ is the probability that the median capacity of a component exceeds a given value $\bar{x}_m$.

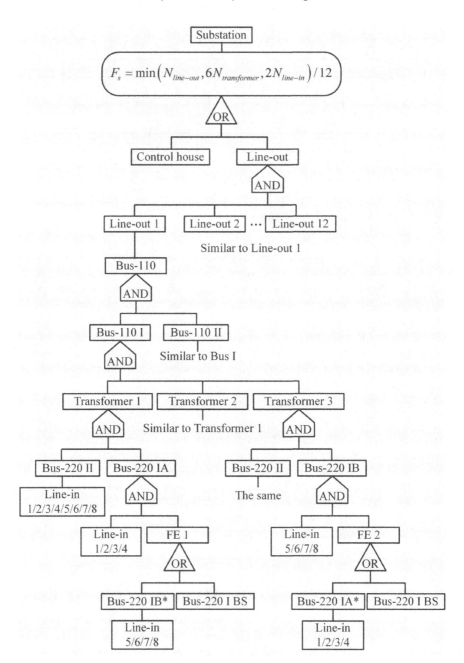

**FIGURE 26.3**    State tree model of the substation.

The fragility parameters, i.e., median capacity and logarithmic standard deviations, and recovery time of different components are shown in Table 26.1 (Li, 2018; Li et al., 2020). Experiential, experimental, and numerical data from literatures are summarized to determine the parameters of the transformer, disconnect switch, circuit breaker, potential transformer, current transformer, and lightning arrester. In contrast, assumed parameters are used for the bus bars, insulators, steel frames, and the control house because few studies have focused on them. To illustrate the use of the framework, the time to replace the components were estimated considering the suggestions of several experienced workers.

**TABLE 26.1**

**Fragility Parameters and Recovery Time of Different Components**

| | 110 kV | | | | 220 kV | | | |
|---|---|---|---|---|---|---|---|---|
| | Median | | | Time | Median | | | Time |
| Equipment | (g) | $\beta_r$ | $\beta_u$ | (day) | (g) | $\beta_r$ | $\beta_u$ | (day) |
| Transformer | 0.43 | 0.68 | 0.49 | 0.50 | 0.59 | 0.47 | 0.30 | 0.60 |
| Disconnect switch | 0.66 | 0.41 | 0.54 | 0.40 | 0.55 | 0.38 | 0.37 | 0.50 |
| Circuit breaker | 0.62 | 0.45 | 0.45 | 0.25 | 0.46 | 0.37 | 0.61 | 0.35 |
| Potential transformer | 0.93 | 0.38 | 0.48 | 0.20 | 0.49 | 0.22 | 0.28 | 0.25 |
| Current transformer | 0.82 | 0.40 | 0.45 | 0.20 | 0.41 | 0.27 | 0.58 | 0.25 |
| Lightning arrester | 0.65 | 0.42 | 0.45 | 0.15 | 0.65 | 0.28 | 0.26 | 0.20 |
| Bus bar | 1.00 | 0.25 | 0.25 | 1.00 | 1.00 | 0.25 | 0.25 | 1.20 |
| Insulator | 0.80 | 0.30 | 0.30 | 0.10 | 0.70 | 0.30 | 0.30 | 0.15 |
| Bus insulator | 0.80 | 0.30 | 0.30 | 0.10 | 0.70 | 0.30 | 0.30 | 0.15 |
| Steel frame | 1.5 | 0.25 | 0.25 | 0.25 | – | – | – | – |
| Control house | 1.0 | 0.25 | 0.25 | 1.00 | – | – | – | – |

## 26.5 QUANTIFICATION OF SYSTEM RELIABILITY

Monte Carlo simulation is used to calculate the reliability of the substation system with the fragility functions of components. The operational state, i.e., failure or success, of a component is marked as zero or one, respectively. In one simulation for a given peak ground acceleration (PGA), random numbers between zero and one are generated for all components. The operational states of the components are determined by comparing these random numbers and the failure probability that is defined by the fragility curve. The system functionality is then evaluated from bottom to top according to the logic defined in the state tree.

The reliability of the substation is defined as the probability of a minimum of $N$ outgoing lines surviving after an earthquake over the total number of outgoing lines. The reliability of the substation with different functionalities are illustrated in Figure 26.4. With a reduction in the substation's functionality $F_s$, the corresponding reliability curves shifted toward the right, indicating that the reliability of the substation is higher if the requirement of functionality is lower. When an earthquake occurs, the residual functionality of the substation is essential to emergency response; therefore, the substation's performance must be assessed based on fully defined functionality.

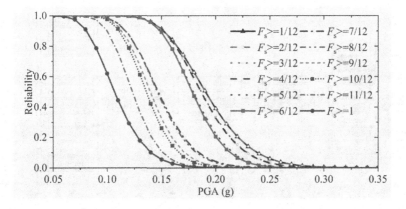

**FIGURE 26.4** Reliability of substation system with different functionalities.

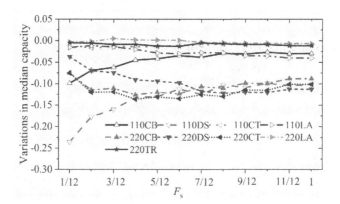

**FIGURE 26.5** Influences on substation's median capacities due to components' median capacities.

Variations in the system's median capacity, i.e., the PGA achieves 50% reliability, with a 30% reduction in one type of equipment are depicted in Figure 26.5. The circuit breaker, disconnect switch, and current transformer were the three most critical pieces of equipment in the substation. Transformers are regarded as the most important piece of equipment in a substation due to their high vulnerability. However, their effects are limited with regard to functionality because there are only three transformers in the substation. In addition, the transformer is connected with other equipment in series. Any piece of equipment of the transformer unit is damaged, its function will be lost.

## 26.6  ASSESSMENT OF SYSTEM RESILIENCE

Seismic resilience quantifies the changes in the system functionality over time after an earthquake. For substation systems, a pre-determined (PD) recovery procedure that suggested by experts or experienced engineers may exist in case of emergency conditions. The whole substation is assumed to be repaired in seven steps, as shown in Figure 26.6. For simplicity, the components in a functional unit are assumed to be repaired sequentially, so as the same groups of functionals units. The recovery curve of the substation is calculated indirectly with seven calculation conditions, i.e., $R1$–$R7$. $Ri$ indicates that the first to the $i$-th functional units are repaired. The total time to repair the substation is the sum of the time to repair all damaged components. Then the time to repair a specific group of functional units is the difference of the time of the adjacent calculation conditions. For example, the difference between $T_{R2}$ and $T_{R1}$ is the time to repair all line-in units.

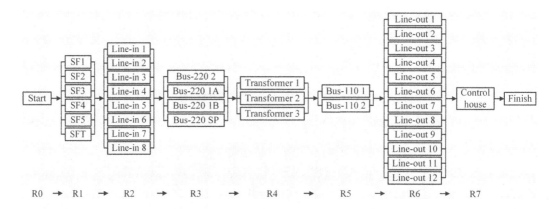

**FIGURE 26.6** Pre-determined (PD) recovery strategy for the substation.

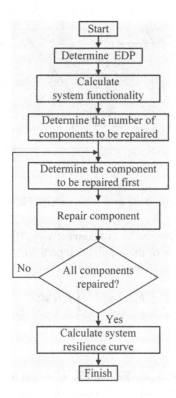

**FIGURE 26.7** Iterative process to determine the recovery sequence.

It is beneficial that the maintenance staff in the substation can be trained before an earthquake with a PD recovery routine. However, the performance of the substation with such a recovery routine may not satisfy the requirements, i.e., the fast restoration of system's functionality. A functionality prioritization (FP) recovery strategy for the substation is determined iteratively using the state tree model, as shown in Figure 26.7. In one the simulation, the damaged component with an operational state of zero can be located using the state tree model. The operational state of each of the $n$ damaged component's is set as one, whereas that of others remains unchanged. Then the functionality of the substation and the time to repair the component is calculated. After $n$ calculations, the component that improves the system functionality most is determined as the one to be repaired first. Such procedures are repeated until all components are repaired. Note that the recovery sequence of the substation is determined at the functional unit level rather than the component level because the functionality of the substation will not be affected unless all components in a functional unit are repaired.

The average recovery curves of the substation with the PD and FP strategy at 0.2g are shown in Figure 26.8. Approximately 83.1% of the functionality of the substation is lost at the beginning of the earthquake. During the recovery process, it is noted that the system functionality may not be improved by every repair sequence. For example, limited improvement in the functionality is observed in the first three days with PD strategy. On the other hand, the repair of the transformer units could significantly improve the substation's functionality because the results of R4 and R3 exhibits substantial difference. The total time to repair the substation are not affected by the recovery strategies, however, faster restoration of the system's functionality is obtained with the FP strategy compared to the PD strategy.

The resilience index $R_{Resi}$ (Cimellaro et al., 2010) is defined in Eq. (26.4). The control time is selected as 72 hours (SPUR, 2009). The resilience indices of the substation are 0.190 and 0.563 with PD and FP strategy, respectively. The loss of system functionality is not resulted from the damage

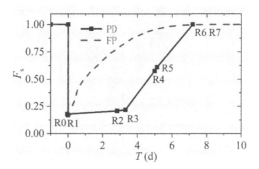

**FIGURE 26.8**    Average recovery curve of the substation with the PD and FP recovery strategies at 0.2g.

of a large number of components, but the loss of connectivity. Therefore, the system functionality is restored step by step with the repair of the critical components.

$$R_{Resi} = \frac{1}{T_{LC}} \int_0^{T_{LC}} F_s(t)dt \tag{26.4}$$

## 26.7  CONCLUSIONS

A new framework based on the state tree model was proposed to assess the seismic reliability and resilience of substation systems. The one-to-one correspondence between the operational states of the components and the system are clearly defined using the state tree method, enabling the quantification of dynamic recovery processes of a damaged system.

The seismic reliability of the substation was found to be lower compared with individual components. The sensitivity analyses identified the circuit breaker, disconnect switch, and current transformer as the three most critical pieces of equipment in the substation, either due to their low seismic capacities or large amounts.

The effects of two different recovery strategies, i.e., PD and FP, on the seismic resilience of the substation were quantified in terms of recovery curves and resilience indices. The system's resilience can be significantly improved if the FP strategy is used.

## ACKNOWLEDGMENTS

This study was sponsored by the Scientific Research Fund of Institute of Engineering Mechanics, CEA (2019B02) and the National Natural Science Foundation of China (51908519). Any opinions, findings and conclusions or recommendations expressed in this paper are those of the authors and do not necessarily reflect the views of the sponsors.

## REFERENCES

Bruneau M, Chang S E, Eguchi R T, et al., (2003). A framework to quantitatively assess and enhance the seismic resilience of communities, Earthquake Spectra, 19(4): 733–752.

Cimellaro G P, Reinhorn A M, Bruneau M, (2010). Framework for analytical quantification of disaster resilience, Engineering Structures, 32(11): 3639–3649.

Electric Power Research Institute (1994). Methodology for developing seismic fragilities. Electric Power Research Institute, Palo Alto, CA.

Espinoza, S, Poulos, A, Rudnick, H, de la Llera, J C, Panteli, M, Mancarella, P, (2020). Risk and resilience assessment with component criticality ranking of electric power systems subject to earthquakes. IEEE Systems Journal, 14(2), 2837–2848.

FEMA (2003). HAZUS-MH Multi-hazard loss estimation methodology, Earthquake Model - Technical Manual. Washington D.C., Federal Emergency Management Agency, Department of Homeland Security.

GB-T 13462-2008 (2008). Economical operation for power transformers. China Zhijian Publishing House, Beijing (In Chinese).

Li J C (2018). Study on probability-based seismic performance assessment method for substation systems. Institute of Engineering Mechanics, China Earthquake Administration, Harbin (In Chinese).

Li J C., Wang T, Shang Q X (2019). Probability-based seismic reliability assessment method for substation systems. Earthquake Engineering and Structural Dynamics, 48(3), 328–346.

Li J C., Wang T, Shang Q X (2020). Probability-based seismic resilience assessment method for substation systems. Structure & Infrastructure Engineering, in press, 1–13. (DOI: 10.1080/15732479.2020.1835998.)

Liu R S., Shu R X., Hu Z X (2019). Study on the seismic resilience of substations. China Earthquake Engineering Journal, 41(4), 827–833 (in Chinese).

San Francisco Planning and Urban Research Association (SPUR) (2009). The resilient city: Defining what San Francisco needs from its seismic mitigation polices. spur.org.

State Grid Corporation of China (2011). General design of power transmission and transformation project. China Electric Power Press, Beijing (in Chinese).

Vanzi I (1996). Seismic reliability of electric power networks: Methodology and application. Structural Safety, 18(4), 311–327.

# 27 Reliability-Based Methods for Inspection Planning and Predictive Maintenance in Fixed Offshore Structures

*Tiago P. Ribeiro*
Tal Projecto, Lda, Lisboa, Portugal

*Ana M.L. Sousa*
Instituto Superior Técnico, Universidade de Lisboa,
Lisboa, Portugal

## CONTENTS

DOI: 10.1201/9781003194613-27

## 27.1   INTRODUCTION AND BACKGROUND

### 27.1.1   WHY USING RELIABILITY-BASED METHODS FOR ANALYZING STRUCTURES IN OCEAN ENVIRONMENT?

Throughout this chapter, we will enter the world of fixed offshore structures, depict its ongoing data-powered and decarbonization-directed fundamental change, and discuss the role of Reliability-Based Methods in it.

Notwithstanding a significant shift towards deep and ultra-deep offshore in new oil and gas explorations in the past two decades, more than 90% of such structures are located in shallow waters (Reddy and Swamidas 2014, Ribeiro 2017, Ribeiro et al. 2020). This led to quite standardized structural concepts, based on braced systems with welded tubular elements, governed by fatigue resistance, except for platforms in some tropical sites. Therefore, the structural behavior of this type of structures is rigid, with a significant mismatch between its Eigen frequencies and ocean actions usual frequency. This is why most of the offshore structures are classified as "fixed."

Currently, the panorama where fixed offshore structures are inserted in is a rapidly changing one. On one hand, these structures are required to function beyond what has been designed for, either due to fields' life extension with Enhanced Oil Recovery and technical developments in reservoir engineering and flow assurance (Sousa, Matos, and Guerreiro 2019) or the expected future trend in converting existing platforms for wind energy production. On the other hand, abandonment is not environmentally or even legally an option and decommissioning bears significant costs and risks (Fowler et al. 2014).

These circumstances pushed regular design and analysis methods, as defined by most codes and as an industry standard, to obsolescence. Current and future analyses are deemed to assess existing structures for its condition and remaining life, especially to fatigue damage, and to establish reinforcement, inspection and maintenance risk-based plans that enable its life extension and conversion for different functions. This can only be attained with the broad and systematic employment of Reliability-Based Methods (Bai et al. 2018, Monsalve-Giraldo et al. 2019, Ribeiro et al. 2020).

A step forward in the analysis of fixed offshore structures with Reliability-Based Methods allows the setting of Predictive Maintenance (PDM) programs, unleashing the potential of increasing data volume that can be routinely gathered.

To understand the potential impact of PDM in fixed offshore structures, one shall regard the common maintenance approaches. These include the reactive approach, usually referred as "*fix and fail*," associated with uneconomical assets' downtime (Ribeiro et al. 2020), and the Preventive Maintenance (PM) approach (Lee et al. 2006, Forsthoffer 2017), comprising frequent material replacements, heavy workforce usage and, even, early deactivation of assets. In both cases, economic inefficiency outstands, as well as the difficulty in dealing with unexpected failure. PDM, on the contrary, matches sensors (Carvajal, Maucec, and Cullick 2018) and inspection data with reliability-based structural analysis. This allows predicting end-of-life scenarios and, therefore, leverages timely repair planning as well as effective maintenance and inspection schedule for attaining the best information with the least cost.

The ongoing Digital Transformation era, which already endowed assets with inexpensive and increasingly reliable sensors, unmanned vehicles for inspection, connections through Industrial Internet of Things (IIoT) and the widespread use of Artificial Intelligence (AI) techniques (Lu et al. 2018, Sousa et al. 2019) is already able to foster PDM to unexpected levels of efficiency in assets management. Yet, for bringing these effective new approaches to reality, assets' structural analysis needs Reliability-Based Methods.

## 27.1.2 Fixed Offshore Structures

### 27.1.2.1 Structural Concepts, Purpose and Materials

Offshore structures are usually employed in energy production. Typical uses include wind farms, wave energy capture, as well as production, storage and control facilities within the oil and gas industry. These systems can be classified as fixed, flexible or floating. On opposition to the latter, the first two are supported in the seabed. Fixed structures (jackets as usually referred) owe this classification to its rigidity, with first modes' Eigen frequencies exceeding wave frequencies. These conditions are very common for offshore structures with a height of no more than 300 m.

Jackets are predominately made of steel tubes. Its conceptual design comprises three or four main legs and braced planes between them. Bracing systems most common layouts include X, K, V or N arrangements.

Governing load cases include seismic action, accidental loading, currents, but mostly wave actions in the splash zone and wind actions in the topside and, especially, in wind energy platforms. This results in a significant vulnerability to fatigue phenomena for most sites, except for tropical waters.

However, failure in a structural part does not necessarily imply collapse. Therefore, offshore structures shall be designed with sufficient redundancy, which can be measured by the collapse probability given the failure of a structural element ($P_{SYS}$). In practical terms designing for redundancy means, for example, favoring four-leg conceptual designs over three-leg layouts.

Other five main aspects that significantly increase the risk of fatigue failure in offshore structures include: the environment aggressiveness with cumulative effects of fatigue and corrosion; maritime transport and installation induced damages; a higher probability for crack initiation in fully welded construction; Hydrogen Induced Cracking (HIC) likely to occur in thick tubes exposed to low temperatures and premature fracture due to the decrease of material toughness, also in a low temperatures scenario.

### 27.1.2.2 Ocean Environment

Reducing the complexity of the ocean environment to mathematical parameters is, evidently, impossible. However, designing and inspecting structures in the ocean environment requires establishing methodologies to infer design actions from measurable parameters. The following summary illustrates how such endeavor can be carried out, accounting for the needs of Reliability-Based Methods in fixed offshore structures.

All waves can be regarded as random processes; however, sea and swell variability are very different. Sea is locally generated by the wind, thus its height and frequency are very heterogeneous. Swell is propagated in time and space, beyond where it was formed by wind action, therefore it is more constant.

Wave data stochastic nature can be modeled with time-series or wave histograms. For such an end, the following parameters are employed: wave height (H), wave period (T) leading to the wave frequency of $f = 1/T$ and $w = 2\pi f$, water depth (d or L), wavelength (L or $\lambda$) given by $L = gT^2/(2\pi)$ for deep waters ($d/L > 0.5$), wave steepness (H/L commonly within the range of 1/20 to 1/18) and wave number ($k = 2\pi/L$). However, real data are not so simple, and conventions are needed. The following are the basics for wave accounting.

For wave height, we use the significant wave height ($H_{1/3}$ or $H_S$), which computes the average height of the third highest of all the measured waves. For wave period we use the significant wave period ($T_S$), which computes the average period of the third-highest period wave, among all measured waves. For period assessing, zero up-crossing method can be applied. Using it, only the waves that crest is above the Mean Sea Level (MSL) and its trough under MSL are accounted for.

As depicted in Figure 27.1, peak-to-peak period, $T_{Ci}$, is measured among any two successive peaks and zero up-crossing period, $T_{Zi}$, is measured between the last peak before the series change

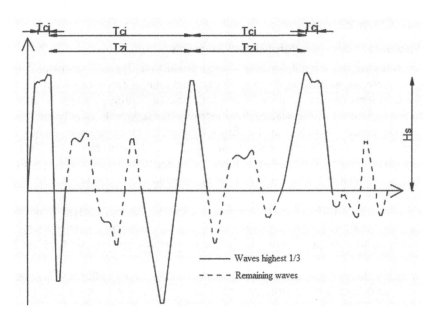

**FIGURE 27.1**    Peak-to-peak and zero up-crossing wave periods.

signal (which means passing through the MSL) and the first peak after changing signal again. Zero up-crossing period $(T_Z)$ is the most common parameter in wave data. However, caution is necessary since some wave spectra formulae use the peak-to-peak period $(T_C$ or $T_P)$.

A graphical summary encompassing the procedures from measuring ocean environment parameters to attaining structural response can be found in Figure 27.2.

Due to its overwhelming complexity, ocean environment can be regarded as a stochastic process. Therefore, it can either be measured on-site, employing the aforementioned parameters or modeled.

Several wave theories can be considered, from the most straightforward Linear (Airy) to more complex ones such as Trochoidal (Boussinesq, Dubreil-Jacotin or Gerstner theories), Cnoidal (Korteweg-de Vries, Keulegan-Patterson or Latoine) or Solitary (Scott-Russel or Boussinesq).

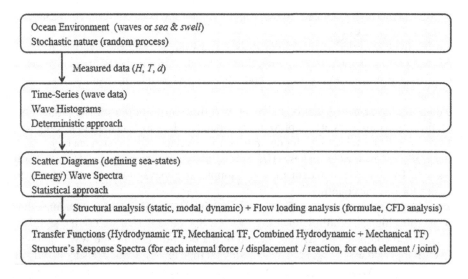

**FIGURE 27.2**    A general outlook from measuring ocean environment parameters to attaining structural response.

However, the most common practice is employing Stokes theories (of 2nd, 3rd or 5th order). In fact, DNVGL-RP-C210 standard prescribes using the 5th-order Stokes theory.

After measuring or modeling sea and swell, deterministic time-series or wave histograms are attained. The same data, for a certain time lapse, can be expressed in these two formats. While time-series are long chains of $H_i$-$T_i$ data, wave histograms are short tables assigning many occurrences $n_i$ to $H_i$-$T_i$ pairs. If data is measured, time series can be rearranged as wave histograms. If data is modeled, wave histograms may be directly synthesized. To build wave histograms, $H_i$ and $T_i$ must be reduced to discrete ranges, instead of continuous variables. That is why the name *Block Histogram* is also common. Usually, measured or modeled data is composed of 2,000,000–6,000,000 waves per year, and shall comprise complete years, to capture seasonality.

A subsequent, and practical step, will be transforming wave histograms into scatter diagrams, transforming occurrences in frequency, provided that wave histogram data sufficiently profuse to represent occurrence probability.

One final step includes computing spectral power density functions named wave spectra. There are Energy Wave Spectra and Height Wave Spectra, both attained by applying the Fourier Transform to wave time-series amplitude and frequency functions.

Since site conditions, encapsulating different sea and swell relations, have a significant impact upon wave spectra, calibration is still needed, and most spectra are semi-empirical. These include Pierson-Moskovitz (P-M or ISSC) spectrum for mature sea states worldwide, JONSWAP spectrum (and TMA Modified JONSWAP) mostly for the North Sea, Ochi (or Ochi-Hubble or Ochi-Hubble Double Peak) spectrum when sea and swell coexist and are both very significant, Information and Telecommunication Technology Center (ITTC) spectrum and Bretschneider spectrum for mature sea states with hurricanes or typhoons for the Gulf of Mexico and the South China Sea.

### 27.1.2.3  Fatigue and Fracture Mechanics

The use of calculation models based on the principles of fracture mechanics (FM) is essential to enable the inclusion of information resulting from inspections. In these circumstances, mathematical models are formulated to describe the fatigue damage attending to critical details' idiosyncrasies.

Designing a numerical method for the simulation of the fatigue failure phenomenon requires, primarily, the idealization of the fracture mechanism. For this purpose, the multidimensionality of the mechanism must be accounted for, considering bidirectional propagation initiated in a micro-crack, usually at the root of a weld. In the case of tubular structures, as most common in the offshore environment, both circumferential and through-thickness propagation should be modeled.

Under these circumstances, profuse FM formulations can be reduced to Eqs. (27.1) and (27.2), with the Stress Intensity Factor for the crack maximum depth, $\Delta K_A$, exceeding the threshold value for crack propagation, $\Delta K_{th}$, and $a_0$ as the crack depth (*a*) initial value, as well as to Eqs. (27.3) and (27.4), with the Stress Intensity Factor for the crack maximum length, $\Delta K_B$, exceeding $\Delta K_{th}$ and $c_0$ as the crack half-length (*c*) initial value (DNV 2015).

$$da/dN = C_A (\Delta K_A)^m \tag{27.1}$$

$$\Delta K_A = \Delta\sigma Y_{(a,c)} \sqrt{\pi a} \tag{27.2}$$

$$dc/dN = C_B (\Delta K_B)^m \tag{27.3}$$

$$\Delta K_B = \Delta\sigma Y_{(a,c)} \sqrt{\pi c} \tag{27.4}$$

In the aforementioned, unstable propagation and, therefore fracture, is found as the Stress Intensity Factors, $\Delta K$ equal the crack unstable propagation limit, $K_{mat}$ (also referred to as $K_{max}$).

Unstable propagation can be conceptualized as the point when crack length reaches its critical dimensions $a_c$ and $c_c$. Once one critical dimension is attained the other will also do, considering its interdependent behavior.

$\Delta K_{th}$, $K_{mat}$ and $m$ are material-specific for the employed steel alloy and environmental conditions. C parameter is an especially influential constant which accounts for the cyclic growth rate of cracks.

In a first phase, the crack should propagate bidirectionally until it reaches the thickness of the tube (T) or the critical depth $a_c$, followed by a second phase in which its circumferential growth will continue until it makes up the perimeter of the section (that is until $c$ reaches half the perimeter) or reaches the critical length, $c_c$.

Stress Intensity Factor is extremely important for quantifying the crack propagation. It contains the mathematical formulations leading to the quantification of the intensity of the stresses at the end of the equivalent crack, which depends, in addition to the geometry of the crack and the elements in which it propagates, on the relative size of the crack and the increase in stresses that its growth represents. Thus, despite the stress resulting from the structural calculation already contemplating macroscopic geometric effects, either by applying Stress Concentration Factors (*SCF* or $K_t$) or by obtaining Hot-Spot stresses through numerical simulations, it is necessary to affect them by geometry functions in the context of calculating the Stress Intensity Factors, $\Delta K_A$ and $\Delta K_B$.

The geometry function expressed in the classical formulation of the Stress Intensity Factor (*K*, also referred as $K_g$), is described as Y or, more precisely, as $Y_{(a)}$ or $Y_{(a,c)}$ since it also depends on the variable crack dimensions. For its quantification it is possible to use the Newman-Raju formulation (Newman and Raju 1981), as depicted in Eq. (27.7) with $\alpha$ as the ratio membrane tension to total stress, as expressed in Eq. (27.8), with *degree of bending* (DOB), $Y_m$ the geometry function for membrane stress and $Y_b$ the geometry function for bending stress.

$$K_g = \sigma_{tot} Y_{(a,c)} \sqrt{\pi a} \qquad (27.5)$$

$$Y = Y_m \alpha + Y_b (1 - \alpha) \qquad (27.6)$$

$$K_g = \sigma_{tot} Y_{(a,c)} \sqrt{\pi c} \qquad (27.7)$$

$$\alpha = 1 - DOB = \sigma_{membrane} / (\sigma_{membrane} + \sigma_{bending}) \qquad (27.8)$$

Considering that the crack starts in a weld makes it necessary to apply an additional geometry function, which aims to adapt the quantification of the stress intensity factor to the effects of the initial imperfection. For this reason, the former definition of the stress intensity factor assumes the nomenclature $K_g$, leaving the common designation of K for the product of the former with a function appropriate to the welding geometry, $M_k$, given by Eq. (27.9).

$$K = M_k K_g \qquad (27.9)$$

$M_k$ functions must accurately describe the intensity of the stresses in the notches created in the welds. For such an end, it is possible to use the Bowness-Lee expressions (Bowness and Lee 2000). Adding the Newman-Raju and Bowness-Lee functions, the geometry functions are written as Eqs. (27.10) and (27.11).

$$Y_A = Y_{Am} \alpha M_{kma} + Y_{Ab} (1 - \alpha) M_{kba} \qquad (27.10)$$

$$Y_B = Y_{Bm} \alpha M_{kmc} + Y_{Bb} (1 - \alpha) M_{kbc} \qquad (27.11)$$

### 27.1.2.4    Structural Analysis

As known from broader structural engineering applications, static analyses owe its name to static loading, while dynamic analyses allow spectral or time-history loading. In offshore structures fatigue design, using dynamic analyses for spectral loading is the widespread procedure. Yet, dynamic analyses require previous modal analyses.

When deterministic fatigue analyses are employed, design actions and resistances are considered and uncertainty is indirectly treated by specifying safety factors. On the other hand, spectral fatigue analyses use scatter histograms, wave theories and wave spectra as inputs, perform modal and dynamic analyses upon the structure while taking load-structure and foundations-structure interactions to synthesize transfer functions, thus computing response spectra of structures.

Furthermore, probabilistic fatigue analyses take uncertainty explicitly into consideration for calculations. Reliability is computed considering each parameter uncertainty, while reliability levels are required to be fulfilled.

To summarize, analysis can be static, modal, dynamic (also referred as modal), spectral (also referred as dynamic or modal) or time-history (also referred as dynamic or modal), concerning structure behavior influence on loading as well as loading conditions along the time. Nomenclature heterogeneity is due to geographical, industry and software differences.

Furthermore, analysis can also be deterministic or probabilistic concerning the way reliability is treated.

S-N data approach and FM approach are both possible when using either deterministic or probabilistic fatigue analyses. The difference is considering the fatigue damage, and corresponding failure criteria, either as a linear sum of each cycle damage, inferred from S-N curves or as the crack growth that leads to sectional fracture.

### 27.1.2.5    Standards

DNVGL-RP-C210 recommendations establish a framework for the use of probabilistic methods in fatigue evidence inspection. However, beyond DNVGL-RP-C210, using DNVGL-RP-0005 (DNV GL 2015a) is necessary to analyze remaining life and plan inspections under DNVGL framework (Ribeiro et al. 2020).

Offshore structures general design standards (such as DNV-OS-C101 (DNV 2015), DNV-OS-C201 and NORSOK N004), in conjunction with execution standards (such as DNV-OS-401 and NORSOK M101), prescribe certain safety and performance levels.

Both the DNV and NORSOK frameworks set Design and Inspection Classes. Those are deemed to be used to assess initial defects in parts and joints as well as to evaluate structural safety.

NORSOK N004 standard (which use must be completed with NORSOK M101) Design Classes are defined as DC1–DC5, while Inspection Classes are A–E. The relation between both groups depends on structural detail type and (high or low) fatigue effects exposure.

NORSOK M101 quantifies how many tests are required for each Inspection Class. DNV-OS-401's philosophy is quite similar to NORSOK M101's. Its Inspection Classes (under the name of Categories) are IC-I, IC-II and IC-III.

For defining acceptability criteria for initial flaws BS 7910 (BSI 2015), DNV-RP-D404 and ISO 5817 can be consulted for visual inspection, EN ISO 17635 for radiographic, electromagnetic and ultrasonic testing, EN ISO 23278 for electromagnetic and liquid penetrant testing (PT), EN ISO 11666 for ultrasonic testing and ISO 10675 for radiographic testing (RT), as well as ISO 6520 and EN 12062 standards regardless of the testing method.

Using visual or electromagnetic inspection techniques, incomplete weld penetration flaws are acceptable until the least of 10 mm and half weld thickness for C, D and E categories, but not for A and B categories.

Using the same techniques, undercuts are acceptable until 0.75 mm for C, D and E categories and until 0.50 mm for A and B categories.

Using RT, pores dimensions are admissible until 3 mm if grouped and the least of 6 mm and one-fourth of material thickness if isolated, for A and B categories. For C, D and E categories those limits rise to 4 mm in the first case and the least of 6 mm and one-third of material thickness in the second.

Regardless of the test method, cracks are not acceptable. This does not mean they will not exist, it just means that they should not exceed methods' detection capacity.

Design standards and execution standards – that, together, define safety criteria and specify fabrication requirements – must comply with fatigue analysis standards assumptions for the later calculation methods (such as S-N curves) to be valid. However, design and execution standards do not fully match. For instance, fatigue design recommendations DNVGL-RP-0005 assume the inexistence of undercuts in type D and harsher S-N fatigue resistance curves, 0.5 mm undercuts in D, E and F curves and 1.0 mm undercuts for F1 and worse (less resistant) curves.

Such conditions must be materialized in the structures' execution. Yet, some of those requirements (for D and upper curves) exceed the most demanding execution and control criteria as in the significant standards, namely ISO 5817.

### 27.1.3 Structural Health Monitoring, Remaining Life Assessment, Inspection Planning and Predictive Maintenance

#### 27.1.3.1 Data Collection

Acquiring data is the first step towards current analyses and forthcoming digital transformation in any industry. It can only be assured by physical means, with very different degrees of complexity, but all deemed to sense physical modifications in the surveyed objects.

Recent developments significantly decreased sensors, computing, physical memory and communications cost (Carvajal et al. 2018), while increasing its capabilities. As a result, its employment is now much more profuse, not only for advanced manufacturing but across industries, including the ones where sensor locations are remote, such as in offshore structures.

Sensors are already so widespread in the industry that a current modern platform is equipped with up to 80,000 sensors (Evensen and Matson 2017). Such gear power allows major Oil & Gas companies to acquire between 1 and 2 TB of data, daily. During an asset's lifetime, 15 petabytes (or 15 million gigabytes) can be gathered (IQPC Middle East 2018). Yet, despite the impressive volume of acquired data that frequently exceeds the capacities for local or online storage, the size is not the only (and possibly not the most significant) challenge of Big Data. Data variety, velocity, value and complexity are other attributes that impose major technological challenges for acquisition, transport and storage (Kaisler et al. 2013). This means that evolving big data analytics will require ever more complex sensor technology at the data network first end.

Different techniques can be employed to acquire information on structural health. Those can be summarized in three types: Quality control during production, periodical inspections or continuous monitoring and deep inspections or transportation to land for investigation. The later applies to floating facilities and implies significant downtime and costs. The first includes both non-destructive testing (NDT) and destructive testing upon samples. While standards and project criteria must be followed, it is common that the fabrication quality exceeds the minimum requirements. Thus, having detailed information about these tests may help to assess fatigue life.

Periodical inspections and continuous monitoring are limited do NDT, to the accessible parts of the structure, to lighter inspection devices, to the sea conditions and to budgetary limits.

Visual inspection, either general (GVI) or close (CVI), is the most common inspection method and can be accompanied by electromagnetic or ultrasonic testing. Otherwise, liquid penetrant and RT require drying the environment locally, which can be very expensive. Underwater inspection can be performed by divers or using Remotely Operated Vehicles (ROV).

Liquid PT allows identifying flaws, cracks, pores and any other superficial discontinuities. It is cheap but does is not able to scan in-depth nor is possible to employ in wet conditions. According to

BS 7910, PT detection capacity (smaller detectable flaw) is 5.0–1.5 mm (length-depth) for cast steel, 5.0–1.0 mm for ground welds and 20.0–4.0 mm for untreated welds.

Electromagnetic testing is mostly used to scan flaws in-depth. It includes magnetic particle inspection (MPI/MT), alternating current field measurement (ACFM), alternating current potential drop (ACPD) and eddy/Focault current (ET/EC). According to BS 7910, MPI detection capacity is 5.0–1.5 mm (length-depth) for ground welds and 20.0–4.0 mm for untreated welds. ET is, currently, the electromagnetic test with wider employment. That is mostly due to the capacity, offered by ET, of waiving the need for painting removal on the tested specimens. This is highly advantageous in the marine environment, given the difficulties in painting with good quality under such conditions. ET detection capacity is very dependent on the flaw morphology, member geometry, equipment and analyst expertise. Nevertheless, a 15.0–3.0 mm (length-depth) detection capacity for most cases can be considered.

ACFM can be performed in wet or dry environments and does not require prior surface cleaning. Its results are impaired by the thick paint layers and its detection capacity is better than the remaining electromagnetic tests, detecting 15.0–2.0 mm (length-depth) flaws.

ACPD is not advantageous compared to ACFM, thus its use decayed with the increase of the first. However, both ACFM and ACPD are usually employed to measure flaws that have been detected by other methods.

Ultrasonic Testing, including Conventional Ultrasonic Testing (UT), Automated Ultrasonic Testing (AUT), Focused Phased Array (FPA) and Time of Flight Diffraction (TOFD) tests are able to detect flaws, pores, cracks and inclusions either on the tested surface, inside the material or in the back surface. Detection capacity is, according to BS7910, 15.0–3.0 mm for UT, 10.0–1.5 mm for AUT and FPA and 10.0–2.0 mm for TOFD.

RT is unpractical to employ in the marine environment, given equipment weight and required isolation, and impossible to use upon hollow sections.

Using either X or $\gamma$ radiation flaws, cracks and pores can be found, regardless of its position within the material. Considering BS7910, RT detection capacity spans different kinds of flaws, with more than 1.2 mm.

Flooded Member Detection (FMD) is a monitoring method which detects indirectly the existence of through-thickness cracks, given its consequence of flooding structural members, previously dry.

Within stochastic analyses, it is necessary to define initial flaws, based on methods' detection capacities and standards' requirements, to incorporate detected flaws or maximum undetectable flaws during service life and to associate a probability of detection (PoD) for each of the aforementioned.

### 27.1.3.2  Evidence-Based Calculations

Bayesian Inference is frequently used as the theory underlying inspections planning in structures subjected to fatigue effects. It provides the background for the calculation procedures that allow using inspections results to continuously improve structural failure probabilities evaluation.

Synthetically, Bayesian Inference can be described as a rational method for estimates updating with additional information. Its simple definition can be explained considering A an uncertain event, whose probability of occurrence can be computed as $P(A \mid \&)$ for a certain state of information $\&$. The probability of another event B occurring given the fact the event A occurred is $P(B \mid A,\&)$. However, if the occurrence of A is unknown, then the probability of B occurring is $P(B \mid \&)$. Under these conditions, Bayes rule states that:

$$P(A \mid B, \&) = P(A \mid \&) \, P(B \mid A, \&)/P(B \mid \&) \tag{27.12}$$

In this manner, Bayesian Inference establishes the mathematical framework for, in every state of information $\&$, using its observations (occurrences and non-occurrences of certain damages) named B, C and beyond, to improve the probability assessment for event A – the fatigue failure.

Within this context, failure probability firstly assessed is as less important as more and better is the additional information gathered in time. Therefore, the first probability assessment can be based on theoretical grounds, past knowledge, expert opinions or bibliography.

Applying the former to condition inspections in fixed offshore structures, information gathered in various moments can be inserted into numerical simulations. Three possible results are considered. Those are crack detection and measurement, crack detection without measurement and the non-detection. The absence of detection leads to considering the lengthiest and deepest undetectable crack. On the other hand, unmeasured detection requires further investigations.

Nevertheless, all possible results are associated with a PoD, depending on the employed means of inspection and the crack dimensions. Furthermore, all measurements are affected by tolerances and plausible errors. More advanced models may also contemplate the probability of false detection.

Adding information is, from a mathematical perspective, updating reliability levels. For doing so, random variables' uncertainty measures must also be updated. This is attained by adding conditions to a conditional probability of fatigue failure.

Hence, the failure probability during life $t$, accounting for the information acquired in $t_I$, is given by Eq. (27.13) if no crack has been detected and Eq. (27.14) otherwise;

$$P_F(t) = P(M(t) < 0 \mid H(t_1) > 0) \tag{27.13}$$

$$P_F(t) = P(M(t) < 0 \mid H(t_1) < 0) \tag{27.14}$$

Limit state function $M(t) = R(t)\text{-}S(t)$ correlates effects and resistances as $H(t_i) = a_d\text{-}d(t_i)$ relates observed crack dimension $d(t_i)$ and crack detectable dimension, $a_d$. Both formulae depict a conditional probability problem, in the sense that the probability of occurrence of a certain event, as a failure, for example, P(F|I), is related with the probability of occurrence of one given observation, which is P(I). Applying the lexicon usually considered in Bayesian inference problems, the probability of an occurrence named F, provided the result I and the information state & is formulated in Eq. (27.15).

$$P(F \mid I) = P(F \cap I)/P(I) \tag{27.15}$$

Expanding the former for various independent results (I, J, K and further, one attains Eq. (27.16).

$$P(F \mid I, \&) = P(F \mid \&) \, P(I \mid F, \&)/P(I \mid \&) \tag{27.16}$$

If one inspection is unable to find cracks, then $a_d$ is expressed by the random variable $A_{di}$, which refers to the smallest detectable crack. Similarly, the damage $d(t_i)$ can be formulated by the crack dimension, $a(t_i)$, at the inspection instant, $t_i$. Thus, the event is expressed by $a(t_i) \leq A_{di}$.

Fatigue failure probability, at the instant $t$, accounting for a no-crack result in $t_I$ can be expressed as Eq. (27.17), leading to Eq. (27.18). In the case of two independent inspections had been performed in $t_1$ and $t_2$, Eq. (27.19) applies.

$$P_F(t) = P(M(t) < 0 \mid H(t_1) > 0) = P(M(t < 0) \cap H(t_1) > 0)/P(H(t_1) > 0) \tag{27.17}$$

$$P_F(t) = P(M(t) < 0 \mid H(t_1) > 0) = P(M(t < 0) \, P(H(t_1) > 0 \mid M(t) < 0)/P(H(t_1) > 0) \tag{27.18}$$

$$P_F(t) = P(M(t) < 0 \mid H(t_1) > 0 \cap H(t_2) > 0) \tag{27.19}$$

Notwithstanding more comprehensive definitions into this book's chapters, let us recall that when relevant parameters cannot be deterministically described, random variables should be considered. Such variables are defined by one expected value and uncertainty measures. Thus, the expected value of x, E(x), can be assumed equal to the mean value of a significant sample, $\mu_x$. The most common absolute dispersion measures are the variance ($\sigma_x^2$) and the standard deviation ($s_x$ or $\sigma_x$). Furthermore, relative measures are defined, such as the (Pearson) Coefficient of Variation, $CoV_x = \sigma_x/\mu_x$, also referred to as CV.

It is important to disambiguate CoV from Cov – the covariance, one different dispersion measure.

Crack growth parameter C shall be considered a random variable, for its extreme impact upon FM models. The same applies to stress ranges $\Delta\sigma$. On the other hand, material parameters m, $K_{IC}$ and $K_{th}$ may be assumed constants.

Even though recognizing the unicity of each case, resulting from inspection methods idiosyncrasies, crack dimensions uncertainty is usually defined by Lognormal (Walbridge 2005) or Exponential (DNV GL 2015b) distributions.

The random variable which expresses the stress range makes use of several other variables. Among those can count the SCF with Lognormal distribution (Walbridge 2005, Sørensen and Ersdal 2008) and CoV = 0.04 (Walbridge 2005), the DOB, having a Normal distribution and CoV = 0.08 (Walbridge 2005), the Newman-Raju functions with Lognormal distribution (Sørensen and Ersdal 2008) and CoV = 0.05 (DNV GL 2015b) or the Bowness-Lee functions with Lognormal distribution (Walbridge 2005) and CoV = 0.10 (DNV GL 2015b).

Crack growth parameter C can be described with a Normal distribution (Walbridge 2005, Sørensen and Ersdal 2008). Its log standard deviation is 0.11 for the base metal in protected underwater structures or emerged structures in the marine environment and 0.22 for welds in the aforementioned circumstances and base metal in unprotected underwater structures (DNV GL 2015b). For adding information through Bayesian Inference, the aforementioned models' parameters are replaced by random variables and several simulations are carried out.

Analyses probabilistic nature is given by using experimental methods like the Monte Carlo Simulations (MCS). Furthermore, First-Order Reliability Method (FORM) and Second-Order Reliability Method (SORM) are possible to employ. However, the latter two are computationally very demanding and eventually unsuitable when the phenomenon to model is too complex or the information too scarce.

Monte Carlo Method can be defined as generating an extensive amount of pseudo-random values, using computer algorithms, with which stochastic simulations are built. Upon such experimental results, statistical analyses can be performed.

To simulate physical phenomena with MCS, random variables values are guided by specifying its expected value and dispersion measures. Generated values for the random variables will, then, be inserted in mathematical models that simulate physical phenomena, so that an extensive pool of results is attained.

As a result, crack growth histories are generated, upon which statistical inference reveals expected values and dispersion measures for fatigue life.

Probabilities of exceedance and non-exceedance, $p_f$, can then be determined considering a suitable statistical distribution.

### 27.1.3.3   Inspection Planning

Assessing structural condition regarding one specific type of damage, as fatigue effects in offshore structures, can be achieved through two broad strategies. One relies on continuous monitoring, by means of frequent inspections, sensors and transducers. Such a strategy is quite common for accessible bridge structures, as part of a Structural Health Monitoring (SHM) approach. The other possibility is endeavoring a program of sparse inspections, whose results will be used in probabilistic analyses. The latter is a cheaper approach, which has been profusely used and validated in the aerospace industry since equipment weight and energy input are significant downsides and downtime is economically penalizing.

However, setting an adequate Inspection Plan for sparse inspections faces significant challenges, namely how to deal with uncertainty. Accounting for aleatory uncertainty, as well as epistemic uncertainty as long as it is not disproportionate, can be facilitated by defining the structural reliability concept, which is inversely proportional to the failure probability.

Then, one can use structural reliability requirements as a tool for defining structural safety criteria. Therefore, the ability for allowing explicit uncertainty accounting is why Reliability-Based calculations and Probabilistic methods are being increasingly regarded as advantageous for fatigue analysis of structures, compared to the deterministic methods, even if the latter is widespread in design codes. Furthermore, probabilistic methods allow assessing structural safety in structures that have already surpassed its service life.

Under these circumstances, Reliability and Risk-based Inspection (RBI) has been proposed as the basis for inspections planning, employing the concept of Remaining Useful Life. That can already be carried out under a regulated framework for offshore structures, applying DNVGL-RP-C210 recommendations (DNV GL 2015b) and for bridge and building structures complying with the Eurocodes, namely with the reliability index control prescribed in Eurocode 3's Part 1–9 for damage tolerant structures.

A critical feature of inspections planning methods is allowing inspection intervals continuous adjustment to the former results, rather than keeping an immutable first program. Yet the goal of assuring that no fatigue failure arises before the developing crack is inspected no less than on two occasions during its growth can be particularly useful.

Nevertheless, some issues arise from continuously adjusting inspections. For instance, using a purely theoretical continuously updated reliability model, if successive good results are attained in inspections, the uncertainty on structural health will decrease and the inspection intervals may be continuously enlarged. Thus, a structure which has shown no damage throughout its service life will not be frequently inspected in its end-of-life.

The former may be a serious hindrance since fatigue failure is known for occurring more frequently either on the beginning of structural life – due to significant damages, material lack of quality, inadequate design or poor construction – or on its end when the cumulative damage sparks rapid crack growth. This has been identified in other industries, including aeronautical, where the term bath-tub effect has been used.

Thus, inspection planning may need additional criteria, other than just compute past inspection results. This may include experience and expertise-based thresholds and recommendations, as material modeling.

## 27.2   THE EMPLOYMENT OF RELIABILITY-BASED METHODS

### 27.2.1   An Outlook

As aforementioned, Reliability-Based Methods are tools for enabling inspections planning and remaining life assessment and extension. Its use in inspections planning enhancement allows employing PDM strategies and conducting Remaining Life Assessments. The latter will be pivotal both for improving safety, economy and environmental risks reduction in current fixed offshore structures and for fostering Energy Transition by, for example, redesigning existing oil and gas offshore platforms to wind farms, based on its current condition.

However, Reliability-Based Methods applications for each of the mentioned purposes are usually case-specific. Moreover, guidelines or recommendations are broad in scope, specifying a philosophy and basic principles, as DNVGL-RP-C210 (DNV GL 2015b), but cannot provide exhaustive procedures.

For the specific case of fixed offshore structures, we recommend the use of a comprehensive methodology, previously proposed in (Ribeiro et al. 2020) and illustrated in Figure 27.3 with a step-by-step flowchart.

This method encompasses a series of analyses, including Global Spectral Analyses, Simplified Probabilistic Analyses, FM models creating and calibration and Probabilistic Analyses with

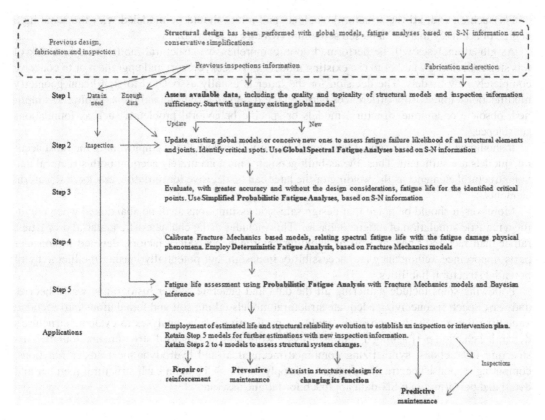

**FIGURE 27.3** Calculation method for assessing fatigue life based on inspection results. (Adapted from Ribeiro et al. 2020.)

Bayesian Inference. It is focused in fatigue damage as the main cause of condition degradation in time, however, extreme climate actions, seismic events, collisions, explosions, equipment vibration and other random loads may cause significant structural damage or accelerate fatigue damage. While there is no possible monitoring or prevention for such events, it is highly recommended that, after being subjected to extreme loading, the structure is fully inspected, not only for direct damages but also for existing fatigue damage disproportionate growth.

## 27.2.2 GLOBAL SPECTRAL ANALYSES

Global Spectral Fatigue Analyses are the basis of the standard structural design standard. Under these circumstances, it would not be possible to address structural behavior and reliability levels without resorting to it in the first place.

From such analyses, it will be possible to assess global behavior and attain transfer functions relating to the ocean environment with stresses in structural parts. Among the global behavior main issues to study one shall highlight the dynamic interaction between wave actions and structures and the non-linear behavior in soil-foundation interactions.

Yet, another particularly important role of Global Spectral Fatigue Analyses to leverage the forthcoming use of Reliability-Based Methods is the preliminary fatigue screening to every parts and detail of the structure, under as many sets of load combinations and material conditions one finds necessary. This knowledge, even if code-based and made under several design assumptions will allow direction the following, much more labor-intensive analyses, to the details where early failure is likely.

For such an end, all the structural elements and joints should be modeled with the same high degree of accuracy to avoid a biased screening (Ribeiro et al. 2020).

As global analyses will be performed upon comprehensive structural models, the first task of this step is deciding whether to use existing models and check, correct and upgrade it or to conceive completely new models. The decision for the latter is usually associated to significant geometry modifications, noteworthy differences between the designed structure and the as-built, too simplified, obsolete or antique structural models or specific behavioral problems, such as foundations subsidence.

It should be taken into account that, for Reliability-Based Methods employ, only the most accurate models are sufficient. Thus, the as-built geometry must accurately account both structural and non-structural elements as the welding of the later can be decisive for initiating cracks in structural elements.

Moreover, it should be noted that design safe-side assumptions shall be abandoned when pursuing a realistic simulation of fatigue behavior. This includes using characteristic material properties, rather than factored ones, as well as avoiding heterogeneous safety factors, deemed to consider parts importance, redundancy and accessibility in design, but potentially biasing the hierarchy of potential structural liabilities.

Following steps include gathering all the data and generate scatter histograms, wave theories and wave spectra, conceiving adequate structural models taking soil and foundations into account, validating the models through static analyses, performing modal analyses to evaluate structure's dynamic behavior, performing dynamic analyses while considering load-structure and foundations-structure interactions, synthetizing combined mechanical and hydrodynamic transfer functions, computing response spectra of the structure, applying the SCFs upon each structural member and detail and performing an SN-data approach for fatigue assessment.

### 27.2.3 SIMPLIFIED PROBABILISTIC ANALYSES

After having selected an adequate amount of structural details where fatigue problems are likelier, further analyses shall be performed to ensure its assessment and comparison. Such analyses must take reliability explicitly into account and employ more realistic quantifications for material properties and loading effects. These are deemed to be performed upon an extensive group of cases. Therefore, expedite methods are required.

For such an end, Simplified Probabilistic Fatigue Analysis, as suggested in DNVGL recommendations (DNV GL 2015b) and defined (DNV GL 2015a) may be considered.

Its formulae include computing damage as a two-parameter Weibull distribution Eq. (27.20).

$$D = (n_0/a_d)q^m\ \Gamma\left(1 + m/h; (S/q)^h\right) \tag{27.20}$$

The former expression includes the S-N curve parameters $S$, $m$ and $log\ a_d$, Weibull distribution's shape parameter, which can be valued as $h = 0.8$ following (DNV GL 2015b) recommendations and the Complementary Incomplete Gamma Function, $\Gamma(a;z)$, can be found in (DNV GL 2015a). Also, scale parameter expression Eq. (27.21) must be merged in the damage expression. It accounts for reference stress range, $\Delta\sigma_0$, the average zero up-crossing frequency, $v_0$ in Hz, and the service life, $T_d$ in seconds.

$$q = \Delta\sigma_0 / \left(ln\ n_0\right)^{\frac{1}{h}} \tag{27.21}$$

$$n_0 = v_0\ T_d \tag{27.22}$$

$$\log\bar{a} = loga - 2S_{logN} \tag{27.23}$$

Reliability index can then be attained from damage expression by computing the limit state function, $g$, as depicted in Eq. (27.26).

$$D = (n_0/\bar{a}) q^m \Gamma(1 + m/h) \tag{27.24}$$

$$g = -\ln D \leftrightarrow g = -\ln(n) + \ln(a) - m \ln(q) - \ln \Gamma(1 + m/h) \tag{27.25}$$

$$g = -z_n + z_a - m z_q - \ln \Gamma(1 + m/h) \tag{27.26}$$

Given the fact that random variables are employed, its expected values, standard deviations and coefficients of variation – $\mu$, $\sigma$ and $CoV$, respectively – shall be estimated beforehand. As a result, limit state function expected value and variance can be assessed using Eqs. (27.27) and (27.28) expressions, allowing a simple estimation for the reliability index in Eq. (27.29).

$$\mu_g = -\mu_{ln(n)} + \mu_{ln(a)} - m \, \mu_{ln(q)} - \ln \Gamma(1 + m/h) \tag{27.27}$$

$$\sigma_g^2 = \sigma_{ln(n)}^2 + \sigma_{ln(a)}^2 + m^2 \sigma_{ln(q)}^2 \tag{27.28}$$

$$\beta = \mu_g / \sigma_g \tag{27.29}$$

In these circumstances, expected component failure can be assessed by comparing obtained reliability levels with a reliability target.

There are several options for setting the values for the aforementioned reliability target. As more extensively depicted in Ribeiro et al. (2020), the most common options are either retrieving the reliability levels implicitly embedded in standards safety factors or inferring reliability thresholds directly from failure consequences. One option for the latter computation, as depicted in Eq. (27.30), is considering an acceptable yearly collapse probability, which relates a critical component failure probability with the global collapse probability given the component failure – usually referred as $P_{SYS}$.

$$P_{f,anual,target} = P_{f,anual,component} / P_{SYS} \tag{27.30}$$

### 27.2.4 Calibration of Physical Models to Deterministic Analyses

Fixed offshore structures design to fatigue – its most concerning collapse mechanism – is usually based on indirect methods, including damage accumulation sums, as the Palmgren-Miner. While this is a profusely disseminated procedure in the design of most civil engineering steel structures it has the obvious disadvantage of not keeping a physical relation between design damage and the actual fatigue mechanism evolution.

Under these circumstances, it is not possible to directly use the design analyses with the monitoring results. FM models are employed for fatigue analysis whenever a physical grounding for the calculations is required.

However, FM models are governed by an extensive group of variables. Some will not vary significantly or will not influence considerably the results but otherwise, they must be considered random variables. Despite this preliminary distinction, random variables will be too numerous, and conceiving an independent model may not be possible due to the unavailability of accurate information. Consequently, FM models, once conceived, may be calibrated to previous analyses results.

Among the parameters whose immutability is an acceptable assumption for given steel and environmental context, one may include material parameters $\Delta K_{th}$, $K_{mat}$ and $m$ (Ribeiro et al. 2020).

On the other hand, calibration should be necessary for crack propagation formulae parameters, including $C$ parameter and initial crack dimensions $a_0$ and $c_0$, as well as Stress Intensity Factor geometry functions. Nevertheless, it shall be stressed out that calibration is constricted by parameters physical meaningfulness and, desirably, by bibliographical backing. For such an end, using (DNV GL 2015b) recommendations and British standard BS 7910 (BSI 2015) is an adequate option.

For calibrating an FM model to Simplified Probabilistic Fatigue Analysis results, the developer must employ one single FM, so that the comparison is possible. Thus, a Deterministic Analysis is due for this step.

FM formulae are very well reported in the literature, yet some developments have recently been proposed, such as Castiglioni and Pucinotti (2009), Fatemi and Shamsaei (2011), Sun et al. (2012), Gao et al. (2014) or Van Do, Lee, and Chang (2015), and can be considered in the making of the FM model.

In its most simple terms, mathematical formulae for crack propagation is synthesized in (27.1) to (27.4) for $\Delta K_A > \Delta K_{th}$ and $\Delta K_B > \Delta K_{th}$.

Macroscopic effects upon the stress fields can be accounted either in simpler linear finite element models, using SCFs, or computing Hot-Spot stresses with volumetric finite element models. Nevertheless, geometric functions will be needed as accounting for local stress peaks in welds is required. Afterwards, $\Delta K_A$ and $\Delta K_B$ Stress Intensity Factors can be determined.

### 27.2.5 Probabilistic Analyses with Physical Models and Bayesian Inference

Evolving from deterministic models to probabilistic ones implies adding inspections and monitoring information into the developed and validated FM framework. Acquired datasets must be reduced to crack detection hypothesis, namely detection with measurement, detection without measurement and the absence of detection, all associated with Probabilities of Detection. The latter results in assuming the lengthiest undetectable crack while detection without measurement does not provide a valuable contribution, except for the earlier stages of crack propagation.

The mathematical formulation for creating the probabilistic models by updating reliability levels has been introduced in Section 27.1.3.2. Therefore, Eqs. (27.13–27.19) can be applied for formulating the conditional probability problem while employing conventional Bayesian inference problems lexicon.

As FM models parameters have now been replaced by random variables, simulations with reliability assessment methods, such as the MCS, FORM and SORM, can be performed.

The choice among MCS, FORM and SORM is dependent on the analysis objective (Du 2005). In a context of information scarcity or too complex modeling FORM and SORM may not be the most adequate option, especially given its computational demands for the most complex models (Leira 2013). This drove some researchers to take advantage of MCS for engineering problems with a profuse number of variables, like the ones involving fatigue and fracture (Walbridge 2005). The downside of developing MCS solutions is its resources expenditure (Papadopoulos and Giovanis 2018).

Looking for recent developments, FORM and SORM solutions are trending for similar problems, especially with the spreading of AI-based solutions (Papadopoulos and Giovanis 2018, Low and Tang 2007, Du and Zhang 2010, Sankararaman, Daigle, and Goebel 2014, Manoj 2016).

For the proposed methodology MCS are regarded as a convenient, yet computationally demanding, method, as each individual simulation comprises a crack growth history until fracture, accounting for the variable's randomness upon the developed FM behavioral model.

The results distribution, obtained from massive simulations, will yield probabilities of exceedance and non-exceedance and associated expected errors.

### 27.2.6 Setting Inspection Plans and Structural Health Monitoring in the Age of Digital Transformation and Additive Manufacturing

The objective in fatigue life extension is to seize the available data and to plan new efficient inspections, to enhance Remaining Life Assessment. Therefore, employing probabilistic methods is needed and, for doing so, Bayesian inference is a very useful tool.

Underlying philosophy can be summarized in creating realistic structural models to determine reliability levels throughout the structural life, employing probabilistic analyses with uncertainty quantifications. By comparing the attained reliability levels with the required ones, the need for further inspections can be continuously evaluated and planned.

Inspection locations and detection techniques adequacy and accuracy can be enhanced with preliminary fatigue analyses for investigating crack expected dimensions and damages locations. As results become available, mathematical models can be updated and reliability levels are computed. For each information addition, inspection plans can be enhanced. The results gathered at the inspections, whether finding cracks or not, are of paramount importance for estimating fatigue life.

Inspection techniques made possible by the current Digital Transformation are enabling Inspection Planning transformation. These consist of ubiquitous, cheaper, and more reliable sensors, even with AI embedded capacities, including image processing to better visual inspection quality, unmanned vehicles availability, as well as IIoT leveraged communication and collaboration.

Therefore, it is expected that monitoring, analysis, condition assessment and decision-making will be made in real-time, with the latter leveraged by well-trained AI systems.

By employing the formerly described methods we will create conditions not only for PDM but also for smarter and more focused repair, which will make use of emerging technologies, such as Additive Manufacturing.

### 27.2.7  Reliability-Based Methods as Tools for Energy Transition

Energy Industry is facing four major changes (World Economic Forum and Accenture 2016). Among those are (generation, transmission and distribution) Assets Life-cycle Management with real-time remote-control or PDM, deemed to extend the life cycle or operating efficiency; Grid Optimization and Aggregation; Integrated Customer Services that go as far as energy generation and management; as well as Personalized Connected Services with the aim of adding value by pushing the electricity commodity to an experience level and fostering customer engagement (Borst 2017). These four grounds of the ongoing change are recommended for investment (World Economic Forum and Accenture 2016), facing an expected profit return of over 100,000,000,000 USD (United States Dollar) within a decade.

The immense potential within Reliability-Based Methods lies, among other aspects, in the possibility for continuous assessment, life extension by calculation and structural integrity assessment for a functional modification scenario.

Thus, the ongoing and accelerating Energy Transition can be largely supported by these methods, not only in the most notorious cases with the conversion of oil and gas offshore platforms into wind farms but, in a broader sense, assisting PDM endeavors and any data-driven Assets Life-cycle Management strategies.

Reliability-Based Methods are suited for design and analysis of structures in a context of data and information abundance. Therefore, such methods are particularly suited for bringing to practice the incoming data-rich developments in autonomous smart machines and cognitive computing with further AI advances and the ability to reach overly complex thought processes in the absence of complete or reliable information (BP 2018).

Furthermore, it is already possible to say without controversy, that Digital Transformation will drive business models, in several industries including Energy, towards a service-driven, technology leveraged, more interactive and collaborative, paradigm (Ustundag and Cevikcan 2018, DNV GL, Vareide, and Ahmed 2015). Within the Oil & Gas industry, Digital transformation is leveraging several ongoing changes that can lead to business models modification. Among the most significant one can count digital asset life-cycle management, more collaborative ecosystems and growing customer engagement (World Economic Forum and Accenture 2017). These are fields where employing Reliability-based analyses can unleash the potential of acquired structures and infrastructures data.

## REFERENCES

Bai, Xiaodong, Yunpeng Zhao, Guohai Dong, and Chunwei Bi. 2018. "Probabilistic Analysis and Fatigue Life Assessment of Floating Collar of Fish Cage Due to Random Wave Loads." *Applied Ocean Research* 81(September): 93–105.

Borst, T. 2017. *Digital Transformation of Energy Systems – A Holistic Approach to Digitization of Utility System Operations through Effective Data Management.*

Bowness, D. and M. M. K. Lee. 2000. "Prediction of Weld Toe Magnification Factors for Semi-Elliptical Cracks in T-Butt Joints." *International Journal of Fatigue* 22(5), 369–387.

BP. 2018. *BP Technology Outlook 2018.*

BSI. 2015. *BS 7910 – Guide to Methods for Assessing the Acceptability of Flaws in Metallic Structures.*

Carvajal, Gustavo, Marko Maucec, and Stan Cullick. 2018. *Intelligent Digital Oil and Gas Fields. Concepts, Collaboration, and Right-Time Decisions.* Cambridge: Gulf Professional Publishing, Elsevier.

Castiglioni, Carlo A. and Raffaele Pucinotti. 2009. "Failure Criteria and Cumulative Damage Models for Steel Components under Cyclic Loading." *Journal of Constructional Steel Research* 65(4), 751–765.

DNV GL, K. Vareide, and N. Ahmed. 2015. *Industry Perspective: Digitalization in the Oil and Gas Sector.*

DNV GL. 2015a. *DNVGL-RP-0005 – Fatigue Design of Offshore Steel Structures.*

DNV GL. 2015b. *DNVGL-RP-C210 – Probabilistic Methods for Planning of Inspection for Fatigue Cracks of Offshore Structures.*

DNV. 2015. *DNV-OS-C101 – Design of Offshore Steel Structures, General (LRFD Method).*

Van Do, Vuong Nguyen, Chin Hyung Lee, and Kyong Ho Chang. 2015. "High Cycle Fatigue Analysis in Presence of Residual Stresses by Using a Continuum Damage Mechanics Model." *International Journal of Fatigue* 70, 51–62.

Du, X. 2005. *First-Order and Second-Order Reliability Methods.* Rolla: University of Missouri.

Du, Xiaoping and Junfu Zhang. 2010. "Second-Order Reliability Method with First-Order Efficiency." in *Proceedings of the ASME Design Engineering Technical Conference.*

Evensen, Ole and John Matson. 2017. *Digital Transformation in Oil and Gas. How Innovative Technologies Modernize Exploration and Production.*

Fatemi, Ali and Nima Shamsaei. 2011. "Multiaxial Fatigue: An Overview and Some Approximation Models for Life Estimation." *International Journal of Fatigue* 33(8), 948–958.

Forsthoffer, Michael S. 2017. *Predictive and Preventive Maintenance.*

Fowler, A. M., P. I. Macreadie, D. O. B. Jones, and D. J. Booth. 2014. "A Multi-Criteria Decision Approach to Decommissioning of Offshore Oil and Gas Infrastructure." *Ocean and Coastal Management* 87: 20–29.

Gao, Huiying, Hong Zhong Huang, Shun Peng Zhu, Yan Feng Li, and Rong Yuan. 2014. "A Modified Nonlinear Damage Accumulation Model for Fatigue Life Prediction Considering Load Interaction Effects." *The Scientific World Journal* 2014, 7, 164378.

IQPC Middle East. 2018. "Oil & gas: Transforming through digital technologies." in *World Digital Refineries Congress.* Kuwait.

Kaisler, Stephen, Frank Armour, J. Alberto Espinosa, and William Money. 2013. "Big Data: Issues and Challenges Moving Forward." in *Proceedings of the Annual Hawaii International Conference on System Sciences*, pp. 995–1004.

Lee, Jay, Jun Ni, Dragan Djurdjanovic, Hai Qiu, and Haitao Liao. 2006. "Intelligent Prognostics Tools and E-Maintenance." *Computers in Industry* 57(6): 476–89.

Leira, Bernt J. 2013. *Optimal Stochastic Control Schemes within a Structural Reliability Framework.* B. J. Leira (Ed.). Springer Briefs in Statistics.

Low, B. K. and Wilson H. Tang. 2007. "Efficient Spreadsheet Algorithm for First-Order Reliability Method." *Journal of Engineering Mechanics* 133(12): 1378–1387.

Lu, Yang, Liping Sun, Xinyue Zhang, Feng Feng, Jichuan Kang, and Guoqiang Fu. 2018. "Condition Based Maintenance Optimization for Offshore Wind Turbine Considering Opportunities Based on Neural Network Approach." *Applied Ocean Research* 74: 69–79.

Manoj, Neethu Ragi. 2016. "First Order Reliability Method: Concepts and Application." Delft University of Technology. https://repository.tudelft.nl/islandora/object/uuid:c4c941fa-a9c1-4bd4-a418-afc54bb6d475?collection=education

Monsalve-Giraldo, J. S., P. M. Videiro, F. J. Mendes de Sousa, C. M. P. M. dos Santos, and L. V. S. Sagrilo. 2019. "Parametric Interpolation Method for Probabilistic Fatigue Analysis of Steel Risers." *Applied Ocean Research* 90(January): 101838.

Newman, J. and I. Raju. 1981. "An Empirical Stress Intensity Factor Equation for the Surface Crack." *Engineering Fracture Mechanics* 15:185–192.

Papadopoulos, Vissarion and Dimitris G. Giovanis. 2018. "Reliability Analysis." in *Mathematical Engineering* (pp. 71–98). Springer, Germany

Reddy, DV and ASJ Swamidas. 2014. *Essentials of Offshore Structures. Framed and Gravity Platforms.* CRC Press.

Ribeiro, Tiago Pinto. 2017. "Extensão Da Vida Útil À Fadiga De Estruturas Offshore Fixas Formadas Por Elementos Tubulares." University of Coimbra.

Ribeiro, Tiago, Constança Rigueiro, Luís Borges, and Ana Sousa. 2020. "A Comprehensive Method for Fatigue Life Evaluation and Extension in the Context of Predictive Maintenance for Fixed Ocean Structures." *Applied Ocean Research* 95: 102050

Sankararaman, Shankar, Matthew J. Daigle, and Kai Goebel. 2014. "Uncertainty Quantification in Remaining Useful Life Prediction Using First-Order Reliability Methods." *IEEE Transactions on Reliability* 63(2):603–619.

Sørensen, J. D. and G. Ersdal. 2008. "Safety and Inspection Planning of Older Installations." *Proceedings of the Institution of Mechanical Engineers, Part O: Journal of Risk and Reliability* 222(3): 403–17.

Sousa, A. L., H. A. Matos, and L. P. Guerreiro. 2019. "Preventing and Removing Wax Deposition inside Vertical Wells: A Review." *Journal of Petroleum Exploration and Production Technology* 9(3): 2091–2107.

Sousa, Ana L., Tiago P. Ribeiro, Susana Relvas, and Ana Barbosa-Póvoa. 2019. "Using Machine Learning for Enhancing the Understanding of Bullwhip Effect in the Oil and Gas Industry." *Machine Learning & Knowledge Extraction* 1(3): 994–1012.

Sun, C., J. Xie, A. Zhao, Z. Lei, and Y. Hong. 2012. "A Cumulative Damage Model for Fatigue Life Estimation of High-Strength Steels in High-Cycle and Very-High-Cycle Fatigue Regimes." *Fatigue and Fracture of Engineering Materials and Structures* 35(7): 638–647.

Ustundag, Alp and Emre Cevikcan. 2018. *Industry 4.0: Managing The Digital Transformation.*

Walbridge, Scott. 2005. "A Probabilistic Study of Fatigue in Post-Weld Treated Tubular Bridge Structures." École Polytechnique Fédérale de Lausanne. Switzerland

World Economic Forum and Accenture. 2016. *Digital Transformation of Industries: Electricity.*

World Economic Forum and Accenture. 2017. *Digital Transformation Initiative: Oil and Gas Industry.*

# 28 Load and Resistance Factor Design Involving Random Variables with Unknown Probability Distributions

*Pei-Pei Li*
Kanagawa University, Yokohama, Japan


*Yan-Gang Zhao*
Kanagawa University, Yokohama, Japan

## CONTENTS

DOI: 10.1201/9781003194613-28

## 28.1 INTRODUCTION

The reliability-based limit state design method, which considers the uncertainty of the load and material properties and employs probabilistic methods to quantitatively solve the safety problems in structural engineering, providing a framework for reasonably solving structural safety design and maintainability problems. Load amplification factor and resistance reduction factor, i.e., load and resistance factor design (LRFD) format, provide a practical method for engineering reliability design (Ellingwood et al. 1982; Galambos and Ellingwood 1982; Ang and Tang 1984). This method has been adopted by various design codes around the world, e.g., the American Society of Civil Engineers (ASCE 2013), the American Association of State Highway and Transportation Officials (AASHTO 2014), the American Concrete Institute (ACI 2014), and the Architectural Institute of Japan (AIJ 2002, 2015). This method amplifies the nominal design loads through proper load factors and reduces the nominal resistance through the related resistance factor. When the factorized resistance is equal to the sum of factored loads at least, the safety of the structure can be ensured (Ugata 2000; Assi 2005). Therefore, satisfactory load and resistance factors (LRFs) should be formulated with the designed structure to meet the required reliability level.

At present, a series of specific LRFs are recommended in many design codes. For example, AIJ's "Recommendations for Limit State Design of Buildings" (AIJ 2002) suggests several sets of LRFs at different target reliability levels $\beta_T = 1, 2, 3$, and 4. Generally, in the process of structural design, practitioners always utilize the LRFs suggested in the design specification without performing complex reliability analysis. However, with the development of performance design, engineers need to estimate the LRFs to achieve more flexible structural design. For example, it is indispensable to estimate LRFs with different target reliability levels, or even the same level of reliability but include different uncertainty characteristics. In this case, it is expected that the design code not only recommends appropriate LRF values but also recommends simple and accurate methods to calculate those factors.

In principle, the LRFs could be assessed through appropriate reliability analysis methods (such as the Monte Carlo simulation (MCS) and first-order reliability method (FORM)). However, using the MCS approach to calculate the LRFs is time-consuming, and using FORM to determine these factors requires not only the determination of design points but also the iterative calculation of derivatives (Hasofer and Lind 1974; Der Kiureghian and Dakessian 1998; Barranco-Cicilia et al. 2009). Therefore, it is necessary to develop a simple and practical calculation approach to evaluate the LRFs.

The method recommended by AIJ (2002, 2015) to evaluate the LRFs requires knowing the probability density function (PDF) of all random variables and converting these variables into corresponding lognormal random variables. Later, Mori et al. (2019) developed an accurate and efficient LRF assessment method. This approach uses the shifted lognormal distribution to approximate the random variables with known probability distributions, and provides a simple and effective separation coefficient calculation technique, thereby avoiding complex iterative calculations. Owing to the absence of statistical evidence, some random variables' probability distributions are usually unknown. Therefore, in this case, a simple, reliable, and efficient approach is needed to calculate the factors of load and resistance.

This chapter discusses some commonly used LRFD methods and their limitations, and then outlines the concept of using the method of moments approach to calculate the LRFs. Ultimately, the detailed process of calculating the LRFs using the first few statistical moments of the basic random variables (i.e., mean, standard deviation, skewness, kurtosis) is introduced, and a simple target mean resistance calculation formula is adopted to avoid iterative calculations. Since this approach only uses the first few statistical moments of the basic random variables, even if the probability distribution of random variables is unknown, the LRFs can be calculated.

## 28.2 REVIEW OF LOAD AND RESISTANCE FACTOR DETERMINATION

### 28.2.1 BASIC CONCEPT OF LOAD AND RESISTANCE FACTORS

For the reliability design of the structural components with the resistance $R$ subjected to a combination of load effects $S_i$ ($i = 1, ..., n$), the LRFD equation is as follows:

$$\phi R_n \geq \sum_{i=1}^{n} \gamma_i S_{ni} \tag{28.1}$$

where $\phi$ represents the resistance factor, $R_n$ denotes the nominal resistance value, $\gamma_i$ represents the load factor, and $S_{ni}$ is the nominal value of the load effect $S_i$.

To achieve the specified reliability, the resistance factor $\phi$ and load factor $\gamma_i$ should be calculated and the design format in Eq. (28.1) should satisfy the following probability conditions:

$$G(X) = R - \sum_{i=1}^{n} S_i \tag{28.2}$$

$$P_f \leq P_{fT} \text{ or } \beta \geq \beta_T \tag{28.3}$$

where $G(X)$ represents the limit-state function, $P_f$ represents the failure probability, $\beta$ represents the reliability index, $P_{fT}$ represents the target failure probability, and $\beta_T$ represents the target reliability index.

Eq. (28.3) indicates that the failure probability $P_f$ related to the limit-state function in Eq. (28.2) should be lower than the specified target level $P_{fT}$, and the reliability index $\beta$ should be larger than the specified target level $\beta_T$.

### 28.2.2 SECOND-MOMENT METHOD FOR DETERMINATION OF LRFS

#### 28.2.2.1 Determination of LRFs under Normal Distribution

If $R$ and $S_i$ are normal random variables that are mutually independent, the second-moment reliability index $\beta_{2M}$ can be used, and the design formula for Eq. (28.3) becomes:

$$\beta_{2M} \geq \beta_T \tag{28.4}$$

where

$$\beta_{2M} = \frac{\mu_G}{\sigma_G} \tag{28.5a}$$

$$\mu_G = \mu_R - \sum_{i=1}^{n} \mu_{S_i} \tag{28.5b}$$

$$\sigma_G = \sqrt{\sigma_R^2 + \sum_{i=1}^{n} \sigma_{S_i}^2} \tag{28.5c}$$

where $\mu_R$ and $\mu_{S_i}$ represent the average value of $R$ and $S_i$, respectively; $\sigma_R$ and $\sigma_{S_i}$ represent the standard deviation of $R$ and $S_i$, respectively; and $\mu_G$ and $\sigma_G$ represent the average value and standard deviation of $G(X)$, respectively.

Substituting Eq. (28.5) into Eq. (28.4), it can be obtained that:

$$\frac{\mu_R - \sum_{i=1}^{n} \mu_{S_i}}{\sqrt{\sigma_R^2 + \sum_{i=1}^{n} \sigma_{S_i}^2}} \geq \beta_T \tag{28.6}$$

Multiplying both sides by $\sigma_G = \sqrt{\sigma_R^2 + \sum_{i=1}^{n} \sigma_{S_i}^2}$ yields:

$$\mu_R - \frac{\sigma_R^2}{\sigma_G} \beta_T \geq \sum_{i=1}^{n} \left( \mu_{S_i} + \beta_T \frac{\sigma_{S_i}^2}{\sigma_G} \right) \tag{28.7}$$

We define

$$\alpha_R = \sigma_R / \sigma_G \text{ and } \alpha_{S_i} = \sigma_{S_i} / \sigma_G \tag{28.8}$$

as the sensitivity factors of $R$ and $S_i$, respectively. Eq. (28.7) can be rewritten as follows:

$$\frac{\mu_R}{R_n} \left( 1 - \alpha_R V_R \beta_T \right) R_n \geq \sum_{i=1}^{n} \frac{\mu_{S_i}}{S_{ni}} \left( 1 + \alpha_{S_i} V_{S_i} \beta_T \right) S_{ni} \tag{28.9}$$

where $V_R$ represents the coefficient of variation (COV) of $R$, and $V_{S_i}$ represents the COV of $S_i$.

Comparing Eq. (28.9) with Eq. (28.1), the calculation formulas of LRFs are formulated as follows:

$$\phi = \left( 1 - \alpha_R V_R \beta_T \right) \frac{\mu_R}{R_n} \tag{28.10a}$$

$$\gamma_i = \left( 1 + \alpha_{S_i} V_{S_i} \beta_T \right) \frac{\mu_{S_i}}{S_{ni}} \tag{28.10b}$$

### 28.2.2.2 Determination of LRFs under Lognormal Distribution

Considering Eq. (28.2), if only a single load is applied and $R$ and $S$ are mutually uncorrelated lognormal random variables with parameters $\lambda_R$, $\zeta_R$, $\lambda_S$, and $\zeta_S$, the reliability index can be accurately determined as:

$$\beta = \frac{\lambda_R - \lambda_S}{\sqrt{\zeta_R^2 + \zeta_S^2}} \tag{28.11}$$

where $\lambda_R$ and $\zeta_R$ are the average value and standard deviation of $\ln R$, and $\lambda_S$ and $\zeta_S$ are the average value and standard deviation of $\ln S$.

The relationship between $\lambda$ and $\zeta$ with $\mu$ and $V$ can be expressed as follows:

$$\zeta^2 = \ln \left( 1 + V^2 \right) \tag{28.12a}$$

$$\lambda = \ln \frac{\mu}{\sqrt{1 + V^2}} \tag{28.12b}$$

Substituting Eq. (28.11) into Eq. (28.4) yields:

$$\frac{\lambda_R - \lambda_S}{\sqrt{\zeta_R^2 + \zeta_S^2}} \geq \beta_T \tag{28.13}$$

Multiply both sides by $\sqrt{\zeta_R^2 + \zeta_S^2}$ yields:

$$\lambda_R - \beta_T \frac{\zeta_R^2}{\sqrt{\zeta_R^2 + \zeta_S^2}} \geq \lambda_S + \beta_T \frac{\zeta_S^2}{\sqrt{\zeta_R^2 + \zeta_S^2}} \tag{28.14}$$

We define $\alpha_R = \zeta_R / \sqrt{\zeta_R^2 + \zeta_S^2}$ and $\alpha_S = \zeta_S / \sqrt{\zeta_R^2 + \zeta_S^2}$ as the sensitivity factors of $R$ and $S$, respectively. Eq. (28.14) can be rewritten as follows:

$$\frac{\mu_R}{R_n} \cdot \frac{1}{\sqrt{1 + V_R^2}} \cdot \exp(-\zeta_R \alpha_R \beta_T) R_n \geq \frac{\mu_S}{S_n} \cdot \frac{1}{\sqrt{1 + V_S^2}} \cdot \exp(\zeta_S \alpha_S \beta_T) S_n \tag{28.15}$$

By comparing Eq. (28.15) with Eq. (28.1), the LRFs are given:

$$\phi = \frac{1}{\sqrt{1 + V_R^2}} \exp(-\zeta_R \alpha_R \beta_T) \frac{\mu_R}{R_n} \tag{28.16a}$$

$$\gamma = \frac{1}{\sqrt{1 + V_S^2}} \exp(\zeta_S \alpha_S \beta_T) \frac{\mu_S}{S_n} \tag{28.16b}$$

Eq. (28.16) can be derived accurately because $R/S$ is a lognormal random variable. However, even if all the fundamental random variables are independent, it is impossible to obtain accurate calculation expressions of LRFs for multiple load effects.

### 28.2.3   FORM FOR DETERMINATION OF LRFs

The second-moment reliability index cannot be utilized to accurately evaluate the failure probability of non-normal random variables $R$ and $S_i$. Under such circumstance, the reliability index can be evaluated via FORM (Hasofer and Lind 1974), and the design format is given by:

$$R^* \geq \sum_{i=1}^{n} S_i^* \tag{28.17}$$

where $R^*$ and $S_i^*$ represent the design points of $R$ and $S_i$ in the original space.

Using the inverse normal transformation (Rackwitz and Flessler 1978; Der Kiureghian and Liu 1986), we obtain:

$$R = F^{-1}[\Phi(U_R)], S_i = F^{-1}[\Phi(U_{S_i})] \tag{28.18}$$

where $F^{-1}[\cdot]$ represents the inverse normal transformation; $\Phi(\cdot)$ represents the standard normal random variables' cumulative distribution function (CDF); and $U_R$ and $U_{S_i}$ represent the transformed standard normal random variables of $R$ and $S_i$, respectively.

The limit-state function of Eq. (28.2) can be rewritten as follows by the inverse normal transformation:

$$G(U_R, U_{S_i}) = F^{-1}[\Phi(U_R)] - \sum_{i=1}^{n} F^{-1}[\Phi(U_{S_i})] \tag{28.19}$$

The separation factors $\alpha_R$ and $\alpha_{S_i}$ are also the directional cosines of $R$ and $S_i$ in standard normal space at their design points, which are defined as follows:

$$\alpha_R = \frac{\partial G^*/\partial u_R}{\sqrt{\left(\partial G^*/\partial u_R\right)^2 + \sum_{i=1}^{n}\left(\partial G^*/\partial u_{S_i}\right)^2}}; \ \alpha_{S_i} = \frac{\partial G^*/\partial u_{S_i}}{\sqrt{\left(\partial G^*/\partial u_R\right)^2 + \sum_{i=1}^{n}\left(\partial G^*/\partial u_{S_i}\right)^2}} \tag{28.20}$$

where

$$\partial G^*/\partial u_R = \frac{\partial G\left(u_R^*, u_{S_i}^*\right)}{\partial u_R}; \ \partial G^*/\partial u_{S_i} = \frac{\partial G\left(u_R^*, u_{S_i}^*\right)}{\partial u_{S_i}} \tag{28.21}$$

where $u_R^*$ and $u_{S_i}^*$ represent the design points of $R$ and $S_i$, which can be determined by an iterative method based on derivatives, but the explicit expressions of $u_R^*$ and $u_{S_i}^*$ are generally not available.

Based on the basic concept of FORM, the design points can also be calculated as follows:

$$u_R^* = -\alpha_R \beta_T, u_{S_i}^* = \alpha_{S_i}\beta_T \tag{28.22}$$

Therefore, the LRFs obtained by FORM are given as:

$$\phi = \frac{F^{-1}[\Phi(-\alpha_R\beta_T)]}{R_n}, \gamma_i = \frac{F^{-1}[\Phi(\alpha_{S_i}\beta_T)]}{S_{ni}} \tag{28.23}$$

Eq. (28.23) gives the general expression of LRFs on the basis of FORM. A flowchart to determine the LRFs using FORM is presented in Figure 28.1. As shown in Figure 28.1, the "design point" should be determined before evaluating the LRFs utilizing FORM, and derivative-based iterations are also required. Therefore, there is a need for a practical method for engineers to determine the LRFs by themselves to produce a flexible and high-performance structural design.

### 28.2.4 Practical Methods for the Determination of LRFs

#### 28.2.4.1 Determination of LRFs via Lognormal Approximation

When $R$ and $S_i$ are uncorrelated lognormal random variables, it is possible to express the inverse normal transformation in Eq. (28.23) as:

$$R^* = \exp[\zeta_R \cdot (-\alpha_R\beta_T) + \lambda_R] = \exp(\lambda_R) \cdot \exp(-\alpha_R \cdot \zeta_R \cdot \beta_T) \tag{28.24a}$$

$$S_i^* = \exp[\zeta_{S_i} \cdot (\alpha_{S_i}\beta_T) + \lambda_{S_i}] = \exp(\lambda_{S_i}) \cdot \exp(\alpha_{S_i} \cdot \zeta_{S_i} \cdot \beta_T) \tag{28.24b}$$

where $\lambda_{S_i}$ represents the mean value of $\ln S_i$, and $\zeta_{S_i}$ represents the standard deviation of $\ln S_i$.

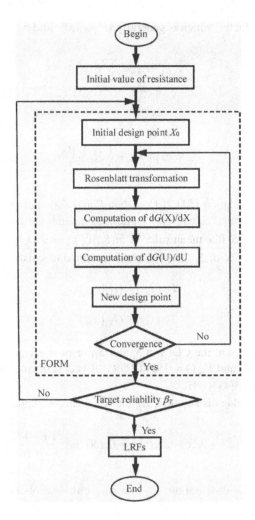

**FIGURE 28.1**   Flowchart of calculating the load and resistance factors via FORM.

For independent lognormal random variables, according to Eq. (28.23), the LRF calculation formulas under multiple load effects are:

$$\phi = \frac{1}{\sqrt{1+V_R^2}} \exp\left(-\zeta_R \alpha_R \beta_T\right) \frac{\mu_R}{R_n} \tag{28.25a}$$

$$\gamma_i = \frac{1}{\sqrt{1+V_{Si}^2}} \exp\left(\zeta_{Si} \alpha_{Si} \beta_T\right) \frac{\mu_{Si}}{S_{ni}} \tag{28.25b}$$

However, the sensitivity factors $\alpha_R$ and $\alpha_{Si}$ must be determined from the design points in FORM. For practical use, Mori and Nonaka (2001) developed empirical formulas to estimate the sensitivity factors:

$$\alpha_R = \frac{\zeta_R}{\sqrt{\zeta_R^2 + \sum_{i=1}^{n}\left(c_{Si}\zeta_{Si}\right)^2}} \cdot \varpi; \; \alpha_{S_i} = \frac{\zeta_{Si}}{\sqrt{\zeta_R^2 + \sum_{i=1}^{n}\left(c_{Si}\zeta_{Si}\right)^2}} \cdot \varpi \tag{28.26}$$

where $\varpi$ is a modification factor, which is generally set as 1.05, and $c_{S_i}$ is given as:

$$c_{S_i} = a_{S_i} \bigg/ \sum_{i=1}^{n} a_{S_i} \qquad (28.27)$$

where

$$a_{S_i} = \exp\left(\lambda_{S_i} + \frac{1}{2}\zeta_{S_i}^2\right) \cdot \frac{\mu_{S_i}}{S_{ni}} \qquad (28.28)$$

According to AIJ recommendation (AIJ 2015), when $R$ and $S_i$ do not obey the lognormal distribution, they can be converted into "equivalent" lognormal random variates. In this specification, the normalized mean value of $ln\tilde{S}_i$ (the mean value of $ln(\tilde{S}_i/\mu_{S_i})$), i.e., $\tilde{\lambda}_{S_i}$ ($\tilde{\mu}_{lnS_i}$), the standard deviation of $ln\tilde{S}_i$, i.e.,$\tilde{\zeta}_{S_i}$ ($\tilde{\sigma}_{lnS_i}$), and COV of $\tilde{S}_i$, i.e., $\tilde{V}_{S_i}$, are determined to satisfy the following conditions:

$$F_{S_i}(s_{50}) = G_{S_i}(s_{50}) \qquad (28.29a)$$

$$F_{S_i}(s_{99}) = G_{S_i}(s_{99}) \qquad (28.29b)$$

where $F_{S_i}(\cdot)$ and $G_{S_i}(\cdot)$ represent the CDFs of $S_i$ and $\tilde{S}_i$, respectively, and $s_{50}$ and $s_{99}$ represent the values satisfying $F_{S_i}(\cdot) = 0.5$ and $F_{S_i}(\cdot) = 0.99$ (0.999 if $S_i$ is described by the Type II extreme value distribution and $\beta_T \geq 2.5$), respectively.

For the Type I extreme value distribution, $\tilde{\lambda}_{S_i}$ and $\tilde{\sigma}_{lnS_i}$ can be explicitly expressed as follows:

$$\tilde{\lambda}_{S_i} = \ln(1 - 0.164V_{S_i}); \tilde{\sigma}_{lnS_i} = 0.430 \cdot \ln\left(\frac{1 + 3.14V_{S_i}}{1 - 0.164V_{S_i}}\right) \qquad (28.30)$$

For the Type II extreme value distribution, $\tilde{\lambda}_{S_i}$ and $\tilde{\sigma}_{lnS_i}$ can be explicitly expressed as follows:

1. $\beta_T \leq 2.5$

$$\tilde{\lambda}_{S_i} = \frac{0.367}{k} - \ln\left[\Gamma\left(1 - \frac{1}{k}\right)\right], \ \tilde{\sigma}_{lnS_i} = \frac{1.82}{k} \qquad (28.31a)$$

2. $\beta_T > 2.5$

$$\tilde{\lambda}_{S_i} = -\frac{0.170}{k} - \ln\left[\Gamma\left(1 - \frac{1}{k}\right)\right], \ \tilde{\sigma}_{lnS_i} = \frac{2.29}{k} \qquad (28.31b)$$

where $k$ is the statistical characteristic coefficient of earthquake disasters for different regions, which can be found in the AIJ recommendation (AIJ 2015).

For the other distributions, the following regression formulae can be used:

$$\tilde{\lambda}_{S_i} = e_0 + e_1 V_{S_i} + e_2 V_{S_i}^2 + e_3 V_{S_i}^3 \qquad (28.32a)$$

$$\tilde{\sigma}_{lnS_i} = s_0 + s_1 V_{S_i} + s_2 V_{S_i}^2 + s_3 V_{S_i}^3 \qquad (28.32b)$$

$$\tilde{V}_{S_i} = \sqrt{\exp(\tilde{\sigma}_{lnS_i}^2) - 1} \qquad (28.32c)$$

where $V_{Si}$ is the COV of the annual maximum load effect before being approximated as the lognormal random variate, and $e_j$ and $s_j$ are provided by the AIJ (2015) and are based on the probability distribution of the annual maximum value.

The reference periods for the serviceability limit-state design and ultimate limit-state design are assumed to be 1 and 50 years, respectively. In summary, when $S_i$ is not a lognormal random variable, only need to replace $\lambda_{S_i}$, $\zeta_{S_i}$, and $V_{S_i}$ with $\tilde{\lambda}_{S_i}$, $\tilde{\zeta}_{S_i}$, and $\tilde{V}_{S_i}$, respectively, and then use Eqs. (28.25–28.28) to evaluate the LRFs. The detailed procedure was presented by Mori (2002) and the AIJ (2002). Additionally, revised version of the method was presented by Mori and Maruyama (2005) and the AIJ (2015).

### 28.2.4.2 Determination of LRFs via Shifted Lognormal Approximation

When the load effect $S_i$ is a shifted lognormal random variable, the CDF of $S_i$ can be expressed as follows (Mori et al. 2019):

$$F_{S_i}(S_i) = \Phi\left(\frac{\ln(S_i - S_{i_0}) - \mu_{\ln Y_i}}{\sigma_{\ln Y_i}}\right) \tag{28.33}$$

where $\mu_{\ln Y_i}$ and $\sigma_{\ln Y_i}$ represent the log-mean and log-standard deviation of the lognormal random variable $Y_i = S_i - S_{i_0}$, and $S_{i_0}$ is the displacement.

The unknown parameters $\mu_{\ln Y_i}$, $\sigma_{\ln Y_i}$, and $S_{i_0}$ can be obtained by setting the three quantiles of the original distribution equal to the corresponding quantiles of the shift lognormal distribution. The three parameters of the shift lognormal distribution can be evaluated through finding the roots of the simultaneous equations as follows:

$$\begin{cases} p_1 = F_{S_i}(S_{i_1}) = \Phi(a_1) \\ p_2 = F_{S_i}(S_{i_2}) = \Phi(a_2) \\ p_3 = F_{S_i}(S_{i_3}) = \Phi(a_3) \end{cases} \Leftrightarrow \begin{cases} \ln(S_{i_1} - S_{i_0}) - \mu_{\ln Y_i} = a_1 \cdot \sigma_{\ln Y_i} \\ \ln(S_{i_2} - S_{i_0}) - \mu_{\ln Y_i} = a_2 \cdot \sigma_{\ln Y_i} \\ \ln(S_{i_3} - S_{i_0}) - \mu_{\ln Y_i} = a_3 \cdot \sigma_{\ln Y_i} \end{cases} \tag{28.34}$$

where $S_{ij}$ $(j = 1, 2, 3)$ represents the quantile of the load effect $S_i$ corresponding to $\Phi(a_j)$.

As shown in Eq. (28.34), these equations are generally difficult to solve directly. We set $a_1 = a$, $a_2 = a + b$, and $a_3 = a + 2b$; the solution obtained under this condition is as follows:

$$S_{i_0} = \frac{S_{i_1} \cdot S_{i_3} - S_{i_2}^2}{S_{i_1} + S_{i_3} - 2S_{i_2}} \tag{28.35a}$$

$$\sigma_{\ln Y_i} = \ln\left\{\left(\frac{S_{i_3} - S_{i_2}}{S_{i_2} - S_{i_1}}\right)^{1/b}\right\} \tag{28.35b}$$

$$\mu_{\ln Y_i} = \ln\left(\frac{(S_{i_2} - S_{i_1})^2}{S_{i_1} + S_{i_3} - 2S_{i_2}}\right) - a \cdot \sigma_{\ln Y_i} \tag{28.35c}$$

Note that $S_{i_1} + S_{i_3} - 2S_{i_2} \neq 0$.

As indicated by Eq. (28.33), $S_i = Y_i + S_{i_0}$; thus, the inverse normal transformation of the load effect $S_i$ can be expressed as follows:

$$S_i = \exp(U_{S_i} \cdot \sigma_{\ln Y_i} + \mu_{\ln Y_i}) + S_{i_0} \tag{28.36}$$

In this circumstance, the LRFs can be evaluated by substituting Eq. (28.36) into Eq. (28.23), as shown below:

$$\phi = \frac{1}{\sqrt{1+V_R^2}} \cdot \exp(\alpha_R \cdot \beta_T \cdot \sigma_{\ln R}) \cdot \frac{\mu_R}{R_n} \tag{28.37a}$$

$$\gamma_i = \left( \frac{1}{\sqrt{V_{Y_i}^2 + 1}} \cdot \exp(-\alpha_{S_i} \cdot \beta_T \cdot \sigma_{\ln Y_i}) + \frac{S_{i0}}{\mu_{S_i}} \right) \cdot \frac{\mu_{S_i}}{S_{ni}} \tag{28.37b}$$

Mori et al. (2019) proposed an accurate and simple calculation method for determining separation factors $\alpha_R$ and $\alpha_{S_i}$ (of $R$ and $S_i$, respectively) in Eq. (28.37), which does not require redundant iterative computations. Use of the shift lognormal distribution can yield good results under a wide range of target reliability levels, but there are still complex computations in estimating the three parameters of the shift lognormal distribution. In addition, this method requires the random variables' probability distribution information that is known in advance.

All the foregoing methods assume that all the fundamental random variables have known PDFs. However, in engineering practice, there are many cases where the distribution of random variables is unknown; thus, there is a practical demand for methods to determine the LRFs for random variables with unknown PDFs.

## 28.3   GENERAL FORMULATION OF METHOD OF MOMENTS FOR DETERMINATION OF LRFs

In general, the limit-state function $G(\mathbf{X})$ in Eq. (28.2) can be standardized utilizing the following formula (Lu et al. 2013):

$$G_S = \frac{G(\mathbf{X}) - \mu_G}{\sigma_G} \tag{28.38}$$

where $G_S$ represents the standardized limit-state function.

Assume that the standardized variable $G_S$ can be expressed by the first few moments of $G(\mathbf{X})$ and the standard normal random variable $U$, as follows:

$$G_S = S^{-1}(U, \mathbf{M}) \tag{28.39a}$$

$$U = S(G_S, \mathbf{M}) \tag{28.39b}$$

where $S^{-1}$ represents the inverse normal transformation, $S$ represents the normal transformation, and $\mathbf{M}$ represents the vector containing the first few moments of $G(\mathbf{X})$.

According to the reliability theory, the failure probability is defined as follows:

$$P_F = \text{Prob}[G(\mathbf{X}) \le 0] = \text{Prob}[G_S \sigma_G + \mu_G \le 0]$$
$$= \text{Prob}\left[ G_S \le -\frac{\mu_G}{\sigma_G} \right] = \text{Prob}[G_S \le -\beta_{2M}] \tag{28.40}$$

Using $F(\cdot)$ to represent the CDF of $G_S$, we can obtain:

$$F(G_S) = \Phi(U) = \Phi[S^{-1}(G_S, \mathbf{M})] \tag{28.41}$$

Accordingly, Eq. (28.40) can be rewritten as:

$$P_F = F(-\beta_{2M}) = \Phi[S^{-1}(-\beta_{2M},\mathbf{M})] \tag{28.42}$$

Thus, the moment-based reliability index can be expressed as:

$$\beta = -\Phi^{-1}(P_F) = -S(-\beta_{2M},\mathbf{M}) \tag{28.43}$$

where $\beta_{2M}$ represents the second-moment reliability index.
   Replace the reliability index given in Eq. (28.3) with the reliability index given in Eq. (28.43) to obtain:

$$\beta = -S(-\beta_{2M},\mathbf{M}) \geq \beta_T \tag{28.44}$$

From Eq. (28.44), it can be derived:

$$\beta_{2M} \geq -S^{-1}(-\beta_T,\mathbf{M}) \tag{28.45}$$

Using $\beta_{2T}$ to denote the right side of Eq. (28.45), we obtain:

$$\beta_{2M} \geq \beta_{2T} \tag{28.46a}$$

$$\beta_{2T} = -S^{-1}(-\beta_T,\mathbf{M}) \tag{28.46b}$$

Eq. (28.46a) is similar to Eq. (28.3), which means that if $\beta_{2M}$ is greater than or equal to $\beta_{2T}$, the associated $\beta$ would be greater than or equal to $\beta_T$, satisfying the required reliability. Therefore, $\beta_{2T}$ is defined as the target second-moment reliability index.
   Without loss of generality, based on the inverse normal transformation, the standardized variable $G_S$ can be expressed by the following formula (Fleishman 1978):

$$G_S = \sum_{i=0}^{n} a_i U^i \tag{28.47}$$

where $a_i$ represents the polynomial coefficients related to the moments of the limit-state function $G(\mathbf{X})$ in Eq. (28.2).
   Correspondingly, $\beta_{2T}$ can be expressed as:

$$\beta_{2T} = -\sum_{i=0}^{n} a_i(-\beta_T)^i \tag{28.48}$$

Because Eq. (28.46a) is also similar to Eq. (28.4), the LRFs of Eq. (28.46a) can be conveniently formulated through replacing $\beta_T$ with $\beta_{2T}$ on the right side of Eq. (28.4). Thereafter, the design formula can be given as:

$$\mu_R(1-\alpha_R V_R \beta_{2T}) \geq \sum_{i=1}^{n} \mu_{S_i}(1+\alpha_{S_i} V_{S_i}\beta_{2T}) \tag{28.49}$$

And the LRFs are:

$$\phi = (1 - \alpha_R V_R \beta_{2T}) \frac{\mu_R}{R_n} \tag{28.50a}$$

$$\gamma_i = (1 + \alpha_{S_i} V_{S_i} \beta_{2T}) \frac{\mu_{S_i}}{S_{ni}} \tag{28.50b}$$

where $\alpha_R$ and $\alpha_{S_i}$ are identical to those in Eq. (28.8), and $\beta_{2T}$ denotes the target second-moment reliability index calculated using Eq. (28.48).

Because the above calculation formula is on the basis of the first few moments of the fundamental random variables, even though the distribution of the fundamental random variables is unknown, the load and resistance factors can be calculated.

## 28.4 THIRD-MOMENT METHOD FOR DETERMINATION OF LRFs

### 28.4.1 DETERMINATION OF LRFs USING THIRD-MOMENT METHOD

If only the first three moments of $G(\mathbf{X})$ given by Eq. (28.2) are known, the following polynomial function of $U$ can be used to express $G_S$ as (Zhao and Lu 2011):

$$G_S = \frac{3}{\alpha_{3G}} \left\{ \exp\left[ \frac{\alpha_{3G}}{3} \left( U - \frac{\alpha_{3G}}{6} \right) \right] - 1 \right\} \tag{28.51}$$

where $\alpha_{3G}$ represents the skewness, i.e., third-order central moment of $G(\mathbf{X})$, which can be estimated as:

$$\alpha_{3G} = \frac{1}{\sigma_G^3} \left( \alpha_{3R} \sigma_R^3 - \sum_{i=1}^{n} \alpha_{3S_i} \sigma_{S_i}^3 \right) \tag{28.52}$$

where $\alpha_{3R}$ represents the skewness of $R$, and $\alpha_{3S_i}$ represents the skewness of $S_i$, respectively.

The application scope of using the inverse normal transformation shown in Eq. (28.51) to evaluate the LRFs is:

$$-1 \le \alpha_{3G} \le 1 \quad \text{and} \quad \alpha_{3G} \le 1/\beta_{2M} \tag{28.53}$$

Then, based on Eqs. (28.51) and (28.43), the third-moment reliability index is given as (Zhao and Lu 2011):

$$\beta_{3M} = -\frac{\alpha_{3G}}{6} - \frac{3}{\alpha_{3G}} \ln\left( 1 - \frac{1}{3} \alpha_{3G} \beta_{2M} \right) \tag{28.54}$$

Substituting Eq. (28.54) into the design format in Eq. (28.3) yields:

$$\beta_{2M} \ge \frac{3}{\alpha_{3G}} \left\{ 1 - \exp\left[ \frac{\alpha_{3G}}{3} \left( -\beta_T - \frac{\alpha_{3G}}{6} \right) \right] \right\} \tag{28.55}$$

By defining the right side of Eq. (28.55) as $\beta_{2T}$, we can get:

$$\beta_{2M} \ge \beta_{2T} \tag{28.56}$$

$$\beta_{2T} = \frac{3}{\alpha_{3G}} \left\{ 1 - \exp\left[ \frac{\alpha_{3G}}{3} \left( -\beta_T - \frac{\alpha_{3G}}{6} \right) \right] \right\} \tag{28.57}$$

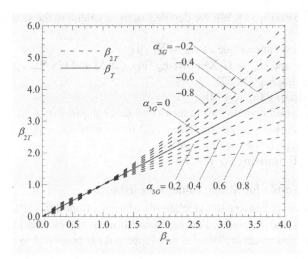

**FIGURE 28.2**  Variation of $\beta_{2T}$ with regard to $\beta_T$.

Eq. (28.56) is similar to Eq. (28.3). Thus, if $\beta_{2M}$ is greater than or equal to $\beta_{2T}$, the associated $\beta$ would be greater than or equal to $\beta_{2T}$, satisfying the required reliability.

Since Eq. (28.56) is also similar to Eq. (28.4), the LRFs related to Eq. (28.56) can be conveniently formulated through replacing $\beta_T$ with $\beta_{2T}$ on the right side of Eq. (28.10), which are given as:

$$\phi = (1 - \alpha_R V_R \beta_{2T}) \frac{\mu_R}{R_n} \tag{28.58a}$$

$$\gamma_i = (1 + \alpha_{Si} V_{Si} \beta_{2T}) \frac{\mu_{Si}}{S_{ni}} \tag{28.58b}$$

Figure 28.2 shows the variation of $\beta_{2T}$ with regard to $\beta_T$. As can be seen from Figure 28.2 that for a negative $\alpha_{3G}$, $\beta_{2T}$ is greater than $\beta_T$, and for a positive $\alpha_{3G}$, $\beta_{2T}$ is smaller than $\beta_T$. When $\alpha_{3G} = 0$ and $\beta_{2T} = \beta_T$, Eq. (28.56) is identical to Eq. (28.4), and Eq. (28.10) can be utilized to calculate the LRFs.

The LRFs can also be evaluated utilizing other forms of third-moment reliability index instead of the third-moment reliability index in Eq. (28.54). Details were presented by Zhao et al. (2006) and Lu et al. (2010).

### 28.4.2  Evaluation Method of Mean Value of Resistance

#### 28.4.2.1  Iteration Method for Mean Value of Resistance

The computation of LRFs according to the third-moment method is carried out under the assumption that $\beta$ is equal to $\beta_T$, so it is necessary to first estimate the average value of resistance under this condition (hereinafter named as the target mean resistance). The calculation formula of target mean resistance is (Takada 2001):

$$\mu_{Rk} = \mu_{Rk-1} + (\beta_T - \beta_{k-1}) \sigma_G \tag{28.59}$$

where $\beta_{k-1}$ represents the third-moment reliability index value for $(k-1)$th iteration, $\mu_{Rk}$ represents the $k$th iteration mean resistance value, and $\mu_{Rk-1}$ represents the $(k-1)$th iteration mean resistance value.

Therefore, combined with Eq. (28.59), the detailed steps to calculate the LRFs are:

1. Give the initial resistance value as $\mu_{R_0} = \sum_{i=1}^{n} \mu_{S_i}$.
2. Compute $\mu_G$, $\sigma_G$, and $\alpha_{3G}$ of $G(\mathbf{X})$ utilizing Eqs. (28.5) and (28.52), respectively, and then determine $\beta_{2T}$ using Eq. (28.57).
3. Estimate $\beta_{3M}$ utilizing Eq. (28.54).
4. Determine $k$th iteration value for the mean resistance $\mu_{Rk}$ utilizing Eq. (28.59).
5. Repeat the calculation process from steps (2) to (4) until $|\beta_T - \beta_{k-1}| < 0.0001$, the target mean resistance can be obtained hereafter.
6. Evaluate the LRFs utilizing Eq. (28.58).

### 28.4.2.2  Simple Formula for Mean Value of Resistance

Iterative calculations are necessary to calculate the aforementioned target mean resistance, which is impractical for users and designers. For these practitioners, the calculation procedure should be as accurate and simple as possible. Thus, a simple formula is proposed to approximate the target mean resistance below.

In the limit state, based on Eqs. (28.5) and (28.56), the following formula can be derived:

$$\mu_R = \sum_{i=1}^{n} \mu_{S_i} + \beta_{2T}\sigma_G \tag{28.60}$$

As shown in Eq. (28.60), $\mu_{S_i}$ is known and can be used to determine $\mu_R$. Since $\sigma_G$ and $\beta_{2T}$ are unknown and can be calculated from $\mu_R$, the initial average resistance value $\mu_{R_0}$ should be set in advance. Combined with the calculation formula of $\beta_{2T}$ given in Eq. (28.57), Eq. (28.60) can be rewritten as follows:

$$\mu_R = \sum_{i=1}^{n} \mu_{S_i} + \beta_{2T}\sigma_G = \sum_{i=1}^{n} \mu_{S_i} + \frac{3}{\alpha_{3G}}\left\{1 - \exp\left[\frac{\alpha_{3G}}{3}\left(-\beta_T - \frac{\alpha_{3G}}{6}\right)\right]\right\} \times \left(\sqrt{1 + (\mu_R V_R)^2 / \sum_{i=1}^{n}\sigma_{S_i}^2}\right)\sqrt{\sum_{i=1}^{n}\sigma_{S_i}^2} \tag{28.61}$$

Since Eq. (28.54) is derived under $|\alpha_{3G}| < 1$ (Zhao and Ang 2003), and $\mu_R$ will increase as the $\beta_T$ increases, the following approximation can be derived through trial and error:

$$\frac{3}{\alpha_{3G}}\left\{1 - \exp\left[\frac{\alpha_{3G}}{3}\left(-\beta_T - \frac{\alpha_{3G}}{6}\right)\right]\right\} \times \left(\sqrt{1 + (\mu_R V_R)^2 / \sum_{i=1}^{n}\sigma_{S_i}^2}\right) \approx \sqrt{\beta_T^{3.5}} \tag{28.62}$$

Therefore, the initial value of the resistance $\mu_{R_0}$ is calculated as:

$$\mu_{R_0} = \sum_{i=1}^{n} \mu_{S_i} + \sqrt{\beta_T^{3.5}\sum_{i=1}^{n}\sigma_{S_i}^2} \tag{28.63}$$

Accordingly, the target mean resistance obtained from iterative computations can be approximated by the following simple formula:

$$\mu_{RT} = \sum_{i=1}^{n} \mu_{S_i} + \beta_{2T_0}\sigma_{G_0} \tag{28.64}$$

where $\mu_{RT}$ represents the target mean resistance, $\beta_{2T_0}$ represents the initial target second-moment reliability index obtained from $\mu_{R_0}$, and $\sigma_{G_0}$ represents the initial standard deviation obtained from $\mu_{R_0}$.

The procedure of calculating the LRFs by means of the current simple approximate formula of the initial resistance is as follows:

1. Give the initial resistance value $\mu_{R_0}$ utilizing Eq. (28.63).
2. Compute the initial standard deviation $\sigma_{G_0}$, skewness $\alpha_{3G_0}$, and target second-moment reliability index $\beta_{2T_0}$ utilizing Eqs. (28.5), (28.52), and (28.57), respectively, and then estimate $\mu_{RT}$ using Eq. (28.64).
3. Determine the standard deviation $\sigma_G$, skewness $\alpha_{3G}$, and target second-moment reliability index $\beta_{2T}$ using Eqs. (28.5), (28.52), and (28.57), respectively, and then compute $\alpha_R$ and $\alpha_{Si}$ using Eq. (28.8).
4. Calculate the LRFs utilizing Eq. (28.58).

## 28.4.3  Numerical Examples

### Example 28.1   LRFD Using Iterative Target Mean Resistance

The statically indeterminate beam is considered in this example, as shown in Figure 28.3, which has been investigated in AIJ (2002). The beam bears three uniformly distributed loads, namely dead load $D$, live load $L$, and snow load $S$. The performance function of this beam is:

$$G(\mathbf{X}) = M_P - (M_D + M_L + M_S)  \tag{28.65}$$

where $M_P$ represents the resistance of the beam, $M_D = (Dl^2)/16$ represents the load effect of the dead load $D$, $M_L = (Ll^2)/16$ represents the load effect of the live load $L$, and $M_S = (Sl^2)/16$ represents the load effect of the snow load $S$, respectively.

Table 28.1 presents the probabilistic characteristics of the member strength and loads. The design working life of the beam is assumed to be 50 years, and among the three loads it bears, the snow load is the main load that changes with time. Here, the LRFs are estimated utilizing the target mean resistance determined by the iterative method to achieve the target reliability index $\beta_T = 2.0$. Due to the annual maximum snow load $S$ is a random variable that obeys Gumbel

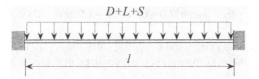

FIGURE 28.3   Statically indeterminate beam of Example 28.1.

## TABLE 28.1
### Probability Characteristics of the Random Variables in Example 28.1

| R or $S_i$ | CDF | Mean/Nominal | Mean | $V_i$ | $\alpha_{3i}$ |
|---|---|---|---|---|---|
| D | Normal | 1.00 | $\mu_D$ | 0.10 | 0.000 |
| L | Lognormal | 0.45 | $0.30\,\mu_D$ | 0.40 | 1.264 |
| S | Gumbel | 0.47 | $1.25\,\mu_D$ | 0.35 | 1.140 |
| Mp | Lognormal | 1.00 | $\mu_{Mp}$ | 0.10 | 0.301 |

distribution, the 50-year maximum snow load $S_{50}$ is also a Gumbel distributed random variable (Melchers 1999). The collected statistical values of the 50-year maximum snow load $S_{50}$ including mean $\mu_{S50} = 2.595\mu_D$, mean/nominal $\mu_{S50}/S_n = 0.972$, COV $V_{S50} = 0.169$, and skewness $\alpha_{3S50} = 1.140$.

**Step 1**. Give the initial mean value of the resistance.

$$\mu_{Mp0} = \sum_{i=1}^{n} \mu_{Ms_i} = 1.0\mu_{M_D} + 0.3\mu_{M_D} + 2.595\mu_{M_D} = 3.895\mu_{M_D}, \mu_{M_D} = \left(\mu_D l^2\right)/16$$

**Step 2**. Compute $\mu_G$, $\sigma_G$, and $\beta_{2M}$ utilizing Eq. (28.5), and calculate $\alpha_{3G}$ utilizing Eq. (28.52).

$$\mu_G = \mu_R - \sum_{i=1}^{n} \mu_{S_i} = \mu_{Mp0} - \sum_{i=1}^{n} \mu_{Ms_i} = 0, \sigma_G = \sqrt{\sigma_R^2 + \sum_{i=1}^{n} \sigma_{S_i}^2} = 0.607\mu_{M_D}$$

$$\alpha_{3G} = \frac{1}{\sigma_G^3}\left(\alpha_{3R}\sigma_R^3 - \sum_{i=1}^{n}\alpha_{3i}\sigma_{S_i}^3\right) = -0.360, \beta_{2M} = \frac{\mu_G}{\sigma_G} = \frac{\mu_R - \sum_{i=1}^{n}\mu_{S_i}}{\sigma_G} = 0$$

**Step 3**. Estimate $\beta_{3M}$ utilizing Eq. (28.54).

$$\beta_{3M} = -\frac{\alpha_{3G}}{6} - \frac{3}{\alpha_{3G}}\ln\left(1 - \frac{1}{3}\alpha_{3G}\beta_{2M}\right) = 0.06003$$

**Step 4**. Determine $\mu_{Mp1}$ utilizing Eq. (28.59).

$$\mu_{Mp1} = \mu_{Mp0} + \left(\beta_T - \beta_{3M}\right)\sigma_G = 5.073$$

**Step 5**. The difference between $\beta_T$ and $\beta_{3M}$ is given as

$$\varepsilon = |\beta_T - \beta_{3M}| = |2.0 - 0.06003| = 1.93997 > 0.0001$$

Because the difference $\varepsilon$ is too large, the foregoing calculation process should be repeated until $|\beta_T - \beta_{k-1}| < 0.001$. The iteration computation is summarized in Table 28.2. As shown, convergence

**TABLE 28.2**

**Iteration Computation for Example 28.1**

| Iter. No. | $\mu_{Mp(k)}$ | $\mu_G$ | $\sigma_G$ | $\beta_{2M}$ | $\alpha_{3G}$ | $\beta_{3M}$ | $\mu_{Mp(k+1)}$ | $\varepsilon$ |
|---|---|---|---|---|---|---|---|---|
| 1 | $3.895\mu_{MD}$ | 0 | $0.607\mu_{MD}$ | 0 | −0.360 | 0.060 | $5.073\mu_{MD}$ | 1.940 |
| 2 | $5.073\mu_{MD}$ | $1.178\mu_{MD}$ | $0.689\mu_{MD}$ | 1.710 | −0.181 | 1.658 | $5.308\mu_{MD}$ | 0.342 |
| 3 | $5.308\mu_{MD}$ | $1.413\mu_{MD}$ | $0.706\mu_{MD}$ | 2.001 | −0.152 | 1.932 | $5.356\mu_{MD}$ | 0.068 |
| 4 | $5.356\mu_{MD}$ | $1.461\mu_{MD}$ | $0.710\mu_{MD}$ | 2.059 | −0.146 | 1.987 | $5.366\mu_{MD}$ | 0.013 |
| 5 | $5.366\mu_{MD}$ | $1.471\mu_{MD}$ | $0.710\mu_{MD}$ | 2.07 | −0.145 | 1.997 | $5.367\mu_{MD}$ | 0.003 |
| 6 | $5.367\mu_{MD}$ | $1.472\mu_{MD}$ | $0.711\mu_{MD}$ | 2.073 | −0.144 | 2.000 | $5.368\mu_{MD}$ | $5 \times 10^{-4}$ |

is achieved after six iterations of computation. Through the third-moment method, the target mean resistance is obtained as $\mu_{MpT} = 5.368\mu_{MD}$.

Calculate $\alpha_{Mp}$ and $\alpha_{MSi}$ utilizing Eq. (28.8), and calculate $\beta_{2T}$ utilizing Eq. (28.57).

$$\alpha_{Mp} = \sigma_{Mp}/\sigma_G = 0.755, \alpha_{MD} = \sigma_{MD}/\sigma_G = 0.141, \alpha_{ML} = \sigma_{ML}/\sigma_G = 0.169$$

$$\alpha_{MS50} = \sigma_{MS50}/\sigma_G = 0.617, \beta_{2T} = \frac{3}{\alpha_{3G}}\left\{1 - \exp\left[\frac{\alpha_{3G}}{3}\left(-\beta_T - \frac{\alpha_{3G}}{6}\right)\right]\right\} = 2.073$$

**Step 6.** Evaluate the LRFs utilizing Eq. (28.58).

$$\phi = \mu_{Mp}(1 - \alpha_{Mp}V_{Mp}\beta_{2T})/R_n = 0.843$$

$$\gamma_{MD} = \mu_{MD}(1 + \alpha_{MD}V_{MD}\beta_{2T})/D_n = 1.029$$

$$\gamma_{ML} = \mu_{ML}(1 + \alpha_{ML}V_{ML}\beta_{2T})/L_n = 0.513$$

$$\gamma_{MS50} = \mu_{MS50}(1 + \alpha_{MS50}V_{MS50}\beta_{2T})/S_n = 1.182$$

Therefore, based on the third-moment method, the LRFD format is:

$$0.84M_{Pn} \geq 1.03M_{Dn} + 0.51M_{Ln} + 1.18M_{Sn}$$

and the design resistance value is

$$\mu_{Mp} \geq 5.37\,\mu_{MD}$$

where $M_{Dn} = (D_n l^2)/16$, $M_{Ln} = (L_n l^2)/16$, and $M_{Sn} = (S_n l^2)/16$.

According to the FORM, the LRFD format is obtained as:

$$0.89M_{Pn} \geq 1.02M_{Dn} + 0.46M_{Ln} + 1.28M_{Sn}$$

and the design resistance value is

$$\mu_{Mp} \geq 5.31\,\mu_{MD}$$

As indicated by this example, while the LRFs estimated utilizing the proposed approach are not the same as those estimated utilizing FORM, the values of the design resistance calculated from these two methods are approximately the same.

## Example 28.2   LRFD Using Simplified Target Mean Resistance

To simplify the computation of the target mean resistance, a simple formula, i.e., Eq. (28.64), is utilized to consider the foregoing limit-state function again. The initial mean resistance value is then determined as follows:

$$\mu_{Mp0} = \sum_{i=1}^{n}\mu_{Msi} + \sqrt{\beta_T^{3.5}\sum_{i=1}^{n}\sigma_{Msi}^2} = 5.461\mu_{MD}, \mu_{Mp0} = (\mu_D l^2)/16$$

Based on the obtained initial mean resistance value, $\sigma_{G_0}$, $\alpha_{3G_0}$, and $\beta_{2T_0}$ can be computed as follows:

$$\sigma_{G_0} = \sqrt{\sigma_{M_{p0}}^2 + \sum_{i=1}^{n} \sigma_{M_{Si}}^2} = 0.718\,\mu_{MD}, \alpha_{3G_0} = \left(\alpha_{3M_p}\sigma_{M_{p0}}^3 - \sum_{i=1}^{n}\alpha_{3M_{Si}}\sigma_{M_{Si}}^3\right)\Big/\sigma_{G_0}^3 = -0.133$$

$$\beta_{2T_0} = \frac{3}{\alpha_{3G_0}}\left\{1 - \exp\left[\frac{\alpha_{3G_0}}{3}\left(-\beta_T - \frac{\alpha_{3G_0}}{6}\right)\right]\right\} = 2.067$$

Using Eq. (28.64), the target mean resistance $\mu_{MpT}$ is determined as:

$$\mu_{MpT} = \sum_{i=1}^{n}\mu_{M_{Si}} + \beta_{2T_0}\sigma_{G_0} = 5.379\mu_{MD}$$

After obtaining the target mean resistance, $\sigma_G$, $\alpha_{3G}$, and $\beta_{2T}$ are calculated as:

$$\sigma_G = 0.711\mu_{MD}, \alpha_{3G} = -0.143, \beta_{2T} = 2.072$$

Considering that $-1.0 < \alpha_{3G} = -0.143 < 0.386$, the skewness is within the scope of the third-moment method. Based on Eq. (28.8), $\alpha_{Mp}$ and $\alpha_{MSi}$ can be calculated as follows:

$$\alpha_{Mp} = \sigma_{Mp}/\sigma_G = 0.756, \alpha_{MD} = \sigma_{MD}/\sigma_G = 0.141$$

$$\alpha_{ML} = \sigma_{ML}/\sigma_G = 0.169, \alpha_{MS50} = \sigma_{MS50}/\sigma_G = 0.617$$

According to Eq. (28.58), the LRFs are calculated as follows:

$$\phi = \mu_{Mp}(1 - \alpha_{Mp}V_{Mp}\beta_{2T})/R_n = 0.843, \gamma_{MD} = \mu_{MD}(1 + \alpha_{MD}V_{MD}\beta_{2T})/D_n = 1.029$$

$$\gamma_{ML} = \mu_{ML}(1 + \alpha_{ML}V_{ML}\beta_{2T})/L_n = 0.513, \gamma_{MS50} = \mu_{MS50}(1 + \alpha_{MS50}V_{MS50}\beta_{2T})/S_n = 1.182$$

Finally, the LRFD format and design resistance value are:

$$0.84M_{Pn} \geq 1.03M_{Dn} + 0.51M_{Ln} + 1.18M_{Sn}$$

$$\mu_{Mp} \geq 5.37\mu_{MD}$$

where $M_{Dn} = (D_n l^2)/16$, $M_{Ln} = (L_n l^2)/16$, and $M_{Sn} = (S_n l^2)/16$.

As shown in previous investigations, the LRFs evaluated utilizing the simplified formula are similar to those evaluated utilizing the iteration method, and the design resistance value of the simple formula under specific design conditions is almost the same as that determined utilizing the iteration method.

## 28.5　FOURTH-MOMENT METHOD FOR DETERMINATION OF LRFs

### 28.5.1　DETERMINATION OF LRFs USING FOURTH-MOMENT METHOD

If the kurtosis effect is not considered when the kurtosis value is large, the values for LRF will be inaccurate. Therefore, this section discusses the calculation approach for the LRFs with consideration of the kurtosis. When the first four moments of $G(\mathbf{X})$ are known, the standardized variable $G_S$ given by Eq. (28.2) can be represented by the following polynomial function of $U$ (Zhao and Lu 2007):

$$G_S = -l_1 + k_1U + l_1U^2 + k_2U^3 \tag{28.66}$$

where the coefficients $l_1$, $l_2$, $k_1$, and $k_2$ can be determined as:

$$l_1 = \frac{\alpha_{3G}}{6(1+6l_2)}, l_2 = \frac{1}{36}\left(\sqrt{6\alpha_{4G} - 8\alpha_{3G}^2 - 14} - 2\right) \qquad (28.67a)$$

$$k_1 = \frac{1 - 3l_2}{\left(1 + l_1^2 - l_2^2\right)}, k_2 = \frac{l_2}{\left(1 + l_1^2 + 12l_2^2\right)} \qquad (28.67b)$$

where $\alpha_{4G}$ represents the kurtosis, i.e., the fourth-order central moment of $G(\mathbf{X})$, and its calculation formula is:

$$\alpha_{4G} = \frac{1}{\sigma_G^4}\left(\alpha_{4R}\sigma_R^4 + 6\sigma_R^2 \sum_{n}^{i=1}\sigma_{S_i}^2 + \sum_{n}^{i=1}\alpha_{4S_i}\sigma_{S_i}^4 + 6\sum_{n-1}^{i=1}\sum_{n}^{j>i}\sigma_{S_i}^2\sigma_{S_j}^2\right) \qquad (28.68)$$

where $\alpha_{4R}$ represents the kurtosis of $R$, and $\alpha_{4Si}$ represents the kurtosis of $S_i$, respectively.

To make Eq. (28.67a) operable, the following requirement should be met:

$$\alpha_{4G} \geq \left(7 + 4\alpha_{3G}^2\right)/3 \qquad (28.69)$$

According to Eqs. (28.43) and (28.66), the fourth-moment reliability index is determined as follows (Lu et al. 2017):

$$\beta_{4M} = \frac{\sqrt[3]{2}p}{\sqrt[3]{-q_0 + \Delta_0}} - \frac{\sqrt[3]{-q_0 + \Delta_0}}{\sqrt[3]{2}} + \frac{l_1}{3k_2} \qquad (28.70)$$

where

$$\Delta_0 = \sqrt{q_0^2 + 4p^3}, p = \frac{3k_1k_2 - l_1^2}{9k_2^2}, q_0 = \frac{2l_1^3 - 9k_1k_2l_1 + 27k_2^2(-l_1 + \beta_{2M})}{27k_2^3} \qquad (28.71)$$

Substituting Eq. (28.70) into the design format in Eq. (28.3) yields:

$$\beta_{2M} \geq l_1 + k_1\beta_T - l_1\beta_T^2 + k_2\beta_T^3 \qquad (28.72)$$

Defining the right side of Eq. (28.72) as $\beta_{2T}$, we obtain:

$$\beta_{2M} \geq \beta_{2T} \qquad (28.73)$$

$$\beta_{2T} = l_1 + k_1\beta_T - l_1\beta_T^2 + k_2\beta_T^3 \qquad (28.74)$$

Since Eq. (28.73) is similar to Eq. (28.3), if $\beta_{2M}$ is greater than or equal to $\beta_{2T}$, the associated $\beta$ would be greater than or equal to $\beta_{2T}$, thus satisfying the required reliability. It can be seen that Eq. (28.73) is also similar to Eq. (28.4), so the LRFs related to Eq. (28.73) can be conveniently formulated by replacing $\beta_T$ with $\beta_{2T}$ on the right side of Eq. (28.10), given as:

$$\phi = (1 - \alpha_R V_R \beta_{2T})\frac{\mu_R}{R_n} \qquad (28.75a)$$

$$\gamma_i = \left(1 + \alpha_{Si} V_{Si} \beta_{2T}\right)\frac{\mu_{Si}}{S_{ni}} \tag{28.75b}$$

In particular, when $\alpha_{4G} = 3$ and $\alpha_{3G}$ is sufficiently small, we obtain $l_2 = k_2 = 0$, $k_1 = 1$, and $l_1 = \alpha_{3G}/6$. Eq. (28.74) becomes

$$\beta_{2T} = \beta_T - \frac{1}{6}\alpha_{3G}\left(\beta_T^2 - 1\right) \tag{28.76}$$

where $\beta_{2T}$ in Eq. (28.76) is similar to $\beta_{2T}$ obtained from Eq. (28.57).

Furthermore, when $\alpha_{4G} = 3$ and $\alpha_{3G} = 0$, $\beta_{2T}$ in Eq. (28.76) becomes $\beta_T$, which is identical to the $\beta_T$ in Eq. (28.4). Under this circumstance, Eq. (28.10) can be utilized to compute the LRFs. Figure 28.4a shows the variation of $\beta_{2T}$ with regard to $\beta_T$ for $\alpha_{3G} = 0$, and Figures 28.4b–d show the variation of $\beta_{2T}$ with regard to $\beta_T$ for $\alpha_{4G} = 2.8$, 3.0, and 3.2, respectively. As shown in these figures, for a negative $\alpha_{3G}$, $\beta_{2T}$ is usually greater than $\beta_T$, and for a positive $\alpha_{3G}$, $\beta_{2T}$ is usually smaller than $\beta_T$. Additionally, for $\alpha_{4G} > 3.0$, $\beta_{2T}$ is usually greater than $\beta_T$, and for $\alpha_{4G} < 3.0$, it is smaller.

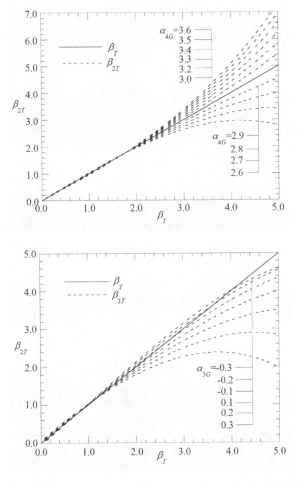

**FIGURE 28.4** Variation of $\beta_{2T}$ with regard to $\alpha_{3G}$ and $\alpha_{4G}$. (a) $\alpha_{3G} = 0.0$, (b) $\alpha_{3G} = 2.8$, (c) $\alpha_{3G} = 3.0$, (d) $\alpha_{3G} = 3.2$.

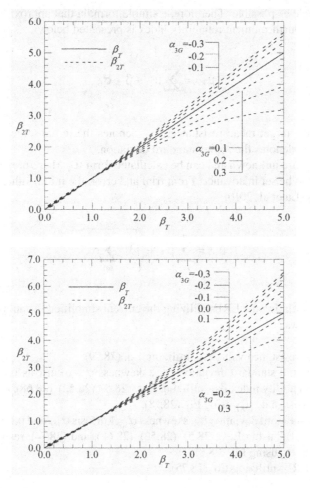

**FIGURE 28.4** (Continued)

## 28.5.2 EVALUATION METHOD OF MEAN VALUE OF RESISTANCE

### 28.5.2.1 Iteration Method for Mean Value of Resistance

Under the assumption that $\beta$ is equal to $\beta_T$, the LRFs are computed according to the fourth-moment method, so it is necessary to first estimate the target mean resistance under this condition. The target mean resistance is calculated as follows:

$$\mu_{Rk} = \mu_{Rk-1} + \left(\beta_T - \beta_{k-1}\right)\sigma_G \tag{28.77}$$

where $\beta_{k-1}$ represents the fourth-moment reliability index value for $(k-1)$th iteration, $\mu_{Rk}$ represents the $k$th iteration mean resistance value, and $\mu_{Rk-1}$ represents the $(k-1)$th iteration mean resistance.

### 28.5.2.2 Simple Formulae for Target Mean Resistance

Iterative calculation is necessary for calculating the abovementioned target mean resistance, which is impractical for users and designers. For these practitioners, the calculation procedure should be as accurate and simple as possible. Therefore, a simple formula that approximates the target mean resistance under the fourth-moment reliability index is presented below:

$$\mu_{RT} = \sum_{i=1}^{n} \mu_{S_i} + \beta_{2T_0} \sigma_{G_0} \qquad (28.78)$$

where $\mu_{RT}$ denotes the target mean resistance, $\beta_{2T_0}$ denotes the initial target second-moment reliability index, and $\sigma_{G_0}$ denotes the initial standard deviation.

Since $\sigma_{G_0}$ and $\beta_{2T_0}$ are unknown and can be calculated form $\mu_{R_0}$, therefore, the initial mean resistance value $\mu_{R_0}$ should be set in advance. From trial and error, the initial value of the resistance $\mu_{R_0}$ can be calculated as (Lu et al. 2010):

$$\mu_{R_0} = \sum_{i=1}^{n} \mu_{S_i} + \sqrt{\beta_T^{3.3} \sum_{i=1}^{n} \sigma_{S_i}^2} \qquad (28.79)$$

The procedure of calculating the LRFs utilizing the current simplified formula of target mean resistance is given as follows:

1. Give the initial resistance value $\mu_{R_0}$ utilizing Eq. (28.79).
2. Compute the initial standard deviation $\sigma_{G_0}$, skewness $\alpha_{3G_0}$, kurtosis $\alpha_{4G_0}$, and target second-moment reliability index $\beta_{2T_0}$ utilizing Eqs. (28.5), (28.52), (28.68), and (28.74), respectively, and then estimate $\mu_{RT}$ using Eq. (28.78).
3. Determine the standard deviation $\sigma_G$, skewness $\alpha_{3G}$, kurtosis $\alpha_{4G}$, and target second-moment reliability index $\beta_{2T}$ using Eqs. (28.5), (28.52), (28.68), and (28.74), respectively, and then compute $\alpha_R$ and $\alpha_{S_i}$ using Eq. (28.8).
4. Calculate the LRFs utilizing Eq. (28.75).

### 28.5.3 NUMERICAL EXAMPLES

### Example 28.3 Accuracy of the Simplified Formula for the Target Mean Resistance

To study the accuracy of using the simplified target mean resistance formula to compute the LRFs, the following limit-state function is considered:

$$G(X) = R - (D + L + S) \qquad (28.80)$$

where $R$ represents the resistance of the unknown PDF, and $V_R = 0.15$, $\alpha_{3R} = 0.453$, $\alpha_{4R} = 3.368$, and $\mu_R/R_n = 1.1$; $D$ represents the dead load of the unknown PDF, and $V_D = 0.1$, $\alpha_{3D} = 0.0$, $\alpha_{4D} = 3.0$, and $\mu_D/D_n = 1$; $L$ represents the live load of the unknown PDF, and $V_L = 0.4$, $\alpha_{3L} = 1.264$, $\alpha_{4L} = 5.969$, and $\mu_L/L_n = 0.45$; and $S$ represents the snow load of the unknown PDF, and $V_S = 0.25$, $\alpha_{3S} = 1.140$, $\alpha_{4S} = 5.4$, and $\mu_S/S_n = 0.47$.

When the mean value of $D$ and $L$ is set as $\mu_D = 1.0$ and $\mu_L/\mu_D = 0.5$, on the basis of the fourth-moment method, the LRFs estimated utilizing the simplified formula in comparison with the factors estimated utilizing iterative computations for $\beta_T = 2.0$, $\beta_T = 3.0$, and $\beta_T = 4.0$ are illustrated in Figures 28.5a–c, respectively. Figure 28.5d shows the target mean resistance values determined from the simplified formula, together with those calculated through iterative calculations. As shown in Figure 28.5, the target mean resistance as well as LRFs estimated by the two methods under the same target reliability index is essentially the same.

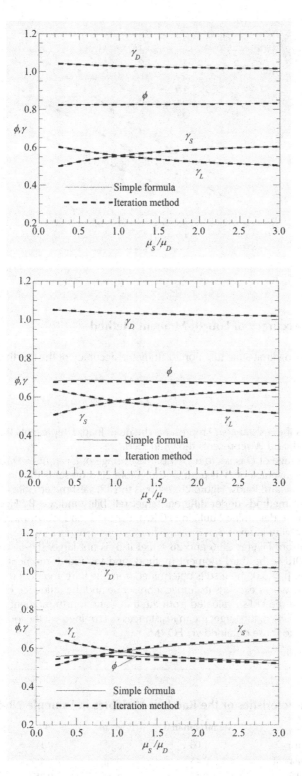

**FIGURE 28.5** Comparison between the simple formula and the iterative method of Example 28.3. (a) Load and resistance factors when $\beta_T = 2.0$, (b) load and resistance factors when $\beta_T = 3.0$, (c) load and resistance factors when $\beta_T = 4.0$, (d) target mean resistance.

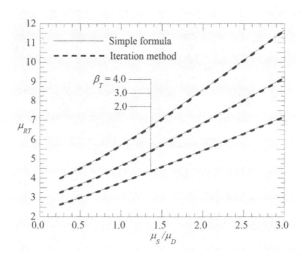

**FIGURE 28.5**  (Continued)

## Example 28.4  Accuracy of Fourth-Moment Method

Consider the following limit-state function to study the accuracy of the fourth-moment method to evaluate the LRFs:

$$G(X) = R - (D + L + S + W) \tag{28.81}$$

where $R$ represents the resistance, $D$ represents the dead load, $L$ represents the live load, $S$ represents the snow load, and $W$ represents the wind load.

Suppose the average of $D$ is set to $\mu_D = 1.0$, the average of $L$ is $\mu_L/\mu_D = 0.5$, the average of $S$ is $\mu_S/\mu_D = 2.0$, and the average of $W$ is $\mu_W/\mu_D = 2.0$. Table 28.3 presents the probabilistic characteristics of the resistance and loads. Figure 28.6 shows the LRFs estimated utilizing FORM, the third- and fourth-moment methods under different target reliability indexes $\beta_T$. Figure 28.7 shows the target mean resistance determined utilizing FORM, the third- and fourth-moment methods under different target reliability indexes $\beta_T$.

It can be seen from Figures 28.6 and 28.7 that if $\beta_T$ is not large, the target mean resistances evaluated from FORM, the third-moment method, and the fourth-moment method are almost identical, and when $\beta_T > 3.0$, the results determined from the third-moment method are unconservative due to the skewness exceeds its application scope and the influence of kurtosis is ignored. Furthermore, even if the LRFs evaluated from the third- and fourth-moment methods differ from these evaluated by FORM, the target mean resistances determined via the fourth-moment method are nearly identical to those obtained via FORM.

**TABLE 28.3**

**Probability Characteristics of the Random Variables in Example 28.4**

| $R$ or $S_i$ | CDF | Mean/Nominal | Mean | $V_i$ | $\alpha_{3i}$ | $\alpha_{4i}$ |
|---|---|---|---|---|---|---|
| $D$ | Normal | 1.00 | $\mu_D$ | 0.1 | 0.000 | 3.000 |
| $L$ | Lognormal | 0.45 | $0.5\mu_D$ | 0.4 | 1.264 | 5.969 |
| $S$ | Gumbel | 0.47 | $2.0\mu_D$ | 0.25 | 1.140 | 5.400 |
| $W$ | Gumbel | 0.60 | $2.0\mu_D$ | 0.2 | 1.140 | 5.400 |
| $R$ | Lognormal | 1.10 | $\mu_R$ | 0.15 | 0.453 | 3.368 |

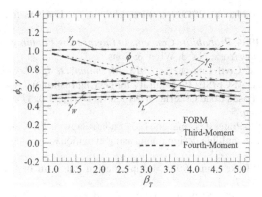

FIGURE 28.6 Load and resistance factors obtained via different methods of Example 28.4.

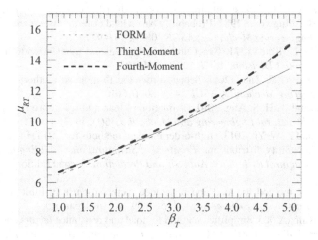

FIGURE 28.7 Target mean resistances obtained via different methods of Example 28.4.

## 28.6 CONCLUSIONS

This chapter illustrates the application of the method of moments in evaluating the LRFs. This method does not require derivative-based iteration or design points, thus it is easier to apply in actual engineering. Although in some cases, the LRFs determined through the third- and fourth-moment methods differ from those determined through FORM, the target mean resistances produced from the proposed methods are basically the same as those produced from FORM. Moreover, since the proposed methods do not rely on the probability distribution of fundamental random variables but on their first few statistical moments, so when the probability distribution of fundamental random variables is unknown, LRFD can also be performed.

## REFERENCES

AASHTO (American Association of State Highway and Transportation Officials). 2014. *AASHTO LRFD Bridge Design Specifications. 7th ed., Customary U.S. Units*. Washington, DC: AASHTO.

ACI (American Concrete Institute). Committee 318. 2014. *Building Code Requirements for Structural Concrete (ACI 318–14) and Commentary (ACI 318R-14)*. Farmington Hills, MI: ACI.

AIJ (Architectural Institute of Japan). 2002. *Recommendations for Limit State Design of Buildings*. Tokyo: AIJ. (in Japanese).

AIJ (Architectural Institute of Japan). 2015. *Recommendations for Limit State Design of Buildings*. Tokyo: AIJ. (in Japanese).

Ang, A. H. S., and W. Tang. 1984. *Probability Concepts in Engineering Planning and Design, Vol II, Decision, Risk and Reliability*. New York, NY: J. Wiley & Sons.

ASCE (American Society of Civil Engineers). 2013. *Minimum Design Loads for Buildings and Other Structures*. Reston, VA: ASCE/SEI (Structural Engineering Institute) 7–10: ASCE.

Assi, I. M. 2005. Reliability-Based Load Criteria for Structural Design: Load Factors and Load Combinations. The University of Jordan.

Barranco-Cicilia, F., E. Castro-Prates de Lima, and L. V. Sudati-Sagrilo. 2009. Structural reliability analysis of limit state functions with multiple design points using evolutionary strategies. *Ingeniería, Investigación y Tecnología* 10(2): 87–97.

Der Kiureghian, A., and T. Dakessian. 1998. Multiple design points in first and second-order reliability. *Structural Safety* 20(1): 37–49.

Der Kiureghian, A., and P. Liu. 1986. Structural reliability under incomplete probability information. *Journal of Engineering Mechanics, ASCE* 112(1): 85–104.

Ellingwood, B., G. MacGregor, T. V. Galambos, and C. A. Cornell. 1982. Probability based load criteria: load factor and load combinations. *Journal of the Structural Division, ASCE* 108(5): 978–997.

Fleishman, A. L. 1978. A method for simulating non-normal distributions. *Psychometrika* 43(4): 521–532.

Galambos, T. V., and B. Ellingwood. 1982. Probability based load criteria: assessment of current design practice. *Journal of Engineering Mechanics, ASCE* 108(5): 957–977.

Hasofer, A. M., and N. C. Lind. 1974. Exact and invariant second moment code format. *Journal of the Engineering Mechanics Division, ASCE* 100(1): 111–121.

Lu, Z. H., D. Z. Hu, and Y. G. Zhao. 2017. Second-order fourth-moment method for structural reliability. *Journal of Engineering Mechanics, ASCE* 143(4): 06016010.

Lu, Z. H., Y. G. Zhao, and A. H. S. Ang. 2010. Estimation of load and resistance factors based on the fourth moment method. *Structural Engineering and Mechanics* 36(1): 19–36.

Lu, Z. H., Y. G. Zhao, and Z. W. Yu. 2013. High-order moment methods for LRFD including random variables with unknown probability distributions. *Frontiers of Structural and Civil Engineering* 7(3): 288–295.

Melchers, R. E. 1999. *Structural Reliability; Analysis and Prediction*. Second Edition. West Sussex, UK: John Wiley & Sons.

Mori, Y. 2002. Practical method for load and resistance factors for use in limit state design. *Journal of Structural and Construction Engineering, AIJ* 559: 39–46. (in Japanese).

Mori, Y., and Y. Maruyama. 2005. Simplified method for load and resistance factors and accuracy of sensitivity factors. *Journal of Structural and Construction Engineering, AIJ* 589: 67–72. (in Japanese).

Mori, Y., and M. Nonaka. 2001. LRFD for assessment of deteriorating existing structures. *Structural Safety* 23(4): 297–313.

Mori, Y., H. Tajiri, and Y. Higashi. 2019. A study on practical method for load and resistance factors using shifted lognormal approximation. *Proceedings of the 9th Jpn Conference on Structural Safety and Reliability*, Vol 9, 8pp. (CR-ROM) (in Japanese).

Rackwitz, R., and B. Flessler. 1978. Structural reliability under combined random load sequences. *Computers and Structures* 9(5): 489–494.

Takada, T. 2001. Discussion on LRFD in AIJ-WG of Limit State Design. Private communication.

Ugata, T. 2000. Reliability analysis considering skewness of distribution-simple evaluation of load and resistance factors. *Journal of Structural and Construction Engineering, AIJ* 529: 43–50. (in Japanese).

Zhao, Y. G., and A. H. S. Ang. 2003. System reliability assessment by method of moments. *Journal of Structural Engineering, ASCE* 129(10): 1341–1349.

Zhao, Y. G., and Z. H. Lu. 2007. Fourth-moment standardization for structural reliability assessment. *Journal of Structural Engineering, ASCE* 133(7): 916–924.

Zhao, Y. G., and Z. H. Lu. 2011. Estimation of load and resistance factors using the third-moment method based on the 3P-lognormal distribution. *Frontiers of Architecture and Civil Engineering in China* 5(3): 315–322.

Zhao, Y. G., Z. H. Lu, and T. Ono. 2006. Estimating load and resistance factors without using distributions of random variables. *Journal of Asian Architecture and Building Engineering* 5(2): 325–332.

# 29 Methods of Moment for System Reliability

*Zhao-Hui Lu*
Beijing University of Technology, Beijing, China

*Yan-Gang Zhao*
Kanagawa University, Yokohama, Japan

## CONTENTS

## 29.1 INTRODUCTION

The reliability of a multicomponent system is essentially a problem involving multiple modes of failure; that is, the failures of different components, or different sets of components constitute distinct and different failure modes of the system. The consideration of multiple modes of failure, therefore, is fundamental to the problem of system reliability. The identification of each individual failure modes and the evaluation of their respective failure probabilities might be problematic in themselves.

This chapter will be concerned with structures with more than one limit states. For simple structures composed of just one element, various limit states such as bending action, shear, buckling, axial stress, deflection, etc., may all need to be considered. Most structures, in practice, are composed of a number of members or elements. Such a composition is hereafter referred to as a structural system.

It is obvious that the reliability of a structural system will be a function of the reliability of its individual members, and the reliability assessment of structural systems therefore needs to account for multiple, perhaps correlated, limit states. Methods to deal with such problems will be the subject of this chapter. Discussion in this chapter will be limited to one-dimensional system such as trusses

DOI: 10.1201/9781003194613-29

and rigid frames. Two-dimensional systems such as plates, slabs, and shells will not be considered, nor will three-dimensional continua such as earth embankments and dams. For these more complex structural systems, the principles given here, however, are also valid and can be used to develop appropriate calculation techniques.

Systems that are composed of multiple components can be classified as series-connected or parallel-connected systems, or combinations thereof. More generally, the failure events, e.g., in the case of multiple failure modes, may also be represented as events in series (union) or in parallel (intersection).

## 29.2   BASIC CONCEPTS OF SYSTEM RELIABILITY

### 29.2.1   SERIES SYSTEMS

Systems that are composed of components connected in series (series systems) are such that the failure of any one or more of these components constitutes the failure of the system. Such systems, therefore, have no redundancy and are also known as "weakest link" systems. In other words, the reliability or safety of the system requires that none of the components fails. A series system may be represented schematically as in Figure 29.1.

Consider a structural system with $k$ possible failure modes and assume that the performance function for failure mode $i$ is given by

$$g_i(\mathbf{X}) = g_i(X_1, X_2, \cdots, X_n), i = 1, 2, \cdots, k \tag{29.1}$$

where $X_1$, $X_2$, ..., $X_n$ are the basic random variables, and $g_i(\cdot)$ is the $i$th performance function.

Define the failure event for failure mode $i$ as

$$E_i = [g_i(\mathbf{X}) \leq 0] \tag{29.2a}$$

Then, the compliment of $E_i$ is the safe event; that is,

$$\bar{E}_i = [g_i(\mathbf{X}) > 0] \tag{29.2b}$$

Since the occurrence of any failure event $E_i$ causes the failure of the structure, the failure event $E$ of the structure is the union of all the possible failure modes, which can be expressed as

$$E = E_1 \cup E_2 \cup \cdots \cup E_k \tag{29.3}$$

In the case of two variables, the above events may be portrayed as in Figure 29.2, which shows three failure modes for the limit state equations $g_i(\mathbf{X}) = 0$, $i = 1, 2, 3$.

In structural reliability theory, the failure probability $P_F$ of a series structural system due to the occurrence of the event $E$ in Eq. (29.3) involves the following integration

$$P_F = \int \cdots \int_{(E_1 \cup E_2 \cup \cdots \cup E_k)} f_{X_1, X_2, \cdots, X_n}(x_1, x_2, \cdots, x_n) dx_1 dx_2 \cdots dx_n \tag{29.4}$$

where $f(\cdot)$ is the pertinent joint probability density function (PDF).

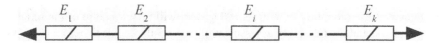

**FIGURE 29.1**   Representations of series systems.

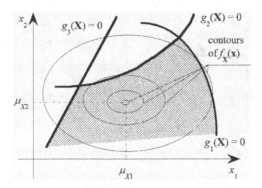

**FIGURE 29.2** Basic problem of system reliability in two dimensions.

## Example 29.1

Consider the static deterministic truss shown in Figure 29.3a. Since the structure is non-redundant, either the failure of member AB or AC will cause the failure of the structure as shown in Figures 29.3b and 29.3c. So the truss is a series system.

Suppose that the strengths of the members AB and AC are $R_1$ and $R_2$, respectively. The failure of members AB and AC will be represented as

$$E_1 = \left[ g_1 = 3R_1 - 5P \leq 0 \right] \tag{29.5a}$$

$$E_2 = \left[ g_2 = 3R_2 - 4P \leq 0 \right] \tag{29.5b}$$

The failure of the system would be the event

$$E_F = E_1 \cup E_2 = [g_1 \leq 0 \cup g_2 \leq 0] \tag{29.6}$$

## Example 29.2

Consider a one-story one-bay elastoplastic frame shown in Figure 29.4a (Ono et al. 1990). The six most likely failure modes obtained from stochastic limit analysis are shown in Figure 29.4b. Since any of the six failure modes will cause the failure of the structure, the frame is a series system. The six failure modes will be represented as

$$E_1 = [g_1 = 2M_1 + 2M_2 - 15S_1 \leq 0] \tag{29.7a}$$

$$E_2 = [g_2 = M_1 + 3M_2 + 2M_3 - 15S_1 - 10S_2 \leq 0] \tag{29.7b}$$

$$E_3 = [g_3 = 2M_1 + M_2 + M_3 - 15S_1 \leq 0] \tag{29.7c}$$

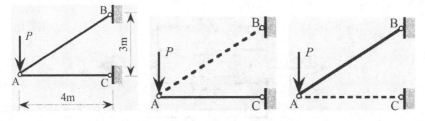

**FIGURE 29.3** A static deterministic truss for Example 29.1. (a) A static deterministic truss; (b) failure of member of AB; (c) failure of member of AC.

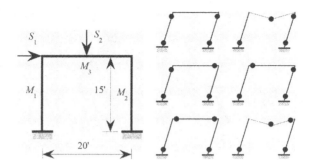

**FIGURE 29.4** One-story one-bay frame for Example 29.2. (a) A one-story one-bay frame; (b) six failure modes.

$$E_4 = [g_4 = M_1 + 2M_2 + M_3 - 15S_1 \le 0] \tag{29.7d}$$

$$E_5 = [g_5 = M_1 + M_2 + 2M_3 - 15S_1 \le 0] \tag{29.7e}$$

$$E_6 = [g_6 = M_1 + M_2 + 4M_3 - 15S_1 - 10S_2 \le 0] \tag{29.7f}$$

The failure of the system would be the event

$$E_F = E_1 \cup E_2 \cup E_3 \cup E_4 \cup E_5 \cup E_6$$
$$= [g_1 \le 0 \cup g_2 \le 0 \cup g_3 \le 0 \cup g_4 \le 0 \cup g_5 \le 0 \cup g_6 \le 0] \tag{29.8}$$

### 29.2.2 PARALLEL SYSTEMS

Systems that are composed of components connected in parallel (parallel systems) are such that the failure of the system requires failures of all the components. In other words, if any one of the components survives, the system remains safe. A parallel system is clearly a redundant system and may be represented schematically as shown in Figure 29.5.

If $E_i$ denotes the failure of component $i$, the failure of a $k$-component parallel system, therefore, is

$$E_F = E_1 \cap E_2 \cap \cdots \cap E_k \tag{29.9}$$

Note that the failure of a series system is the union of the component failures, whereas the safety of parallel system is the union of the survival events of the components. Also, the safety of a series system is the intersection of the safety of the components, whereas the failure of a parallel system is the intersection of the component failures.

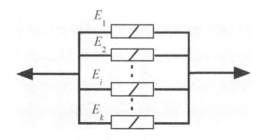

**FIGURE 29.5** Representations of parallel systems.

## Example 29.3

Consider the idealized parallel system shown in Figure 29.6a. If the elements are brittle, with different fracture strength $R_1$ and $R_2$, then the failure of elements 1 and 2 will be represented as Eqs. (29.10a) and (29.10b), respectively.

$$E_1 = \left[ g_1 = R_1 - \frac{1}{2}S \leq 0 \right] \qquad (29.10a)$$

$$E_2 = \left[ g_2 = R_2 - \frac{1}{2}S \leq 0 \right] \qquad (29.10b)$$

The failure of the system would be the event

$$E_F = E_1 \cap E_2 = \left[ \left( R_1 - \frac{1}{2}S \right) \cap \left( R_2 - \frac{1}{2}S \right) \leq 0 \right] \qquad (29.11)$$

However, since both the elements carry the load $S$, as element 1 fails, element 2 has to support the entire load $S$ by itself, so the failure of element 2 becomes

$$E_2 = \left[ g_2 = R_2 - S \leq 0 \right] \qquad (29.12)$$

The failure of the system in this failure sequence would be the event

$$E_F = \left[ \left( R_1 - \frac{1}{2}S \leq 0 \right) \cap \left( R_2 - S \leq 0 \right) \right] \qquad (29.13)$$

Similarly, when element 2 fails first, element 1 has to support the whole load $S$ by itself, and then the failure of the system in this failure sequence would be the event

$$E_F = \left[ \left( R_1 - S \leq 0 \right) \cap \left( R_2 - \frac{1}{2}S \leq 0 \right) \right] \qquad (29.14)$$

As there is no assurance as to which of the elements will fail first, both the two failure sequences should be considered. The failure of the system, therefore, would be the event

$$E_F = \left[ \left( R_1 - \frac{1}{2}S \leq 0 \right) \cap \left( R_2 - S \leq 0 \right) \right] \cup \left[ \left( R_2 - \frac{1}{2}S \leq 0 \right) \cap \left( R_1 - S \leq 0 \right) \right] \qquad (29.15)$$

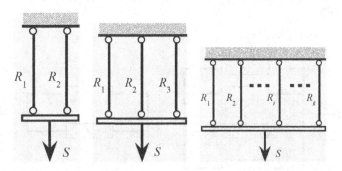

**FIGURE 29.6** Parallel systems for Example 29.3. (a) With two elements; (b) with three elements; (c) with $k$ elements.

Then consider the idealized parallel system shown in Figure 29.6b with different brittle fracture strength $R_1$, $R_2$, and $R_3$. According to the analysis above, there are six failure sequences, i.e., 1-2-3, 1-3-2, 2-1-3, 2-3-1, 3-2-1, and 3-1-2. Since there is no assurance as to which of the elements will fail first, all of the six failure sequences should be considered. The failure of the system would be the event

$$E_F = \left[\left(R_1 - \frac{1}{3}S \le 0\right) \cap \left(R_2 - \frac{1}{2}S \le 0\right) \cap (R_3 - S \le 0)\right] \cup \left[\left(R_1 - \frac{1}{3}S \le 0\right) \cap \left(R_3 - \frac{1}{2}S \le 0\right) \cap (R_2 - S \le 0)\right]$$

$$\cup \left[\left(R_2 - \frac{1}{3}S \le 0\right) \cap \left(R_3 - \frac{1}{2}S \le 0\right) \cap (R_1 - S \le 0)\right] \cup \left[\left(R_2 - \frac{1}{3}S \le 0\right) \cap \left(R_1 - \frac{1}{2}S \le 0\right) \cap (R_3 - S \le 0)\right]$$

$$\cup \left[\left(R_3 - \frac{1}{3}S \le 0\right) \cap \left(R_1 - \frac{1}{2}S \le 0\right) \cap (R_2 - S \le 0)\right] \cup \left[\left(R_3 - \frac{1}{3}S \le 0\right) \cap \left(R_2 - \frac{1}{2}S \le 0\right) \cap (R_1 - S \le 0)\right]$$

(29.16)

It can be observed that Eq. (29.16) is quite cumbersome. In order to deal this problem more briefly, introduce a new set of random variables $R^{(1)}$, $R^{(2)}$, and $R^{(3)}$, where $R^{(1)}$ is the smallest of $(R_1, R_2, R_3)$, $R^{(2)}$ is the second smallest of $(R_1, R_2, R_3)$, and $R^{(3)}$ is the third smallest of $(R_1, R_2, R_3)$, i.e., the largest of $(R_1, R_2, R_3)$, then the failure sequence of the system will be $R^{(1)}$-$R^{(2)}$-$R^{(3)}$. Therefore, the failure of the system would be the event

$$E_F = \left[\left(R^{(1)} - \frac{1}{3}S \le 0\right) \cap \left(R^{(2)} - \frac{1}{2}S \le 0\right) \cap \left(R^{(3)} - \frac{1}{3}S \le 0\right)\right]$$

(29.17)

Generally, for the parallel system shown in Figure 29.6c with $k$ elements, introduce a new set of random variables $R^{(i)}$, $i = 1, 2, \ldots, n$, where $R^{(i)}$ is the $i$th order smallest of $(R_i, i = 1, 2, \ldots, k)$. The failure of the system would be the event

$$E_F = \left[\left(R^{(1)} - \frac{1}{k}S \le 0\right) \cap \left(R^{(2)} - \frac{1}{k-1}S \le 0\right) \cap \cdots \cap \left(R^{(i)} - \frac{1}{k-i+1}S \le 0\right) \cap \cdots \cap \left(R^{(k)} - S \le 0\right)\right]$$

(29.18)

### 29.2.3 COMBINED SERIES-PARALLEL SYSTEMS

Structural systems may be composed of a combination of series- and parallel-connected components. However, it should be noted that not all systems could be decomposed into such series and parallel components. A general system may, nevertheless, be represented as a combination of failure or safe events in series and/or in parallel. The pertinent events may not be simply the failures or survivals of the individual components. Combined systems may be schematically represented as shown in Figure 29.7.

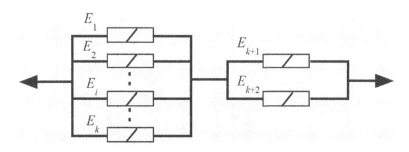

**FIGURE 29.7** Combined systems.

## Example 29.4

Consider a simple parallel-chain system shown in Figure 29.8. Assume that the individual components of the system will fail through tensile fracture or compressive buckling, and thus may be assumed to be brittle; i.e., once failure occurs, the strength of a component is reduced to zero. In this case, the system is a non-series system. Suppose that the brittle elements have different fracture strength $R_1$, $R_2$, $R_3$, and $R_4$. The system is series composed of $R_1$ and a sub-system $R_{234}$ that is composed of $(R_2, R_3, R_4)$. Suppose the failures of $R_1$ and the sub-system $R_{234}$ are presented by event $E_1$ and $E_{234}$, then the failure of the system would be the event

$$E_F = [E_1 \cup E_{234}] \tag{29.19}$$

For element $R_1$, the failure is readily expressed as

$$E_1 = (R_1 - S \le 0) \tag{29.20}$$

Note the sub-system $R_{234}$ is parallel composed of $R_4$ and another sub-system $R_{23}$ that is composed of $(R_2, R_3)$. So the failure of sub-system $R_{234}$ would be the event

$$E_{234} = \left[ \min(R_4, R_{23}) - \frac{1}{2} S \le 0 \right] \cap \left[ \max(R_4, R_{23}) - S \le 0 \right] \tag{29.21}$$

where $R_{23}$ is the strength of the sub-system $R_{23}$. Note again the sub-system $R_{23}$ is series composed of $R_2$ and $R_3$. So the strength of the sub-system $R_{23}$ would be the minimum of $R_2$ and $R_3$, i.e.,

$$R_{23} = \min(R_2, R_3) \tag{29.22}$$

Finally, the failure of the system would be the event

$$
\begin{aligned}
E_F &= [E_1 \cup E_{234}] = (R_1 - S \le 0) \cup \left[ \min(R_4, R_{23}) - \frac{1}{2} S \le 0 \right] \cap \left[ \max(R_4, R_{23}) - S \le 0 \right] \\
&= (R_1 - S \le 0) \cup \left\{ \min[R_4, \min(R_2, R_3)] - \frac{1}{2} S \le 0 \right\} \cap \left\{ \max[R_4, \min(R_2, R_3)] - S \le 0 \right\} \\
&= (R_1 - S \le 0) \cup \left[ \min(R_2, R_3, R_4) - \frac{1}{2} S \le 0 \right] \cap \left\{ \max[R_4, \min(R_2, R_3)] - S \le 0 \right\}
\end{aligned}
\tag{29.23}
$$

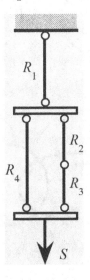

**FIGURE 29.8** A combined system for Example 29.4.

## 29.3 SYSTEM RELIABILITY BOUNDS

The calculation of the failure probability of a structural system is generally difficult through integration such as Eq. (29.4). Approximations are always necessary. One of the approximations is to develop upper and lower bounds on the probability of failure of a system.

### 29.3.1 UNIMODAL BOUNDS

Suppose the failure of a series system with $k$ failure modes is presented by an event $E$, and the $i$th failure mode is presented by the event $E_i$, then the probability of survival and failure is given by

$$P_S = P(\bar{E}_1 \cap \bar{E}_2 \cap ... \cap \bar{E}_k), \text{ and } P_F = 1 - P_S$$

First, consider events $E_i$ and $E_j$ with failure mode $i$ and $j$. The relationship between $E_i$ and $E_j$ is shown in Figure 29.9, one can easily understand that

$$P(E_j|E_i) \geq P(E_j), P(\bar{E}_j|\bar{E}_i) \geq P(\bar{E}_j)$$

Since

$$P(\bar{E}_i \cap \bar{E}_j) \geq P(\bar{E}_j|\bar{E}_i)P(\bar{E}_i)$$

We have

$$P(\bar{E}_i \cap \bar{E}_j) \geq P(\bar{E}_j)P(\bar{E}_i)$$

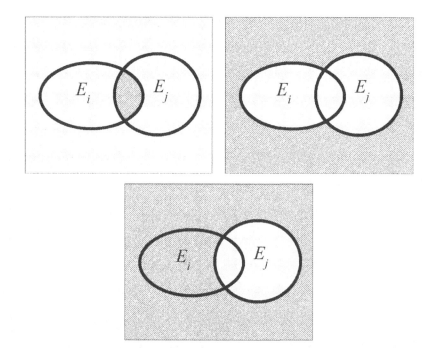

**FIGURE 29.9** Relationship between events $E_i$ and $E_j$. (a) $E_i \cap E_j$; (b) $\bar{E}_i \cap \bar{E}_j$; (c) $\bar{E}_j$.

For $k$ events, this can be generalized to yield

$$P_s = P\left(\bar{E}_1 \cap \bar{E}_2 \cap \cdots \cap \bar{E}_k\right) \geq \prod_{i=1}^{k} P\left(\bar{E}_i\right) \tag{29.24}$$

Conversely, since

$$\bar{E}_1 \cap \bar{E}_2 \cap \ldots \cap \bar{E}_k \subset \min_{1 \leq i \leq k}(\bar{E}_i)$$

We have

$$P_S \leq \min_{1 \leq i \leq k}\left[P\left(\bar{E}_i\right)\right] \tag{29.25}$$

If we denote the reliability against the $i$th failure mode as $P_{Si}$, and the probability of failure corresponding to the $i$th failure mode as $P_{fi}$, then we have

$$P_{Si} = P\left(\bar{E}_i\right), \quad P_{fi} = P(E_i) = 1 - P_{Si}$$

According to Eqs. (29.14a–b), the reliability of the system $P_S$ is then bounded as follows.

$$\prod_{i=1}^{k} P_{Si} \leq P_S \leq \min_{1 \leq i \leq k}\left(P_{Si}\right) \tag{29.26a}$$

Conversely, the corresponding bounds for the failure probability $P_F$ would be (e.g., Cornell 1967, Ang and Amin 1968).

$$\max_{1 \leq i \leq k}\left(P_{fi}\right) \leq P_F \leq 1 - \prod_{i=1}^{k}\left(1 - P_{fi}\right) \tag{29.26b}$$

where $P_{fi}$ is the failure probability of the $i$th failure mode.

The above bounds in Eqs. (29.26a–b) are often referred to as the "first-order" or "uni-modal" bounds on $P_F$ and $P_S$, in the sense that the lower and upper probability bounds involve single-mode probabilities.

Since only the failure probability of a single failure mode is considered, the correlation of the failure modes is neglected. The above wide bound estimation method is simple to evaluate. For many practical structural systems, especially for complex systems, the first-order bound given by Eq. (29.26b) is, however, too wide to be meaningful (Grimmelt and Schueller 1982). Better bounds have been developed but at the expense of more computation.

## 29.3.2 BIMODAL BOUNDS

The first-order bounds described above can be improved by taking into account the correlation between pairs of potential failure modes. The resulting improved bounds will necessary require the probabilities of joint events such as

$$E_i \cap E_j \text{ or } \bar{E}_i \cap \bar{E}_j$$

and thus may be called "bi-modal" or "second-order" bounds.

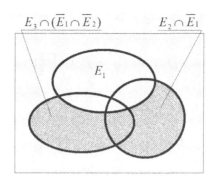

**FIGURE 29.10**   Decomposition of $E$.

Suppose the failure of a system with $k$ failure modes is represented by the event $E$, and the $i$th failure mode is presented by the event $E_i$, then the failure event is given by

$$
\begin{aligned}
E &= E_1 \cup E_2 \cup \cdots \cup E_k \\
&= E_1 + \left(E_2 \cap \bar{E}_1\right) + \left[E_3 \cap \left(\bar{E}_1 \cap \bar{E}_2\right)\right] + \cdots + \left[E_k \cap \left(\bar{E}_1 \cap \bar{E}_2 \cap \cdots \cap \bar{E}_{k-1}\right)\right]
\end{aligned}
\tag{29.27}
$$

where $E_2 \cap \bar{E}_1$ and $E_3 \cap (\bar{E}_1 \cap \bar{E}_2)$ are illustrated in Figure 29.10.

According to the De Morgan's rule

$$
\bar{E}_1 \cap \bar{E}_2 \cap \cdots \cap \bar{E}_{i-1} = \overline{E_1 \cup E_2 \cup \cdots \cup E_{i-1}}
$$

for $i = 2, 3, \ldots, k$, we have

$$
E_i \cap \left(\bar{E}_1 \cap \bar{E}_2 \cap \cdots \cap \bar{E}_{i-1}\right) = E_i \cap \overline{E_1 \cup E_2 \cup \cdots \cup E_{i-1}}
$$

Note that

$$
\overline{E_1 \cup E_2 \cup \ldots \cup E_{i-1}} + (E_1 \cup E_2 \cup \ldots \cup E_{i-1}) = S
$$

where $S$ is a certain event.

Then we have

$$
P\left[E_i \cap \left(\bar{E}_1 \cap \bar{E}_2 \cap \ldots \cap \bar{E}_{i-1}\right)\right] = P(E_i) - P\left[(E_i \cap E_1) \cup (E_i \cap E_2) \cup \cdots \cup (E_i \cap E_{i-1})\right]
$$

Therefore,

$$
P\left[E_i \cap \left(\bar{E}_1 \cap \bar{E}_2 \cap \ldots \cap \bar{E}_{i-1}\right)\right] \geq P(E_i) - \sum_{j=1}^{i-1} P\left(E_i \cap E_j\right)
\tag{29.28}
$$

Substituting Eq. (29.28) into Eq. (29.27) leads to

$$
P(E) \geq P(E_1) + \sum_{i=2}^{k} \max \left\{ \left[ P(E_i) - \sum_{j=1}^{i-1} P\left(E_i \cap E_j\right) \right], 0 \right\}
\tag{29.29}
$$

On the other hand, since

$$\bar{E}_1 \cap \bar{E}_2 \cap ... \cap \bar{E}_{i-1} \subset \min_{1 \le j \le i-1} \left( \bar{E}_j \right)$$

We have

$$P\left[E_i \cap \left(\bar{E}_1 \cap \bar{E}_2 \cap ... \cap \bar{E}_{i-1}\right)\right] \le P(E_i) - \max_{1 \le j \le i-1} P\left(E_i \cap E_j\right) \tag{29.30}$$

Substituting Eq. (29.30) into Eq. (29.27), one obtains that

$$P[E] \le \sum_{i=1}^{k} \left[P(E_i)\right] - \sum_{i=2}^{k} \left[\max_{1 \le j \le i-1} P\left(E_i \cap E_j\right)\right] \tag{29.31}$$

Then the probability of failure of the system $P_F$ is bounded according to Eqs. (29.29) and (29.31) as follows (Ditlevsen 1979).

$$P_{f1} + \sum_{i=2}^{k} \max\left(P_{fi} - \sum_{j=1}^{i-1} P_{fij}, 0\right) \le P_F \le \sum_{i=1}^{k} P_{fi} - \sum_{i=2}^{k} \max_{j \le i}\left(P_{fij}\right) \tag{29.32}$$

where $P_{fij}$ is referred to the probability of the intersection $E_i \cap E_j$ as $P_{fij}$, i.e., the joint probability of the simultaneous occurrences of the $i$th and $j$th failure modes. The left- and right-hand sides of Eq. (29.32) are, respectively, the lower bound and upper bound of the failure probability of a series structural system with $k$ potential failure modes. It can be observed that because the joint probability of simultaneous failures of every pair of failure modes must be evaluated, the resulting bounds of Eq. (29.32) are narrower than those from Eq. (29.26b).

## 29.4 MOMENT APPROACH FOR SYSTEM RELIABILITY

From the previous section, as a function of the failure probabilities of the individual modes, the failure probability of a system may be estimated using unimodal bound techniques. However, the bounds would be wide for a complex system. Bimodal bound technique (Ditlevsen 1979; Feng 1989; Zhao et al. 2007) has been developed to improve the bound width of unimodal bound technique, but mutual correlations and the joint failure probability matrices among the failure modes have to be determined. Since the number of potential failure modes for most practical structures is very large, the determination of the mutual correlations and the joint failure probability matrices is quite cumbersome and difficult. The failure probability of a system may also be estimated approximately using the probabilistic network evaluation technique (PNET) developed by Ang and Ma (1981), where the mutual correlations matrix among the failure modes have also to be computed. Other methods have been proposed or discussed (Moses 1982; Bennett and Ang 1986; Thoft-Christensen and Murotsu 1986; Miao and Ghosn 2011; Chang and Mori 2013). In this section, a computationally more effective method using moment approximations is introduced and examined for system reliability of both series and non-series systems.

### 29.4.1 PERFORMANCE FUNCTION FOR A SYSTEM

Consider a structural system with multiple modes of potential failure; e.g., $E_1, E_2, ..., E_k$. For a series structural system, each of the failure modes, $E_i$, can be defined by a performance function $gi = g_i(X)$ such that $E_i = (g_i < 0)$ and the failure probability of the system is then:

$$P_F = P\left[g_1 \le 0 \cup g_2 \le 0 \cup \cdots \cup g_k \le 0\right] \tag{29.33}$$

Conversely, the safety of a system is the event in which none of the $k$ potential failure modes occurs; again in the case of a series system, this means

$$P_S = P[g_1 > 0 \cap g_2 > 0 \cap \cdots \cap g_k > 0] = P[\min(g_1, g_2, \cdots g_k) > 0] \qquad (29.34)$$

Thus, the performance function of a series system, $G$, can be expressed as the minimum of the performance functions corresponding to all the potential failure modes (Zhao and Ang 2003); that is,

$$G(\mathbf{X}) = \min[g_1, g_2, \cdots, g_k] \qquad (29.35)$$

where $g_i = g_i(\mathbf{X})$ is the performance function of the $i$th failure mode.

Eq. (29.35) can also be derived using probability integration method and the procedure that have been developed by Li et al. (2007), where it is referred to as the equivalent extreme-value events.

Similarly, for a parallel structural system, each of the failure modes, $E_i$, can be defined by a performance function $gi = g_i(\mathbf{X})$ such that $E_i = (g_i < 0)$ and the failure probability of the system is then:

$$P_F = P[g_1 \leq 0 \cap g_2 \leq 0 \cap \cdots \cap g_k \leq 0] = P[\max(g_1, g_2, \cdots, g_k) \leq 0] \qquad (29.36)$$

Thus, the performance function of a parallel system, $G$, can be expressed as the maximum of the performance functions corresponding to all the potential failure modes; that is,

$$G(\mathbf{X}) = \max[g_1, g_2, \cdots, g_k] \qquad (29.37)$$

For a combined series-parallel system, the performance function of the system will generally involve combinations of maximum and minimum of the component performance functions. For example, suppose each of the failure modes, $E_i$, can be defined by a performance function $g_i = g_i(\mathbf{X})$ such that $E_i = (g_i < 0)$ the performance function corresponding to the combined system shown in Figure 29.11 will be given as

$$G(\mathbf{X}) = \min\{\max(g_1, g_2), \max[\min(g_3, g_5), \min(g_4, g_6)]\} \qquad (29.38)$$

In summary, the performance function of a series system and a parallel system can be expressed as the minimum and the maximum, respectively, of the performance functions corresponding to all the potential failure modes. And the performance function of a combined series-parallel system generally involves combinations of maximum and minimum of the component performance functions. Since the system performance function $G(\mathbf{X})$ will not be smooth even though the performance function of a component is smooth, one may doubt whether the PDF of the system performance function $Z = G(\mathbf{X})$ is smooth or not. We will investigate this problem through the following example.

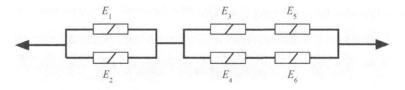

**FIGURE 29.11**  Combined systems.

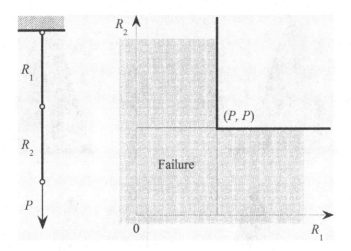

**FIGURE 29.12** A series system of two members for Example 29.5.

## Example 29.5 A Simple Series System

Consider a series system with only two elements as shown in Figure 29.12, the performance function is defined as

$$G(\mathbf{X}) = \min(R_1 - P, R_2 - P) \tag{29.39}$$

where $R_1$, $R_2$ are independent random variables with mean and standard deviation of $\mu_{R1} = 200$, $\mu_{R2} = 300$, $\sigma_{R1} = 20$, $\sigma_{R2} = 60$, $P$ is a load with deterministic value of 100. The following four cases are investigated:

Case 1, $R_1$ – Normal, $R_2$ – Lognormal;
Case 2, $R_1$ – Lognormal, $R_2$ – Lognormal;
Case 3, $R_1$ – Normal, $R_2$ – Weibull;
Case 4, $R_1$ – Gumbel, $R_2$ – Gamma.

Figure 29.13 presents the histograms of the performance function with 1,000,000 for the four cases. It can be observed that the histograms have good behaviors and the PDF should be smooth.

### 29.4.2 METHODS OF MOMENT FOR SYSTEM RELIABILITY

In this section, methods of moment (Zhao and Ono 2001) will be applied to system reliability evaluation. If the central moments of the system performance function, as described in Eq. (29.35) for a series system, Eq. (29.37) for a parallel system, and that for series-parallel system like Eq. (29.38) can be obtained, the failure probability of a system, which is defined as $P[G(\mathbf{X}) < 0]$, can be expressed as a function of the central moments. Because the first two moments are generally inadequate, high-order moments will invariably be necessary.

If the first three or four moments of $G(\mathbf{X})$ are obtained, the reliability analysis becomes a problem of approximating the distribution of a specific random variable with its known first three or four moments.

Approximating the distribution of a random variable using its moments of finite order is a well-known problem in statistics, and various approximations such as the Pearson, Johnson and Burr systems, the Edgeworth and Cornish-Fisher expansions were developed (Stuart and Ord 1987). Their applications in structural reliability have been examined by Winterstein (1988), Grigoriu (1983), and

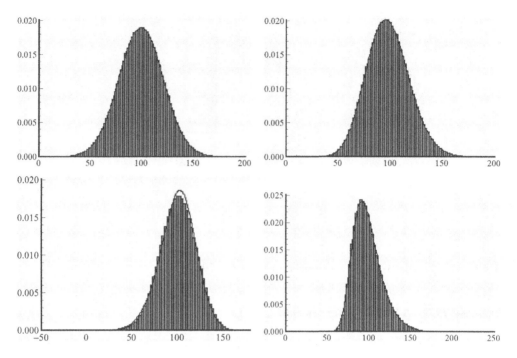

**FIGURE 29.13**  The histograms of the performance functions in Example 29.5. (a) Case 1; (b) case 2; (c) case 3; (d) case 4.

Hong and Lind (1996). In the present chapter, the fourth-moment standardization function will be used. For the standardized performance function,

$$Z_s = \frac{Z - \mu_G}{\sigma_G} \tag{29.40}$$

The fourth-moment standardization function is expressed as (Zhao and Lu 2007)

$$Z_s = S(U, \mathbf{M}) = -l_1 + k_1 U + l_1 U^2 + k_2 U^3 \tag{29.41}$$

where

$$l_1 = \frac{\alpha_{3G}}{6(1 + 6l_2)} \tag{29.42a}$$

$$l_2 = \frac{1}{36}\left(\sqrt{6\alpha_{4G} - 8\alpha_{3G}^2 - 14} - 2\right) \tag{29.42b}$$

$$k_1 = \frac{1 - 3l_2}{\left(1 + l_1^2 - l_2^2\right)} \tag{29.42c}$$

$$k_2 = \frac{l_2}{\left(1 + l_1^2 + 12l_2^2\right)} \tag{29.42d}$$

Since

$$F(Z_s) = \Phi(U) = \Phi[S^{-1}(Z_s, \mathbf{M})] \tag{29.43}$$

The PDF of $Z = G(\mathbf{X})$ can be obtained as

$$f_Z(z) = \frac{1}{\sigma}\phi(u) \Big/ \frac{dz_s}{u}$$
(29.44)

where

$$u = S^{-1}\left(\frac{z - \mu_G}{\sigma_G}\right)$$
(29.45)

Therefore, the PDF of the performance function is expressed as

$$f_Z(z) = \frac{\phi(u)}{\sigma\left(k_1 + 2l_1u + 3k_2u^2\right)}$$
(29.46)

where

$$u = S^{-1}\left(\frac{z - \mu_G}{\sigma_G}\right) = \frac{1}{D} - Dp - l$$
(29.47a)

$$D = \sqrt[3]{2}\left(\sqrt{q^2 + 4p^3} - q\right)^{-1/3}$$
(29.47b)

$$q = l(2l^2 - k_1/k_2 - 3) + \frac{z - \mu_G}{k_2\sigma_G}$$
(29.47c)

$$p = k_1/(3k_2) - l^2, \quad l = l_1/(3k_2)$$
(29.47d)

The probability of failure is expressed as

$$P_f = \text{Prob}[G \leq 0] = \text{Prob}[Z_s\sigma_G + \mu_G \leq 0]$$

$$= \text{Prob}\left[Z_s \leq -\frac{\mu_G}{\sigma_G}\right] = \text{Prob}[Z_s \leq -\beta_{2M}]$$

That is,

$$P_f = F(-\beta_{2M}) = \Phi[S^{-1}(-\beta_{2M}, \mathbf{M})]$$
(29.48)

Then the corresponding reliability index is expressed as

$$\beta = -\Phi^{-1}(P_f) = -S^{-1}(-\beta_{2M}, \mathbf{M})$$
(29.49)

Therefore,

$$\beta_{4M} = \frac{\sqrt[3]{2}p}{\sqrt[3]{-q_0 + \Delta_0}} - \frac{\sqrt[3]{-q_0 + \Delta_0}}{\sqrt[3]{2}} + \frac{l_1}{3k_2}$$
(29.50a)

$$\Delta_0 = \sqrt{q_0^2 + 4p^3}$$
(29.50b)

$$q_0 = \frac{2l_1^3 - 9k_1k_2l_1 + 27k_2^2(-l_1 + \beta_{2M})}{27k_2^3}$$
(29.50c)

The first few moments of a performance function can be evaluated using the generalized multivariate dimension-reduction method proposed by Xu and Rahman (2004), in which the $n$-dimensional performance function is approximated by the summation of a series of, at most, $D$-dimensional functions ($D < n$). Here, the bivariate- and one-dimension reduction ($D = 2$ and 1) are introduced.

For a performance function $Z = G(\mathbf{X})$, using the inverse Rosenblatt transformation, the $k$th moments about zero, of $Z$ can be defined as (Zhao and Ono 2000)

$$\mu_{kG} = E\{[G(\mathbf{X})]^k\} = \int_{-\infty}^{\infty} \cdots \int_{-\infty}^{\infty} \{G[\mathbf{x}]\}^k f_{\mathbf{X}}(\mathbf{x}) d\mathbf{x} = \int_{-\infty}^{\infty} \cdots \int_{-\infty}^{\infty} \{G[T^{-1}(\mathbf{u})]\}^k \phi(\mathbf{u}) d\mathbf{u} \qquad (29.51)$$

Let $L(\mathbf{u}) = G[T^{-1}(\mathbf{u})]$ in Eq. (29.51). By the bivariate dimension-reduction method (Xu and Rahman 2004)

$$L(\mathbf{u}) \cong L_2 - (n-2)L_1 + \frac{(n-1)(n-2)}{2} L_0 \qquad (29.52)$$

where

$$L_0 = L(0, ..., 0, ..., 0) \qquad (29.53a)$$

$$L_1 = \sum_{i=1}^{n} L_i(0, ..., u_i, ..., 0) = \sum_{i=1}^{n} L_i(u_i) \qquad (29.53b)$$

$$L_2 = \sum_{i<j} L_{ij}(0, ..., u_i, ..., u_j, ..., 0) = \sum_{i<j} L_{ij}(u_i, u_j) \qquad (29.53c)$$

where $i, j = 1, 2, ..., n$ and $i < j$. It is noted that $L_1$ is a summation of $n$ one-dimensional functions and $L_2$ is a summation of $[n(n-1)]/2$ two-dimensional functions.

Substituting Eqs. (29.53a–c) in Eq. (29.52) helps reduce the $n$-dimensional integral of Eq. (29.51) into a summation of, at most, two-dimensional integrals

$$\mu_{kG} = E\left(\{G[T^{-1}(U)]\}^k\right) \cong E\left(\{L(U)\}^k\right) = \sum_{i<j} \mu_{k-L_{ij}} - (n-2)\sum_{i} \mu_{k-L_i} + \frac{(n-1)(n-2)}{2} L_0^k \qquad (29.54)$$

where

$$\mu_{k-L_{ij}} = \int_{-\infty}^{\infty}\int_{-\infty}^{\infty} \left[L_{ij}(u_i, u_j)\right]^k \phi(u_i)\phi(u_j) du_i du_j \qquad (29.55a)$$

$$\mu_{k-L_i} = \int_{-\infty}^{\infty} \left[L_i(u_i)\right]^k \phi(u_i) du_i \qquad (29.55b)$$

$$L_0^k = [L(0, ..., 0, ..., 0)]^k \qquad (29.55c)$$

Using the Gauss-Hermite integration, the one-dimensional integral in Eq. (29.55b) can be approximated by the following equation.

$$\mu_{k-L_i} = \sum_{r=1}^{m} P_r \left[ L_i(u_{ir}) \right]^k \tag{29.56}$$

Similarly, the two-dimensional integral in Eq. (29.55a) can be approximated by

$$\mu_{k-L_{ij}} = \sum_{r_1=1}^{m} \sum_{r_2=1}^{m} P_{r_1} P_{r_2} \left[ L_{ij}(u_{ir_1}, u_{jr_2}) \right]^k \tag{29.57}$$

The estimating points $u_r$ and the corresponding weights $P_r$ can be readily obtained as

$$u_r = \sqrt{2} x_r, \quad P_r = \frac{w_r}{\sqrt{\pi}} \tag{29.57}$$

where $x_r$ and $w_r$ are the abscissas and weights, respectively, for Hermite integration with the weight function $\exp(-x^2)$, as given in Abramowitz and Stegun (1972). For a seven-point estimate ($m = 7$) in standard normal space, we have the following:

$$u_0 = 0, P_0 = 16/35 \tag{29.58a}$$

$$u_{1+} = -u_{1-} = 1.1544054, P_1 = 0.2401233 \tag{29.58b}$$

$$u_{2+} = -u_{2-} = 2.3667594, P_2 = 3.07571 \times 10^{-2} \tag{29.58c}$$

$$u_{3+} = -u_{3-} = 3.7504397, P_3 = 5.48269 \times 10^{-2} \tag{29.58d}$$

Finally, the mean, standard deviation, the skewness, and the kurtosis of a performance function $G(\mathbf{X})$ with $n$ random variables can be obtained as

$$\mu_G = \mu_{1G} \tag{29.59a}$$

$$\sigma_G = \sqrt{\mu_{2G} - \mu_{1G}^2} \tag{29.59b}$$

$$\alpha_{3G} = \left( \mu_{3G} - 3\mu_{2G}\mu_{1G} + 2\mu_{1G}^3 \right)/\sigma_G^3 \tag{29.59c}$$

$$\alpha_{4G} = \left( \mu_{4G} - 4\mu_{3G}\mu_{1G} + 6\mu_{2G}\mu_{1G}^2 - 3\mu_{1G}^4 \right)/\sigma_G^4 \tag{29.59d}$$

### Example 29.6   A Simple Series System

Consider the series system with only two elements as shown in Example 29.5 again. The first four moments of the performance functions are readily obtained using the point estimate method and are listed in Table 29.1, from which the PDF defined by the first four moments can be obtained using the Eq. 29.46. The PDFs of the performance functions are also illustrated in Figure 29.13. From Figure 29.13, it can be observed that the histograms of the system performance function can be generally approached by the PDF in Eq. 29.46, that is, the system reliability can be approximate by the methods of moment.

**TABLE 29.1**

**The Parameters and Moments Results of the Four Cases for Example 29.6**

| Case | Distributions | | Moment by MCS | | | |
|---|---|---|---|---|---|---|
| | $R_1$ | $R_2$ | $\mu_G$ | $\sigma_G$ | $\alpha_{3G}$ | $\alpha_{4G}$ |
| Case 1 | normal | Lognormal | 99.31 | 19.77 | −0.0161 | 3.009 |
| Case 2 | Lognormal | Lognormal | 99.30 | 19.65 | 0.265 | 3.136 |
| Case 3 | Normal | Weibull | 97.93 | 21.39 | −0.400 | 4.252 |
| Case 4 | Gumbel | Gamma | 98.99 | 19.39 | 0.948 | 4.794 |

## Example 29.7    A Simple Problem of System Reliability

Consider the performance function defined as the minimum value of the following eight linear performance functions as shown in Figure 29.14.

$$G(\mathbf{X}) = \min\{g_1, g_2, g_3, g_4, g_5, g_6, g_7, g_8\} \tag{29.60}$$

where

$$g_1 = 2X_1 - 2X_2 + 8 \qquad (\beta_F = 2.828)$$
$$g_2 = 2.6X_1 - 2X_2 + 9.3 \qquad (\beta_F = 2.835)$$
$$g_3 = 1.4X_1 - 2X_2 + 7.2 \qquad (\beta_F = 2.949)$$
$$g_4 = 4X_1 - 2X_2 + 14 \qquad (\beta_F = 3.131)$$
$$g_5 = 0.7X_1 - 2X_2 + 6.8 \qquad (\beta_F = 3.209)$$
$$g_6 = -0.5X_1 - 2X_2 + 8 \qquad (\beta_F = 3.881)$$
$$g_7 = -2X_1 - 2X_2 + 11 \qquad (\beta_F = 3.889)$$
$$g_8 = -1.5X_1 - 2X_2 + 10 \qquad (\beta_F = 4.000)$$

where $X_1$ and $X_2$ are independent standard normal random variables.

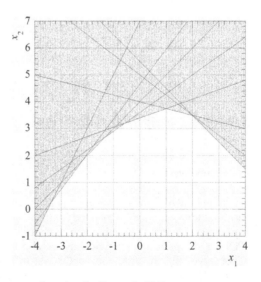

**FIGURE 29.14**   The performance function for Example 29.7.

The example is a series system problem to illustrate the numerical details of the procedure for the assessment of system reliability using the moment method. Since the gradients of the performance function are not convenient to obtain, it is not convenient to solve the example directly using the conventional procedure of first-order reliability method (FORM) described in Chapter 5. The FORM results would be obtained by first locating all the design points and then apply one of the methods to compute the probability on unions. If Ditlevsen's bounds (Ditlevsen 1979) are used, the correlation matrix and the joint failure probability are obtained as

$$[C]=\begin{pmatrix} 1. & 0.9916 & 0.984784 & 0.948683 & 0.901002 & 0.514496 & 0. & 0.141421 \\ 0.9916 & 1. & 0.954035 & 0.981615 & 0.837324 & 0.399267 & -0.129339 & 0.0121942 \\ 0.984784 & 0.954035 & 1. & 0.879292 & 0.962682 & 0.655687 & 0.173785 & 0.311308 \\ 0.948683 & 0.981615 & 0.879292 & 1. & 0.717581 & 0.21693 & -0.316228 & -0.178885 \\ 0.901002 & 0.837324 & 0.962682 & 0.717581 & 1. & 0.835555 & 0.433816 & 0.556876 \\ 0.514496 & 0.399267 & 0.655687 & 0.21693 & 0.835555 & 1. & 0.857493 & 0.921635 \\ 0. & -0.129339 & 0.173785 & -0.316228 & 0.433816 & 0.857493 & 1. & 0.989949 \\ 0.141421 & 0.0121942 & 0.311308 & -0.178885 & 0.556876 & 0.921635 & 0.989949 & 1. \end{pmatrix}$$

$$[P_{ij}]=10^{6}\begin{pmatrix} 2341.99 & 1943.37 & 1441.39 & 740.392 & 486.883 & 10.6655 & 0.11787 & 0.385213 \\ 1943.37 & 2291.28 & 1170.3 & 847.979 & 364.894 & 5.07798 & 0.0172665 & 0.085113 \\ 1441.39 & 1170.3 & 1594.02 & 458.581 & 583.31 & 19.272 & 0.590553 & 1.36889 \\ 740.392 & 847.979 & 458.581 & 871.061 & 125.313 & 0.571774 & 0.0000783872 & 0.00124889 \\ 486.883 & 364.894 & 583.31 & 125.313 & 665.987 & 33.8198 & 2.93761 & 4.87381 \\ 10.6655 & 5.07798 & 19.272 & 0.571774 & 33.8198 & 52.0139 & 13.0526 & 15.9204 \\ 0.11787 & 0.0172665 & 0.590553 & 0.0000783872 & 2.93761 & 13.0526 & 50.3291 & 28.8034 \\ 0.385213 & 0.085113 & 1.36889 & 0.00124889 & 4.87381 & 15.9204 & 28.8034 & 31.6712 \end{pmatrix}$$

The bounds of the failure probability are given as [0.00272, 0.00301], which means that the system reliability index is in the range of [2.747, 2.779].

Using the seven-point estimates described above, the first four moments of $G(\mathbf{X})$ are approximately obtained as $\mu_G = 6.5288$, $\sigma_G = 2.19674$, $\alpha_{3G} = -0.1369$, and $\alpha_{4G} = 3.1762$. The second-moment reliability index is readily obtained as $\beta_{2M} = 2.972$.

Using Eqs. (29.50a–c) and (29.42a–c), the fourth-moment reliability index is computed as follows.

$$l_2 = \frac{1}{36}\left(\sqrt{6\alpha_{4G} - 8\alpha_{3G}^2 - 14} - 2\right) = \frac{1}{36}\left(\sqrt{6\times3.1762 - 8\times0.1369^2 - 14} - 2\right) = 0.00598$$

$$l_1 = \frac{\alpha_{3G}}{6(1 + 6l_2)} = \frac{-0.1369}{6(1 + 6\times0.00598)} = -0.022$$

$$k_1 = \frac{1 - 3l_2}{\left(1 + l_1^2 - l_2^2\right)} = \frac{1 - 3\times0.00598}{(1 + 0.022^2 - 0.00598^2)} = 0.9816$$

$$k_2 = \frac{l_2}{\left(1 + l_1^2 + 12l_2^2\right)} = \frac{0.00598}{(1 + 0.022^2 + 12\times0.00598^2)} = 0.00597$$

$$p = \frac{3k_1 k_2 - l_1^2}{9k_2^2} = \frac{3\times0.9816\times0.00597 - 0.022^2}{9\times0.00597^2} = 53.268$$

$$q_0 = \frac{2l_1^3 - 9k_1 k_2 l_1 + 27k_2^2(-l_1 + \beta_{2M})}{27k_2^3}$$

$$= \frac{2(-0.022)^3 - 9\times0.9816\times0.00597\times(-0.022) + 27\times0.00597^2(0.022 + 2.972)}{27\times0.00597^3} = 699.532$$

$$\Delta_0 = \sqrt{q_0^2 + 4p^3} = \sqrt{699.532^2 + 4 \times 53.268^3} = 1045.9$$

$$\beta_{4M} = \frac{\sqrt[3]{2}p}{\sqrt[3]{-q_0 + \Delta_0}} - \frac{\sqrt[3]{-q_0 + \Delta_0}}{\sqrt[3]{2}} + \frac{l_1}{3k_2}$$

$$= \frac{\sqrt[3]{2} \times 53.268}{\sqrt[3]{-699.532 + 1045.9}} - \frac{\sqrt[3]{-699.532 + 1045.9}}{\sqrt[3]{2}} + \frac{-0.022}{3 \times 0.00597} = 2.7531$$

The corresponding probability of failure is equal to 0.00295.

Using the method of Monte Carlo simulation (MCS) with 1,000,000 samples, the probability of failure for this performance function is obtained as 0.00307 with corresponding reliability index of $\beta = 2.740$ and the coefficients of variation of $P_F$ is 1.803%. It is found that both the results of the third- and fourth-moment approximations are in close agreement with the MCS results, whereas the second-moment approximation overestimated the reliability index by 8.5%.

It can also be observed that, for this example, the probability of failure obtained using the fourth-moment method is closer to the result of MCS than that of FORM. Furthermore, the methods of moment can be easily conducted without shortcomings associated with the design points, and do not require iteration or computation of derivatives, and do not conduct reliability analysis for each mode and need not to compute the correlation matrix and joint probability.

### Example 29.8 Two-Story One-Bay Truss Structure

The example is an elastoplastic truss structure with two stories and one bay as shown in Figure 29.15, which is also a series system. The statistics of the member strengths and loads are as follows: mean values are of $\mu_{T1} = \mu_{T2} = 90$kip, $\mu_{T3} = 9$kip, $\mu_{T4} = \mu_{T5} = 48$kip, $\mu_{T6} = \mu_{T7} = 21$kip, $\mu_{T8} = 15$kip, $\mu_{T9} = \mu_{T10} = 30$kip, $\mu_{F1} = 11$kip, $\mu_{F2} = 3.6$kip; and coefficients of variation are of $V_{T1} = \ldots = V_{T10} = 0.15$, $V_{F1} = 0.3$, $V_{F2} = 0.2$. The performance functions corresponding to the eight most likely failure modes are given in Eqs. (29.51a–i) (Ono et al. 1990), with the respective

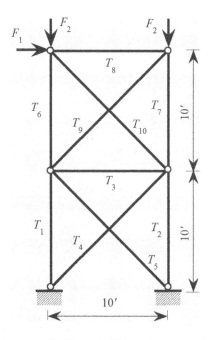

**FIGURE 29.15** Two-story one-bay truss for Example 29.8.

FORM reliability indices listed in parentheses showing that none of the modes are significantly dominant.

$$g_1 = 0.7071T_4 + 0.7071T_5 - 2.2F_1 \quad (\beta_F = 3.409) \tag{29.51a}$$

$$g_2 = T_6 + 0.7071T_{10} - 1.2F_1 - F_2 \quad (\beta_F = 3.497) \tag{29.51b}$$

$$g_3 = T_3 + 0.7071T_5 + 0.7071T_{10} - 2.2F_1 \quad (\beta_F = 3.264) \tag{29.51c}$$

$$g_4 = T_8 + 0.7071T_{10} - 1.2F_1 \quad (\beta_F = 3.333) \tag{29.51d}$$

$$g_5 = T_6 + T_7 - 1.2F_1 \quad (\beta_F = 3.814) \tag{29.51e}$$

$$g_6 = T_3 + 0.7071T_5 - 1.2F_1 - F_2 \quad (\beta_F = 3.484) \tag{29.51f}$$

$$g_7 = 0.7071T_9 + 0.7071T_{10} - 1.2F_1 \quad (\beta_F = 3.846) \tag{29.51h}$$

$$g_8 = T_1 + 0.7071T_5 - 3.4F_1 - F_2 \quad (\beta_F = 3.793) \tag{29.51i}$$

Using the five-point estimates, the first four moments of the system performance function $G(\mathbf{X}) = \min\{g_1, g_2, g_3, g_4, g_5, g_6, g_7, g_8\}$ are approximately obtained as $\mu_G = 22.316$, $\sigma_G = 5.379$, $\alpha_{3G} = -0.408$, and $\alpha_{4G} = 3.566$. With these first four moments of the system performance function, the moment-based reliability indices are obtained as $\beta_{2M} = 4.148$ and $\beta_{4M} = 3.246$. Using MCS with 1,000,000 samples, the probability of failure for this system is obtained as $7.86 \times 10^{-4}$ with corresponding reliability index of $\beta = 3.161$. The coefficient of variation (COV) of this MCS estimate is 3.57%. For this example, the fourth-moment reliability index errors about 2.7%. The second-moment reliability index has a significant error of about 31%.

## 29.5 SYSTEM RELIABILITY ASSESSMENT OF DUCTILE FRAME STRUCTURES

### 29.5.1 INTRODUCTION

The evaluation of system reliability for ductile frame structures has been an active area of research in the past 50 years. Several commonly used assumptions are summarized in the following (e.g., Ditlevsen and Bjerager 1986; Zhao and Ono 1998; Mahadevan and Raghothamachar 2000).

1. The constitutive relationship of the frame members is assumed elastic-perfectly plastic. The failure of a section means the imposition of a plastic hinge and an artificial moment equal to the plastic moment capacity of the section is applied at the hinge.
2. The frame member does not lose stability until it reaches the limit state of bending capacity.
3. Structural uncertainties are represented by considering only moment capacities of the frame members as random variables.
4. The effects of axial forces on the reduction of moment capacities of the frame members are neglected. Geometrical second-order and shear effects are also neglected.

Using the assumptions above, based on the upper-bound theorem of plasticity (Livesley 1975), failure of ductile frame structure is defined as the formation of a kinematically admissible mechanism due to the formation of plastic hinges at a certain number of sections. Because of the uncertainties in the moment capacities of the structures and the load vector, frame structures may fail in different modes, and the general definition of performance function for ductile frame is the minimum of the performance functions corresponding to all the potential failure modes as

$$G(\mathbf{M},\mathbf{P}) = \min\{G_1(\mathbf{M},\mathbf{P}), G_2(\mathbf{M},\mathbf{P}), \ldots, G_n(\mathbf{M},\mathbf{P})\} \tag{29.52}$$

where $\mathbf{M}$ is the vector of moment capacities, $\mathbf{P}$ is the load vector, and $G_i(\mathbf{M}, \mathbf{P})$ is the performance function corresponding to $i$th failure mode.

To obtain the performance function defined by Eq. (29.52), the potential failure modes should be identified. However, there are generally an astronomically large number of potential failure paths in large-scale structures, and in most cases, only small fraction of them contributes significantly to the overall failure probability. These may be referred to as significant failure sequences, and the estimation of system failure probability based on these sequences is expected to be close to the true answer. Several different approaches have been developed to identify the significant failure modes (or sequences) of a ductile frame structure, such as incremental load approach (e.g., Moses 1982; Feng 1988), β-unzipping method (e.g., Thoft-Christensen and Murotsu 1986; Thoft-Christensen 1990; Yang et al. 2014), truncated enumeration method (e.g., Melchers and Tang 1984; Nafday 1987), branch and bound method (e.g., Murotsu et al. 1984; Mahadevan and Raghothamachar 2000; Lee and Song 2011), stable configuration approach (Ang and Ma 1981; Bennett and Ang 1986), and mathematical programming techniques (Zimmerman et al. 1993), etc. Although some of these methods present elegant approaches for identifying the significant failure modes, they tend to be uneconomical or even unresolvable when applied to large structures.

If the potential failure modes are known or can be identified, computational procedures can be used for system reliability estimation such as bounding techniques (e.g., Cornell 1967; Ditlevsen 1979; Feng 1989; Song 1992; Penmetsa and Grandhi 2002; Zhao et al. 2007), the PNET (Ang and Ma 1981), and direct or smart MCSs (Melchers 1994). However, the calculation of the failure probability is generally difficult for a system because of the large number of potential failure modes for most of practical structures, the difficulty in obtaining the sensitivity of the performance function, and the complications in computing mutual correlation matrix and joint failure probability matrix among failure modes as described in Section 29.3.

Considering the difficulties in both failure mode identification and failure probability computation when using traditional methods, we will first define a performance function independent of failure modes and compute the corresponding probability of failure.

### 29.5.2 PERFORMANCE FUNCTION INDEPENDENT OF FAILURE MODES

In order to define a failure mode dependent performance function, one can first observe the limit state of function under only one load $P$. Because the structure will become a kinematically admissible mechanism when the load $P$ is equal to the ultimate load, the performance function can be described as

$$G(\mathbf{M}, P) = u_p(\mathbf{M}) - P \tag{29.53}$$

where $u_p(\mathbf{M})$ is the ultimate load. The value of $u_p(\mathbf{M})$ is dependent on $\mathbf{M}$, and a different failure mode may happen with different component of $\mathbf{M}$.

In limit analysis, the ultimate load is generally described as the production of load $P$ and a load factor, therefore, Eq. (29.53) can be described as follows:

$$G(\mathbf{M}, P) = \lambda(\mathbf{M}, P)P - P \tag{29.54}$$

where $\lambda(\mathbf{M}, P)$ is the load factor.

In Eq. (29.54), for a specific load $P$ with certain direction, the amount of $P$ does not influence the shape of the limit state surface $G(\mathbf{M}, \mathbf{P}) = 0$. Therefore, the same reliability analysis results will be obtained if it is written as follows:

$$G(\mathbf{M}, P) = \lambda(\mathbf{M}, P) - 1 \tag{29.55}$$

For multiple loads, the limit state function is defined in a load space, and it cannot be dealt directly with as Eq. (29.55). One can divide the load space into various load path and consider one load path. For example, for a frame structure with two load $\mathbf{P} = [P_1, P_2]$. Considering a load path $\mathbf{P}(\theta_i)$ where $\theta_i$ is defined as $\tan\theta_i = P_2/P_1$, since the limit analysis is based on the principal of the proportional loading, the structure will become a kinematically admissible mechanism when the load increased along this load path. The utmost value of $P_1$ or $P_2$ can be described as the production of the load factor and $P_1$ or $P_2$ itself. The performance function can be written as follows, as already described:

$$G[\mathbf{M}, \mathbf{P}(\theta_i)] = \lambda[\mathbf{M}, \mathbf{P}(\theta_i)]P_1 - P_1 \text{ or } G[\mathbf{M}, \mathbf{P}(\theta_i)] = \lambda[\mathbf{M}, \mathbf{P}(\theta_i)]P_2 - P_2 \qquad (29.56)$$

where $\lambda[\mathbf{M}, \mathbf{P}(\theta_i)]$ is the load factor and $\mathbf{P}(\theta_i)$ is the load path.

In Eq. (29.56), for a specific load path, the amount of $P_1$ or $P_2$ does not influence the shape of the limit state surface $G[\mathbf{M}, \mathbf{P}(\theta_i)] = 0$. Therefore, the same reliability analysis results will be obtained if it is written as follows:

$$G[\mathbf{M}, P(\theta_i)] = \lambda[\mathbf{M}, \mathbf{P}(\theta_i)] - 1 \qquad (29.57)$$

Different failure modes will happen with different $\mathbf{M}$ and different load path $\mathbf{P}(\theta_i)$, Eq. (29.57), however, always holds true with any load path. With given $(\mathbf{M}, \mathbf{P})$, the loading path is certain and only one value of $\lambda(\mathbf{M}, \mathbf{P})$ can be obtained, therefore, one needs not to specify the loading path and Eq. (29.57) is written as

$$G(\mathbf{M}, \mathbf{P}) = \lambda(\mathbf{M}, \mathbf{P}) - 1 \qquad (29.58)$$

Eq. (29.58) is the general form of performance function of ductile frame structures (Zhao and Ono 1998). Because the ultimate load is obtained as the minimum load of formulation a kinematically admissible mechanism, limit state surface expressed in Eqs. (29.52) and (29.58) are the same, although their performance functions are different.

Using the performance function expressed in Eq. (29.58), the failure probability can be estimated from first- or second-order reliability method (FORM/SORM) and the response surface approach (RSA) without the identification procedure of failure modes and the correlation computation among the large number of failure modes (Zhao and Ono 1998; Leu and Yang 2000). However, the procedures need iteration and sometimes have local convergence problems. It is, therefore, very necessary to check the results of different fitting points. Besides, it becomes difficult to deal with complex large-scale structures especially with uniform distributed loads due to the strong nonlinearity of the performance function. In order to avoid iteration and overcome the fitting point dependence described above, we will use the methods of moment described in Section 29.4, in which, the first four moments of the performance function will be utilized to compute the probability of failure without derivation-based iteration.

Previous study (Zhao and Ono 1998) has revealed that the failure mode independent performance function in Eq. (29.58) was essentially the inner connotative surface of the limit state surfaces of all the potential failure modes of the structural system. The failure mode independent performance function in Eq. (29.58) is generally implicit, and for practical structures, it may be strongly nonlinear and the distribution of the performance function may have strong non-normality. Since the methods of moment are good for weak non-normality of the distributions of the performance functions, to improve the computational efficiency of methods of moment in application to the system reliability analysis of frame structures, it is necessary to have a failure mode independent performance function with distributions of weak non-normality.

It has been pointed out that the Box-Cox transformation (Box and Cox 1964; Zhang and Pandey 2013) is useful to bring the distributions of the transformed random variables or performance functions closer to the normality. Since the load factor $\lambda(\mathbf{M}, \mathbf{P})$ is greater than 0, according to the Box-Cox transformation, the performance function expressed in Eq. (29.58) can be equivalently expressed as:

$$G^*(\mathbf{M},\mathbf{P}) = \begin{cases} \dfrac{[\lambda(\mathbf{M},\mathbf{P})]^q - 1}{q}, & q \neq 0 \\ \ln[\lambda(\mathbf{M},\mathbf{P})], & q = 0 \end{cases} \tag{29.59}$$

where $q$ is an undetermined coefficient.

For simplicity, let $q = 0$, the performance function for the system reliability analysis of ductile frame structures can be expressed as

$$G^*(\mathbf{M},\mathbf{P}) = \ln[\lambda(\mathbf{M},\mathbf{P})] \tag{29.60}$$

Although the performance function of Eq. (29.60) has a different form than that of Eq. (29.58), they correspond to the same limit state physically. Since Eq. (29.60) can be considered as a logarithmic transformation of Eq. (29.58), it can reduce the variation of the performance function through reducing the measure scale. That is, the random fluctuations in Eq. (29.60) are generally much weaker than that in Eq. (29.58). In addition, the transformation is generally an effective way to make distributions less skewed as will be demonstrated in the following numerical examples. Eq. (29.60) can therefore be used as the failure mode independent performance function with weak non-normality for system reliability evaluation of ductile frame structures.

### Example 29.9  System Reliability of a Six-Story Two-Bay Ductile Frame Structure under Lateral Concentrated Loads and Vertical Uniform Loads

Consider a six-story two-bay plane building frame subjected to lateral concentrated loads and vertical uniform loads as shown in Figure 29.16, with the probabilistic information of member strength ($R_b$ and $R_c$) and loads ($L_1$, $L_2$, and $L_3$) listed in Table 29.2.

According to Eq. (29.60), the proposed failure mode independent performance functions for this example can be expressed as

$$G^*(\mathbf{M},\mathbf{P}) = \ln\left[\lambda\left(R_b, R_c, L_1, L_2, L_3\right)\right] \tag{29.69}$$

Using bivariate-dimension reduction-based point estimate method, the first four moments of the performance function $G^*(\mathbf{M}, \mathbf{P})$ are approximately $\mu_G = 0.611$, $\sigma_G = 0.141$, $\alpha_{3G} = -0.029$, and $\alpha_{4G} = 2.981$. With these first four moments of the proposed failure mode independent performance function for the system reliability evaluation of the ductile frame structure, the moment-based reliability indices are $\beta_{4M} = 4.305$ with $P_f = 8.332 \times 10^{-6}$. The total number of calls of structural limit analysis is 182 in the proposed method. For comparison, the probability of failure is also conducted with subset simulation (SS) (Au and Beck 2001; Li and Cao 2016), which is instrumental in improving the computational efficiency of MCS. Here, the conditional probability of SS is 0.1 and the number of samples in each simulation level $N = 5000$ is used. The probability of failure is obtained as $9.261 \times 10^{-5}$ with corresponding reliability index of 4.282, in which the total number of calls for structural limit analysis in SS is 23,000.

It can be observed that: (1) the reliability results of the proposed method and SS are in close agreement with the exact values; and (2) the computational effort of the proposed method is about 0.72% of that of SS.

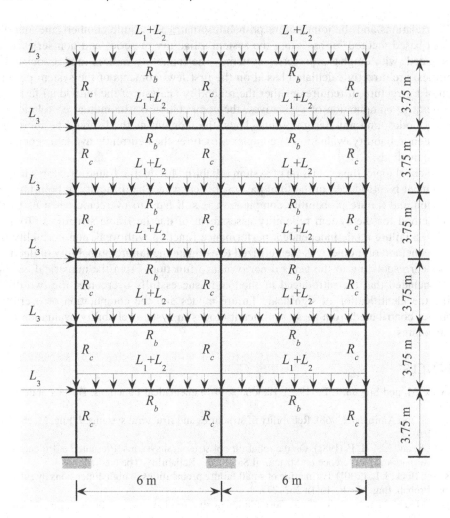

**FIGURE 29.16** A six-story two-bay frame structure for Example 29.9.

**TABLE 29.2**
**Random Variables for Example 29.9**

| Basic Variables | Description | Distribution | Mean | Coefficient of Variation (COV) |
|---|---|---|---|---|
| $R_b$ | Moment capacity of beams | Lognormal | 150 kN·m | 0.05 |
| $R_c$ | Moment capacity of columns | Lognormal | 200 kN·m | 0.05 |
| $L_1$ | Dead load | Lognormal | 20 kN/m | 0.05 |
| $L_2$ | Live load | Gumbel | 12 kN/m | 0.30 |
| $L_3$ | Lateral load | Weibull | 12 kN | 0.40 |

## 29.6 SUMMARY

The concept of system reliability is described in this chapter. In order to improve the narrow bounds of the failure probability of a series structural system, a point estimation method is introduced for calculating the joint probability of every pair of failure modes of the system. Since the number of potential failure modes for most practical structures is very large, the determination of the mutual correlations and the joint failure probability matrices is quite cumbersome and difficult. A moment-based method for assessing the system reliability of series and non-series structures is investigated, with emphasis on series systems. The reliability indices (and associated failure probabilities) are directly calculated based on the first few moments of the system performance function of a structure. It requires neither the reliability analysis of the individual failure modes nor the iterative computation of derivatives, the computation of the mutual correlations among the failure modes, and any design points. Thus, the moment method should be more effective for the system reliability evaluation of complex structures than currently available computational (non-MCS) methods.

As a practical topic, the evaluation of system reliability for ductile frame structures using methods of moment is introduced in the final part of this chapter. The difficulty in both failure mode identification and failure probability computation is still hard to overcome when using the traditional method for the system reliability assessment of ductile frame structures. To avoid the difficulties, a failure mode independent performance function with weak non-normality is introduced. The method of moments is then utilized to estimate the failure probability of ductile frame structures corresponding to the general performance function. From the numerical examples, it can be concluded that the introduced methodology successfully overcomes the two difficulties including the identification of significant failure modes and the computation of overall failure probabilities contributed from these significant modes in system reliability evaluation of ductile frame structures.

## REFERENCES

Abramowitz, M. and Stegun, I. E. (1972). Handbook of Mathematical Functions. Dover Publications: New York.

Ang, A. H.-S. and Amin, M. (1968). Reliability of structures and structural systems. J. Eng. Mech., 94(EM2): 671–691.

Ang, A. H. -S. and Ma, H. F. (1981). On the reliability of structural systems. Presented at Proceedings of the 3rd International Conference on Structural Safety and Reliability, The Netherlands.

Au, S. K. and Beck, J. L. (2001). Estimation of small failure probabilities in high dimensions by subset simulation. Probab. Eng. Mech., 16(4): 263–277.

Bennett, R. M. and Ang, A. H.-S. (1986). Formulation of structural system reliability. J. Eng. Mech., 112(11): 1135–1151.

Box, G. E. P. and Cox, D. R. (1964). An analysis of transformations. J. Roy. Statist. Soc. Ser. B, 26(2): 211–252.

Chang, Y. and Mori, Y. (2013). A study on the relaxed linear programming bounds method for system reliability. Struct. Saf., 41: 64–72.

Cornell, C. A. (1967). Bounds on the reliability of structural systems. J. Struct. Div., 93(1): 181–200.

Ditlevsen, O. (1979). Narrow reliability bounds for structural systems. J. Struct. Mech., 8(4): 453–482.

Ditlevsen, O. and Bjerager, P. (1986). Methods of structural system reliability. Struct. Saf., 3(3): 195–229.

Feng, Y. S. (1988). Enumerating significant failure modes of a structural system by using criterion methods. Comput. Struct., 30(5): 1153–1157.

Feng, Y. S. (1989). A method for computing structural system reliability with high accuracy. Comput. Struct., 33(1): 1–5.

Grigoriu, M. (1983). Approximate analysis of complex reliability problems. Struct. Saf., 1: 277–288.

Grimmelt, M. J. and Schueller, G. I. (1982). Benchmark study on methods to determine collapse failure probabilities of redundant structures. Struct. Saf., 1: 93–106.

Hong, H. P. and Lind, N. C. (1996). Approximation reliability analysis using normal polynomial and simulation results. Struct. Saf., 18(4): 329–339.

Lee, Y. and Song, J. (2011). Risk analysis of fatigue-induced sequential failures by branch-and-bound method employing system reliability bounds. J. Eng. Mech., 137(12): 807–821.

Leu, L. J. and Yang, S. S. (2000). System reliability analysis of steel frames considering the effect of spread of plasticity. 8th ASCE Specialty Conference on Probabilistic Mechanics and Structural Reliability, pp. 19–22.

Li, H. S. and Cao, Z. J. (2016). MatLab codes of subset simulation for reliability analysis and structural optimization. Struct. Multidiscip. Optim., 54(2): 391–410.

Li, J., Chen, J. B. and Fan, W. L. (2007). The equivalent extreme-value event and evaluation of the structural system reliability[J]. Struct. Saf., 29(2): 112–131.

Livesley, R. K. (1975). Matrix Methods of Structural Analysis, 2. Pergamon Press: New York.

Mahadevan, S. and Raghothamachar, P. (2000). Adaptive simulation for system reliability analysis of large structures. Comput. Struct., 77(6): 725–734.

Melchers, R. E. (1994). Structural system reliability assessment using directional simulation. Struct. Saf., 16(1–2): 23–37.

Melchers, R. E. and Tang, L. K. (1984). Dominant failure modes in stochastic structural systems. Struct. Saf., 2(2): 128–143.

Miao, F. and Ghosn, M. (2011). Modified subset simulation method for reliability analysis of structural systems. Struct. Saf., 33(4–5): 251–260.

Moses, F. (1982). System reliability developments in structural engineering. Struct. Saf., 1: 3–13.

Murotsu, Y., Okada, H., Taguchi, K., Grimmelt, M. and Yonezawa, M. (1984). Automatic generation of stochastically dominant failure modes of frame structures. Struct. Saf., 2(1): 17–25.

Nafday, A. M. (1987). Failure mode identification for structural frames. J. Struct. Eng., 113(7): 1415–1432.

Ono, T., Idota, H. and Dozuka, A. (1990). Reliability evaluation of structural systems using higher-order moment standardization technique. J. Struct. Constr. Eng., (418): 71–79 (In Japanese).

Penmetsa, R. and Grandhi, R. (2002). Structural system reliability quantification using multipoint function. AIAA J., 40(12): 2526–2531.

Song, B. F. (1992). A numerical Integration method in affine space and a method with high accuracy for computing structural system reliability. Comput. Struct., 42(2): 255–262.

Stuart, A. and Ord, J. K. (1987). Kendall's Advanced Theory of Statistics. Charles Griffin & Company Ltd.: London, 1: 210–275.

Thoft-Christensen, P. (1990). Consequence modified β-unzipping of plastic structures. Struct. Saf., 7(2): 191–198.

Thoft-Christensen, P. and Murotsu, Y. (1986). Application of Structural Systems Reliability Theory. Springer: Berlin.

Winterstein, S. R. (1988). Nonlinear vibration models for extremes and fatigue. J. Eng. Mech., 114: 1772–1790.

Xu, H. and Rahman, S. (2004). A generalized dimension-reduction method for multidimensional integration in stochastic mechanics. Int. J. Numer. Methods Eng., 61(12): 1992–2019.

Yang, L., Liu, J. and Yu, B. (2014). Adaptive dynamic bounding method for reliability analysis of structural system. China Civ. Eng. J., 47(4): 38–46. (in Chinese)

Zhang, X. and Pandey, M. D. (2013). Structural reliability analysis based on the concepts of entropy, fractional moment and dimensional reduction method. Struct. Saf., 43: 28–40.

Zhao, Y. G. and Ang, A. H.-S. (2003). System reliability assessment by method of moments. J. Struct. Eng., 129(10): 1341–1349.

Zhao, Y. G. and Lu, Z. H. (2007). Fourth-moment standardization for structural reliability assessment. J. Struct. Eng., 133(7): 916–924.

Zhao, Y. G. and Ono, T. (1998). System reliability evaluation of ductile frame structure. J. Struct. Eng., 124(6): 678–685.

Zhao, Y. G. and Ono, T. (2000). New point-estimates for probability moments. J. Eng. Mech., 126(4): 433–436.

Zhao, Y. G. and Ono, T. (2001). Moment methods for structural reliability. Struct. Saf., 23(1): 48–85.

Zhao, Y. G., Zhong, W. Q. and Ang, A. H.-S. (2007). Estimating joint failure probability of series structural systems. J. Eng. Mech., 133(5): 588–596.

Zimmerman, J. J., Ellis, J. H. and Corotis, R. B. (1993). Stochastic optimization models for structural reliability analysis. J. Struct. Eng., 119(1): 223–239.

# 30 Reliability-Based Design Optimization Using Transfer Learning

*Tayyab Zafar*
University of Electronic Science and Technology
of China, Chengdu, China
National University of Sciences and Technology, Islamabad, Pakistan

*Zhonglai Wang, Shui Yu*
University of Electronic Science and Technology
of China, Chengdu, China

## CONTENTS

## 30.1 INTRODUCTION

Dynamic uncertainties exist in various stages of the design process including environmental, loading, and operational conditions (Yu et al. 2020). Due to these uncertainties, the strength of a structure may degrade with time. Therefore, uncertainties must be considered in designing a structure to enhance structural performance and reliability (Singh, Mourelatos and Li 2010). The reliability-based design optimization (RBDO) is introduced in the literature (Kharmanda, El Hami and Souza De Cursi 2013) for the incorporation of reliability in the design optimization method. RBDO attempts to reduce the cost of a structure while considering the structural safety and hence maintaining a balance between reliability and cost (Lee and Jung 2008). A reliability assessment (SORA) and sequential optimization process (Du and Chen 2004) is introduced to transform the nested loops into the inverse of reliability analysis and deterministic optimization sequence. A single loop-based method is suggested to change the probabilistic optimization problem into a deterministic optimization (Liang, Mourelatos and Tu 2008). Another approach to decompose the RBDO problem to several cycles of deterministic design optimization is introduced in Chen et al. (2013). Generally, in the RBDO problems, time-independent reliability is considered; however, the reliability of a structure changes with time due to dynamic uncertainties (Wang et al. 2020; Yuan et al. 2020; Hyeon Ju and Chai Lee 2008). The time-dependent reliability (TDR) can be well-defined as, the probability that the structure executes its task properly for a specified time interval under the given conditions (Wang et al. 2014). Considering the TDR significantly increases the computational efforts. Several approaches are proposed in the literature for the reduction of computational efforts in TDR-based design optimization (TDRBDO). The sequential optimization approach in Du and Chen (2004) is extended to time-dependent problems involving stochastic processes using an equivalent most probable point (Hu and Du 2016). Composite limit state-based method (Singh and Mourelatos 2010a) is developed to optimize the lifecycle cost; however the method is computationally prohibitive. Another out-crossing rate based approach is proposed (Kuschel and Rackwitz 2000) for mono-level TDRBDO problem. To handle the high dimensional problems, a line search strategy and a nonlinear point algorithm is proposed (Jensen et al. 2012). Generally, these methods are efficient numerically; however, the TDR accuracy for these methods cannot be ensured. For example, the above-mentioned sequential and mono-level methods may not be so accurate in estimating the TDR which may result in suboptimal design (Agarwal 2004). To obtain more reliable results with minimal computational efforts, a double-loop optimization methodology is introduced in the literature (Dubourg, Sudret and Bourinet 2011). In the double loop method, two nested loops are used to embed the TDR analysis in a TDRBDO problem. A sequential Kriging surrogate model-based method is recommended in Hawchar, El Soueidy and Schoefs (2018). The Kriging surrogate model based process is suggested in which performance function is approximated by a global Kriging surrogate model is made in an amplified reality region. In TDRBDO problems, TDR analysis is required to be performed repeatedly, therefore reducing the computational efforts for reliability analysis would significantly impact the optimization process. Hence, reliability analysis is still one of the active areas of the research. The TDR estimation methods are generally categorized into three types as (1) analytical methods (2) sampling-based methods (3) surrogate model-based methods. Generally, analytical methods (van Noortwijk et al. 2007; Singh and Mourelatos 2010b) not so appropriate for highly non-linear and multimodal problems due to their high computational requirements. Sampling-based methods (Singh, Mourelatos and Nikolaidis 2011) have improved efficiency as compared to analytical method, however, their efficiency still dependent on non-linearity of limit state functions (LSFs). To further reduce the computational efforts for highly non-linear problems, surrogate model-based methods are introduced. Wang and Wang develop a nested great comeback surface framework (Wang and Wang 2012). The proposed framework builds a response surface on random variable space to predict time corresponding to greater response of system using Efficient Global Optimization (EGO). Hence, the TDR is converted to time-independent reliability. To improve its computational efficiency, Hu and Du proposed a new method in which the random variable and time samples are drawn simultaneously (Hu and Du 2015). In these methods, double loop-based method is proposed to estimate the TDR; however, these methods are not computationally efficient for problems involving stochastic processes. A single loop Kriging model is established in Hu and Mahadevan (2016). The key model of the suggested method is to predict the number of failing trajectories accurately without having to consider the extreme value of the LSF.

These methods are used to analyze the reliability of a structure for a given design point. For these methods, experiments or simulations need to be performed for a complete period of time. However, the acquisition of time-dependent LSF information is a challenging task for long simulations. In this work, prediction-based TDRBDO (P-TDRBDO) is proposed that has never been addressed. In the proposed scheme, the total time is broken down into two sub-intervals. Let's call them for instance the present and future interval (PI) and (FI), respectively. The stochastic process samples (SPS) in the two sub-intervals are transformed and are made similar. These samples are then utilized to estimate and predict the TDR for the whole time interval. A transformation matrix ™ using a transfer learning-based technique is calculated to transmute the SPS. A Kriging model of the given LSF in the PI only made in the global space by means of an adaptive sampling procedure. Random variables samples as well as transformed SPS are utilized to build the surrogate model. The development of the Kriging surrogate model is decoupled from the reliability prediction and design optimization process. After building the surrogate model, the method then attempts to search for the optimal design of the P-TDRBDO problem. The cost function is taken as an objective function whereas to ensure the optimal design reliability, the dynamic LSF is taken as a constraint to the optimization problem. To search for the optimal design of the P-TDRBDO problem, two nested loops are used. In the outer loop, the design is augmented using a gradient-based method. In the inner loop, the TDR is predicted for the whole time interval using the global Kriging surrogate model initially built in the PI only. The research work contributions are summarized as follows: (1) P-TDRBDO is introduced in the paper which has never been addressed in the literature. (2) A transfer learning-based method is proposed to find the TM. (3) The global Kriging surrogate model is decoupled from the optimization process. (4) The future interval LSF information is not required to build the Kriging surrogate model.

The rest of paper is structured as follows: P-TDRBDO problem is formulated in Section 30.2. Section 30.3 explains the planned method in detail. Section 30.4 exhibits the results and discussion of the proposed method using several examples. The research is concluded in Section 30.5.

## 30.2  PREDICTED TIME-DEPENDENT RELIABILITY-BASED DESIGN OPTIMIZATION (P-TDRBDO)

Traditionally, the TDRBDO problem is defined as the constrained optimization problem with dynamic probabilistic constraints and given as follows:

$$\begin{array}{c} \underset{D,\mu_{X_D}}{\text{minimize}:} f\left(D,\mu_{X_D}\right) \\ \text{subject to}: \Pr\left\{g_i\left(D,X,Y(t)\right)\geq 0,\ \forall t\in[0,T]\right\}>[R_i] \\ i=1,2,\ldots,n_g, \mu_{X_D}^{(L)}\leq\mu_{X_D}\leq\mu_{X_D}^{(U)} \end{array} \tag{30.1}$$

where $f\left(D,\mu_{X_D}\right)$ represents the objective function, $D$ is the deterministic variables, and $X$ is random variables. The random variables are further divided into two categories $X=[X_D,X_R]$ where $X_D$ represents the design random variables and $X_R$ represents the random variables with known means. The lower and upper bound of random design variables are represented by $\left[\mu_{X_D}^{(L)},\mu_{X_D}^{(U)}\right]$, respectively. The $i$th LSF of the TDRBDO problem is given by $g_i\left(D,X,Y(t)\right)$ where $Y(t)$ represents the stochastic process in the given time interval $[0,T]$ and $n_g$ represents the total number of LSFs. During the optimization process, the TDR of a given $i$th LSF should be greater than a pre-defined threshold $R_i$. Mathematically, TDR is defined as

$$R(0,\ T)=\Pr\left\{g_i\left(D,X,Y(t)\right)\geq 0,\ \forall t\in[0,T]\right\} \tag{30.2}$$

Generally, two types of optimization algorithms are used to find the solution of TDRBDO, namely evolutionary algorithms and gradient based optimization algorithms (Hawchar, El Soueidy and Schoefs 2018). Evolutionary algorithms such as genetic algorithms, particle swarm optimization, etc. can perform well for complex problems; however, these algorithms are computationally prohibitive as they have to evaluate complex LSFs many times. Therefore, gradient-based algorithms are typically used as they are efficient in finding the optimal solution. However, they still have to evaluate the LSF multiple times.

To further increase the computational efficiency of gradient-based algorithms, surrogate model-based methods are introduced. Usually, in these surrogate model-based methods, a double loop procedure is typically used to build a surrogate model based on extreme values. In the first loop, the extreme value of the LSF is found using EGO. In the second loop, the Kriging surrogate model is built using the extreme values. This strategy has two drawbacks (1) the error in finding the extreme values would propagate to the surrogate model development and thus affecting the reliability accuracy, (2) the method is computationally expensive especially in the case of long-run simulations. To overcome these issues, a single loop surrogate model-based method is introduced in the literature (Hu and Mahadevan 2016).

### 30.2.1 Development of Kriging Surrogate Model

Let the input random parameters $[D,X,Y(t)]$ be represented by $S=[D,X,Y(t)]$. The vector $D=[d_1,d_2,...,d_{n_D}]$ consists of $d_{n_D}$ deterministic variables, $X=[x_{d_1},x_{d_2},...,x_{dn},x_{r_1},x_{r_2},...,x_{rn}]$ consists of $d_n$ and $r_n$ number of design and random variables, respectively, and $Y(t)=[y_1(t),y_2(t),...y_{n_y}(t)]$ consists of $n_y$ number of stochastic processes. Hence, $S$ can be represented as $S=[d_1,...,d_{n_D},x_{d_1},...,x_{nXd},x_{r_1},...,x_{nXr},y_1(t),...,y_{n_y}(t)]=[s_1,...,s_{nD},s_{nD+1},...,s_{nd+nXd},$ $s_{nd+nXd+1},...,s_{nd+nXd+nXr},s_{nd+nXd+nXr+1},...,s_{ns}]$ with $n_s=n_d+n_{Xd}+n_{Xr}+n_y$. To build an accurate surrogate model with minimal computational efficiency, an adaptive sampling technique is used. An initial model $\hat{g}_i(S)$ for the $i^{th}$ LSF $g_i(S)$ is developed using $n_{in}$ input samples $s_j=[s_1,s_2,...,s_{n_{in}}]$ and their corresponding responses $G_i=[g_i(s_1),g_i(s_2),...,g_i(s_{n_{in}})]$ where $j=1,2,...,n_{in}$ which is given as

$$\hat{g}_i(s_j)=\Theta_i(s_j)^T \xi_i+\epsilon_i(s_j) \tag{30.3}$$

where $\Theta_i(s_j)=[\Theta_{i1}(s_j),\Theta_{i2}(s_j),...,\Theta_{iz}(s_j)]$ represents a vector of $iz$ basis functions, $\xi_i=[\xi_1,\xi_2,...,\xi_{iz}]$ represents the vector of regression coefficients. The product of two terms is known as the prediction trend. $\epsilon_i$ represents a stationary Gaussian process with zero mean and covariance function given as

$$\mathrm{cov}(\epsilon_i(s_{ji}),\epsilon_i(s_{jj}))=\sigma_\epsilon^2 R(s_{ji},s_{jj}),\ ji,jj=1,2,...,n_{nin} \tag{30.4}$$

where $R(s_{ji},s_{jj})$ is the correlation function. For any given samples, the vector of unknown coefficients $\xi_i$ and $\sigma_\epsilon^2$ can be given as

$$\xi_i=(\vartheta_i^T R^{-1}\vartheta_i)^{-1}\vartheta_i^T R^{-1}G_i \tag{30.5}$$

$$\sigma_{\epsilon_i}^2=\frac{1}{n_{in}}(G_i-\vartheta_i\xi_i)^T R^{-1}(G_i-\vartheta_i\xi_i) \tag{30.6}$$

where, $\vartheta_i=[\Theta_i(s_j)]_{n_{in}\times iz}$ and $R=[R(s_{ji},s_{jj},\theta_i)]_{n_{in}\times n_{in}}$ is the correlation matrix with an unknown parameter $\theta_i$. The optimal value of $\theta_i$ is found using maximum log-likelihood function (Lophaven, Nielsen and Søndergaard 2002) and can be given as

$$\theta_i^{opt}=\underset{\theta_i}{\mathrm{argmin}}\left(\sigma_i^2|R|^{\frac{1}{n_{in}}}\right) \tag{30.7}$$

$n_{MCS}$ samples are generated and utilized to refine the initially built surrogate model. The model is used to predict the response for any new sample $s_a$. The response for these samples are given as

Gaussian random variable with mean $\mu_{s_a}$ and variance $\sigma^2_{s_a}$. The mean and error for the given new sample point $s_a$ is given as

$$\hat{\mu}_{s_a} = \Theta_i(s_a)^T \xi_i + r_{n_{in}}(s_a) R^{-1}(G_i - \Theta_i \xi_i) \tag{30.8}$$

$$\hat{\sigma}^2_{s_a} = \left(1 - \left[\Theta_i(s_a) \; r_{n_{in}}(s_a)\right] \begin{bmatrix} 0 & \vartheta_i^T \\ \vartheta_i & R \end{bmatrix} \begin{bmatrix} \Theta_i(s_a) \\ r_{n_{in}}(s_a) \end{bmatrix} \right) \tag{30.9}$$

where $r_{n_{in}}(s_a)$ is the correlation vector between $s_{n_{in}}$ samples and the new sample $s_a$. The response is predicted for all the $n_{MCS}$ samples. A sample point is selected among the MCS samples to refine the initially built surrogate model based on the $U$ function (Hu and Mahadevan 2016) which is given as

$$U_i(s_a) = \frac{\hat{\mu}_{s_a}}{\hat{\sigma}_{s_a}} \tag{30.10}$$

Based on the $U$ function value, it can be determined that the sign of response prediction is accurate or not. If the U value is greater than two, it means that the sign is predicted accurately whereas a value less than two represents a false prediction. The Monte Carlo simulation (MCS) samples are categorized into two groups, namely $\Lambda_g$ and $\Lambda_o$. Sample points with U value greater than two are grouped into $\Lambda_g$ and all other samples are grouped into $\Lambda_o$. A sample point $s_b$ is selected considering two important criteria: (1) the sample with minimum U value is selected among the $\Lambda_o$ as it would most affect the built model accuracy.

$$s_b = \min_s U(s_o) \tag{30.11}$$

(2) The maximum correlation value of the sample point $s_b$ and the initial point must be less than 0.95.

$$\max\left\{\rho\left(s_i^{new}, s_{n_{in}}\right)\right\} < 0.95 \tag{30.12}$$

The response for the selected sample point $s_b$ is then recorded using the original LSF and then added to initial samples. The process is continued until a stopping criterion for each LSF is satisfied. The stopping criterion can be given as

$$\varepsilon_i^{max} = \max_{n_{fio}^* \in [0, n_{io}]} \frac{n_{fio} + n_{fio}^*}{\left| n_{fig}^* - n_{fio}^* \right|} \times 100\% \tag{30.13}$$

Here, $n_{fig}$ and $n_{fio}$ denotes the number of failing trajectories in the two groups $\Lambda_g$ and $\Lambda_o$, respectively, whereas $n_{fio}^*$ is the actual number of failing trajectories.

## 30.2.2 Time-Dependent Reliability Prediction

The future performance conditions of a structure usually depend severely on the previous conditions and there usually exists a veritable well of relevant information that can be exploited to predict the future performance condition of a component to improve computational efficiency. Transfer learning is a branch of machine learning which attempts to transfer knowledge from one domain to another (Pan and Yang 2009). Domain adaptation, a paradigm of transfer learning aims to find a latent space where the knowledge from two domains has a similar representation. The domain

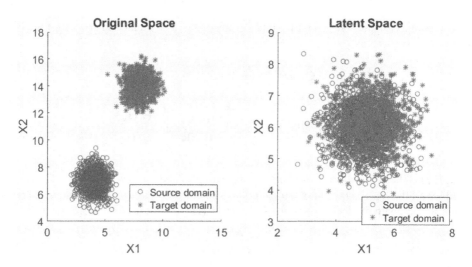

**FIGURE 30.1** Concept of domain adaptation.

for which complete information is available is known as the source domain whereas the domain for which partial information is available is known as the target domain. Figure 30.1 shows the basic concept of domain adaptation learning. It can be seen from the figure that in the original space the two domain samples have different statistical properties. A surrogate model built using source domain samples would not work for the target domain. To overcome this, the two domains' samples are transformed in a way that they have similar representations. Therefore, the surrogate model built using source domain samples can be utilized to build for the target domain as well.

To find this latent space, two criteria are proposed in the literature. These criteria are named as maximum mean discrepancy (MMD) (Pan et al. 2010) and Hilbert Schmidt independence criteria (HSIC) (Yan, Kou and Zhang 2017). MMD can deal with the problems where a discrete change is involved. However, mechanical system problems often deal with continuous change in their operating or environmental conditions. Hence, HSIC is utilized to transfer LSF information from one-time interval to another. For this purpose, the given time interval $[0, T]$ is broken down into two sub-intervals $[0,\tau] \cup [\tau,T]$. It is assumed that the complete information (random parameters and LSF) is available for $[0,\tau]$ only whereas partial information (input parameters) is available for $[\tau,T]$. The reliability is required for complete time-interval $[0,T]$. Based on predicted TDR, the optimal design needs to be found.

## 30.3 THE PROPOSED METHOD FOR P-TDRBDO

To solve the P-TDRBDO problem, an integrated framework is proposed. The framework integrates the transfer learning and the Kriging surrogate model. The proposed frameworks treat the time-dependent LSFs as probabilistic constraints. The dynamic constraints are converted to deterministic constraints by estimating the TDR for a given design point. To predict the TDR using Kriging surrogate, a TM is found using transfer learning. The SPS is generated in the PI and the FI. The SPS samples are utilized to find the TM. Subsequently, transformed SPS is used to build the Kriging model. Surrogate model for each of the LSF is built in the PI $[0,\tau]$ only. The TDR for the given time $[0, T]$ is predicted using the same surrogate model. It should be noted that the Kriging surrogate model is decoupled from the optimization process and built independently. Once, the surrogate model is built, the optimization process utilizes the surrogate model to predict the TDR, which greatly enhances the computational efficiency of the proposed method. To find the optimal design a gradient based optimization algorithm is used. The basic flowchart of the proposed method is shown in Figure 30.2.

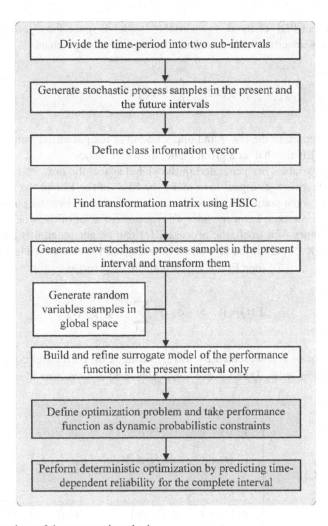

**FIGURE 30.2**   Flowchart of the proposed method.

One of the challenging issues in building the Kriging surrogate model is the variation in the input parameters (mean or standard deviation) value of the design random variables. Different values of the parameters would result in different probabilistic density functions (PDF). Hence, the accuracy of the surrogate model built for a specific value of the input parameter would be compromised for a different input parameter. Therefore, the random variables need to be generated in global space so that the change in the input parameters does not affect the surrogate model accuracy. Hence, to build the Kriging surrogate model, the input design random variables samples $x_d$ are represented by $x_d'$, which are generated in the global space.

The input design random variables samples with varying mean are generated by finding the inverse of some CDF function

$$x_d' = F^{-1}(p) \tag{30.14}$$

where the CDF function $F(p)$ can be found by

$$F(p) = \int_{-\infty}^{x_d} f(x_d)\,dx_d \tag{30.15}$$

where $p(x_d)$ represent PDF of the design variable in global space for a generalized value of the parameter $P$ which is uncertain with a uniform distribution $u(p)$ given as

$$f(x_d) = \int_{p^{(l)}}^{p^{(u)}} \varphi(x_d|p) u(p) dp \tag{30.16}$$

where $p^{(l)}$ and $p^{(u)}$ represent the lower and upper bound of the parameter value of the design variable $x_d$ and $\varphi(x_d \mid p)$ is the PDF at a given input parameter value.

Once the design variables are generated in the global space, the next step is to find the TM for SPS using transfer learning. The goal is to make the SPS (of the PI and the FI) as much as similar while preserving some statistical properties such as variance. For the purpose, SPS is generated using the EOLE method. Let's consider $Y(t)$ represent a set of the stochastic process where $t \in [0, T]$. The trajectory of a stochastic process $Y(t)$ can be generated using Expansion Optimal Linear Estimator (EOLE) [42]. If the given time interval $[0, T]$ is discretized into $p$ time points with $(t_i)_{i=1,2,...,p} = (t_1 = t_0, ..., t_{p=}T)$ then EOLE expansion can be given as

$$Y(t) = \mu_Y(t) + \sigma_Y(t) \sum_{r=1}^{s} \frac{\xi_r}{\lambda_r} f_r^T \rho_Y(t) \tag{30.17}$$

where, the process mean is represented by $\mu_Y(t)$, variance by $\sigma_Y^2(t)$ and correlation function by $C(t_i, t_j)$. $s$ is truncation order, $\lambda$ and $f$ are eigenvalues and eigenvectors of auto-correlation function $C$ given by

$$C(t_i, t_j) = \begin{bmatrix} \rho_Y(t_1, t_1) & \rho_Y(t_1, t_2) & \cdots & \rho_Y(t_1, t_p) \\ \rho_Y(t_2, t_1) & \rho_Y(t_2, t_2) & \cdots & \rho_Y(t_2, t_p) \\ \vdots & \vdots & \ddots & \vdots \\ \rho_Y(t_p, t_1) & \rho_Y(t_p, t_2) & \cdots & \rho_Y(t_p, t_p) \end{bmatrix} \tag{30.18}$$

The correlation coefficient is represented by $\rho_Y(t)$ where its $j$th element be $\rho_Y(t, t_j)$, $j = 1, 2, ..., s < p$ and $\xi_r$ is the set of standard normal variables.

To find the relationship between SPS, the given time interval $[0, T]$ is divided into two sub-intervals $[0, \tau] \cup [\tau, T]$. The SPS is generated in the PI as well as the future interval. Let us represent the SPS in the present and the FI as $y_p = \{Y(t_p), t_p \in [0, \tau], \forall p = 1, 2, ..., n_{ls}\}$ and $y_f = \{Y(t_f), t_f \in [\tau, T], \forall f = 1, 2, ..., n_{ls}\}$, respectively. HSIC can be defined as

$$\text{HSIC}(\chi) = (n_{ls} - 1)^{-2} \text{ tr}(\Sigma_{y_p} H \Sigma_{y_f} H) \tag{30.19}$$

Here, $\chi = y_p \times y_f = \{(y_{p1}, y_{f1}), (y_{p2}, y_{f2}), ..., (y_{pn_{ls}}, y_{pn_{ls}})\}, \Sigma_{y_p}, \Sigma_{y_f} \in \mathcal{R}^{n_{ls} \times n_{ls}}$ is the covariance matrix of $yp$ and $yf$, $H = I - n_{ls}^{-1} 1_{n_{ls}} 1_{n_{ls}}^T \in \mathcal{R}^{n_{ls} \times n_{ls}}$ is a centering matrix, $I$ is the identity matrix and $1_{n_{ls}}$ is a matrix of ones of size $n_{ls} \times n_{ls}$. A class information vector is defined to treat them equally irrespective of their class (stationary or non-stationary process)

$$r = \begin{cases} \phi & \text{For stationary stochastic process} \\ t & \text{For non-stationary stochastic process} \end{cases} \tag{30.20}$$

where $r$ is of $n_{ls} \times 1$ length, $\phi$ is a vector whose value is equal to 1 and 0 if the sample belongs to the present and the interval, respectively. In case of a non-stationary stochastic process $r = t$, where $t$ is the time of the stochastic process sample time $(t \in [0,T])$. Once the vector $r$ is defined, a TM $\mathbb{M} \in \mathcal{R}^{n_{ls} \times n_{ls}}$ can be found which transforms the samples to represent them as much as similar. To find the TM, a vector $\vartheta = [y_p \ y_f]^T$ is defined which contains $2n_{ls}$ SPS from the PI and the FI. The defined vector $\vartheta$ is then augmented with the vector defined $\hat{\vartheta} = \begin{bmatrix} \vartheta \\ r \end{bmatrix}$ (augmentation can also generally be represented as $\hat{y}(t) = \begin{bmatrix} y(t) \\ r \end{bmatrix}$. The purpose of this augmentation step is to learn the class-specific hidden space. A mapping function $\gamma$ can be utilized to project the augmented vector $\hat{\vartheta}$ to a hidden space. The exact form of the $\gamma(\hat{\vartheta})$ need not to be known a covariance matrix rather $\Sigma_{\hat{\vartheta}} = \gamma(\hat{\vartheta})^T \gamma(\vartheta)$ can be utilized by the Kernel trick (Shawe-Taylor and Cristianini 2004). A projection matrix $\mathbb{M}$ can be applied to project $\gamma(\vartheta)$ to a hidden space which and hence the projected samples $Z = \hat{\mathbb{M}} \gamma(\vartheta)$. The purpose of this representation is to express each projection direction as a linear combination of all the samples $\hat{\mathbb{M}} = \gamma(\vartheta) \mathbb{M}$ where $\mathbb{M}$ is the TM to be found. Hence, the transformed samples can be represented as

$$Z = \mathbb{M}^T \gamma(\vartheta)^T \gamma(\vartheta) = \mathbb{M}^T \Sigma_{\hat{\vartheta}} \tag{30.21}$$

Hence, Eq. (30.19) in terms of transformation matrices can be represented as

$$\text{tr}(\Sigma_z \mathbf{H} \Sigma_r \mathbf{H}) = \text{tr}\left(\Sigma_{\hat{\vartheta}} \mathbf{W} \mathbf{W}^T \Sigma_{\hat{\vartheta}} \mathbf{H} \Sigma_r \ \mathbf{H}\right) \tag{30.22}$$

where the matrices $\Sigma_z$ and $\Sigma_r$ represents covariance matrices of the transformed SPS and the class information vector, respectively. One of the challenging issues in finding the TM is to preserve some of the important statistical properties such as variance. For this purpose, the trace of the covariance matrix of the transformed samples can be maximized. The covariance matrix is given by

$$\text{cov}(Z) = \text{cov}\left(\mathbb{M}^T \Sigma_{\hat{\vartheta}}\right) = \mathbb{M}^T \Sigma_{\hat{\vartheta}} \mathbf{H} \Sigma_{\hat{\vartheta}} \mathbb{M} \tag{30.23}$$

Hence, the learning problem can be defined as

$$\max_{\mathbf{W}} - \text{tr}\left(\mathbb{M}^T \Sigma_{\hat{\vartheta}} \mathbf{H} \Sigma_r \mathbf{H} \Sigma_{\hat{\vartheta}} \mathbb{M}\right) + \alpha \ \text{tr}(\mathbb{M}^T \Sigma_{\hat{\vartheta}} \mathbf{H} \Sigma_{\hat{\vartheta}} \mathbb{M}) \tag{30.24}$$

$$\text{s.t. } \mathbb{M}^T \mathbb{M} = \mathbf{I}$$

Here, $\alpha > 0$ is a trade-off parameter. The required TM $\mathbb{M}$ consists of $m$ leading eigenvectors of $\Sigma_{\hat{\vartheta}}(-\mathbf{H} \Sigma_r \mathbf{H} + \alpha \mathbf{H}) \Sigma_{\hat{\vartheta}}$ where $m$ is the number of stochastic processes. In the research work, $m = 1$ is taken. To find the covariance matrices $\Sigma_z$ and $\Sigma_r$ a proper covariance function which satisfies Mercer's theorem (Sun 2005) should be used.

Once the design variables samples are generated in the global space and the SPS are transformed, the input sample vector can be represented as $s' = [d, x'_d, x_r, z(t)]$. The variables $x'_d$ are generated in the global space whereas $z(t)$ represents the transformed SPS. The Kriging model is built using the input sample vector $s'$ and the corresponding LSF information $g(s)$ in the PI only by employing the procedure defined in Section 30.2. Once the surrogate model is accurate enough, the built surrogate model is then utilized in the proposed framework to predict the TDR for any design point instead of the original LSF.

In the proposed framework, the surrogate model is built before the optimization process. Therefore, any optimization algorithm such as an evolutionary algorithm or gradient-based algorithm can be used without any additional computational efforts. However, to show the effectiveness of the proposed framework, a gradient-based optimization algorithm such as SQP in Matlab is employed to search for the optimal design of a given P-TDRBDO problem. An arbitrary design is provided as the initial point of the algorithm. For each design point, the reliability is predicted using the built surrogate model. To predict the TDR for any design point, generate $n_{MCS}$ trajectories of the stochastic process. Transform the trajectories using the TM and predict the TDR using the built surrogate model. The reliability for the complete interval is estimated as

$$R(0,T) = 1 - \frac{n_f}{n_{MCS}}$$

(30.25)

where, $n_f$ is the number of failing trajectories.

Summary of the proposed method is given as follows

Step 1: Generate the design variable samples in the global space.
Step 2: Divide the complete interval $[0, T]$ into two sub-intervals namely the PI$[0, \tau]$ and the FI$[\tau, T]$.
Step 3: Find the TM using Eq. 30.24.
Step 4: Transform the SPS using the TM.
Step 5: Build and refine the Kriging surrogate model in the PI only using the input samples $s'$ and the LSF information $g(s)$.
Step 6: Define the optimization problem and provide an initial design point.
Step 7: Predict the TDR of the complete interval $[0, T]$ for the design point.
Step 8: Check the optimization algorithm convergence.
Step 9: If the algorithm does not converge, refine the design point, and repeat Steps 7 and 8.

## 30.4  RESULTS AND DISCUSSION

To demonstrate the effectiveness of the proposed method, one mathematical and two engineering examples are presented here. The mathematical example is presented to show the predicted reliability accuracy, whereas the two engineering examples are presented to show the efficiency of the proposed method.

### 30.4.1  EXAMPLE 1

In the P-TDRBDO problem, TDR is required to be predicted for each design point, and hence, it comprises a large part of the optimization problem. Here, a mathematical example is presented here to validate the proposed method's accuracy and efficiency. The LSFs of the problem is provided as

$$g_1(X, Y(t)) = (X_1 - X_2 + 0.2Y(t) - 12)^2 - 80$$

(30.26)

$$g_2(X, Y(t)) = 25 - X_1^2 X_2 - 5X_1 + (X_2 + 1)Y(t)$$

(30.27)

where $X_1, X_2$ are random variables and $Y(t)$ represents the stochastic process. Table 30.1 shows the input parameter values of the variables and the stochastic process. The reliability is required for a time period $[0, 10]$. The time interval $[0,10]$ can be divided as $[0, 5] \cup [5, 10]$. It should be noted that the complete interval is divided into half of the total length, however, other values such as $[0, 1], [0, 2],...,[0, 9]$ can also be taken. This is due to the fact that the PI length does not affect the proposed method accuracy rather establishing the accurate relationship between the intervals.

**TABLE 30.1**
**Parameter Values for Mathematical Example**

| Variable | Mean | Standard Deviation | Autocorrelation Function | Distribution Type |
|---|---|---|---|---|
| $X_1$ | 5 | 0.5 | – | Normal |
| $X_2$ | 2 | 0.2 | – | Normal |
| $Y(t)$ | 6 | 0.6 | $\cos\left(2\pi\left(t_2 - t_1\right)\right)$ | Gaussian Process |

**TABLE 30.2**
**Result Comparison for Reliability Prediction**

| Method | $n_{sa}$ | Reliability $[0, 10]$ | | Error $\delta$ | |
|---|---|---|---|---|---|
| | | LSF 1 | LSF 2 | LSF 1 | LSF 2 |
| Proposed method | 20 | 0.9634 | 0.9956 | 0.3001 | 0.0201 |
| MCS | $1 \times 10^6$ | 0.9664 | 0.9954 | – | – |

To predict the TDR for complete the Kriging surrogate models are built using the LSFs defined for $[0, 5]$ only. The TDR is predicted for the complete interval $[0,10]$ using the same surrogate models. $n_{ls} = 100$ SPS are generated and utilized to calculate the TM. The matrix is used to transform the SPS. To build the initial surrogate model, $n_0 = 6$ samples are utilized. It should be noted that these samples are generated in the $[0, 5]$ interval only. Corresponding LSF values are evaluated for these samples. The SPS is transformed using the TM. The transformed SPS is then utilized to build the initial surrogate model. The surrogate model is refined until the stopping criterion is satisfied. Once the surrogate model is accurate enough, TDR is predicted for the complete interval $[0, 10]$. The results of the proposed method are tabulated in Table 30.2. A total of $n_{sa} = 20$ samples are needed to build and refine the surrogate model. From the results, it is clear that the proposed method can predict the TDR accurately with high efficiency.

## 30.4.2 EXAMPLE 2

A simple beam under stochastic loading shown in Figure 30.3 is considered here as an engineering example. The beam is subjected to a uniformly distributed load $q(t)$ and a concentrated loading $F(t)$ under its weight. The weight of the beam can be considered as a uniformly distributed load $\rho_{st}wh$ where $w$ & $h$ represents the width and height of the beam. The beam is required to be designed for 30 years. It is assumed that the LSF information is available for $[0, 15]$ only.

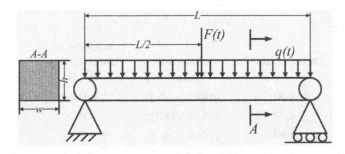

**FIGURE 30.3** A simply supported beam.

The P-TDRBDO problem can be defined as

$$\underset{D,\mu_{X_D}}{\text{minimize}}: \quad \mu_w \mu_h$$

$$\text{subject to}: \quad \Pr\{g(D,X_D,X_R,Y(t)) \geq 0, \ \forall t \in [0,30]\} > [R] \qquad (30.28)$$

$$0.04m \leq \mu_w \leq 0.15m, \ 0.15m \leq \mu_w \leq 0.25m, \ \mu_h \leq 4\mu_w$$

where $R = 0.95$. The LSF is given as

$$g(D,X_d,X_r,Y(t)) = 1 - 4\left(\frac{F(t)l}{4} + \frac{q(t)l^2}{8} + \frac{\rho_{st}whl^2}{8}\right)/wh^2\sigma_u$$

where $D = [\rho_{st}, l, \rho_u]$ is the deterministic variables vector, $X_D = [w, h]$, is random design variable vector, $X_r = [\sigma_u]$ representing the random variables with given parameters, and $Y(t) = [q(t), F(t)]$ represent stochastic processes. Distribution parameters of random variables and stochastic processes are given in Table 30.3. $n_{ls} = 100$ SPS are generated and utilized to calculate the TM. The random variables samples are generated in the global space, and the SPS is transformed to build and refine the surrogate model. Once the surrogate model is constructed, the optimal results are found using the built surrogate model.

The results of the proposed method are presented in Table 30.4 and are compared with the PSO-t-IRS (Li and Chen 2019) and the Time-variant Reliability-baseddesign Optimization using Simultaneously refined Kriging (TROSK) method (Hawchar, El Soueidy and Schoefs 2018). It can

**TABLE 30.3**

**Distribution Parameters for Example 2**

| Variable | Mean | Standard Deviation | Autocorrelation Function | Distribution Type |
|---|---|---|---|---|
| $w$ (m) | $\mu_w$ | 0.01 | NA | Lognormal |
| $h$ (m) | $\mu_h$ | $4 \times 10^{-3}$ | NA | Lognormal |
| $\sigma_u$ (Pa) | $2.4 \times 10^8$ | $2.4 \times 10^7$ | NA | Lognormal |
| $F(t)$N | 6000 | 600 | $\exp\left(-((t_2 - t_1)^2/0.8^2)\right)$ | Gaussian Process |
| $q(t)$N/m | 900 | 90 | $\cos(\pi(t_2 - t_1))$ | Gaussian Process |
| L (m) | 5 | – | NA | Deterministic |
| $\rho_{st}$ (KN/m³) | 78.5 | – | NA | Deterministic |

**TABLE 30.4**

**Results for Example 2**

| Method | Objective Value $f$ | Optimal Point $\mu_w, \mu_h$ | Reliability for Optimal Design $R(0,30)$ | Number of Function Evaluations |
|---|---|---|---|---|
| Proposed Method | 0.0084 | (0.0457, 0.1840) | 0.9500 | 109 |
| TROSK | 0.0083 | (0.0456, 0.1825) | 0.9283 | 112 |
| PSO-t-IRS | 0.0085 | (0.0461, 0.1843) | 0.9500 | 134 |
| MCS | – | – | 0.9502 | $1 \times 10^6$ |

be seen from the results that the number of function evaluations for the TRSOK method and the proposed method are comparable; however, the TROSK method does not satisfy the reliability requirement. On the other hand, the PSO-t-IRS method can meet the reliability requirement, however, its number of evaluations are high. Moreover, the PSO-t-IRS attempts to build the $n_t$ the number of Kriging models using the standard normal variables $\xi$ instead of $y(t)$ directly, which makes the method computationally extensive. It should be noted that to estimate the TDR using other methods, complete LSF information $[0, 30]$ is used. However, the proposed method can predict the TDR for the complete interval using LSF information in the PI $[0, 15]$.

### 30.4.3  Example 3

A two bar frame shown in Figure 30.4 is used as another example to demonstrate the proposed method's effectiveness. The radius of the two bars is represented as $r_1$ and $r_2$. The material strength is represented by $\rho_{M_1}$ and $\rho_{M_2}$ whereas the length of the two bars is $L_1$ and $L_2$, respectively. At the point $O_3$ the frame is subjected to a dynamic force $P(t)$.

The P-TDRBDO problem can be given as

$$\underset{D,\mu_{X_D}}{\text{minimize}}: \quad \pi\mu_{r_1}^2\mu_{l_1} + \pi\mu_{r_2}^2\sqrt{\mu_{l_1}^2 + \mu_{l_2}^2}$$

$$\text{subject to}: \quad \Pr\{g_i(D,X_D,X_R,Y(t)) \geq 0, \ \forall t \in [0,10]\} > [R_i] \tag{30.29}$$

$$0.07m \leq \mu_{n,2} \leq 0.25m, \ R_i = [0,99, \ 0.999]$$

where the LSFs are defined as

$$g_1(D,X_D,X_R,Y(t)) = 1 - \frac{P(t)\sqrt{L_1^2 + L_2^2}}{\pi L_2 r_1^2 \rho_{M_1}} \tag{30.30}$$

$$g_2(D,X_D,X_R,Y(t)) = 1 - \frac{4P(t)L_1}{\pi L_2 r_2^2 \rho_{M_2}} \tag{30.31}$$

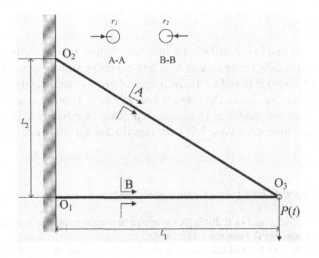

**FIGURE 30.4**  A two bar frame.

**TABLE 30.5**
**Input Parameter of Two Bar Frame**

| Variable | Mean | Standard Deviation | Distribution Type |
|---|---|---|---|
| $r_1$ (m) | $\mu_{r_1}$ | $1 \times 10^{-3}$ | Normal |
| $r_2$ (m) | $\mu_{r_2}$ | $1 \times 10^{-3}$ | Normal |
| $L_1$ (m) | 0.4m | $1 \times 10^{-3}$ | Normal |
| $L_2$ (m) | 0.3m | $1 \times 10^{-3}$ | Normal |
| $\sigma_{M_1}$ (Pa) | $1.7 \times 10^8$ Pa | $1.7 \times 10^7$ | Lognormal |
| $\sigma_{M_2}$ (m) | $1.7 \times 10^8$ Pa | $1.7 \times 10^7$ | Lognormal |
| $P(t)$ (N) | $2.2 \times 10^6$ | $2.0 \times 10^5$ | Gaussian Process |

**TABLE 30.6**
**Results for Example 2**

| Method | Objective Value $f$ | Optimal Point $\mu_w, \mu_h$ | Reliability for Optimal Design $R_1(0,10)$ | Reliability for Optimal Design $R_2(0,10)$ | Number of Function Evaluations |
|---|---|---|---|---|---|
| Proposed Method | 0.0281 | $(0.1043, 0.0961)$ | 0.9911 | 0.9990 | 521 |
| TROSK | 0.0270 | $(0.1013, 0.0947)$ | 0.9646 | 0.9956 | 508 |
| PSO-t-IRS | 0.0286 | $(0.1049, 0.0971)$ | 0.9931 | 0.9990 | 542 |
| MCS | – | – | 0.9930 | 0.9990 | $2 \times 10^6$ |

where $D = []$ is the deterministic variables vector, $X_D = [r_1, r_2]$, is random design variable vector, $X_r = [\rho_{M_1}, \rho_{M_2}, L_1, L_2]$ representing the random variables with given parameters, and $Y(t) = [F(t)]$ represent the stochastic process. Distribution parameters of random variables and stochastic processes are given in Table 30.5. $n_{ls} = 100$ SPS are generated to find the TM. The results are provided in Table 30.6. It can be seen from the results that the proposed method can effectively predict the TDR with significant accuracy and efficiency.

## 30.5 CONCLUSION

TDR plays an important role in the RBDO of a structure. Acquisition of accurate LSF is a challenging task in real life, especially in the case of long and extensive simulations. In this research work, a novel P-TDRBDO problem is introduced. The proposed method predicts the reliability of a design point using the LSF information in the present interval only. Transfer learning-based method is integrated with the surrogate modeling technique to predict the reliability. The method can handle multiple non-linear and time-dependent LSFs with significant accuracy and improved efficiency.

## REFERENCES

Agarwal, H. 2004. *Reliability based design optimization: formulations and methodologies.* University of Notre Dame.

Chen, Z., Qiu H., Gao L., Su L., and Li P. 2013. "An adaptive decoupling approach for reliability-based design optimization." *Computers & Structures* 117: 58–66.

Du, X., and Chen W. 2004. "Sequential optimization and reliability assessment method for efficient probabilistic design." *Journal of Mechanical Design* 126 (2): 225–233.

Dubourg, V., Sudret B., and Bourinet J. -M. 2011. "Reliability-based design optimization using kriging surrogates and subset simulation." *Structural and Multidisciplinary Optimization* 44 (5): 673–690.

Hawchar, L., El Soueidy C.- P., and Schoefs F. 2018. "Global kriging surrogate modeling for general time-variant reliability-based design optimization problems." *Structural and Multidisciplinary Optimization* 58 (3): 955–968.

Hu, Z., and Du X. 2015. "Mixed efficient global optimization for time-dependent reliability analysis." *Journal of Mechanical Design* 137 (5).

Hu, Z., and Du X. 2016. "Reliability-based design optimization under stationary stochastic process loads." *Engineering Optimization* 48 (8): 1296–1312.

Hu, Z., and Mahadevan S. 2016. "A single-loop kriging surrogate modeling for time-dependent reliability analysis." *Journal of Mechanical Design* 138 (6).

Hyeon Ju, B., and Chai Lee B. 2008. "Reliability-based design optimization using a moment method and a kriging metamodel." *Engineering Optimization* 40 (5): 421–438.

Jensen, H., Kusanovic D., Valdebenito M., and Schuëller G. 2012. "Reliability-based design optimization of uncertain stochastic systems: gradient-based scheme." *Journal of Engineering Mechanics* 138 (1): 60–70.

Kharmanda, G., El Hami A., and Souza De Cursi E. 2013. "Reliability-based Design Optimization (RBDO)." *Multidisciplinary Design Optimization in Computational Mechanics*: 425–458.

Kuschel, N., and Rackwitz R. 2000. "Optimal design under time-variant reliability constraints." *Structural Safety* 22 (2): 113–127.

Lee, T. H., and Jung J. J. 2008. "A sampling technique enhancing accuracy and efficiency of metamodel-based RBDO: Constraint boundary sampling." *Computers & Structures* 86 (13-14): 1463–1476.

Li, J., and Chen J. 2019. "Solving time-variant reliability-based design optimization by PSO-t-IRS: A methodology incorporating a particle swarm optimization algorithm and an enhanced instantaneous response surface." *Reliability Engineering & System Safety* 191: 106580.

Liang, J., Mourelatos Z. P., and Tu J. 2008. "A single-loop method for reliability-based design optimisation." *International Journal of Product Development* 5 (1-2): 76–92.

Lophaven, S. N., Nielsen H. B., and Søndergaard J. 2002. *DACE: a Matlab kriging toolbox.* Vol. 2: Citeseer.

Pan, S. J., Tsang I. W., Kwok J. T., and Yang Q. 2010. "Domain adaptation via transfer component analysis." *IEEE Transactions on Neural Networks* 22 (2): 199–210.

Pan, S. J., and Yang Q. 2009. "A survey on transfer learning." *IEEE Transactions on Knowledge and Data Engineering* 22 (10): 1345–1359.

Shawe-Taylor, J., and Cristianini N. 2004. *Kernel methods for pattern analysis.* Cambridge University Press.

Singh, A., and Mourelatos Z. P. 2010a. "Design for Lifecycle Cost Using Time-Dependent Reliability." *Journal of Mechanical Design* 132 (9).

Singh, A., and Mourelatos Z. P. 2010b. "On the time-dependent reliability of non-monotonic, non-repairable systems." *SAE International Journal of Materials and Manufacturing* 3 (1): 425–444.

Singh, A., Mourelatos Z. P., and Li J. 2010. "Design for lifecycle cost using time-dependent reliability." *Journal of Mechanical Design* 132 (9).

Singh, A., Mourelatos Z., and Nikolaidis E. 2011. "Time-dependent reliability of random dynamic systems using time-series modeling and importance sampling." *SAE International Journal of Materials and Manufacturing* 4 (1): 929–946.

Sun, H. 2005. "Mercer theorem for RKHS on noncompact sets." *Journal of Complexity* 21 (3): 337–349.

van Noortwijk, J. M., van der Weide J. A., Kallen M. -J., and Pandey M. D. 2007. "Gamma processes and peaks-over-threshold distributions for time-dependent reliability." *Reliability Engineering & System Safety* 92 (12): 1651–1658.

Wang, Z., Almeida Jr J. H. S., St-Pierre L., Wang Z., and Castro S. G. 2020. "Reliability-based buckling optimization with an accelerated Kriging metamodel for filament-wound variable angle tow composite cylinders." *Composite Structures* 254: 112821.

Wang, Z., Mourelatos Z. P., Li J., Baseski I., and Singh A. 2014. "Time-dependent reliability of dynamic systems using subset simulation with splitting over a series of correlated time intervals." *Journal of Mechanical Design* 136 (6).

Wang, Z., and Wang P. 2012. "A nested extreme response surface approach for time-dependent reliability-based design optimization." *Journal of Mechanical Design* 134 (12).

Yan, K., Kou L., and Zhang D. 2017. "Learning domain-invariant subspace using domain features and independence maximization." *IEEE transactions on cybernetics* 48 (1): 288–299.

Yu, S., Zhang Y., Li Y., and Wang Z. 2020. "Time-variant reliability analysis via approximation of the first-crossing PDF." *Structural and Multidisciplinary Optimization*: 1–15.

Yuan, K., Xiao N. -C., Wang Z., and Shang K. 2020. "System reliability analysis by combining structure function and active learning kriging model." *Reliability Engineering & System Safety* 195: 106734.

# Index

Note: Locators in *italics* represent figures and **bold** indicate tables in the text.

Printed in the United States
by Baker & Taylor Publisher Services